University Texts in the Mathematical Sciences

Textbooks in this series cover a wide variety of courses in mathematics, statistics and computational methods. Ranging across undergraduate and graduate levels, books may focus on theoretical or applied aspects. All texts include frequent examples and exercises of varying complexity. Illustrations, projects, historical remarks, program code and real-world examples may offer additional opportunities for engagement. Texts may be used as a primary or supplemental resource for coursework and are often suitable for independent study.

Nazibuddin Ahmed · Dipunja Gohain · Rajdeep Bordoloi

Classical Dynamics

Nazibuddin Ahmed
Department of Mathematics
Cotton University
Guwahati, Assam, India

Rajdeep Bordoloi
Department of Mathematics
Thong Nokbe College
Karbi Anglong, Assam, India

Dipunja Gohain
Department of Mathematics
Girijananda Chowdhury University
Guwahati, Assam, India

Thong Nokbe College
Dokmoka, Karbi Anglong, India

ISSN 2731-9318 ISSN 2731-9326 (electronic)
University Texts in the Mathematical Sciences
ISBN 978-981-95-6393-7 ISBN 978-981-95-6394-4 (eBook)
https://doi.org/10.1007/978-981-95-6394-4

Mathematics Subject Classification: 35F21, 70-03, 35J05, 70-00, 70F15

Responsible Editor: Shamim Ahmad

This Springer imprint is published by the registered company Springer Nature Singapore Pte Ltd.
The registered company address is: 152 Beach Road, #21-01/04 Gateway East, Singapore 189721, Singapore

Preface

This book is designed for graduate and postgraduate students in Mathematics, Mathematical Physics, and Engineering. The book also seems to be fruitful to researchers in Applied Mathematics. Each chapter begins with clear statements of definitions, principles, and theorems. These are followed by graded sets of solved and supplementary problems. The solved problems serve to illustrate and amplify the theories and principles.

Vector treatment is highlighted, but the scalar treatment is not disregarded. Theorems and principles are solved by the vector method, whereas, in the case of solved problems, both vector and scalar treatments are utilized. It is assumed that the students have a fair working knowledge of Differential and Integral Calculus, Differential Equations, Vector Algebra and Calculus.

The problems considered in the book are based on some laws of Physics, for instance, Newton's laws of motion, laws of conservation of momentum, and energy. Equations are written in vector form as such the readers understand the originality of the problems. Physical quantities involved in the book are properly defined with respective SI units. The problems are mainly taken from University question papers and some standard books mentioned in the reference section. A large number of illustrated problems are framed by the author himself.

A few references are given in this book. The students who are new to this subject will find plenty to occupy them without looking further afield. The solved problems in the book will be very helpful for the readers. The book consists of nine chapters.

Chapter 1 of the book deals with the basic concepts of the subject. Here the discussion on Newton's laws of motion, Lines of force, Work, Power, Principle of Energy, Conservative force and some of its properties, Linear and angular impulses, Torque and Angular momentum, and principles of linear and angular momentum are carried out in details.

The study of central forces and central orbits are presented in Chap. 2. In this chapter, components of velocity and acceleration in radial and transverse directions are derived. Besides the definition of a central force field, some of its important properties are discussed. Differential equations of the central orbit are obtained in (r, θ) and (p, r) systems. Important aspects of apses in central orbit are highlighted.

Chapter 3 is devoted to the discussion of the law of inverse square and two-body problems. Here, Newton's law of gravitation is presented in vector form. Gravitational potential is derived using a vector tool. The equation to the orbit under the law of inverse is derived. Kepler's laws of planetary motion are described with their physical significance. The chapter is incorporated with the inclusion of the two-body problem. Kepler's equation is also derived.

The motion of a particle in three dimensions is discussed in Chap. 4. Here, the general orthogonal curvilinear coordinates are introduced. Unit vectors in terms of orthogonal curvilinear coordinates are obtained. Components of velocity and acceleration in Cartesian, cylindrical polar, and spherical polar coordinates are derived.

The study of motion in a system of particles is covered in Chap. 5. Centre of mass, the center of gravity, conservation of linear and angular momentum, the relation between external torque and angular momentum, work, the principle of conservation of energy, associated theorem, impulse, virtual displacement, virtual work, D'Alembert principle, and some pertinent examples are all included in this chapter.

Chapter 6 presents an introduction to the dynamics of rigid bodies. Moments and products of inertia are defined. Expression for the moment of inertia is obtained in vector form. The parallel axes theorem and six constant theorems are proved by adopting the vector tool. The principle of angular momentum is derived in a convenient manner. General equations of motion of a rigid body are derived on the basis of D'Alembert's principle. The motion of the center of inertia and motion about the center of inertia are discussed. Conservation of energy and the principle of energy are derived in a convenient and systemic way.

The motion of a rigid body under impulsive forces is focused on in Chap. 7. Here the general equations of motion of a rigid body under impulsive forces are derived. Applications of the principle of virtual works are discussed in this chapter. Carnot's theorem, Bertrand's theorem, and Kelvin's theorem are proved using the tool of vectors.

Chapter 8 is concerned with the study of rigid body motion in space with one point fixed. This chapter is highlighted with the introduction of Inertia Matrix Inertia Tensor. The kinetic energy of rotation is expressed in terms of the Inertia Tensor. Euler's dynamical equations for both the general case and particular case are derived. Euler's geometrical equations are discussed in detail.

Chapter 9 is devoted to a discussion on Lagrange's Equations. Here, the generalized coordinates, generalized velocities, generalized forces, and generalized momentum are introduced systematically. Lagrange's equations are derived by adopting the vector method. This chapter is added with a good number of solved and supplementary problems.

The theory of Hamilton is described in Chap. 10. This chapter includes the introduction of the Hamiltonian, derivation of Hamilton's equations, a glimpse of the calculus of variations, and the solution of the Brachistochrone problem. Like other chapters, this chapter also ended with a graded set of solved and supplementary problems.

We would like to express our gratitude to Mrs. Habiba Ahmed, Shahil Ahmed, Nasiba Ahmed, Nazima Ahmed, Dr. Nilakshi Goswami, Dr. Suman Agarwalla, Dr. Kangkan Choudhury, Dr. Mridusmita Bormudoi, Dr. Richa Deb Dowerah, and Dr. Masuma Khanam, for their encouragement and moral support in the preparation of the book.

Guwahati, India Nazibuddin Ahmed
Guwahati, India Dipunja Gohain
Karbi Anglong, India Rajdeep Bordoloi

Competing Interests The authors have no competing interests to declare that are relevant to the content of this manuscript.

Contents

About the Authors

Nazibuddin Ahmed is Adjunct Professor, Department of Mathematics, Cotton University. He got retired from Gauhati University as Professor and Head, Department of Mathematics, Gauhati University in September, 2023. Recently, he retired from Gauhati University, and at present, he is appointed as an Adjunct Professor at Cotton University. He completed his M.Sc., Ph.D., and D.Sc. from the Department of Mathematics, Gauhati University, Guwahati, Assam. His field of specialization is Magneto-Fluid Dynamics, Heat and Mass Transfer in hydromagnetic flows. He has published more than 200 research papers in many reputed international journals like *Heat Transfer* (Wiley), *ZAMM* (Wiley), *Journal of Heat Transfer* (ASME), *Canadian Journal of Physics*, *Journal of Nanofluid* (ASP), *Scientific Reports* (Springer nature), *International Journal of Modern Physics-B* (World Scientific), etc. He is also the author of the book: *Thermal and Solutal Convection in Some Hydromagnetic Flows* published by Springer. During his academic career, he has done two major projects: one is from UGC, which is remarked as "Excellent" in the evaluation report, and another is from CSIR.

Dipunja Gohain completed his M.Sc. and Ph.D. from Gauhati University, Guwahati, Assam. He cleared CSIR UGC NET 2019 with JRF and also GATE 2019. His fields of interest are fluid dynamics, MHD, heat and mass transfer, and nanofluids. He has publications in various international and national journals like *ZAMM* (Wiley), *Heat Transfer* (Wiley), *Journal of Nanofluid* (ASP), and *International Journal of Modern Physics B* (World Scientific). His thirst for knowledge makes him very passionate about researching new things and new trends.

Rajdeep Bordoloi is Assistant Professor at Thong Nokbe College, Karbi Anglong, India in Mathematics in 2018 and successfully cleared the CSIR-UGC NET with JRF in Mathematics in 2019. He was awarded the Ph.D. degree in 2023 in the field of Fluid Dynamics from Gauhati University. His research primarily focuses on fluid dynamics, hybrid nanofluids, and related applied mathematics problems. He has more than 13 research publications in esteemed international journals, including *Heat Transfer* (Wiley), *International Journal of Modern Physics B* (World Scientific),

ZAMM (Wiley), *Modern Physics Letters B* (World Scientific), *Special Topics and Reviews in Porous Media* (Begell House), *Latin American Applied Research* and so on. In addition to his research publications, he has contributed to academic publishing as the editor of an edited book published by Sibsagar University.

Symbols

$\mathbf{F}$	Force
$\mathbf{g}$	Acceleration due to gravity
H	Hamiltonian, J
L	Lagrangian, J
M or m	Mass, kg
T	Kinetic Energy, J
V	Potential Energy, J
Ω	Angular Momentum Vector
ω	Angular Velocity
Λ	Torque or Moment

$$\nabla \equiv \left(\frac{\partial}{\partial x}, \frac{\partial}{\partial y}, \frac{\partial}{\partial z}\right)$$

Chapter 1
Force, Work, Power, and Energy

1.1 Basic Concepts

1.1.1 Mechanics

It is a branch of Physics that is concerned with the motion or changes in position of a physical object due to the application of forces. Mechanics is subdivided into three branches: *Kinematics*, *Kinetics*, and *Statics*.

- Kinematics is concerned with the geometry of motion.
- Kinetics is concerned with the physical causes of the motion.
- Statics is concerned with conditions under which no motion takes place.

1.1.2 Matter, Mass, and Force

Matter is defined to be that which can be perceived by the senses. A body that is incapable of any rotation or moves without any rotation is defined as a particle. In fact, a particle is a portion of matter.

- The mass of a body is the measure of the quantity of matter contained by it, and it is associated with a particle. Units of mass are grams, kilograms, etc.
- A force is that which changes or tends to change the state of rest, or uniform motion, of a body. Units of force are Dyne, Newton, etc.
- Length, mass, and time are often called dimensions from which other physical quantities are constructed.

N. Ahmed et al., *Classical Dynamics*, University Texts in the Mathematical Sciences,
https://doi.org/10.1007/978-981-95-6394-4_1

1.1.3 Velocity and Acceleration of a Particle Along a Curve

Let a particle move along a curve C. Suppose that P and Q are the positions of the particle at times t and $t + \delta t$, where P and Q have position vectors $\mathbf{r}$ and $\mathbf{r} + \delta\mathbf{r}$ relative to a fixed origin O (see Fig. 1.1). We note that $\mathbf{r} = \mathbf{r}(t)$ and $\mathbf{r} + \delta\mathbf{r} = \mathbf{r}(t + \delta t)$. Then the velocity of the particle at P is given by

$$\mathbf{v} = \lim_{\delta t \to 0} \frac{\mathbf{PQ}}{\delta t} = \lim_{\delta t \to 0} \frac{\delta \mathbf{r}}{\delta t} = \frac{d\mathbf{r}}{dt} \quad (1.1.1)$$

which is a vector tangent to C at P. If $\mathbf{r} = \mathbf{r}(t) = x\hat{i} + y\hat{j} + z\hat{k}$, $\mathbf{v}$ can be written as

$$\mathbf{v} = \frac{d\mathbf{r}}{dt} = \dot{x}\hat{i} + \dot{y}\hat{j} + \dot{z}\hat{k}, \quad (1.1.2)$$

where $(\cdot)$ denotes differentiation with respect to t.

The speed of the particle at P is given by

$$v = |\mathbf{v}| = \left|\frac{d\mathbf{r}}{dt}\right| = \left|\dot{x}\hat{i} + \dot{y}\hat{j} + \dot{z}\hat{k}\right| = \sqrt{\dot{x}^2 + \dot{y}^2 + \dot{z}^2} = \frac{ds}{dt}, \quad (1.1.3)$$

where $s = \overset{\frown}{AP}$ is the arc length measured from some initial point A to P. Let $\mathbf{v} + \delta\mathbf{v}$ be the velocity of the particle at Q. We have $\mathbf{v} = \mathbf{v}(t)$ and $\mathbf{v} + \delta\mathbf{v} = \mathbf{v}(t + \delta t)$.

Just as the velocity $\mathbf{v}$ of the particle at P is the time rate of change of the position vector $\mathbf{r}$, the acceleration $\mathbf{a}$ of the particle at P is defined as the time rate of change of the velocity vector $\mathbf{v}$. So, we can write

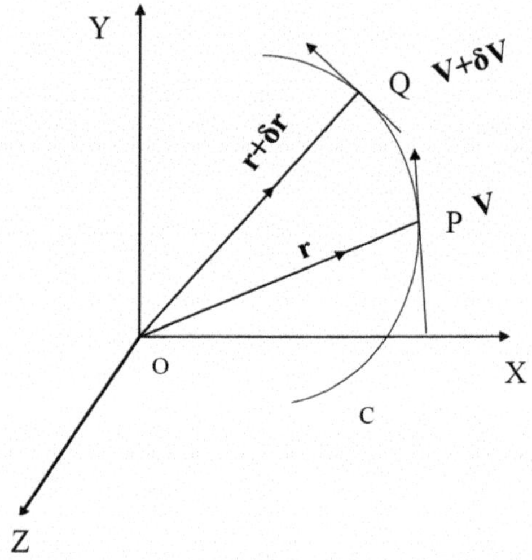

Fig. 1.1 Velocity of a particle moving along a curve

$$\mathbf{a} = \lim_{\delta t \to 0} \frac{\mathbf{v}(t+\delta t) - \mathbf{v}(t)}{\delta t} = \frac{d\mathbf{v}}{dt}. \tag{1.1.4}$$

In (x, y, z) system, $\mathbf{r} = x\hat{i} + y\hat{j} + z\hat{k}$ and

$$\mathbf{a} = \frac{d\mathbf{v}}{dt} = \frac{d}{dt}\left(\frac{d\mathbf{r}}{dt}\right) = \frac{d^2\mathbf{r}}{dt^2} = \ddot{x}\hat{i} + \ddot{y}\hat{j} + \ddot{z}\hat{k}. \tag{1.1.5}$$

The magnitude of the acceleration $\mathbf{a}$ is given by

$$a = |\mathbf{a}| = \sqrt{\ddot{x}^2 + \ddot{y}^2 + \ddot{z}^2}. \tag{1.1.6}$$

1.1.4 Momentum

If a particle of mass m moves with velocity $\mathbf{v}$ at time t, then its momentum $\mathbf{p}$ at that time is given by

$$\mathbf{p} = m\,\mathbf{v}. \tag{1.1.7}$$

1.1.5 Newton's Laws of Motion

I. *Everybody continues in its state of rest, or of uniform motion in a straight line, except in so far as it is compelled by impressed force to change that state.*

II. *The rate of change of momentum is proportional to the impressed force and takes place in the direction in which the force acts.*
If a particle of mass m moves with velocity $\mathbf{v}$ at time t under the action of a force $\mathbf{F}$ then following this law,

$$\mathbf{F} = \lambda \frac{d}{dt}(m\,\mathbf{v}), \tag{1.1.8}$$

where λ is some constant. If m is independent of time t,
Equation (1.1.8) gives

$$\mathbf{F} = \lambda m \frac{d\mathbf{v}}{dt} = \lambda m\,\mathbf{a}. \tag{1.1.9}$$

Equation (1.1.9) yields

$$\mathbf{F} = \lambda m\,\mathbf{a}. \tag{1.1.10}$$

Let the unit of force be so chosen that F = 1 when a = 1, $m = 1$ and hence (1.1.10) gives $\lambda = 1$ and as a consequence, (1.1.9) takes the form

$$\mathbf{F} = m\mathbf{a} = m\frac{d\mathbf{v}}{dt}. \tag{1.1.11}$$

If m depends on time t, we must have, instead,

$$\mathbf{F} = \frac{d}{dt}(m\mathbf{v}). \tag{1.1.12}$$

III. *To every action there is an equal and opposite reaction.*

1.1.6 Lines of Force

A curve C drawn in a force field **F** is said to be a line of force if the tangent vector drawn at any point on the curve is the direction of the force at that point. In Fig. 1.2, let $\mathbf{r} = \mathbf{r}(t)$ define the line of force for the force field **F** at time t. By definition of the line of force, $\frac{d\mathbf{r}}{ds}$, which is a unit vector tangential to the line of force, is parallel to the force field. Therefore, there exists a non-zero scalar λ such that

$$\frac{d\mathbf{r}}{ds} = \lambda\mathbf{F}. \tag{1.1.13}$$

In (x, y, z) system,

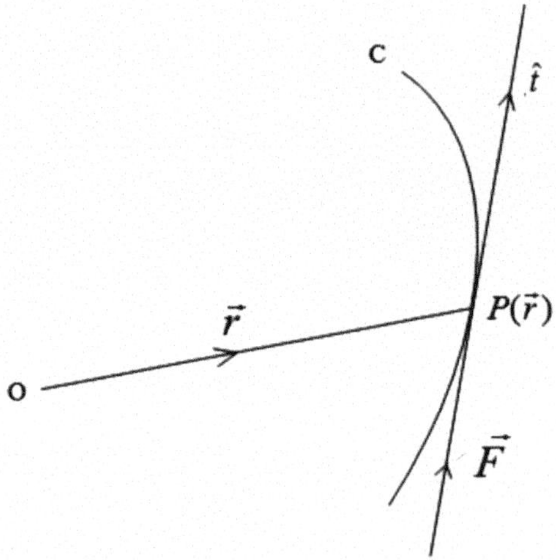

Fig. 1.2 Geometry of lines of force in a vector field

$$\frac{d\mathbf{r}}{ds} = \hat{i}\frac{dx}{ds} + \hat{j}\frac{dy}{ds} + \hat{k}\frac{dz}{ds}; \quad \mathbf{F} = \hat{i}X + \hat{j}Y + \hat{k}Z. \tag{1.1.14}$$

Therefore, Eq. (1.1.13) gives

$$\frac{dx}{X} = \frac{dy}{Y} = \frac{dz}{Z} = \lambda ds. \tag{1.1.15}$$

Thus, the lines of force are given by

$$\frac{dx}{X} = \frac{dy}{Y} = \frac{dz}{Z}. \tag{1.1.16}$$

1.1.7 Relative Velocity and Acceleration

If two particles P and Q move with velocities $\mathbf{v}_1$ and $\mathbf{v}_2$ and acceleration $\mathbf{a}_1$ and $\mathbf{a}_2$ the vectors $\mathbf{v}_{P/Q} = \mathbf{v}_1 - \mathbf{v}_2$ and $\mathbf{a}_{P/Q} = \mathbf{a}_1 - \mathbf{a}_2$ are respectively defined as relative velocity and relative acceleration of P with respect of Q.

1.1.8 Units of Mass and Force

SI unit of mass is kilogram (kg), and the standard unit of mass in the cgs system is gram (g). The SI unit of force is Newton (N), and 1 N is that amount of force which produces an acceleration of $\frac{1\text{m}}{\text{s}^2}$ acting on a mass of 1 kg. i.e., 1 N $=$ 1 kg $\times 1$ m/s^2 $=$ 1 kgm/s^2 The cgs unit of force is dyne, and 1 dyne $=$ 1 gms^{-2}. That is, 1 dyne is the amount of force that produces an acceleration of $\frac{1\,\text{cm}}{\text{s}^2}$ acting on a mass of 1 gm.

1.1.9 Inertial Frames of Reference

Newton's laws are postulated under the assumption that all observations are taken with respect to a coordinate system fixed in space. This coordinate system is termed a frame of reference. It can be established that, if Newton's laws hold in one frame of reference, they also hold in any other frame of reference moving at constant velocity relative to it. All such frames of reference are called inertial frames of reference or Newtonian frames of reference.

1.1.10 Work

Let a particle move along a curve C under the action of a force $\mathbf{F}$. In Fig. 1.3, suppose that P is the position of the particle at time t with position vector $\mathbf{r}$ relative to some

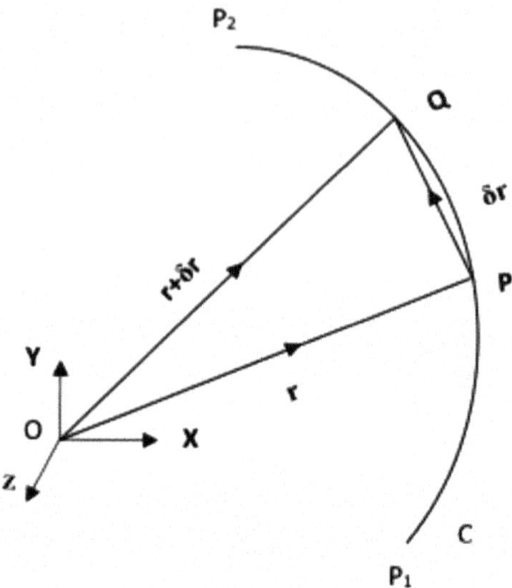

Fig. 1.3 Illustration of work done by a force

origin O. Let Q be the position of the same particle at time $t + \delta t$ and suppose that $\mathbf{OQ} = \mathbf{r} + \delta\mathbf{r}$ and so the displacement of the particle in the δt is $\mathbf{PQ} = \delta\,\mathbf{r}$.

$\delta W = \mathbf{F} \cdot \delta\,\mathbf{r}$ is defined to be the work done on the particle by the force $\mathbf{F}$ in displacing the particle from the point P to Q. The total work done by the force $\mathbf{F}$ in moving the particle from point A to B along the curve C is given by

$$W = \int_C \mathbf{F} \cdot d\mathbf{r} = \int_A^B \mathbf{F} \cdot d\mathbf{r}. \tag{1.1.17}$$

1.1.11 Power

The work done per unit of time on a particle by a force is called the power applied to the particle. Let the symbols W and $\widehat{P}$ stand for work and power, respectively. By definition, we have

$$\widehat{P} = \frac{dW}{dt}. \tag{1.1.18}$$

We note that $\delta W = \mathbf{F} \cdot \delta\mathbf{r}$ and in limiting case, $dW = \mathbf{F} \cdot d\mathbf{r}$, or $\frac{dW}{dt} = \mathbf{F} \cdot \mathbf{v}$, where $\mathbf{v} = \frac{d\mathbf{r}}{dt}$ is the velocity of the particle.

Theorem 1.1 *The total work done in moving a particle along a curve C from the point A to the point B is given by W=KE at B-KE at A =Change in KE.*

Proof Let **F** be the force acting on the particle. We have

$$W = \int_A^B \mathbf{F} \cdot d\mathbf{r}. \tag{1.1.19}$$

Let A and B be the positions of the particle at times t_1 and t_2 respectively. Let $\mathbf{v}_1, \mathbf{v}_2$ denote the velocities of the particle at A and B, respectively.

By Newton's 2nd law of motion,

$$\mathbf{F} = \frac{d}{dt}(m\mathbf{v}). \tag{1.1.20}$$

By the use of (1.1.21), (1.1.20) reduces to

$$\begin{aligned} W &= \int_A^B \frac{d}{dt}(m\mathbf{v}) \cdot d\mathbf{r} = \int_{t_1}^{t_2} \frac{d}{dt}(m\mathbf{v}) \cdot \frac{d\mathbf{r}}{dt} dt \\ &= m \int_{t_1}^{t_2} v \frac{dv}{dt} dt = \frac{1}{2}mv_2^2 - \frac{1}{2}mv_1^2. \end{aligned} \tag{1.1.21}$$

Thus, $W =$ K.E. at B − K.E. at A = Change in KE.
This establishes the proof of the theorem. □

1.1.12 Conservative Force

A force **F** acting on a particle is said to be conservative if ∃ a differentiable scalar point function V (say) such that

$$\mathbf{F} = -\nabla V. \tag{1.1.22}$$

In Eq. (1.1.22), V is called the **force potential** or **potential energy** of the particle.

Theorem 1.2 *In a conservative force field, the work done in moving a particle along curve C from A to B is given by* $W =$ *P.E. at A* − *P.E. at B* $= V_1 - V_2$.

Proof Let **F** be the conservative force acting on a particle of mass m. Let V_1, V_2 denote the potential energies of the particle at A and B respectively. We have

$$W = \int_A^B \mathbf{F} \cdot d\mathbf{r}. \tag{1.1.23}$$

Given $\mathbf{F}$ is conservative with force potential V(say)

$$\therefore \quad \mathbf{F} = -\nabla V. \tag{1.1.24}$$

In(x, y, z) system,

$$d\mathbf{r} = \hat{i}dx + \hat{j}dy + \hat{k}dz, \tag{1.1.25}$$

where $\hat{i}$, $\hat{j}$, $\hat{k}$ are the unit vectors along the +ve directions of X-axis, Y-axis and Z-axis respectively. Further,

$$\nabla V = \hat{i}\frac{\partial V}{\partial x} + \hat{j}\frac{\partial V}{\partial y} + \hat{k}\frac{\partial V}{\partial z}. \tag{1.1.26}$$

$$\therefore \quad \nabla V \cdot d\mathbf{r} = \frac{\partial V}{\partial x}dx + \frac{\partial V}{\partial y}dy + \frac{\partial V}{\partial z}dz, \tag{1.1.27}$$

$$\text{i.e.,} \quad \mathbf{F} \cdot d\mathbf{r} = -dV. \tag{1.1.28}$$

Using (1.1.28) in (1.1.23), we obtain

$$W = -\int_A^B dV = -V\Big]_{V_1}^{V_2} = V_1 - V_2 = \text{P.E. at A} - \text{P.E. at B}. \tag{1.1.29}$$

Hence the theorem. □

Note: The above theorem yields that in a conservative force field, the work done in moving a particle from A to B is independent of the path joining the points A and B. Based on this phenomenon, we can give an alternative definition of conservative force as follows: *A force* **F** *acting on a particle is said to be conservative, if the work done by* **F** *in moving the particle from A to B is independent of the path joining A and B.*

Theorem 1.3 *A force* $\boldsymbol{F}$ *is conservative if and only if* $\nabla \times \boldsymbol{F} = \mathbf{0}$.

Proof Suppose that $\mathbf{F}$ is conservative with force potential V.

$$\therefore \quad \mathbf{F} = -\nabla V, \quad \text{and so} \quad \nabla \times \mathbf{F} = \nabla \times (-\nabla V) = \mathbf{0}.$$

Fig. 1.4 Surface bounded by a closed curve

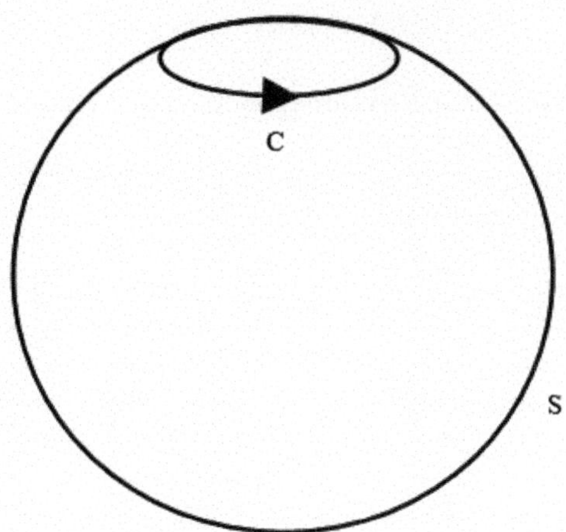

Converse part: Suppose that

$$\nabla \times \mathbf{F} = \mathbf{0}. \tag{1.1.30}$$

Consider a surface S bounded by curve C as seen in Fig. 1.4. Let $\hat{n}$ denote a unit vector normal to the surface S at a point P on S.

By Stoke's theorem,

$$\oint_C \mathbf{F}.d\mathbf{r} = \iint_S \nabla \times \mathbf{F} \cdot \hat{n} ds = 0. \tag{1.1.31}$$

Let A be a fixed point and P be any point on C. Join A and P by the curves and

$\therefore \quad C = C_1 \cup (-C_2)$

(1.1.31) gives

$$\oint_{C_1 \cup (-C_2)} \mathbf{F} \cdot d\mathbf{r} = 0$$

$$\text{i.e.,} \quad \int_{C_1} \mathbf{F} \cdot d\mathbf{r} = \int_{C_2} \mathbf{F} \cdot d\mathbf{r},$$

$$\text{i.e.,} \quad \int_A^P \mathbf{F} \cdot d\mathbf{r} = -V_p,$$

$$\text{i.e.,} \quad \mathbf{F} \cdot d\mathbf{r} = -dV_p = -\nabla V_p \cdot d\mathbf{r},$$

$$\text{i.e.,}\quad (\mathbf{F} + \nabla V_p) \cdot d\mathbf{r} = 0. \tag{1.1.32}$$

In (1.1.32), $d\mathbf{r}$ is arbitrary vector and hence,
we must have

$$\mathbf{F} = -\nabla V_p = -\nabla V, \quad V = -V_p. \tag{1.1.33}$$

That is there exists a scalar point function V such that

$$\mathbf{F} = -\nabla V. \tag{1.1.34}$$

Equation (1.1.34) yields that **F** is conservative. □

Theorem 1.4 *In a conservative force field, the total energy is constant.*

Proof Let W be the work done by a conservative force **F** in moving a particle along a curve C from a fixed point A to any point P on C. It follows from Theorem 1.1, that

$$W = \text{K.E. at P} - \text{K.E. at A} = T - T_1. \tag{1.1.35}$$

As **F** is conservative,

$$\therefore \quad W = \text{P.E. at A} - \text{PE at P} = V_1 - V \quad (\text{Ref. 1.1.14}). \tag{1.1.36}$$

Equations (1.1.35) and (1.1.36) give

$$T + V = T_1 + V_1. \tag{1.1.37}$$

This shows that Total energy at P = Total energy at A.

We note that P is any point on C and A is a fixed point on the same curve C. Equation (1.1.37) establishes the fact that total energy is the same for all points on C. That is to say, that total energy is constant. Thus, in a conservative force field, the total energy is conserved. □

1.1.13 Impulse

Let A and B be the positions of a particle at times t_1 and t_2 and $\mathbf{v}_1$ and $\mathbf{v}_2$ be the velocities of the particle at A and B respectively. Let **F** denote the force acting on the particle. Then the integral

$$\mathbf{I} = \int_{t_1}^{t_2} \mathbf{F}\, dt \tag{1.1.38}$$

is called the impulse of **F** on the particle.

Theorem 1.5 *The impulse is equal to the change in momentum in the direction of the force.*

Proof Let **I** be the impulse of a force **F** applied to a particle of mass m from time $t = t_1$ to $t = t_2$. Let $\mathbf{v}_1$ and $\mathbf{v}_2$ be the velocities of the particle at times $t = t_1$ and $t = t_2$ respectively.

By definition,

$$\mathbf{I} = \int_{t_1}^{t_2} \mathbf{F}\, dt. \tag{1.1.39}$$

By Newton's 2nd law of motion, we have

$$\mathbf{F} = \frac{d}{dt}(m\mathbf{v}). \tag{1.1.40}$$

Using (1.1.40) in (1.1.39), we obtain

$$\mathbf{I} = \int_{t_1}^{t_2} \frac{d}{dt}(m\mathbf{v})\, dt = m \int_{\mathbf{v}_1}^{\mathbf{v}_2} d\mathbf{v} = m\mathbf{v}_2 - m\mathbf{v}_1. \tag{1.1.41}$$

Thus, **I** = Momentum at time t_2 − Momentum at time t_1 = Change in momentum. □

1.1.14 Torque and Angular Momentum

Let **r** denote the position vector of a particle at time t referred to point O moving with velocity **v** under the action of a force **F**. The moment or torque of **F** about O is denoted by $\boldsymbol{\Lambda}$ and defined by

$$\boldsymbol{\Lambda} = \mathbf{r} \times \mathbf{F}. \tag{1.1.42}$$

The angular momentum or moment of momentum of the particle about O is denoted by $\boldsymbol{\Omega}$ and defined by

$$\boldsymbol{\Omega} = \mathbf{r} \times m\mathbf{v}, \tag{1.1.43}$$

where m is the mass of the particle.

Theorem 1.6 *The rate of change of angular momentum is equal to torque.*

Proof Let $\mathbf{r}$ denote the position vector of a particle at time t moving with velocity $\mathbf{v}$. Let $\mathbf{F}$ be the force acting on the particle.

By definition,

$$\boldsymbol{\Lambda} = \mathbf{r} \times \mathbf{F}, \tag{1.1.44}$$

and

$$\boldsymbol{\Omega} = \mathbf{r} \times m\mathbf{v}. \tag{1.1.45}$$

Now

$$\frac{d\boldsymbol{\Omega}}{dt} = \frac{d}{dt}(\mathbf{r} \times m\mathbf{v}) = \frac{d\mathbf{r}}{dt} \times m\mathbf{v} + \mathbf{r} \times \frac{d}{dt}(m\mathbf{v}),$$

$$\text{i.e.,} \quad \frac{d\boldsymbol{\Omega}}{dt} = \mathbf{v} \times m\mathbf{v} + \mathbf{r} \times \mathbf{F} = \mathbf{r} \times \mathbf{F} = \boldsymbol{\Lambda}. \tag{1.1.46}$$

It confirms the proof of the theorem. □

1.1.15 Angular Impulse

The time integral $\mathbf{J} = \int_{t_1}^{t_2} \boldsymbol{\Lambda} dt$ is defined to be the angular impulse of the torque $\boldsymbol{\Lambda}$ applied on a particle about a point O(say) from time $t = t_1$ and $t = t_2$.

Theorem 1.7 *Angular impulse=Change in angular momentum*

Proof By definition and the usual meanings of notations,

$$\mathbf{J} = \int_{t_1}^{t_2} \boldsymbol{\Lambda} dt = \int_{t_1}^{t_2} \frac{d\boldsymbol{\Omega}}{dt} dt = \int_{\boldsymbol{\Omega}_1}^{\boldsymbol{\Omega}_2} d\boldsymbol{\Omega} = \boldsymbol{\Omega}_2 - \boldsymbol{\Omega}_1 = \text{Change in angular momentum.} \tag{1.1.47}$$

This confirms the proof of the theorem. □

1.1.16 Continuous Force and Impulsive Force

A force acting on a body is said to be a continuous force if it acts continuously on the body. For example, the gravitational force. A force is said to be an impulsive force if it acts on the body for an instant only. For example, the blow of a bat on a ball.

1.1.17 Principle of Conservation of Momentum

Theorem 1.8 *If the net external force acting on a particle is zero, its momentum remains unchanged.*

Proof Let a particle of mass m be moving with velocity $\mathbf{v}$ at time t under the action of a force $\mathbf{F}$. By Newton's 2nd law of motion, we have

$$\frac{d}{dt}(m\mathbf{v}) = \mathbf{F} \tag{1.1.48}$$

If $\mathbf{F} = \mathbf{0}$, Eq. (1.1.48) gives $\frac{d}{dt}(m\mathbf{v}) = \mathbf{0}$, which yields, $m\mathbf{v} =$ a constant vector. That is, the momentum of the particle is conserved if the external force acting on the particle is zero. This phenomenon is termed as the *principle of conservation of momentum*. □

1.1.18 Principle of Conservation of Angular Momentum

Theorem 1.9 *If the net external torque applied on a particle is zero, then the angular momentum remains constant.*

Proof Let $\mathbf{\Lambda}$ denote the torque about a point O applied to a moving particle of mass m. Suppose that $\mathbf{\Omega}$ is the angular momentum of the particle about O.

The angular momentum $\mathbf{\Omega}$ and external torque $\mathbf{\Lambda}$ are connected by the relation,

$$\frac{d\mathbf{\Omega}}{dt} = \mathbf{\Lambda}. \tag{1.1.49}$$

If $\mathbf{\Lambda} = \mathbf{0}$, Eq. (1.1.49) reduces to $\dfrac{d\mathbf{\Omega}}{dt} = \mathbf{0}$ which gives, $\mathbf{\Omega} =$ a constant vector. That is the angular momentum is conserved. □

1.2 Partial Derivatives

Consider a function $f : \mathbb{R}^2 \to \mathbb{R}$ defined by $f = f(x,\ y)$. Let us give an increment Δx to x and let Δf be the corresponding change in f.

$$\therefore \qquad f + \Delta f = f(x + \Delta x, y)$$

$$\text{i.e.,} \qquad \Delta f = f(x + \Delta x, y) - f(x, y)$$

$$\therefore \qquad \frac{\Delta f}{\Delta x} = \frac{f(x + \Delta x, y) - f(x, y)}{\Delta x}.$$

This gives

$$\lim_{\Delta x \to 0} \frac{\Delta f}{\Delta x} = \lim_{\Delta x \to 0} \frac{f(x + \Delta x, y) - f(x, y)}{\Delta x}.$$

If this limit exists, then it is denoted by $\frac{\partial f}{\partial x}$ or $f_x(x,\ y)$ and is called the partial derivative of f with respect to x.

$$\therefore \qquad \frac{\partial f}{\partial x} = f_x(x, y) = \lim_{\Delta x \to 0} \frac{f(x + \Delta x, y) - f(x, y)}{\Delta x}.$$

Similarly, we define

$$\therefore \qquad \frac{\partial f}{\partial y} = f_y(x, y) = \lim_{\Delta y \to 0} \frac{f(x, y + \Delta x) - f(x, y)}{\Delta y}$$

to be the partial derivative of f w.r.t. y.

The above results may be generalized for any number of independent variables.

Illustrations

1. Let $f = ax^2 + 2hxy + by^2$, then

$$\frac{\partial f}{\partial x} = 2ax + 2hy,$$

$$\frac{\partial f}{\partial y} = 2hx + 2by.$$

2. Let $f = xy + yz + zx$, then

$$\frac{\partial f}{\partial x} = y + z,$$
$$\frac{\partial f}{\partial y} = x + z,$$
$$\frac{\partial f}{\partial z} = y + x.$$

3. Let $f = \tan^{-1} \frac{y}{x}$, then

$$f_x = \frac{1}{1+\frac{y^2}{x^2}}\left(-\frac{y}{x}\right)$$
$$= -\frac{x^2}{x^2+y^2}\frac{y}{x}$$
$$= -\frac{y}{x^2+y^2}.$$
$$f_y = \frac{1}{1+\frac{y^2}{x^2}}\left(\frac{1}{x}\right)$$
$$= \frac{x}{x^2+y^2}.$$

1.3 Successive Derivatives

For the function $f = f(x, y)$, the usual notations for 2nd order partial derivatives are as follows:

$$\frac{\partial}{\partial x}\left(\frac{\partial f}{\partial x}\right) = \frac{\partial^2 f}{\partial x^2} = f_{xx},$$
$$\frac{\partial}{\partial x}\left(\frac{\partial f}{\partial y}\right) = \frac{\partial^2 f}{\partial x \partial y} = f_{xy},$$
$$\frac{\partial}{\partial y}\left(\frac{\partial f}{\partial x}\right) = \frac{\partial^2 f}{\partial y \partial x} = f_{yx},$$
$$\frac{\partial}{\partial y}\left(\frac{\partial f}{\partial y}\right) = \frac{\partial^2 f}{\partial y^2} = f_{yy}.$$

Illustrations

1. Let $f = x^3 + 3x^2y + 3xy^2 + y^3$, then

$$\begin{aligned}
f_x &= 3x^2 + 6xy + 3y^2,\\
f_y &= 3x^2 + 6xy + 3y^2,\\
f_{xx} &= 6x + 6y,\\
f_{xy} &= 6x + 6y,\\
f_{yx} &= 6x + 6y,\\
f_{yy} &= 6x + 6y.
\end{aligned}$$

2. Let $f = x\cos y + y\cos x$, then

$$\begin{aligned}
f_x &= \cos y - y\sin x,\\
f_y &= -x\sin y + \cos x,\\
f_{xx} &= -y\cos x,\\
f_{xy} &= -\sin y - \sin x,\\
f_{yx} &= -\sin y - \sin x,\\
f_{yy} &= -x\cos y.
\end{aligned}$$

1.4 Homogeneous Function

A function $f(x_1, x_2, \ldots, x_n)$ is said to be a homogeneous function of degree n in x_1, x_2, ..., x_n if

$$f(x_1t, x_2t, \ldots, x_nt) = t^n f(x_1, x_2, \ldots, x_n).$$

Illustrations

1. Let $f(x, y) = ax^2 + 2hxy + by^2$, then

$$\begin{aligned}
f(xt, yt) &= ax^2t^2 + 2hxyt^2 + by^2t^2\\
&= t^2(ax^2 + 2hxy + by^2)\\
&= t^2 f(x, y).
\end{aligned}$$

Therefore, f is a homogeneous function of degree 2 in x and y.

2. Let $f(x, y, z) = yz + zx + xy$, then

$$\begin{aligned} f(xt, yt, zt) &= yzt^2 + zxt^2 + xyt^2 \\ &= t^2(yz + zx + xy) \\ &= t^2 f(x, y, z). \end{aligned}$$

Therefore, f is a homogeneous function of degree 2 in x, y, and z.

3. Let $f(x, y, z) = \sqrt{x} + \sqrt{y} + \sqrt{z}$, then

$$\begin{aligned} f(xt, yt, zt) &= \sqrt{st} + \sqrt{yt} + \sqrt{zt} \\ &= t^{\frac{1}{2}}\left(\sqrt{x} + \sqrt{y} + \sqrt{z}\right) \\ &= t^{\frac{1}{2}} f(x, y, z). \end{aligned}$$

Therefore, f is a homogeneous function of degree $\frac{1}{2}$ in x, y, and z.

1.5 Total Differential Coefficient

Let $f = f(x_1, x_2, \ldots, x_n)$ where $x_1 = \phi_1(t)$, $x_2 = \phi_2(t)$, …, $x_n = \phi_n(t)$, then

$$\frac{df}{dt} = \frac{\partial f}{\partial x_1}\frac{dx_1}{dt} + \frac{\partial f}{\partial x_2}\frac{dx_2}{dt} + \cdots + \frac{\partial f}{\partial x_n}\frac{dx_n}{dt}.$$

1.6 Differential

Let $f = f(x_1, x_2, \ldots, x_n)$. Then

$$df = \frac{\partial f}{\partial x_1}dx_1 + \frac{\partial f}{\partial x_2}dx_2 + \cdots + \frac{\partial f}{\partial x_n}dx_n$$

is called the differential of f.

1.7 Euler's Theorem on Homogeneous Function

Theorem 1.10 *If $f = f(x_1, x_2, \ldots, x_n)$ be a homogeneous function of degree n in x_1, x_2,…, x_n. Then*

$$x_1\frac{\partial f}{\partial x_1}+x_2\frac{\partial f}{\partial x_2}+\cdots+x_n\frac{\partial f}{\partial x_n}=nf.$$

Proof By the definition of a homogeneous function,

$$f(x_1t, x_2t, \ldots, x_nt)=t^n f(x_1, x_2, \ldots, x_n). \tag{1.7.1}$$

Equation (1.7.1) gives

$$f(u_1, u_2, \ldots, u_n)=t^n f(x_1, x_2, \ldots, x_n), \tag{1.7.2}$$

where $u_1 = x_1t$, $u_2 = x_2t$, ..., $u_n = x_nt$.
Differentiating (1.7.2) w.r.t. t, we get

$$\frac{df}{dt}=nt^{n-1}f,$$

$$\text{i.e.,}\quad \frac{\partial f}{\partial u_1}\frac{du_1}{dt}+\frac{\partial f}{\partial u_2}\frac{du_2}{dt}+\cdots+\frac{\partial f}{\partial u_n}\frac{du_n}{dt}=nt^{n-1}f,$$

$$\text{i.e.,}\quad \frac{\partial f}{\partial u_1}x_1+\frac{\partial f}{\partial u_2}x_2+\cdots+\frac{\partial f}{\partial u_n}x_n=nt^{n-1}f. \tag{1.7.3}$$

Putting $t = 1$ in (1.7.3), we get

$$\frac{\partial f}{\partial x_1}x_1+\frac{\partial f}{\partial x_2}x_2+\cdots+\frac{\partial f}{\partial x_n}x_n=nf,$$

$$\text{i.e.,}\quad x_1\frac{\partial f}{\partial x_1}+x_2\frac{\partial f}{\partial x_2}+\cdots+x_n\frac{\partial f}{\partial x_n}=nf.$$

Hence proved. □

1.8 Illustrations

Example 1.1 A particle moves so that its position vector is given by $\mathbf{r} = \cos\omega t\,\hat{i} + \sin\omega t\,\hat{j}$, where ω is constant. Show that
(i) $\mathbf{v}$ is $\perp^r$ $\mathbf{r}$,
(ii) $\mathbf{a}$ is directed towards O,
(iii) $\mathbf{r}\times\mathbf{v}$=a constant vector.

Solution Let P be the position of the particle at time t with position vector $\mathbf{r}$ relative to O.
Given

$$\mathbf{r} = \cos \omega t \, \hat{i} + \sin \omega t \, \hat{j}. \tag{1.8.1}$$

$$\therefore \quad \mathbf{v} = \frac{d\mathbf{r}}{dt} = -\hat{i}\omega \sin \omega t + \hat{j}\omega \cos \omega t \tag{1.8.2}$$

and

$$\mathbf{a} = \frac{d\mathbf{v}}{dt} = -\omega^2 \hat{i} \cos \omega t - \hat{j}\omega^2 \sin \omega t \tag{1.8.3}$$

Equations (1.8.1) and (1.8.2) give

$$\mathbf{r} \cdot \mathbf{v} = -\omega \sin \omega t \cos \omega t + \omega \sin \omega t \cos \omega t = 0. \tag{1.8.4}$$

It shows that **v** is $\perp^r$ to **r**.
Equation (1.8.3) can be written as

$$\mathbf{a} = -\omega^2 \left(\hat{i} \cos \omega t + \hat{j} \sin \omega t \right) = -\omega^2 \mathbf{r} = \omega^2 (-\mathbf{r}) = \omega^2 (-\mathbf{OP}) = \omega^2 \mathbf{PO}. \tag{1.8.5}$$

It shows that **a** acts along **PO**.

That is, **a** is directed towards O.

Again

$$\mathbf{r} \times \mathbf{v} = \begin{vmatrix} \hat{i} & \hat{j} & \hat{k} \\ \cos \omega t & \sin \omega t & 0 \\ -\omega \sin \omega t & \omega \cos \omega t & 0 \end{vmatrix} = \omega \hat{k} = \text{a constant vector.} \tag{1.8.6}$$

Example 1.2 A rod AB of length a rests against a vertical wall OA. The foot B of the rod is pulled away with constant speed v. Find the velocity and speed of the midpoint of the rod at the instant when B is at a distance b ($< a$) from the wall.

Solution Let AB be the position of the rod at time t. Take the X-axis horizontally along **OB** and the Y-axis vertically upwards along **OA** (see Fig. 1.5). Let $\hat{i}$, $\hat{j}$ denote the unit vectors along **OX** and **OY** respectively. Suppose that OB $= x$ and OA $= y$.

Let P be the midpoint of AB with position vector **r**.

$$\therefore \quad \mathbf{r} = \frac{x}{2}\hat{i} + \frac{y}{2}\hat{j}. \tag{1.8.7}$$

Given,

$$\frac{d}{dt}\left(x\hat{i}\right) = v\hat{i},$$

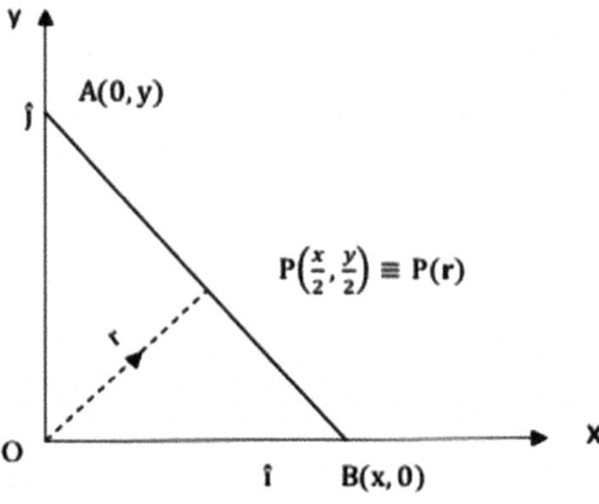

Fig. 1.5 Illustration for Example 1.2

$$\text{i.e.,}\quad \dot{x} = v. \tag{1.8.8}$$

We have

$$AB = a, \quad \text{i.e.,} \quad AB^2 = a^2, \quad \text{i.e.,} \quad x^2 + y^2 = a^2,$$

$$\text{i.e.,}\quad y^2 = a^2 - x^2. \tag{1.8.9}$$

(1.8.9) gives

$$2y\dot{y} = -2x\dot{x},$$

$$\text{i.e.,}\quad y\dot{y} = -xv$$

$$\text{i.e.,}\quad \dot{y} = -\frac{xv}{y} = \frac{-xv}{\sqrt{a^2 - x^2}}. \tag{1.8.10}$$

Now

$$\mathbf{q} = \text{Velocity of P} = \frac{d\mathbf{r}}{dt} = \frac{\dot{x}}{2}\hat{i} + \frac{\dot{y}}{2}\hat{j} = \frac{v}{2}\hat{i} - \frac{xv}{2\sqrt{a^2 - x^2}}\hat{j}. \tag{1.8.11}$$

Let

$$\mathbf{q}_0 = \mathbf{q}\big]_{x=b} = \frac{v}{2}\hat{i} - \frac{bv}{2\sqrt{a^2 - b^2}}\hat{j}. \tag{1.8.12}$$

Equation (1.8.12) gives

$$q_0^2 = \frac{v^2}{4} + \frac{b^2v^2}{4\left(a^2 - b^2\right)} = \frac{a^2v^2}{4\left(a^2 - b^2\right)}. \tag{1.8.13}$$

$$\therefore \quad q_0 = \left|\mathbf{q}_0\right| = \frac{av}{2\sqrt{a^2 - b^2}}. \tag{1.8.14}$$

Example 1.3 Three straight lines are equally inclined to one another in a plane and three equal particles P, Q, and R move along them with velocities **u**, **v**, **w** respectively, show that the magnitude of the velocity of P relative to the centroid of the particles is $\frac{1}{3}\left[4u^2 + v^2 + w^2 + 2uv + 2wu - vw\right]^{1/2}$.

Solution Let m be the mass of each of the particles. Let $\mathbf{r}_1$, $\mathbf{r}_2$, $\mathbf{r}_3$ denote the position vectors of P, Q, and R respectively relative some origin O (see Fig. 1.6). Let G ($\mathbf{r}$) be centroid of the particles.

$$\therefore \quad \mathbf{r} = \frac{m\mathbf{r}_1 + m\mathbf{r}_2 + m\mathbf{r}_3}{m + m + m} = \frac{\mathbf{r}_1 + \mathbf{r}_2 + \mathbf{r}_3}{3}. \tag{1.8.15}$$

Now,

$$\begin{aligned} \dot{\mathbf{r}}_1 &= \text{velocity of P} = \mathbf{u}, \\ \dot{\mathbf{r}}_2 &= \text{velocity of Q} = \mathbf{v}, \\ \text{and} \quad \dot{\mathbf{r}}_3 &= \text{velocity of R} = \mathbf{w}. \end{aligned}$$

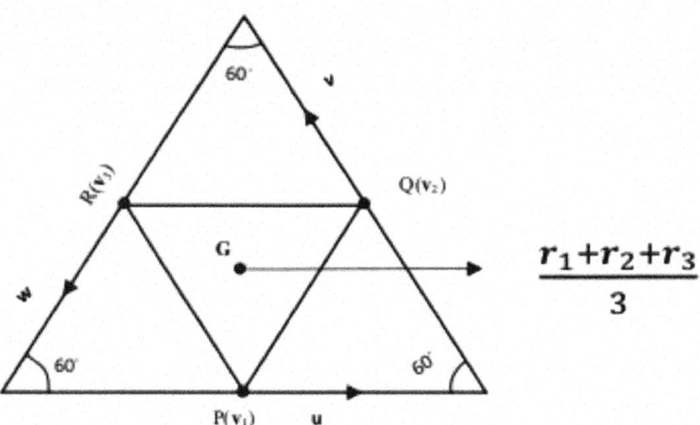

Fig. 1.6 Illustration for Example 1.3

$$\therefore \quad \mathbf{q} = \text{velocity of G} = \dot{\mathbf{r}} = \frac{1}{3}(\dot{\mathbf{r}}_1 + \dot{\mathbf{r}}_2 + \dot{\mathbf{r}}_3) = \frac{1}{3}(\mathbf{u} + \mathbf{v} + \mathbf{w})\,. \tag{1.8.16}$$

Let $\mathbf{q}_0 = \mathbf{q}_{P/G}$ = velocity of P − velocity of G.

$$\therefore \quad \mathbf{q}_0 = \mathbf{u} - \mathbf{q} = \mathbf{u} - \frac{1}{3}(\mathbf{u} + \mathbf{v} + \mathbf{w}) = \frac{1}{3}(2\mathbf{u} - \mathbf{v} - \mathbf{w})\,. \tag{1.8.17}$$

Equation (1.8.17) gives

$$q_0^2 = \frac{1}{9}\left[4u^2 + v^2 + w^2 - 4\mathbf{u}\cdot\mathbf{v} + 2\mathbf{v}\cdot\mathbf{w} - 4\mathbf{w}\cdot\mathbf{u}\right],$$

$$\text{i.e.,} \quad q_0^2 = \frac{1}{9}\left[4u^2 + v^2 + w^2 - 4uv\,\cos 120^\circ + 2vw\,\cos 120^\circ - 4wu\,\cos 120^\circ\right],$$

$$\text{i.e.,} \quad q_0^2 = \frac{1}{9}\left[4u^2 + v^2 + w^2 + 2uv - vw + 2wu\right],$$

$$\therefore \quad q_0 = \frac{1}{3}\left[4u^2 + v^2 + w^2 - 2uv - vw + 2wu\right]^{\frac{1}{2}}. \tag{1.8.18}$$

Example 1.4 An impulse **I** changes the velocity of a part of mass m from $\mathbf{v}_1$ to $\mathbf{v}_2$; show that the KE gained is $\frac{1}{2}\mathbf{I}\cdot(\mathbf{v}_1 + \mathbf{v}_2)$.

Solution

$$\begin{aligned} \mathbf{I} &= \text{Impulse applied to the particle} \\ &= \text{Change in momentum} = m\mathbf{v}_2 - m\mathbf{v}_1\,. \end{aligned} \tag{1.8.19}$$

Now

$$T = \text{KE gained} = \text{KE just after impulse} - \text{KE just before impulse},$$

$$\text{i.e.,} \quad T = \frac{1}{2}mv_2^2 - \frac{1}{2}mv_1^2 = \frac{1}{2}m\left(v_2^2 - v_1^2\right) = \frac{1}{2}m\,(\mathbf{v}_2 + \mathbf{v}_1)\cdot(\mathbf{v}_2 - \mathbf{v}_1)\,,$$

$$\text{i.e.,} \quad T = \frac{1}{2}(\mathbf{v}_2 + \mathbf{v}_1)\cdot(m\mathbf{v}_2 - m\mathbf{v}_1) = \frac{1}{2}(\mathbf{v}_2 + \mathbf{v}_1)\cdot m\,(\mathbf{v}_2 - \mathbf{v}_1)\,,$$

$$\text{i.e.,} \quad T = \frac{1}{2}(\mathbf{v}_2 + \mathbf{v}_1)\cdot\mathbf{I} = \frac{1}{2}\mathbf{I}\cdot(\mathbf{v}_1 + \mathbf{v}_2)\,. \tag{1.8.20}$$

Example 1.5 Two particles of masses m_1 and m_2 move so that their relative velocity is **V** and the velocity of their center of mass is $\mathbf{V}_0$. Show that the total KE is $\frac{1}{2}M\mathbf{V}_0^2 + \frac{1}{2}\mu\mathbf{V}^2$ where $M = m_1 + m_2$ and $\mu = \frac{m_1 m_2}{m_1+m_2}$.

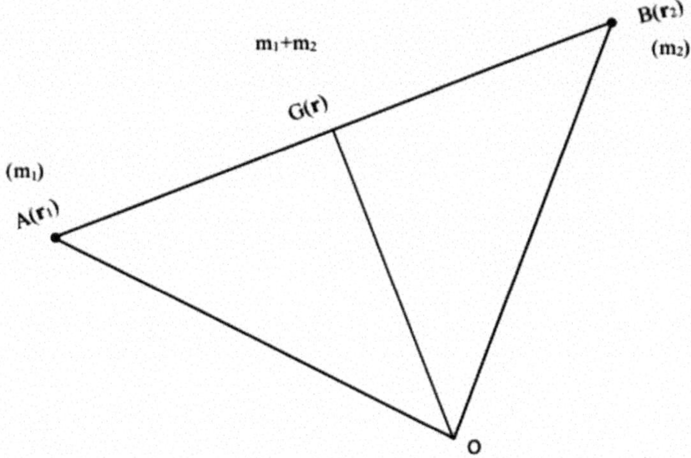

Fig. 1.7 Illustration for Example 1.5

Solution Let A and B denote the positions of the particles of masses m_1 and m_2 respectively at time t as described in Fig. 1.7. Suppose that $\mathbf{r}_1$ and $\mathbf{r}_2$ are the position vectors of the points A and B relative to some origin O. Let G be the center of mass with position vector $\mathbf{r}$.

$$\therefore \quad \mathbf{r} = \frac{m_1\mathbf{r}_1 + m_2\mathbf{r}_2}{m_1 + m_2} = \frac{m_1\mathbf{r}_1 + m_2\mathbf{r}_2}{M}. \tag{1.8.21}$$

Now,

$$\mathbf{V}_0 = \text{velocity of C.M.} = \dot{\mathbf{r}} = \frac{m_1\dot{\mathbf{r}}_1 + m_2\dot{\mathbf{r}}_2}{M} = \frac{m_1\mathbf{v}_1 + m_2\mathbf{v}_2}{M}, \tag{1.8.22}$$

where $\mathbf{v}_1 = \dot{\mathbf{r}}_1 =$ velocity of m_1, $\mathbf{v}_2 = \dot{\mathbf{r}}_2 =$ velocity of m_2.

Equation (1.8.22) gives,

$$m_1\mathbf{v_1} + m_2\mathbf{v_2} = M\mathbf{V}_0. \tag{1.8.23}$$

Further,

$$\mathbf{V} = \text{Velocity of } m_2 \text{ relative to } m_1 = \mathbf{v}_2 - \mathbf{v}_1. \tag{1.8.24}$$

Solving (1.8.23) and (1.8.24), we obtain

$$\mathbf{v}_1 = \mathbf{V}_0 - \frac{m_2}{M}\mathbf{V}, \tag{1.8.25}$$

and

$$\mathbf{v}_2 = \mathbf{V}_0 + \frac{m_1}{M}\mathbf{V}. \tag{1.8.26}$$

Therefore, the total KE is given by

$$T = \frac{1}{2}m_1\mathbf{v}_1^2 + \frac{1}{2}m_2\mathbf{v}_2^2 = \frac{1}{2}m_1\left(\mathbf{V}_0 - \frac{m_2}{M}\mathbf{V}\right)^2 + \frac{1}{2}m_2\left(\mathbf{V}_0 + \frac{m_1}{M}\mathbf{V}\right)^2,$$

$$\text{i.e., } 2T = (m_1 + m_2)\,\mathbf{V}_0^2 + \left(m_1\frac{m_2^2}{M^2} + m_2\frac{m_1^2}{M^2}\right)\mathbf{V}^2 = M\mathbf{V}_0^2 + \frac{m_1 m_2\,(m_1 + m_2)}{M^2}\mathbf{V}^2,$$

$$\text{i.e., } 2T = M\mathbf{V}_0^2 + \frac{m_1 m_2}{m_1 + m_2}\mathbf{V}^2 = M\mathbf{V}_0^2 + \mu\mathbf{V}^2,$$

$$\text{i.e., } T = \frac{1}{2}M\mathbf{V}_0^2 + \frac{1}{2}\mu\mathbf{V}^2.$$

Example 1.6 Show that $\mathbf{F} = f(r)\,\hat{r}$ is conservative.

Solution $\mathbf{F} = f(r)\,\hat{r} = f(r)\,\frac{\mathbf{r}}{r} = \varphi(r)\,\mathbf{r};\quad \varphi(r) = \frac{f(r)}{r}$

$$\begin{aligned}\therefore\quad \nabla \times \mathbf{F} &= \nabla \times \varphi(r)\,\mathbf{r}\\ &= \nabla\varphi(r) \times \mathbf{r} + \varphi(r)\,\nabla \times \mathbf{r}\\ &= \varphi'(r)\,\nabla r \times \mathbf{r} + \varphi(r)\,\mathbf{0}\\ &= \varphi'(r)\,\hat{r} \times \mathbf{r} = \varphi'(r)\,\frac{\mathbf{r}}{r} \times \mathbf{r} = \mathbf{0}.\end{aligned}$$

Example 1.7 Show that the force $\mathbf{F}$ given by $\mathbf{F} = \left(2xy + z^3\right)\hat{\imath} + x^2\hat{\jmath} + 3xz^2\hat{k}$ is conservative and find the corresponding force potential.

Solution Given $\mathbf{F} = \left(2xy + z^3\right)\hat{\imath} + x^2\hat{\jmath} + 3xz^2\hat{k}$.

$$\therefore\quad \nabla \times \mathbf{F} = \begin{vmatrix} \hat{\imath} & \hat{\jmath} & \hat{k}\\ \frac{\partial}{\partial x} & \frac{\partial}{\partial y} & \frac{\partial}{\partial z}\\ 2xy + z^3 & x^2 & 3xz^2 \end{vmatrix}$$

$$= \hat{\imath}\,(0 - 0) + \hat{\jmath}\,\left(3z^2 - 3z^2\right) + \hat{k}\,(2x - 2x) = \mathbf{0}. \tag{1.8.27}$$

Therefore, $\mathbf{F}$ is conservative.

Let V denote the force potential.

$$\therefore\quad \mathbf{F} = -\nabla V. \tag{1.8.28}$$

This gives

$$\frac{\partial V}{\partial x} = -X = -2xy - z^3, \tag{1.8.29}$$

$$\frac{\partial V}{\partial y} = -Y = -x^2, \tag{1.8.30}$$

$$\frac{\partial V}{\partial z} = -Z = -3xz^2. \tag{1.8.31}$$

Integrating (1.8.29) w.r.t. x (1.8.30) w.r.t. y and (1.8.31) w.r.t. z, we obtain

$$V = -x^2y - z^3x + f_1(y, z), \tag{1.8.32}$$

$$V = -x^2y + f_2(z, x), \tag{1.8.33}$$

$$V = -z^3x + f_3(x, y). \tag{1.8.34}$$

Equation (1.8.32)–(1.8.34) give

$$V = -x^2y - z^3x.$$

Example 1.8 A particle of mass m moves along the X-axis under the influence of a conservative force field having potential $V(x)$. If the particle is located at positions x_1 and x_2 at respective times $t_1$and t_2, prove that if E is the total energy $t_2 - t_1 = \sqrt{\frac{m}{2}}\int\limits_{x_1}^{x_2}\frac{dx}{\sqrt{E-V(x)}}$. Find also the impulse of the force acting on the particle in the time interval $t_2 - t_1$

Solution By the law of conservation of energy,

$$T + V = E,$$

$$\text{i.e.,}\quad \frac{1}{2}mv^2 + V(x) = E,$$

$$\text{i.e.,}\quad mv^2 = 2[E - V(x)],$$

$$\text{i.e.,}\quad v^2 = \frac{2}{m}[E - V(x)],$$

$$\text{i.e.,}\quad v = \sqrt{\frac{2}{m}[E - V(x)]},$$

$$\text{i.e.,}\quad dx = \sqrt{\frac{2}{m}}\sqrt{E - V(x)}\,dt,$$

$$\text{i.e.,}\quad \frac{dx}{\sqrt{E - V(x)}} = \sqrt{\frac{2}{m}}\,dt,$$

$$\text{i.e.,}\quad \int_{x_1}^{x_2} \frac{dx}{\sqrt{E - V(x)}} = \sqrt{\frac{2}{m}}\int_{t_1}^{t_2} dt = \sqrt{\frac{2}{m}}\,(t_2 - t_1),$$

$$\text{i.e.,}\quad t_2 - t_1 = \sqrt{\frac{m}{2}}\int_{x_1}^{x_2} \frac{dx}{\sqrt{E - V(x)}}.$$

2nd part: The impulse of the force =Change in momentum,

$$\text{i.e.,}\quad \text{Impulse} = mv_2 - mv_1 = mv(x_2) - mv(x_1) = m\left[v(x_2) - v(x_1)\right],$$

$$\text{i.e.,}\quad \text{Impulse} = \sqrt{2m}\left[\sqrt{E - V(x_2)} - \sqrt{E - V(x_1)}\right].$$

Example 1.9 A particle moves on the xy plane under a force $\mathbf{F} = X\hat{i} + Y\hat{j}$ per unit mass (X, Y being functions of x, y in any position). Write the vector equation of motion of the particle and show that the differential equation of its path is $\frac{d}{dx}\left[\frac{\dot{Y} - \dot{X}\frac{dy}{dx}}{\frac{d^2y}{dx^2}}\right] = 2X$.

Solution Let $\mathbf{r}$ be position vector of the particle of mass m at time t relative to some origin.

By Newton's 2nd law of motion,

$$m\ddot{\mathbf{r}} = m\mathbf{F},$$

$$\text{i.e.,}\quad \ddot{\mathbf{r}} = \mathbf{F}, \tag{1.8.35}$$

which is the vector equation of motion of the particle.

Let

$$\mathbf{r} = \hat{i}x + \hat{j}y.$$

$$\therefore\quad \ddot{\mathbf{r}} = \hat{i}\,\ddot{x} + \hat{j}\,\ddot{y},$$

and hence Eq. (1.8.35) gives

$$\hat{i}\,\ddot{x} + \hat{j}\,\ddot{y} = X\hat{i} + Y\hat{j}. \tag{1.8.36}$$

Equation (1.8.36) yields

$$\left.\begin{array}{l} \ddot{x} = X \\ \ddot{y} = Y \end{array}\right\}. \tag{1.8.37}$$

Equation (1.8.37) is the equation of motion of the particle in the (x, y) system.

Now,

$$\frac{dy}{dx} = \frac{\frac{dy}{dt}}{\frac{dx}{dt}} = \frac{\dot{y}}{\dot{x}}$$

and

$$\frac{d^2y}{dx^2} = \frac{d}{dx}\frac{\dot{y}}{\dot{x}} = \frac{\dot{x}\,\ddot{y}\frac{1}{\dot{x}} - \dot{y}\,\ddot{x}\frac{1}{\dot{x}}}{\dot{x}^2} = \frac{\dot{x}\,\ddot{y} - \dot{y}\,\ddot{x}}{\dot{x}^3}.$$

We have

$$\frac{Y - X\frac{dy}{dx}}{\frac{d^2y}{dx^2}} = \frac{\ddot{y} - \ddot{x}\frac{\dot{y}}{\dot{x}}}{\frac{\dot{x}\ddot{y} - \dot{y}\ddot{x}}{\dot{x}^3}} = \dot{x}^2,$$

$$\therefore \quad \frac{d}{dx}\left[\frac{Y - X\frac{dy}{dx}}{\frac{d^2y}{dx^2}}\right] = \frac{d}{dx}\dot{x}^2 = 2\dot{x}\frac{d\dot{x}}{dx} = 2\dot{x}\,\ddot{x}/\dot{x} = 2\,\ddot{x} = 2X.$$

Example 1.10 In $(r,\ \theta,\ \varphi)$ coordinate system, a force $\mathbf{F}$ is given by

$$\mathbf{F} = \left(2r^{-3}\cos\theta,\ r^{-3}\sin\theta,\ 0\right).$$

Show that $\mathbf{F}$ is solenoidal and conservative. Obtain the force potential and lines of force.

Solution

$$\nabla\cdot\mathbf{F} = \frac{1}{h_1h_2h_3}\left[\frac{\partial}{\partial\lambda}(h_2h_3F_1) + \frac{\partial}{\partial\mu}(h_3h_1F_2) + \frac{\partial}{\partial\nu}(h_1h_2F_3)\right] \tag{1.8.38}$$

and

$$\nabla\times\mathbf{F} = \frac{1}{h_1h_2h_3}\begin{vmatrix} h_1\hat{a}_1 & h_2\hat{a}_2 & h_3\hat{a}_3 \\ \frac{\partial}{\partial\lambda} & \frac{\partial}{\partial\mu} & \frac{\partial}{\partial\nu} \\ h_1F_1 & h_2F_2 & h_3F \end{vmatrix}, \tag{1.8.39}$$

where

$$\mathbf{F} = F_1\hat{a}_1 + F_2\hat{a}_2 + F_3\hat{a}_3,$$
$$ds^2 = h_1^2 d\lambda^2 + h_2^2 d\mu^2 + h_3^2 d\nu^2,$$
$$d\mathbf{r} = h_1\hat{a}_1 d\lambda + h_2\hat{a}_2 d\mu + h_3\hat{a}_3 d\nu,$$

and $\hat{a}_1,\ \hat{a}_2,\ \hat{a}_3$ are the unit vectors in $\lambda,\ \mu,\ \nu$ directions.

In $(r,\ \theta,\ \varphi)$ system,

$$\lambda = r,\quad \mu = \theta,\quad \nu = \varphi;$$
$$\hat{a}_1 = \hat{r},\quad \hat{a}_2 = \hat{\theta},\quad \hat{a}_3 = \hat{\varphi};$$
$$h_1 = 1,\quad h_2 = r,\quad h_3 = r\sin\theta;$$
$$F_1 = F_r,\quad F_2 = F_\theta,\quad F_3 = F_\varphi.$$

Here $F_r = 2r^{-3}\cos\theta,\ F_\theta = r^{-3}\sin\theta,\ F_\varphi = 0.$
Equation (1.8.38) gives

$$\begin{aligned}\nabla\cdot\mathbf{F} &= \frac{1}{r^2\sin\theta}\left[\frac{\partial}{\partial r}\left(r.r\sin\theta.2r^{-3}\cos\theta\right) + \frac{\partial}{\partial\theta}\left(r\sin\theta.1.r^{-3}\sin\theta\right)\right]\\ &= \frac{1}{r^2\sin\theta}\left[\frac{\partial}{\partial r}\left(2r^{-1}\sin\theta\cos\theta\right) + \frac{\partial}{\partial\theta}\left(r^{-2}\sin^2\theta\right)\right]\\ &= \frac{1}{r^2\sin\theta}\left[2\sin\theta\cos\theta\,(-1)\,r^{-2} + r^{-2}.2\sin\theta\cos\theta\right]\\ &= 0.\end{aligned} \tag{1.8.40}$$

Equation (1.8.40) shows that **F** is solenoidal. Equation (1.8.39) yields

$$\nabla\times\mathbf{F} = \frac{1}{1.r.r\sin\theta}\begin{vmatrix}\hat{r} & r\hat{\theta} & r\sin\theta\hat{\varphi}\\ \dfrac{\partial}{\partial r} & \dfrac{\partial}{\partial\theta} & \dfrac{\partial}{\partial\varphi}\\ F_r & rF_r & 0\end{vmatrix}$$

$$= \frac{1}{r^2\sin\theta}\begin{vmatrix}\hat{r} & r\hat{\theta} & r\sin\theta\hat{\varphi}\\ \dfrac{\partial}{\partial r} & \dfrac{\partial}{\partial\theta} & \dfrac{\partial}{\partial\varphi}\\ 2r^{-3}\cos\theta & rr^{-3}\sin\theta & 0\end{vmatrix}$$

$$= \frac{1}{r^2 \sin\theta} \begin{vmatrix} \hat{r} & r\hat{\theta} & r\sin\theta\hat{\varphi} \\ \dfrac{\partial}{\partial r} & \dfrac{\partial}{\partial \theta} & \dfrac{\partial}{\partial \varphi} \\ 2r^{-3}\cos\theta & r^{-2}\sin\theta & 0 \end{vmatrix}$$

$$= \frac{r\sin\theta\hat{\varphi}}{r^2\sin\theta}\left[\sin\theta(-2r^{-3}) - 2r^{-3}(-\sin\theta)\right]$$

$$= \mathbf{0}. \tag{1.8.41}$$

Thus **F** is conservative.

Example 1.11 A force is given by

$$\mathbf{F} = (X, Y, Z);\;\; X = \frac{-2xyz}{\left(x^2+y^2\right)^2},\quad Y = \frac{\left(x^2-y^2\right)z}{\left(x^2+y^2\right)^2},\quad Z = \frac{y}{x^2+y^2}$$

(i) Show that **F** is solenoidal.
(ii) Show that **F** is conservative.
(iii) Obtain the force potential corresponding to **F**.
(iv) Obtain the equations of lines of force.
(v) Obtain the equipotential energy surfaces.

Solution 1st Part: Given

$$X = \frac{-2xyz}{\left(x^2+y^2\right)^2},\quad Y = \frac{\left(x^2-y^2\right)z}{\left(x^2+y^2\right)^2},\quad Z = \frac{y}{x^2+y^2},$$

$$\text{i.e.}\;\; X = \frac{-2xyz}{r^4},\quad Y = \frac{\left(x^2-y^2\right)z}{r^4},\quad Z = \frac{y}{r^2};\;\; r^2 = x^2+y^2.$$

We have

$$\frac{\partial r}{\partial x} = \frac{x}{r},\quad \frac{\partial r}{\partial y} = \frac{y}{r},\quad \frac{\partial r}{\partial z} = 0.$$

$$\frac{\partial X}{\partial x} = -2yz\frac{r^2-4x^2}{r^6},\quad \frac{\partial Y}{\partial y} = -2yz\frac{r^2+2x^2-2y^2}{r^6},\quad \frac{\partial Z}{\partial z} = 0.$$

Thus

$$\nabla\cdot\mathbf{F} = \frac{\partial X}{\partial x} + \frac{\partial Y}{\partial y} + \frac{\partial Z}{\partial z} = -\frac{2yz}{r^6}\left(r^2-4x^2+r^2+2x^2-2y^2\right),$$

$$\text{i.e.,}\;\; \nabla\cdot\mathbf{F} = -\frac{2yz}{r^6}\left(2r^2-2x^2-2y^2\right),$$

$$\text{i.e.,}\quad \nabla\cdot\mathbf{F}=-\frac{2yz}{r^6}\left(2r^2-2r^2\right)=0,$$

i.e., **F** is solenoidal.

2nd Part:

$$\boldsymbol{\xi}=\nabla\times\mathbf{F}=\begin{vmatrix}\hat{i} & \hat{j} & \hat{k}\\ \frac{\partial}{\partial x} & \frac{\partial}{\partial y} & \frac{\partial}{\partial z}\\ X & Y & Z\end{vmatrix}=\hat{i}\left(\frac{\partial Z}{\partial y}-\frac{\partial Y}{\partial z}\right)+\hat{j}\left(\frac{\partial X}{\partial z}-\frac{\partial Z}{\partial x}\right)+\hat{k}\left(\frac{\partial Y}{\partial x}-\frac{\partial X}{\partial y}\right),$$

$$\text{i.e.,}\quad \boldsymbol{\xi}=\hat{i}\xi_x+\hat{j}\xi_y+\hat{k}\xi_z;$$

where

$$\xi_x=\frac{\partial Z}{\partial y}-\frac{\partial Y}{\partial z},\quad \xi_y=\frac{\partial X}{\partial z}-\frac{\partial Z}{\partial x},\quad \xi_z=\frac{\partial Y}{\partial x}-\frac{\partial X}{\partial y}.$$

Now

$$\xi_x=\frac{\partial Z}{\partial y}-\frac{\partial Y}{\partial z}=\frac{r^2-2y^2}{r^4}-\frac{x^2-y^2}{r^4}=\frac{1}{r^4}\left(r^2-x^2-y^2\right)=0,$$

$$\xi_y=\frac{\partial X}{\partial z}-\frac{\partial Z}{\partial x}=-\frac{2xy}{r^4}+\frac{2xy}{r^4}=0,$$

$$\xi_z=\frac{\partial Y}{\partial x}-\frac{\partial X}{\partial y}=\frac{2xz}{r^6}\left[r^2-2\left(x^2-y^2\right)\right]+\frac{2xz}{r^6}\left(r^2-4y^2\right),$$

$$\text{i.e.,}\quad \xi_z=\frac{2xz}{r^6}\left[r^2-2\left(x^2-y^2\right)+r^2-4y^2\right]=\frac{2xz}{r^6}\left(2r^2-2x^2-2y^2\right),$$

$$\text{i.e.,}\quad \xi_z=\frac{2xz}{r^6}\left(2r^2-2r^2\right)=0.$$

Thus

$$\boldsymbol{\xi}=\nabla\times\mathbf{F}=\xi_x\hat{i}+\xi_y\hat{j}+\xi_z\hat{k}=\mathbf{0}.$$

Hence **F** is irrotational, i.e. **F** is conservative.

3rd Part: Let V be the force potential.

$$\therefore\quad \mathbf{F}=-\nabla V,$$

$$\text{i.e.,}\quad \mathbf{F}\cdot d\mathbf{r}=-\nabla V\cdot d\mathbf{r}=-dV,$$

i.e., $-dV = Xdx + Ydy + Zdz,$

i.e., $$-dV = -\frac{2xyz}{\left(x^2+y^2\right)^2}dx + \frac{\left(x^2-y^2\right)}{\left(x^2+y^2\right)^2}dy + \frac{y}{x^2+y^2}dz,$$

i.e., $$-dV = \frac{-2xyzdx + \left(x^2-y^2\right)zdy + y\left(x^2+y^2\right)dz}{\left(x^2+y^2\right)^2},$$

i.e., $$-dV = \frac{y\left(x^2+y^2\right)dz + \left(x^2+y^2-2y^2\right)zdy - 2xyzdx}{\left(x^2+y^2\right)^2},$$

i.e., $$-dV = \frac{y\left(x^2+y^2\right)dz + \left(x^2+y^2\right)zdy - 2y^2zdy - 2xyzdx}{\left(x^2+y^2\right)^2},$$

i.e., $$-dV = \frac{\left(x^2+y^2\right)(ydz+zdy) - 2y^2zdy - 2xyzdx}{\left(x^2+y^2\right)^2},$$

i.e., $$-dV = \frac{\left(x^2+y^2\right)d(yz) - 2yz(xdx+ydy)}{\left(x^2+y^2\right)^2},$$

i.e., $$-dV = \frac{\left(x^2+y^2\right)d(yz) - yzd\left(x^2+y^2\right)}{\left(x^2+y^2\right)^2},$$

i.e., $$-dV = d\left(\frac{yz}{x^2+y^2}\right),$$

i.e., $$V = -\frac{yz}{x^2+y^2}.$$

4th part: Force lines are given by

$$\frac{dx}{X} = \frac{dy}{Y} = \frac{dz}{Z},$$

i.e., $$\frac{r^4dx}{-2xyz} = \frac{r^4dy}{\left(x^2-y^2\right)z} = \frac{r^2dz}{y},$$

i.e., $$\frac{dx}{-2xyz} = \frac{dy}{\left(x^2-y^2\right)z} = \frac{dz}{yr^2},$$

$$\text{i.e.,}\quad \frac{dx}{-2xyz} = \frac{dy}{\left(x^2 - y^2\right)z} = \frac{dz}{y\left(x^2 + y^2\right)} = \frac{xdx + ydy + zdz}{0}. \tag{1.8.42}$$

4th ratio of (1.8.42) gives

$$xdx + ydy + zdz = 0,$$

$$\text{i.e.,}\quad x^2 + y^2 + z^2 = b^2,$$

which represents a family of concentric spheres with center at the origin.
1st and 2nd ratios of (1.8.42) give

$$\frac{dx}{-2xy} = \frac{dy}{x^2 - y^2},$$

$$\text{or}\quad \left(x^2 - y^2\right)dx + 2xydy = 0. \tag{1.8.43}$$

In Eq. (1.8.43),

$$M = x^2 - y^2,\quad N = 2xy,$$

$$\frac{\partial M}{\partial y} = -2y,\quad \frac{\partial N}{\partial x} = 2y.$$

Now,

$$\frac{\dfrac{\partial M}{\partial y} - \dfrac{\partial N}{\partial x}}{N} = -\frac{2}{x},$$

$$\therefore\quad \text{IF} = e^{\int P dx} = e^{\int -\frac{2}{x} dx} = e^{-2\log x} = \frac{1}{x^2}.$$

Equation (1.8.43) reduces to

$$\left(1 - \frac{y^2}{x^2}\right)dx + \frac{2y}{x}dy = 0,$$

which is exact and hence its solution is

$$x + \frac{y^2}{x} = 2c,$$

$$\text{i.e.,}\quad x^2 + y^2 = 2cx,$$

$$\text{i.e.,}\quad (x - c)^2 + y^2 = c^2,$$

which represents a family of right circular cylinders touching the z-axis.

Thus, the force lines are given by

$$\left.\begin{array}{l} x^2 + y^2 + z^2 = b^2 \\ (x - c)^2 + y^2 = c^2 \end{array}\right\}.$$

5th Part: Equi-potential energy surfaces are given by

$$V = \text{Constant},$$

$$\text{i.e.,}\quad \frac{yz}{x^2 + y^2} = \text{Constant},$$

$$\text{i.e.,}\quad \frac{x^2 + y^2}{yz} = \text{Constant} = \lambda \ (\text{say}),$$

$$\text{i.e.,}\quad x^2 + y^2 = \lambda yz,$$

which represents a family of cones with vertex at the origin.

Example 1.12 A shell of mass m is ejected from a gun of mass M by an explosion which generates K.E. E. Prove that the initial velocity of the shell is $\sqrt{\frac{2ME}{(M+m)m}}$.

Solution Let v be muzzle velocity of the shell and V be the recoil velocity of the gun.

By the law of conservation of linear momentum,

$$mv + M(-V) = m.0 + M.0,$$

$$\text{i.e.,}\quad mv = MV,$$

$$\text{i.e.,}\quad \frac{v}{M} = \frac{V}{m} = k\,(say)$$

$$\therefore\quad v = Mk \text{ and } V = mk.$$

As the energy E is generated,

$$E = \frac{1}{2}mv^2 + \frac{1}{2}MV^2,$$

$$\text{i.e.,}\quad 2E = mv^2 + MV^2,$$

$$\text{i.e.,}\quad 2E = mM^2k^2 + Mm^2k^2 = mM(M+m)k^2,$$

$$\text{i.e.,}\quad k^2 = \frac{2E}{mM(M+m)},$$

$$\text{i.e.,}\quad k = \sqrt{\frac{2E}{mM(M+m)}}.$$

Therefore, the initial velocity of the shell is given by

$$v = Mk = M\sqrt{\frac{2E}{mM(M+m)}} = \sqrt{\frac{2EM}{m(M+m)}}.$$

Example 1.13 A gun of mass M fires a shell of mass m horizontally, and the energy of the explosion is such as would be sufficient to project the shell vertically to a height h. Show that the velocity of recoil of the gun is $\left[\frac{2m^2gh}{M(m+M)}\right]^{\frac{1}{2}}$.

Solution Let v be the muzzle velocity of the shell and V denote the recoil velocity of the gun.

By the principle of conservation of linear momentum,

$$mv = MV,$$

$$\text{i.e.,}\quad \frac{v}{M} = \frac{V}{m} = k\,(say)$$

$$\text{i.e.,}\quad v = Mk,\quad V = mk. \tag{1.8.44}$$

If E is the energy generated due to the explosion, then

$$E = mgh. \tag{1.8.45}$$

Further

$$E = \frac{1}{2}mv^2 + \frac{1}{2}MV^2,$$

$$\text{i.e.,}\quad 2E = mv^2 + MV^2,$$

$$\text{i.e.,}\quad 2mgh = mv^2 + MV^2,$$

$$\text{i.e.,}\quad 2mgh = mM^2k^2 + Mm^2k^2 = mM(m+M)k^2,$$

$$\text{i.e.,}\quad k^2 = \frac{2gh}{M\,(m+M)}$$

$$\text{i.e.,}\quad k = \sqrt{\frac{2gh}{M\,(m+M)}}. \tag{1.8.46}$$

The velocity of recoil of the gun is given by

$$V = mk = m\sqrt{\frac{2gh}{M\,(m+M)}} = \sqrt{\frac{2m^2gh}{M\,(m+M)}}$$

which is the required result.

Example 1.14 A bullet of mass m with velocity u strikes a block of mass M, which is free to move in the direction of the motion of the bullet and is embedded in it. Show that the loss of KE is $\frac{1}{2}\ \frac{mM}{m+M}u^2$.

Solution It is given that a block of mass M is struck by a bullet of mass m, as illustrated in Fig. 1.8. Let v be the velocity of the combined body after the impact.

By the law of conservation of momentum, we have

$$mu + M\,0 = (m+M)\ v,$$

$$\text{i.e.,}\quad v = \frac{mu}{m+M}. \tag{1.8.47}$$

Loss in KE = KE of the system before impact − KE of the system after impact

$$= \frac{1}{2}mu^2 - \frac{1}{2}\,(m+M)\,v^2 = \frac{1}{2}mu^2 - \frac{1}{2}\,(m+M)\,\frac{m^2u^2}{(m+M)^2}$$

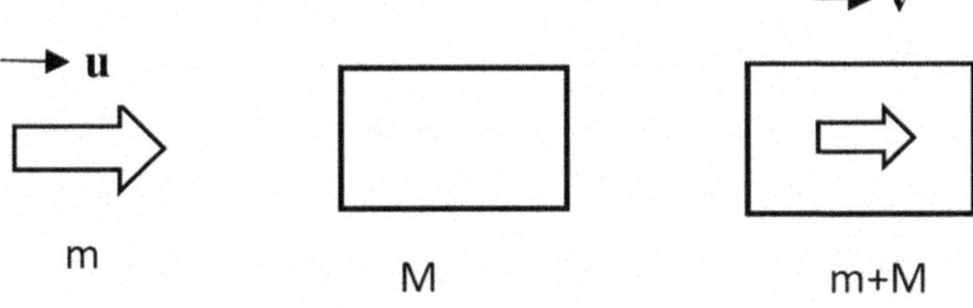

Fig. 1.8 Illustration for Example 1.14

$$= \frac{1}{2}\frac{mM}{m+M}u^2 = \frac{1}{2}mu^2\left[1 - \frac{m}{m+M}\right] = \frac{1}{2}mu^2\frac{M}{m+M}.$$

Example 1.15 A small block of mass M is moving with velocity V when it is struck by a bullet of mass m moving with velocity v in the same direction. If the bullet becomes embedded in the block, show that there is a loss of KE $\frac{Mm(V-v)^2}{2(m+M)}$ and find the impulse on the block.

Solution Let u denote the velocity of the block after impact and let V be the velocity of the block of mass M as seen in Fig. 1.9, therefore, by the principle of conservation of linear momentum, we have

$$(m+M)\,u = mv + MV; \tag{1.8.48}$$

$$\text{or} \quad u = \frac{mv + MV}{m+M}. \tag{1.8.49}$$

Loss of KE = KE of the system just before impact−KE of the system after impact Therefore,

$$\begin{aligned}
\text{Loss in K.E.} &= \frac{1}{2}mv^2 + \frac{1}{2}MV^2 - \frac{1}{2}(m+M)\,u^2 \\
&= \frac{1}{2}\left[mv^2 + MV^2 - (m+M)\left(\frac{mv+MV}{m+M}\right)^2\right] \\
&= \frac{1}{2}\left[mv^2 + MV^2 - \frac{(mv+MV)^2}{m+M}\right] \\
&= \frac{1}{2\,(m+M)}\left[m\,(m+M)\,v^2 + M\,(m+M)\,V^2 - (mv+MV)^2\right] \\
&= \frac{1}{2\,(m+M)}\left[mMv^2 + mMV^2 - 2mMvV\right] \\
&= \frac{mM}{2\,(m+M)}(V-v)^2.
\end{aligned} \tag{1.8.50}$$

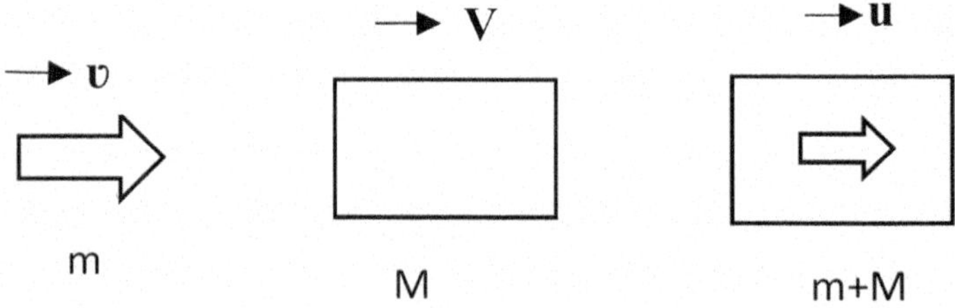

Fig. 1.9 Illustration for Example 1.15

Impulse on the block = Change in momentum

= Momentum after impact − momentum before impact.

$$\text{i.e., Impulse on the block} = (m+M)u - MV$$
$$= (m+M)\frac{mv+MU}{m+M} - MU$$
$$= mv.$$

1.9 Exercise-I

1. A shell lying in a smooth horizontal tube suddenly explodes and breaks into two portions of masses m and m'. If s is the distance apart in the tube of the masses after time t, show that the work done by the explosion is $\frac{1}{2}\,\frac{mm'}{m+m'}\,\frac{s^2}{t^2}$.
2. Show that the force **F** given by $\mathbf{F} = \left(y^2 - 2xyz^3\right)\hat{\imath} + \left(3 + 2xy - x^2z^3\right)\hat{\jmath} + \left(6z^3 - 3x^2yz^2\right)\hat{k}$ is conservative and find its corresponding force potential.
   Answer: $V = x^2yz^3 - xy^2 - \frac{3z^4}{2}$
3. Show that the force **F** given by $\mathbf{F} = -\lambda r^5\mathbf{r}$ is conservative and hence obtain the force potential.
   Answer: $V = \lambda\frac{r^7}{7}$
4. Show that $\mathbf{F} = -\lambda r^3\mathbf{r}$ is conservative and find the force potential.
   Answer: $V = \lambda\frac{r^5}{5}$
5. A particle moves in a force field $\mathbf{F} = r^3\mathbf{r}$, where is the position of the particle. Prove that the angular momentum of the particle is conserved.
   Hints: $\mathbf{\Lambda} = \mathbf{r}\times\mathbf{F} = \mathbf{0}.\ \therefore\ \frac{d\mathbf{\Omega}}{dt} = \mathbf{0}$, or $\mathbf{\Omega} = \mathbf{C}$, **C** being a constant vector.
6. Show that the force $\mathbf{F} = \left(x^2y - z^3\right)\hat{\imath} + \left(3xyz + xz^2\right)\hat{\jmath} + \left(2x^2yz + yz^4\right)\hat{k}$ is non conservative.
   Hints: $\nabla\times\mathbf{F}\neq\mathbf{0}$
7. A force **F** is specified by

$$\mathbf{F} = \frac{x\hat{\jmath} - y\hat{\imath}}{x^2+y^2}.$$

   Show that **F** is solenoidal. Test whether the force **F** is conservative. If so, find the force potential. Also, obtain the equations of the line of the force.
8. A force $\mathbf{F} = (x + 2y + az)\,\hat{\imath} + (bx - 3y - z)\,\hat{\jmath} + (4x + cy + 2z)\,\hat{k}$is conservative. Find a, b, c.
   Answer: $a = 4,\ b = 2,\ c = -1$
9. A force **F** is given by $\mathbf{F} = \left(\frac{3xz}{r^5}, \frac{3yz}{r^5}, \frac{3z^2-r^2}{r^5}\right);\ r^2 = x^2+y^2+z^2$. Prove that **F** is solenoidal and conservative. Show that the force potential is given by $V = \frac{\cos\theta}{r^2}$.

10. Discuss the nature of the force $\mathbf{F} = \left(\frac{ax-by}{r^2}, \frac{ay+bx}{r^2}, 0\right)$; $r^2 = x^2 + y^2$.
Answer: (**I**). Force is Solenoidal, (**II**). Force is of Potential kind.
11. A force field is given by $\mathbf{F} = \left(-\frac{2xyz}{r^4}, \frac{(x^2-y^2)}{r^4}, \frac{y}{r^2}\right)$; $r^2 = x^2 + y^2$. Show that the force is conservative and solenoidal.
12. A particle moves with velocity $\mathbf{r} = \hat{i} - \hat{j} + \hat{k}$ under the influence of a constant force $\mathbf{F} = 2\hat{i} - 3\hat{j} + 5\hat{k}$. Obtain the instantaneous power applied to the particle.
Answer: 10 units.
13. A particle of mass m moves along a space curve defined by $\mathbf{r} = a \cos \omega t\, \hat{i} + b \sin \omega t\, \hat{j}$. Find (**I**) the angular momentum above the origin, (**II**) the torque about the origin.
Answer (I): $2m\, ab\, \omega\, \hat{k}$ (**II**): **0**
14. A particle moves in a force field given by $\mathbf{F} = \varphi(r)\,\mathbf{r}$. Prove that the angular momentum of the particle about the origin is constant.
15. Obtain the magnitude of the impulse developed by a mass of 200g which changes its velocity form $\left(5\hat{i} - 3\hat{j} + 7\hat{k}\right)$ m/s to $\left(2\hat{i} + 3\hat{j} + \hat{k}\right)$ m/s.
Answer: 1.8 Ns
16. The angular momentum of a particle is given as a function of time t by $\mathbf{\Omega} = 6t^2\hat{i} - (2t+1)\,\hat{j} + \left(12t^3 - 8t^2\right)\hat{k}$. Find the torque at time $t = 0$.
Answer: $-2\hat{j}$
17. A small mass M is moving with velocity V which is struck by a bullet of mass m moving with velocity v in the same direction. If the bullet is embedded in the block, show that the impulse on the block is mv.
18. A shell lying in a smooth horizontal tube suddenly explodes and breaks into two portions of masses m and m'. Show that the KE generated is given by $T = \frac{1}{2}\frac{mm'}{m+m'}v^2$, where v is the magnitude of the average velocity of either portion of the shell.
19. A gun of mass M fires a shell of mass m horizontally and the energy of explosion is such as would be sufficient to project the shell vertically to a height h. Show that the muzzle velocity of the shell is $\left[\frac{2Mgh}{m+M}\right]^{\frac{1}{2}}$.
20. A force $\mathbf{F}$ is given by $\mathbf{F} = \left(\frac{3xz}{r^5}, \frac{3yz}{r^5}, \frac{3z^2-r^2}{r^5}\right)$; $r^2 = x^2 + y^2 + z^2$
I. Show that **F** is solenoidal.
II. Show that **F** is conservative.
III. Obtain the force potential corresponding to **F**.
IV. Obtain the equations of lines of force.
V. Obtain the equipotential energy surfaces.

Chapter 2
Central Forces and Central Orbits: Orbital Motion

2.1 Polar Coordinates

2.1.1 *Definition*

Let P(x, y) be a point on XY plane. Join OP and suppose that OP $= r$ and $\angle$XOP $= \theta$. Draw $\perp^r$ PM from P on OX.

From Fig. 2.1, we have $x =$ OM=OP $\cos\theta = r\cos\theta$, $y =$ PM=OP $\sin\theta = r\sin\theta$.

Then (r, θ) are called the polar coordinates of P referred to O as a pole, and the directed line OX as the initial line.

The Cartesian coordinates (x, y) and the polar coordinates (r, θ) of the point P are connected by the relations:

$$\left.\begin{aligned} x &= r\cos\theta \\ y &= r\sin\theta \end{aligned}\right\}. \tag{2.1.1}$$

Equation (2.1.1) gives

$$\left.\begin{aligned} r &= \sqrt{x^2 + y^2} \\ \theta &= \tan^{-1}\tfrac{y}{x} \end{aligned}\right\}. \tag{2.1.2}$$

2.1.2 *Velocity and Acceleration in Polar Coordinates*

Let us consider the motion of a particle on a plane XY (say).

Let the position of the particle at time t be specified by polar coordinates (r, θ) referred to O as a pole and **OX** as an initial line. Let (x, y) denote the Cartesian coordinates of P.

N. Ahmed et al., *Classical Dynamics*, University Texts in the Mathematical Sciences,
https://doi.org/10.1007/978-981-95-6394-4_2

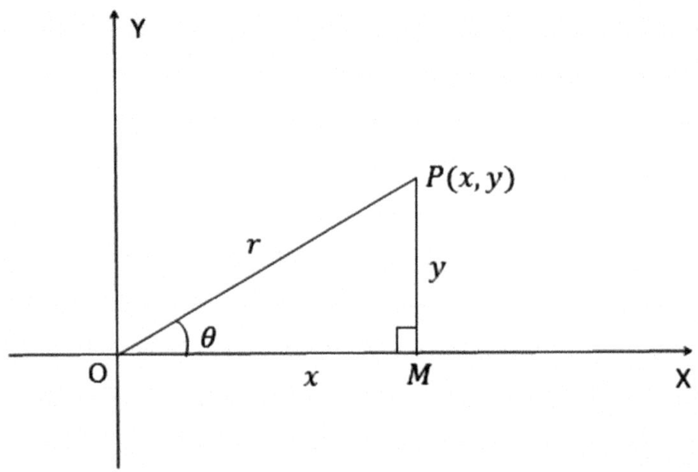

Fig. 2.1 Polar representation of complex number

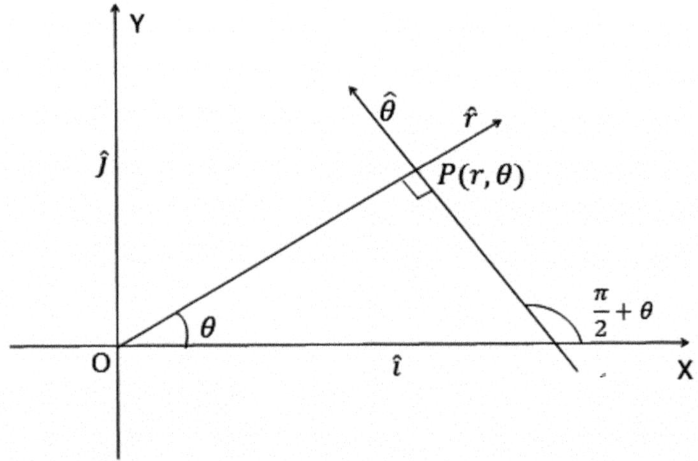

Fig. 2.2 Illustration of velocity and acceleration in (r, θ) system

$$\therefore \quad \left.\begin{aligned} x &= r\ \cos\theta \\ y &= r\ \sin\theta \end{aligned}\right\}. \tag{2.1.3}$$

In Fig. 2.2, let $\mathbf{r}$ be the position vector of P relative to O. Let $\hat{\imath}$, $\hat{\jmath}$ be the unit vectors along OX and OY, respectively. Let $\hat{r}$, $\hat{\theta}$ denote the unit vectors along OP and perpendicular to OP in the sense of θ increasing.

Since $\mathbf{r}$ is the position vector of $P\ (x,\ y)$ relative to O,

$$\therefore \quad \mathbf{r} = x\hat{\imath} + y\hat{\jmath} = \hat{\imath}\ r\ \cos\theta + \hat{\jmath}\ r\ \sin\theta,$$

$$\text{i.e.,}\quad \frac{\mathbf{r}}{r} = \hat{\imath}\ \cos\theta + \hat{\jmath}\ \sin\theta$$

$$\text{i.e.,}\quad \hat{r} = \hat{\imath}\ \cos\theta + \hat{\jmath}\ \sin\theta. \tag{2.1.4}$$

Similarly

$$\hat{\theta} = \hat{\imath}\ \cos\left(\frac{\pi}{2} + \theta\right) + \hat{\jmath}\ \sin\left(\frac{\pi}{2} + \theta\right) = -\hat{\imath}\ \sin\theta + \hat{\jmath}\ \cos\theta. \tag{2.1.5}$$

Equation (2.1.4) gives

$$\dot{\hat{r}} = \hat{\imath}\,(-\sin\theta)\,\dot{\theta} + \hat{\jmath}\ \cos\theta\ \dot{\theta} = \left(-\hat{\imath}\ \sin\theta + \hat{\jmath}\ \cos\theta\right)\dot{\theta} = \hat{\theta}\dot{\theta}. \tag{2.1.6}$$

Equation (2.1.5) yields

$$\dot{\hat{\theta}} = -\hat{\imath}\ \cos\theta\,\dot{\theta} - \hat{\jmath}\ \sin\theta\ \dot{\theta} = -\dot{\theta}\left(\hat{\imath}\ \cos\theta + \hat{\jmath}\ \sin\theta\right) = -\dot{\theta}\ \hat{r}. \tag{2.1.7}$$

We have

$$\mathbf{r} = \mathbf{OP} = \mathrm{OP}\,\hat{r} = r\,\hat{r}.$$

If $\mathbf{v}$ is the velocity of the particle at P, then

$$\mathbf{v} = \frac{d}{dt}\mathbf{r} = \frac{d\mathbf{r}}{dt},$$

$$\text{i.e.,}\quad \mathbf{v} = \frac{d}{dt}\left(r\,\hat{r}\right) = r\,\dot{\hat{r}} + \dot{r}\ \hat{r} = r\ \dot{\theta}\ \hat{\theta} + \ \dot{r}\ \hat{r} = \dot{r}\ \hat{r} + r\dot{\theta}\ \hat{\theta}. \tag{2.1.8}$$

From Eq. (2.1.8), we obtain

$$v_r = \text{Radial component of velocity} = \dot{r}$$

and

$$v_\theta = \text{Transverse component of velocity} = r\dot{\theta}.$$

Let $\mathbf{f}$ be the acceleration of the particle at time t.

$$\therefore\quad \mathbf{f} = \frac{d\mathbf{v}}{dt} = \frac{d}{dt}\left(\dot{r}\,\hat{r} + r\dot{\theta}\,\hat{\theta}\right)$$

$$= \ddot{r}\,\hat{r} + \dot{r}\,\dot{\hat{r}} + \dot{r}\,\dot{\theta}\,\hat{\theta} + r\,\ddot{\theta}\,\hat{\theta} + r\,\dot{\theta}\,\dot{\hat{\theta}}$$

$$= \ddot{r}\,\hat{r} + \dot{r}\,\dot{\theta}\,\hat{\theta} + \dot{r}\,\dot{\theta}\,\hat{\theta} + r\,\ddot{\theta}\,\hat{\theta} + r\,\dot{\theta}\,(-\dot{\theta}\,\hat{r})$$

$$= \ddot{r}\,\hat{r} + 2\dot{r}\dot{\theta}\hat{\theta} + r\,\ddot{\theta}\,\hat{\theta} - r\,\dot{\theta}^2\,\hat{r}$$

$$= \left(\ddot{r} - r\dot{\theta}^2\right)\,\hat{r} + \left(2\dot{r}\dot{\theta} + r\ddot{\theta}\right)\hat{\theta}$$

$$= \left(\ddot{r} - r\dot{\theta}^2\right)\,\hat{r} + \frac{1}{r}\left[2r\dot{r}\dot{\theta} + r^2\ddot{\theta}\right]\hat{\theta}$$

$$= \left(\ddot{r} - r\dot{\theta}^2\right)\hat{r} + \frac{1}{r}\left[\dot{\theta}\frac{d}{dt}r^2 + r^2\frac{d}{dt}\dot{\theta}\right]\hat{\theta}$$

$$= \left(\ddot{r} - r\dot{\theta}^2\right)\,\hat{r} + \frac{1}{r}\frac{d}{dt}\left(r^2\dot{\theta}\right)\,\hat{\theta}$$

$$\therefore \quad f_r = \text{Radial acceleration} = \ddot{r} - r\dot{\theta}^2$$

and

$$f_\theta = \text{Cross radial acceleration} = \frac{1}{r}\,\frac{d}{dt}\left(r^2\dot{\theta}\right).$$

2.2 Central Forces

2.2.1 *Definition*

A force acting on a particle P is said to be a central force with a fixed point O as the center if

- it is always directed from P toward or away from O,
- its magnitude is a function of r where $r = \text{OP}$.

Symbolically, a force **F** is said to be a central force iff

$$\mathbf{F} = f\,(r)\,\hat{r} = f\,(r)\,\frac{\mathbf{r}}{r} = \varphi\,(r)\,\mathbf{r};\quad \varphi\,(r) = \frac{f\,(r)}{r},\quad \mathbf{r} = \mathbf{OP}.$$

If $\varphi\,(r) > 0$, then the force is one of repulsion from O, whereas it is one of attraction if $\varphi\,(r) < 0$.

2.2.2 Some Properties of Central Force Field

I. *A central force field is conservative.*
Proof: Let $\mathbf{F}$ be a central force with the center of force O, acting on a particle P at time t. Let $\mathbf{r}$ be the position vector of P relative to O.
The force $\mathbf{F}$ may be expressed as

$$\mathbf{F} = \varphi(r)\,\mathbf{r}, \quad \text{where} \quad r = |\mathbf{r}|$$

Now,

$$\begin{aligned}\nabla \times \mathbf{F} = \nabla \times \varphi(r)\,\mathbf{r} &= \nabla\varphi(r) \times \mathbf{r} + \varphi(r)\,\nabla \times \mathbf{r} \\ &= \varphi'(r)\nabla r \times \mathbf{r} \\ &= \varphi'(r)\,\hat{r} \times \mathbf{r} \\ &= \varphi'(r)\,\frac{\mathbf{r}}{r} \times \mathbf{r} \\ &= \mathbf{0}\end{aligned}$$

Therefore, $\mathbf{F}$ is conservative. Thus, a central force is conservative.

II. *A central orbit is a plane curve.*
Proof: In Fig. 2.3, let C be a central orbit with the fixed point O as the center of force. Let P be the position of the particle in its orbit at time t.
Let $\mathbf{F}$ be the central force acting on the particle at P with position vector $\mathbf{r}$ relative to O.

$$\therefore \quad \mathbf{F} = \varphi(r)\,\mathbf{r}$$

The equation of motion of the particle of mass m is

$$m\,\ddot{\mathbf{r}} = \mathbf{F} = \varphi(r)\,\mathbf{r}. \tag{2.2.1}$$

Equation (2.2.1) gives

$$\mathbf{r} \times m\,\ddot{\mathbf{r}} = \mathbf{0},$$

$$\text{i.e.,} \quad \mathbf{r} \times \ddot{\mathbf{r}} = \mathbf{0},$$

$$\text{i.e.,} \quad \mathbf{r} \times \frac{d\mathbf{v}}{dt} + \mathbf{v} \times \mathbf{v} = \mathbf{0},$$

$$\text{i.e.,} \quad \mathbf{r} \times \frac{d\mathbf{v}}{dt} + \frac{d\mathbf{r}}{dt} \times \mathbf{v} = \mathbf{0},$$

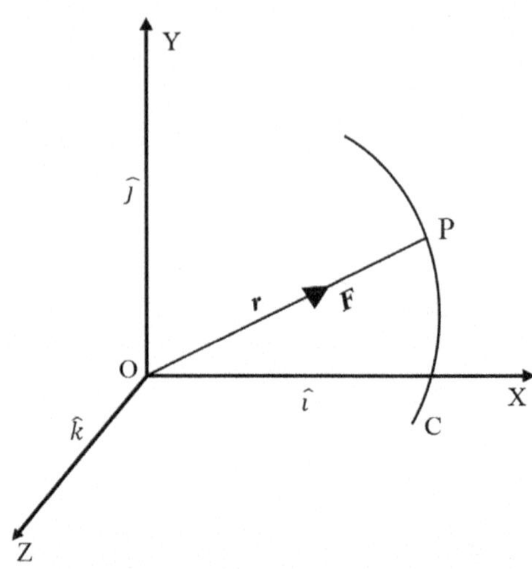

Fig. 2.3 Central orbit with O as the center of force

$$\text{i.e.,}\quad \frac{d}{dt}(\mathbf{r}\times\mathbf{v}) = \mathbf{0}.$$

$$\therefore\quad \mathbf{r}\times\mathbf{v} = \text{a constant vector} = \mathbf{h}\ (\text{say}). \tag{2.2.2}$$

Equation (2.2.2) gives

$$\mathbf{r}\cdot(\mathbf{r}\times\mathbf{v}) = \mathbf{r}\cdot\mathbf{h},$$

$$\text{i.e.,}\quad \mathbf{r}\cdot\mathbf{h} = 0. \tag{2.2.3}$$

Introduce a tri-rectangular coordinate system with point O as the origin. Let (α, β, γ) be the coordinates of P.

$$\therefore\quad \mathbf{r} = \alpha\hat{\imath} + \beta\hat{\jmath} + \gamma\hat{k} \quad \text{and let} \quad \mathbf{h} = \lambda\hat{\imath} + \mu\hat{\jmath} + \upsilon\hat{k}.$$

Equation (2.2.3) yields

$$\lambda\alpha + \mu\beta + \upsilon\gamma = 0. \tag{2.2.4}$$

Therefore, the point P lies on the fixed plane $\lambda x + \mu y + \upsilon z = 0$ which passes through the center of force.
Since the point P is an arbitrary point on C. Therefore, the orbit C entirely lies on the plane given by $\lambda x + \mu y + \upsilon z = 0$.
In other words, the orbit C is a plane curve, i.e., a central orbit is a plane curve. □

Fig. 2.4 Area swept out by radius vector

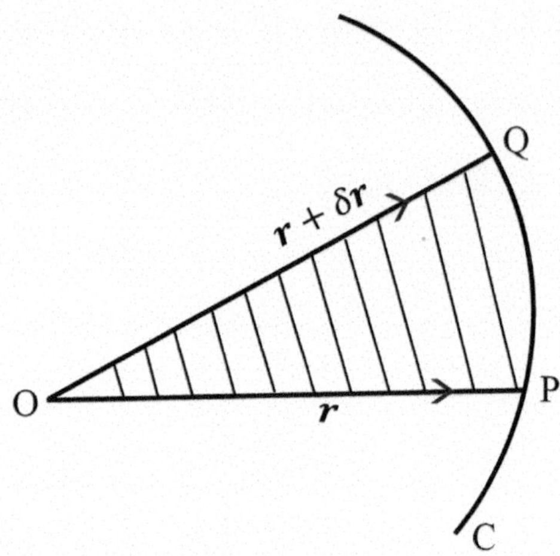

III. *For a particle moving under a central force field, the areal velocity of the radius vector joining the center of force and the particle is constant.*

Proof: Let C be a central orbit with the center of force O.
Let P and Q be the positions of the particle on C at time t and $t + \delta t$, respectively. Suppose that $\mathbf{OP} = \mathbf{r}$ and $\mathbf{OQ} = \mathbf{r} + \delta\,\mathbf{r}$. As Q is very close to P, the arc PQ may be taken as a straight line segment to the first order of approximation. That is the shaded area in Fig. 2.4 may be considered as a triangular area.
Let $\delta \mathrm{A} = \Delta \mathrm{OPQ}$
Now

$$\delta \mathrm{A} = \Delta \mathrm{OPQ} = \frac{1}{2}\,|\mathbf{OP} \times \mathbf{OQ}\,| = \frac{1}{2}\,|\mathbf{r} \times (\mathbf{r} + \delta\mathbf{r})| = \frac{1}{2}\,|\,\mathbf{r} \times \delta\mathbf{r}\,|,$$

$$\text{i.e.,}\quad \frac{\delta \mathrm{A}}{\delta t} = \frac{1}{2}\left|\mathbf{r} \times \frac{\delta\mathbf{r}}{\delta t}\right|. \tag{2.2.5}$$

Letting $\delta t \to 0$ in (2.2.5), we obtain

$$\frac{d\mathrm{A}}{dt} = \frac{1}{2}\left|\mathbf{r} \times \frac{d\mathbf{r}}{dt}\right| = \frac{1}{2}\,|\mathbf{r} \times \mathbf{v}| = \frac{1}{2}\,|\mathbf{h}| = \frac{h}{2} = \text{a constant.}$$

Thus, the areal velocity of the radius vector joining the center of force and the particle is constant.
Therefore, the radius vector joining the center of force and the particle sweeps out equal areas in equal time t. □

2.2.3 Physical and Geometrical Interpretation of h

From the property (**III**) of Sect. 2.2.2, it follows that

$$\frac{dA}{dt} = \frac{h}{2},$$
$$\text{i.e.,}\quad h = 2\frac{dA}{dt}.$$

This shows that h is twice the areal velocity of the radius vector joining the center force and the particle moving under a central force field.

In Fig. 2.5, let P be the position of the particle at time t with position vector $\mathbf{r}$ relative to the center of force O. If m is the mass of the particle moving with velocity $\mathbf{v}$, then the angular momentum $\mathbf{\Omega}$ of the particle about O is given by

$$\mathbf{\Omega} = \mathbf{r} \times m\mathbf{v} = m\,(\mathbf{r} \times \mathbf{v}\,) = m\mathbf{h}.$$

$$\therefore \quad \Omega = |\mathbf{\Omega}| = |m\mathbf{h}| = mh.$$

If $m = 1,\ \ \Omega = h$. Thus, h is the magnitude of the angular momentum about O, of a particle of unit mass moving under a central force field.

2.2.4 Differential Equation of a Central Orbit

A particle moves on a plane with an acceleration that is always directed to a fixed point O in the plane to obtain the differential equation of its path.

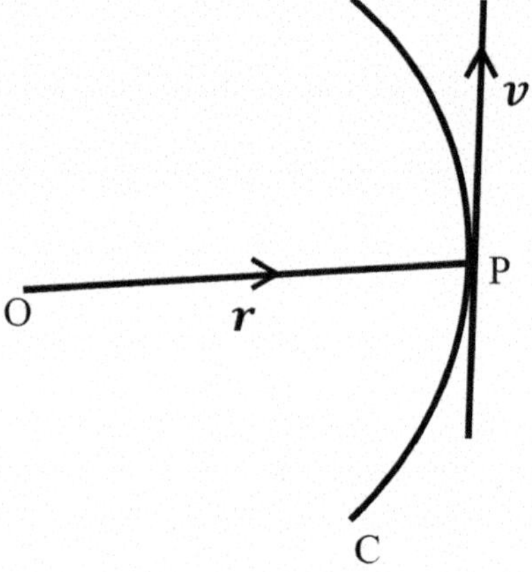

Fig. 2.5 Graphical interpretation for angular momentum

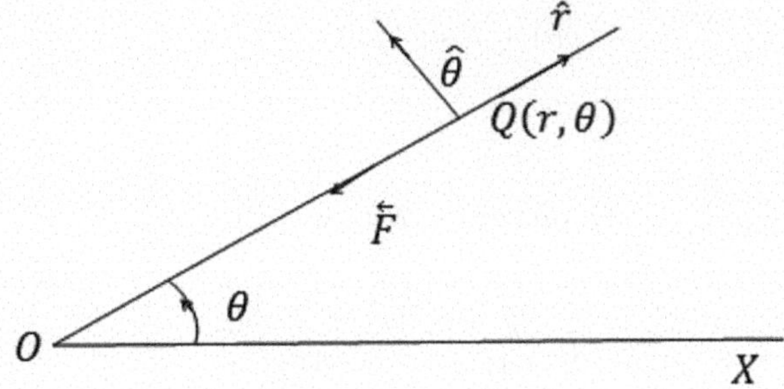

Fig. 2.6 Motion of a particle under central force directed toward a fixed point O

Derivation: Let Q be the position of the particle at time t. Let $\mathbf{r}$ be the position vector of Q relative to O.

In Fig. 2.6, let the point O referred as a pole and a fixed straight line OX through O as the initial line. Let the polar coordinates of Q be (r, θ). Let P be the magnitude of the acceleration directed towards O.

Therefore, the central force $\mathbf{F}$ acting on the particle at Q is specified by

$$\mathbf{F} = mP\left(-\hat{r}\right) = -mP\hat{r},$$

m is the mass of the particle.

The vector equation of motion of the particle based on Newton's 2nd law is

$$m\,\ddot{\mathbf{r}} = \mathbf{F} = -mP\,\hat{r},$$

$$\text{i.e.,}\quad \ddot{\mathbf{r}} = -P\,\hat{r},$$

$$\text{i.e.,}\quad \left(\ddot{r} - r\,\dot{\theta}^2\right)\hat{r} + \frac{1}{r}\frac{d}{dt}\left(r^2\dot{\theta}\right)\hat{\theta} = -P\,\hat{r}. \tag{2.2.6}$$

By equating the coefficients of $\hat{r},\ \hat{\theta}$ in (2.2.6), we obtain the following differential equations:

$$\ddot{r} - r\,\dot{\theta}^2 = -P, \tag{2.2.7}$$

$$\frac{1}{r}\frac{d}{dt}\left(r^2\dot{\theta}\right) = 0. \tag{2.2.8}$$

Equation (2.2.8) gives

$$r^2\,\dot{\theta} = h. \tag{2.2.9}$$

Consider the substitution, $r = \frac{1}{u}$.

Equation (2.2.9) gives

$$\dot{\theta} = hu^2. \tag{2.2.10}$$

Now,

$$\dot{r} = \frac{dr}{dt} = \frac{d}{dt}\left(\frac{1}{u}\right) = -\frac{1}{u^2}\frac{du}{d\theta}\,\dot{\theta} = -\frac{1}{u^2}\frac{du}{d\theta}\,hu^2 = -h\,\frac{du}{d\theta}\,.$$

$$\therefore\quad \ddot{r} = -h\frac{d}{d\theta}\left(\frac{du}{d\theta}\right)\dot{\theta} = -h\frac{d^2u}{d\theta^2}hu^2 = -h^2u^2\frac{d^2u}{d\theta^2}. \tag{2.2.11}$$

Using (2.2.10) and (2.2.11) in (2.2.7), we obtain

$$h^2u^2\frac{d^2u}{d\theta^2} + h^2u^3 = P,$$

i.e.,

$$\frac{d^2u}{d\theta^2} + u = \frac{P}{h^2u^2}.$$

This equation is termed as the differential equation of a central orbit in (r, θ) system.

2.2.5 *(p, r) Equation of a Central Orbit*

Derivation: For a central orbit, we have the following relations

$$vp = h, \tag{2.2.12}$$

$$v^2 = h^2\left[\left(\frac{du}{d\theta}\right)^2 + u^2\right], \tag{2.2.13}$$

$$\frac{d^2u}{d\theta^2} + u = \frac{P}{h^2u^2}. \tag{2.2.14}$$

Eliminating v from (2.2.12) and (2.2.13), we obtain

$$\frac{1}{p^2} = \left(\frac{du}{d\theta}\right)^2 + u^2. \tag{2.2.15}$$

Differentiating (2.2.15) w.r.t. u, we derive

$$-2p^{-3}\frac{dp}{d\theta}=2\left(\frac{du}{d\theta}\right)\frac{d^2u}{d\theta^2}+2u\frac{du}{d\theta},$$

$$\text{i.e.,}\quad -\frac{1}{p^3}\frac{dp}{d\theta}=\frac{du}{d\theta}\frac{d^2u}{d\theta^2}+u\frac{du}{d\theta},$$

$$\text{i.e.,}\quad -\frac{1}{p^3}\frac{dp}{dr}\frac{dr}{d\theta}=\frac{du}{d\theta}\left(\frac{d^2u}{d\theta^2}+u\right),$$

$$\text{i.e.,}\quad -\frac{1}{p^3}\frac{dp}{dr}\frac{d}{d\theta}\left(\frac{1}{u}\right)=\frac{du}{d\theta}\frac{P}{h^2u^2},$$

$$\text{i.e.,}\quad -\frac{1}{p^3}\frac{dp}{dr}\left(-\frac{1}{u^2}\right)\frac{du}{d\theta}=\frac{du}{d\theta}\frac{P}{h^2u^2},$$

$$\text{i.e.,}\quad \frac{1}{p^3u^2}\frac{dp}{dr}=\frac{P}{h^2u^2},$$

$$\text{i.e.,}\quad \frac{1}{p^3}\frac{dp}{dr}=\frac{P}{h^2},$$

$$\text{i.e.,}\quad \frac{h^2}{p^3}\frac{dp}{dr}=P.$$

This equation is termed as the (p, r) equation of a central orbit.

Example 2.1 A particle moves in an ellipse under a force that is always directed towards its focus; determine the law of force and velocity at every point of its path.

Solution The (r, θ) equation of the ellipse (Fig. 2.7) referred to the focus S as the pole and SX as the initial line is

$$\frac{1}{r}=1+e\cos\theta,$$

$$\text{i.e.,}\quad lu=1+e\cos\theta,$$

$$\text{i.e.,}\quad u=\frac{1+e\cos\theta}{l}. \tag{2.2.16}$$

The differential equation of a central orbit is

$$\frac{P}{h^2u^2}=\frac{d^2u}{d\theta^2}+u. \tag{2.2.17}$$

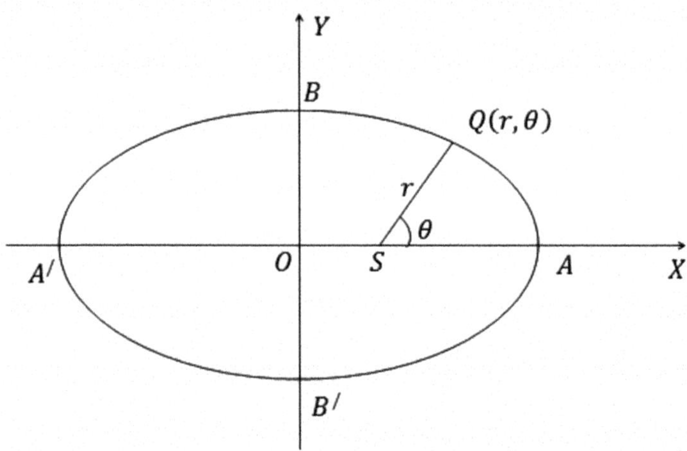

Fig. 2.7 Illustration for Example 2.1

Equation (2.2.16) gives

$$\frac{du}{d\theta} = \frac{1}{l}(-e \sin\theta) = -\frac{e \sin\theta}{l},$$

and

$$\frac{d^2u}{d\theta^2} = -\frac{e}{l}\cos\theta.$$

Therefore, Eq. (2.2.17) yields

$$\frac{P}{h^2u^2} = -\frac{e\cos\theta}{l} + u = -\frac{e\cos\theta}{l} + \frac{1 + e\cos\theta}{l} = \frac{1}{l},$$

$$\text{i.e.,}\quad P = \frac{h^2u^2}{l} = \frac{\frac{h^2}{l}}{r^2} = \frac{\mu}{r^2},$$

where $\mu = \frac{h^2}{l}$.

$$\therefore\quad P \propto \frac{1}{r^2},$$

which is the law of force.

Now $\mu = \frac{h^2}{l}$,

$$\text{i.e.}\quad h^2 = \mu l. \tag{2.2.18}$$

The velocity of the particle at Q (r, θ) on the ellipse is given by

$$v^2 = h^2\left[\left(\frac{du}{d\theta}\right)^2 + u^2\right],$$

$$\text{or}\quad v^2 = \mu l\left[\frac{e^2\sin^2\theta}{l^2} + \frac{(1+e\cos\theta)^2}{l^2}\right] = \frac{\mu}{l}\left[e^2 + 2e\cos\theta + 1\right],$$

$$\text{or}\quad v^2 = \frac{\mu}{l}\left[e^2 + 1 + 2\left(\frac{l}{r} - 1\right)\right] = \frac{\mu}{l}\left[e^2 - 1 + \frac{2l}{r}\right],$$

$$\text{or}\quad v^2 = \frac{\mu}{l}\left[\frac{2l}{r} - \left(1 - e^2\right)\right]. \tag{2.2.19}$$

For an ellipse,

$$b^2 = a^2\left(1 - e^2\right) \text{ and } l = \frac{b^2}{a}.$$

$$\therefore\quad al = a^2\left(1 - e^2\right),$$

$$\text{i.e.,}\quad 1 - e^2 = \frac{l}{a}. \tag{2.2.20}$$

Therefore, Eq. (2.2.19) reduces to

$$v^2 = \frac{\mu}{l}\left[\frac{2l}{r} - \frac{l}{a}\right] = \mu\left[\frac{2}{r} - \frac{1}{a}\right].$$

2.2.6 *Apses*

An apse is a point in a central orbit at which the radius vector is drawn from the center of force to the moving particle, attains its maximum or minimum value.

In Fig. 2.8, let C be a central orbit, and Q be the position of an apse. Let r be the radius vector and p be the perpendicular distance of the tangent at Q from the center of force O.

By the principle of differential calculus, for an apse,

$$\frac{dr}{d\theta} = 0,$$

Fig. 2.8 Apse in a central orbit

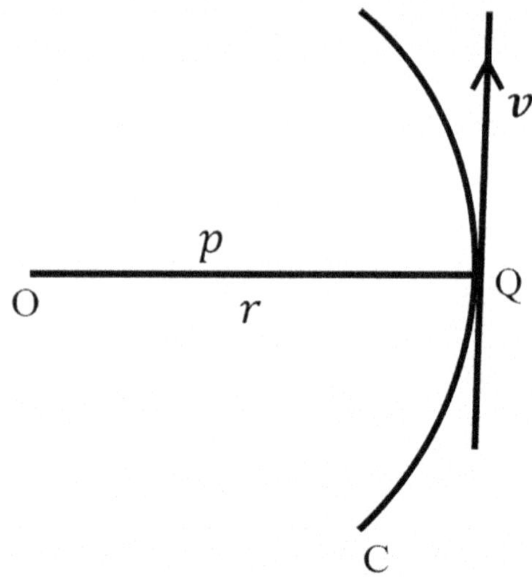

$$\text{i.e.,} \quad \frac{d}{d\theta}\left(\frac{1}{u}\right) = 0,$$

$$\text{i.e.,} \quad -\frac{1}{u^2}\frac{du}{d\theta} = 0,$$

$$\text{i.e.,} \quad \frac{du}{d\theta} = 0. \tag{2.2.21}$$

We have

$$v^2 = h^2\left[\left(\frac{du}{d\theta}\right)^2 + u^2\right]. \tag{2.2.22}$$

Therefore, at an apse,

$$v^2 = h^2u^2 = \frac{h^2}{r^2}. \tag{2.2.23}$$

Again,

$$vp = h,$$

$$\text{i.e.,} \quad v^2 = \frac{h^2}{p^2}. \tag{2.2.24}$$

Thus, at an apse,

$$\frac{h^2}{p^2} = \frac{h^2}{r^2},$$

i.e., $p = r.$

Hence, at an apse, the radius vector is equal to the perpendicular distance of the center of forces from the tangent to the orbit. Therefore, at an apse, **v** is $\perp^r$ to the radius vector.

2.2.7 *Some Important Deductions*

I. **Velocity in a Circle**: Let v be the velocity of a particle moving on a circle with the center of force at the center of the circle. Let P be the acceleration of the particle directed towards O. Let a denote the radius of the circular orbit.
From Fig. 2.9, we have

$$v_r = \dot{r} = 0, \quad v_\theta = r\dot{\theta} = a\omega.$$

$$\therefore \quad v^2 = v_r^2 + v_\theta^2 = a^2\omega^2 \text{ and so } \omega = \frac{v}{a}.$$

As P is the acceleration directed towards O,

$$\therefore \quad \ddot{r} - r\dot{\theta}^2 = \text{acceleration along } \mathbf{OQ} = -P,$$

$$\text{i.e.,} \quad a\omega^2 = P \implies P = a\frac{v^2}{a^2} = \frac{v^2}{a} \implies v^2 = Pa,$$

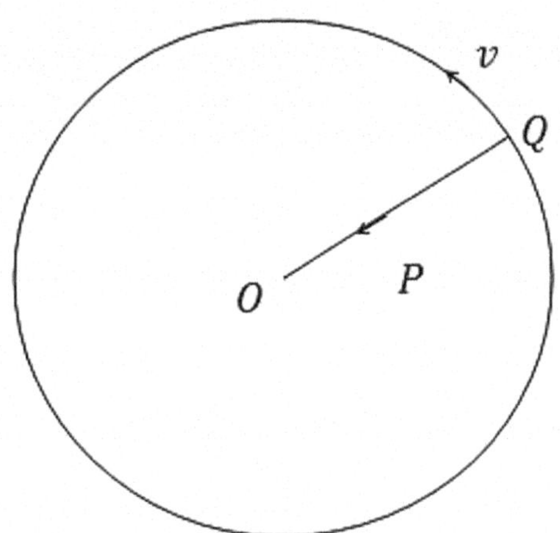

Fig. 2.9 Velocity of a particle moving on a circle

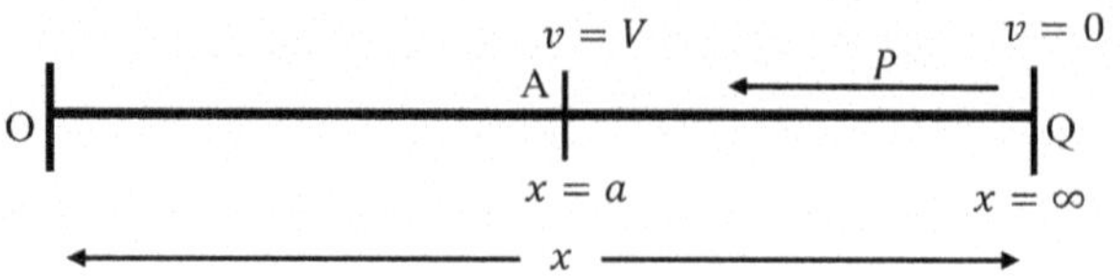

Fig. 2.10 Velocity of a particle fall from infinity

$$\text{i.e.,}\quad v^2 = Pr$$

II. **Velocity of Fall from Infinity**: Let a particle fall from ∞ under a central acceleration P to a point A at a distance a from the center of force O. Let Q be the position of the particle at time t and suppose that OQ $= x$ (see Fig. 2.10). Equations of motion of the particle is

$$\ddot{x} = -P,$$

$$\text{i.e.,}\quad v dv = -P dx,$$

$$\therefore\quad \int_{v=0}^{V} v\,dv = -\int_{\infty}^{a} P dx,$$

$$\text{i.e.,}\quad V^2 = -2\int_{\infty}^{a} P\,dx,$$

$$\text{i.e.,}\quad V^2 = -2\int_{\infty}^{a} P\,dr = 2\int_{a}^{\infty} P\,dr.$$

III. **Velocity of Fall to the Point of Projection**: In Fig. 2.11, let a particle fall from the center of repulsion under a force P to a point A at a distance a from the center of force O. Equation of motion of the particle is

$$\ddot{x} = P,$$

$$\text{i.e.,}\quad v dv = P dx.$$

$$\therefore\quad \int_{v=0}^{V} v dv = \int_{x=0}^{a} P dx,$$

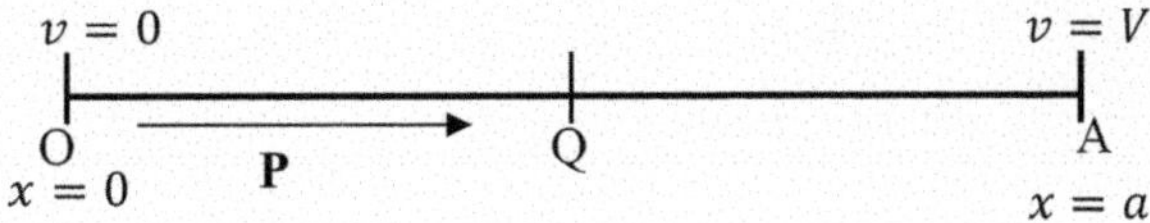

Fig. 2.11 Motion of particle from center O to point A

$$\text{i.e.,}\quad V^2 = 2\int_0^a P dx,$$

$$\text{i.e.,}\quad V^2 = 2\int_0^a P dr.$$

2.3 Illustrations

Example 2.2 Prove that $r^2\dot{\theta} = h$ for a particle moving under a central force field.

Solution It is already established that

$$\mathbf{r} \times \mathbf{v} = \mathbf{h}. \tag{2.3.1}$$

Further

$$\mathbf{v} = v_r\hat{r} + v_\theta\hat{\theta} = \dot{r}\hat{r} + r\dot{\theta}\hat{\theta}. \tag{2.3.2}$$

Eliminating $\mathbf{v}$ from (2.3.1) and (2.3.2) we get

$$\mathbf{r} \times \left(\dot{r}\,\hat{r} + r\,\dot{\theta}\,\hat{\theta}\right) = \mathbf{h},$$

$$\text{i.e.,}\quad r\,\hat{r} \times \left(\dot{r}\,\hat{r} + r\dot{\theta}\,\hat{\theta}\right) = \mathbf{h},$$

$$\text{i.e.,}\quad r^2\dot{\theta}\,\hat{r} \times \hat{\theta} = \mathbf{h},$$

$$r^2\,\dot{\theta}\,\hat{k} = h\,\hat{k}, \tag{2.3.3}$$

$\hat{k}$ is a unit vector $\perp^r$ to the plane of the orbit measured in the anti-clockwise sense.

Equation (2.3.3) gives

$$r^2\,\dot{\theta} = h. \tag{2.3.4}$$

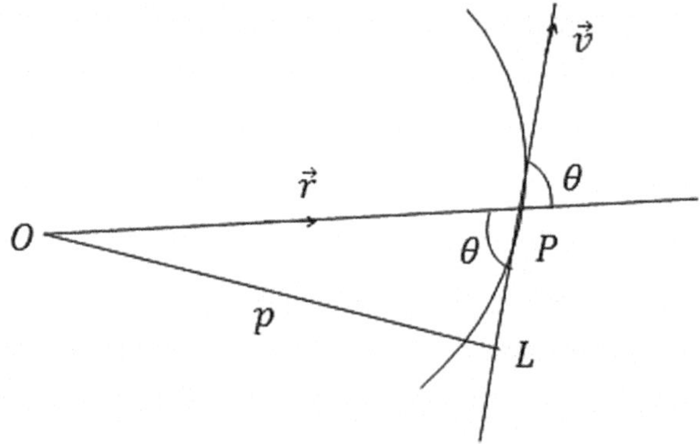

Fig. 2.12 Illustration of Example 2.3

Example 2.3 With usual rotations, prove that $vp = h$ in case of a central orbit.

Solution Let O be the center of force and P be the position of the particle at time t. Let $\mathbf{v}$ be the velocity of the particle at P and let $\mathbf{r}$ denote the position vector of P relative to O.

In Fig. 2.12, let p be the $\perp^r$ distance of the point O from the tangent to the central orbit at the point P. Let θ be the angle between $\mathbf{r}$ and $\mathbf{v}$. In the case of a central orbit,

$$\mathbf{r} \times \mathbf{v} = \mathbf{h},$$

$$\text{i.e.,} \quad r\,v\,\sin\theta\,\hat{k} = h\,\hat{k},$$

$$\text{i.e.,} \quad r\,v\,\sin\theta = h,$$

$$\text{i.e.,} \quad r\,v\,\frac{p}{r} = h,$$

$$\text{i.e.,} \quad v\,p = h.$$

Example 2.4 If $u = \frac{1}{r}$, show that $v^2 = h^2\left[\left(\frac{du}{d\theta}\right)^2 + u^2\right]$.

Solution Let P denote the position of a particle moving under a central force field with the fixed point O as the center of force.

In Fig. 2.13, O is referred as a pole and a straight line **OX** as the initial line, let the polar coordinates of P be (r, θ). If $\mathbf{v}$ is the velocity of the particle at P, then

$$\mathbf{v} = \dot{r}\,\hat{r} + r\,\dot{\theta}\,\hat{\theta},$$

$$\text{i.e.,} \quad v^2 = \dot{r}^2 + r^2\dot{\theta}^2. \tag{2.3.5}$$

For a central orbit, $r^2\dot{\theta} = h$,

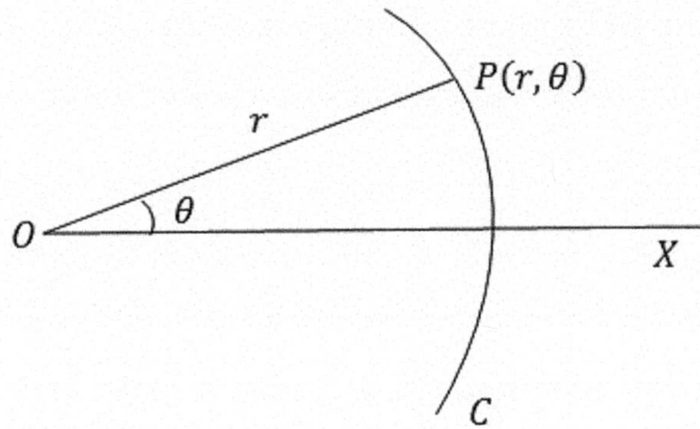

Fig. 2.13 Illustration of Example 2.4

$$\text{i.e.,}\quad \dot{\theta} = \frac{h}{r^2}. \tag{2.3.6}$$

By the use of (2.3.6), in (2.3.5), we obtain

$$v^2 = \dot{r}^2 + \frac{h^2}{r^2}. \tag{2.3.7}$$

Given $r = \frac{1}{u}$.

Now,

$$\dot{r} = \frac{d}{dt}\left(\frac{1}{u}\right) = -\frac{1}{u^2}\frac{du}{d\theta}\dot{\theta} = -\frac{1}{u^2}\frac{du}{d\theta}hu^2 = -h\frac{du}{d\theta}. \tag{2.3.8}$$

By use of Eqs. (2.3.8), (2.3.7) reduces to $v^2 = h^2\left[\left(\frac{du}{d\theta}\right)^2 + u^2\right]$.

Example 2.5 Find the law of force to the pole when the path is the cardioid $r = a\,(1 - \cos\theta)$, and prove that, if **F** be the force at the apse, and **v** the velocity, $3v^2 = 4aF$.

Solution The equation of path (Central orbit) is

$$r = a\,(1 - \cos\theta)\,,$$

$$\therefore\quad u = \frac{1}{a\,(1 - \cos\theta)} = \frac{1}{a\cdot 2\sin^2\frac{\theta}{2}} = \frac{\csc^2\frac{\theta}{2}}{2a},$$

$$\therefore\quad \frac{du}{d\theta} = \frac{1}{2a}2\csc\frac{\theta}{2}\left(-\csc\frac{\theta}{2}\cot\frac{\theta}{2}\right)\frac{1}{2} = -\frac{1}{2a}2\csc^2\frac{\theta}{2}\cot\frac{\theta}{2} = -u\cot\frac{\theta}{2},$$

$$\therefore \quad \frac{d^2u}{d\theta^2} = -\left[-\frac{u}{2}\csc^2\frac{\theta}{2} + \cot\frac{\theta}{2}\frac{du}{d\theta}\right] = \frac{u}{2}\csc^2\frac{\theta}{2} + u\cot^2\frac{\theta}{2}. \tag{2.3.9}$$

The differential equation of a central orbit is

$$\frac{d^2u}{d\theta^2} + u = \frac{P}{h^2u^2}. \tag{2.3.10}$$

On the use of (2.3.9) in (2.3.10), it is derived that

$$\frac{u}{2}\csc^2\frac{\theta}{2} + u\cot^2\frac{\theta}{2} + u = \frac{P}{h^2u^2},$$

$$\text{i.e.,} \quad \frac{P}{h^2u^2} = \frac{u}{2}\csc^2\frac{\theta}{2} + u\ \csc^2\frac{\theta}{2} = \frac{3}{2}u\ \csc^2\frac{\theta}{2} = \frac{3}{2}2au^2 = 3au^2,$$

$$\text{i.e.,} \quad P = 3ah^2u^4 = \frac{3ah^2}{r^4}. \tag{2.3.11}$$

$$\therefore \quad P \propto \frac{1}{r^4},$$

which is the law of force.

2nd Part: At an apse,

$$\frac{du}{d\theta} = 0,$$

$$\text{i.e.,} \quad u\cot\frac{\theta}{2} = 0,$$

$$\text{i.e.,} \quad \cot\frac{\theta}{2} = 0,$$

$$\text{i.e.,} \quad \theta = \pi.$$

Therefore, at an apse

$$r = a(1+1) = 2a.$$

The $v \sim p$ relation in the case of a central orbit is

$$Vp = h. \tag{2.3.12}$$

At the apse, $V = v, \quad p = r = 2a$.

Therefore, Eq. (2.3.12) gives $2av = h$. Hence, Eq. (2.3.11) gives

$$P = \frac{3a}{r^4}4a^2v^2 = \frac{12a^3v^2}{r^4}. \tag{2.3.13}$$

At the apse, $P = F, \quad r = 2a$.

Thus, Eq. (2.3.13) gives

$$F = \frac{12a^3v^2}{16a^4} = \frac{3v^2}{4a},$$

i.e., $3v^2 = 4aF$.

Example 2.6 If a body moves under a central force in a medium that exerts a resistance equal to k times the velocity per unit of mass, prove that $\frac{d^2u}{d\theta^2} + u = \frac{P}{h^2u^2} \; e^{2kt}$ where h is twice the initial moment of momentum about the center of force.

Solution Referring to the center of force O as the pole and **OX** as the initial line, let the position Q of the particle at time t be specified by the polar coordinates $(r, \; \theta)$ (see Fig. 2.14).

The equations of motion of the particle are

$$\ddot{r} - r\dot{\theta}^2 = -P - k\,v\cos\phi, \tag{2.3.14}$$

$$\frac{1}{r}\frac{d}{dt}\left(r^2\dot{\theta}\right) = -kv\sin\phi. \tag{2.3.15}$$

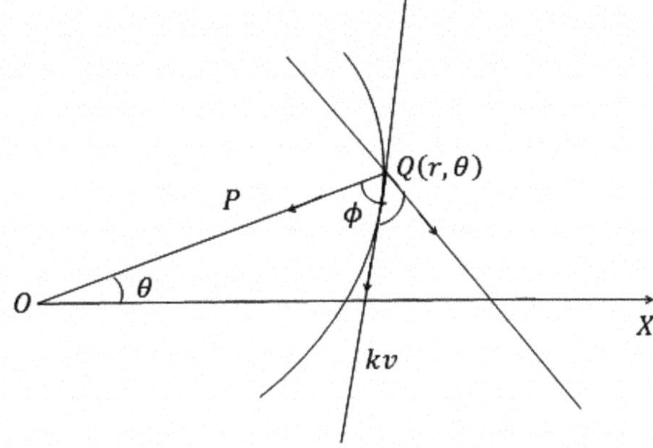

Fig. 2.14 Illustration of Example 2.6

Equation (2.3.15) gives

$$\frac{1}{r}\frac{d}{dt}\left(r^2\dot{\theta}\right) = -kv\frac{rd\theta}{ds} = -k\frac{ds}{dt}\frac{rd\theta}{ds} = -kr\frac{d\theta}{dt} = -kr\dot{\theta},$$

i.e., $\frac{d}{dt}\left(r^2\dot{\theta}\right) = -kr^2\dot{\theta},$

i.e., $\frac{d\left(r^2\dot{\theta}\right)}{r^2\dot{\theta}} = -kdt,$

i.e., $\log\left(r^2\dot{\theta}\right) = -kt + \log C,$

i.e., $\log\frac{r^2\dot{\theta}}{C} = -kt,$

i.e., $r^2\dot{\theta} = Ce^{-kt}.$ (2.3.16)

For unit mass,

$$\Omega = \frac{1}{2}\left|\mathbf{r}\times\mathbf{v}\right| = \frac{1}{2}\left|r\hat{r}\times\left(\dot{r}\hat{r} + r\dot{\theta}\hat{\theta}\right)\right| = \frac{1}{2}\left|\, r^2\dot{\theta}\,\hat{z}\,\right| = \frac{1}{2}\left(r^2\,\dot{\theta}\right).$$

$\therefore\quad r^2\dot{\theta} = 2\Omega.$ (2.3.17)

Therefore, Eq. (2.3.16) gives

$2\Omega = Ce^{-kt}.$ (2.3.18)

Initially, $h = 2\Omega$ at $t = 0$.

Therefore, Eq. (2.3.18) implies $h = C$.

So, Eq. (2.3.18) leads

$$2\Omega = he^{-kt},$$

i.e., $r^2\dot{\theta} = he^{-kt}.$ (2.3.19)

From Eq. (2.3.14), we obtain

$$\ddot{r} - r\dot{\theta}^2 = -P - kv\frac{dr}{ds} = -P - k\frac{ds}{dt}\frac{dr}{ds} = -P - k\frac{dr}{dt} = -P - k\dot{r}. \tag{2.3.20}$$

Let $r = \frac{1}{u}$.

$$\therefore \quad \dot{r} = -\frac{1}{u^2}\frac{du}{d\theta}\frac{d\theta}{dt} = -\frac{1}{u^2}\frac{du}{d\theta}\dot{\theta} = -\frac{1}{u^2}\frac{du}{d\theta}\frac{he^{-kt}}{r^2} = -he^{-kt}\frac{du}{d\theta}. \tag{2.3.21}$$

$$\therefore \quad \ddot{r} = -h\left[e^{-kt}\frac{d^2u}{d\theta^2}\dot{\theta} + \frac{du}{d\theta}e^{-kt}(-k)\right] = -he^{-kt}\left[\frac{d^2u}{d\theta^2}he^{-kt}u^2 - k\frac{du}{d\theta}\right],$$

$$\text{i.e.,} \quad \ddot{r} = -he^{-kt}\left[hu^2e^{-kt}\frac{d^2u}{d\theta^2} - k\frac{du}{d\theta}\right]. \tag{2.3.22}$$

Using (2.3.19), (2.3.21) and (2.3.22) in (2.3.20), we obtain

$$-he^{-kt}\left[hu^2e^{-kt}\frac{d^2u}{d\theta^2} - k\frac{du}{d\theta}\right] - \frac{h^2}{u}e^{-2kt}u^4 = -P + khe^{-kt}\frac{du}{d\theta},$$

$$\text{i.e.,} \quad -h\left[hu^2e^{-kt}\frac{d^2u}{d\theta^2} - k\frac{du}{d\theta}\right] - h^2u^3e^{-kt} = -Pe^{kt} + kh\frac{du}{d\theta},$$

$$\text{i.e.,} \quad h^2u^2e^{-kt}\frac{d^2u}{d\theta^2} + kh\frac{du}{d\theta} - h^2u^3e^{-kt} = -Pe^{-kt} + kh\frac{du}{d\theta},$$

$$\text{i.e.,} \quad h^2u^2e^{-kt}\frac{d^2u}{d\theta^2} - h^2u^3e^{-kt} = -Pe^{kt},$$

$$\text{i.e.,} \quad h^2u^2\frac{d^2u}{d\theta^2} + h^2u^3 = Pe^{2kt},$$

$$\text{i.e.,} \quad \frac{d^2u}{d\theta^2} + u = \frac{P}{h^2u^2}e^{2kt}.$$

Example 2.7 A particle moves with a central acceleration P in a medium of which the resistance is $k(\text{velocity})^2$; show that the equation of its path is $\frac{d^2u}{d\theta^2} + u = \frac{P}{h^2u^2}e^{2ks}$, where s is the arc described and h is twice the initial moment of momentum about the center of force.

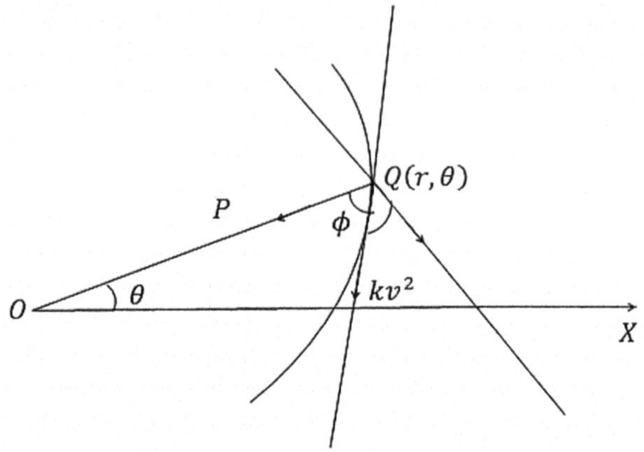

Fig. 2.15 Illustration of Example 2.7

Solution Similar to Example 2.6, and from Fig. 2.15, the equations of motion are

$$\ddot{r} - r\dot{\theta}^2 = -P - kv^2 \cos\phi, \tag{2.3.23}$$

$$\frac{1}{r}\frac{d}{dt}\left(r^2\dot{\theta}\right) = -kv^2 \sin\phi. \tag{2.3.24}$$

Equation (2.3.24) gives

$$\frac{1}{r}\frac{d}{dt}\left(r^2\dot{\theta}\right) = -kv^2\frac{rd\theta}{ds},$$

$$\text{i.e.,}\quad \frac{1}{r}\frac{d}{ds}\left(r^2\dot{\theta}\right)\dot{s} = -kvv\frac{rd\theta}{ds} = -kv\frac{ds}{dt}\frac{rd\theta}{ds} = -kv\frac{rd\theta}{dt},$$

$$\text{i.e.,}\quad \frac{1}{r}\frac{d}{ds}\left(r^2\dot{\theta}\right)\dot{s} = -k\dot{s}\frac{rd\theta}{dt},$$

$$\text{i.e.,}\quad \frac{d}{ds}\left(r^2\dot{\theta}\right) = -kr^2\dot{\theta},$$

$$\text{i.e.,}\quad \frac{d\left(r^2\dot{\theta}\right)}{r^2\dot{\theta}} = -kds,$$

$$\text{i.e.,}\quad r^2\dot{\theta} = Ce^{-ks}. \tag{2.3.25}$$

Initially $r^2\dot{\theta} = h$ when $t = 0$, i.e., $r^2\dot{\theta} = h$ when $s = 0$.

Therefore, Eq. (2.3.25) gives $C = h$.
Thus, Eq. (2.3.25) reduces to

$$r^2\dot{\theta} = he^{-ks}. \tag{2.3.26}$$

On substitution of $r = \frac{1}{u}$, Eq. (2.3.26) gives

$$\dot{\theta} = hu^2e^{-ks}. \tag{2.3.27}$$

Now

$$\dot{r} = \frac{dr}{dt} = \frac{d}{dt}\left(\frac{1}{u}\right) = -\frac{1}{u^2}\frac{du}{d\theta}\dot{\theta} = -\frac{1}{u^2}\frac{du}{d\theta}\cdot hu^2e^{-ks} = -h\ e^{-ks}\frac{du}{d\theta}, \tag{2.3.28}$$

$$\ddot{r} = -h\left[e^{-ks}\frac{d^2u}{d\theta^2}\dot{\theta} - \frac{du}{d\theta}e^{-ks}k\dot{s}\right] = -he^{-ks}\left[\frac{d^2u}{d\theta^2}\dot{\theta} - k\dot{s}\frac{du}{d\theta}\right],$$

$$\text{i.e., } \ddot{r} = -he^{-ks}\left[\frac{d^2u}{d\theta^2}hu^2e^{-ks} - k\dot{s}\frac{du}{d\theta}\right]. \tag{2.3.29}$$

Using (2.3.27)–(2.3.29) in (2.3.23), we obtain

$$-he^{-ks}\left[hu^2e^{-ks}\frac{d^2u}{d\theta^2} - k\dot{s}\frac{du}{d\theta}\right] - \frac{1}{u}h^2u^4e^{-2ks} = -P - kv^2\cos\phi,$$

$$\begin{aligned}\text{i.e., } he^{-ks}\left[h^2u^2e^{-ks}\tfrac{d^2u}{d\theta^2} - k\dot{s}\tfrac{du}{d\theta}\right] + h^2u^3e^{-2ks} &= P + kv^2\cos\phi \\ &= P + kv\tfrac{ds}{dt}\tfrac{dr}{ds} = P + kv\dot{r} = P + k\dot{s}\left(-he^{-ks}\tfrac{du}{d\theta}\right),\end{aligned}$$

$$\text{i.e., } h\left[h^2u^2e^{-ks}\frac{d^2u}{d\theta^2} - k\dot{s}\frac{du}{d\theta}\right] + h^2u^3e^{-ks} = Pe^{ks} - hk\dot{s}\frac{du}{d\theta},$$

$$\text{i.e., } h^2u^2e^{-ks}\frac{d^2u}{d\theta^2} - hk\dot{s}\frac{du}{d\theta} + h^2u^3e^{-ks} = Pe^{ks} - hk\dot{s}\frac{du}{d\theta},$$

$$\text{i.e., } h^2u^2e^{-ks}\frac{d^2u}{d\theta^2} + h^2u^3e^{-ks} = Pe^{ks},$$

$$\text{i.e., } h^2u^2\frac{d^2u}{d\theta^2} + h^2u^3 = Pe^{2ks},$$

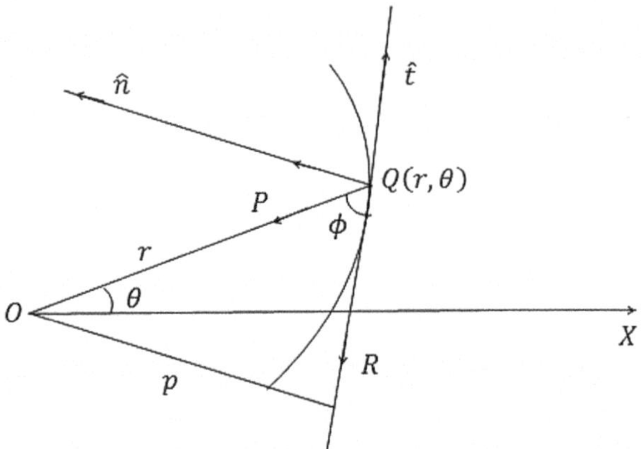

Fig. 2.16 Diagram illustrating Example 2.8

i.e., $\dfrac{d^2u}{d\theta^2} + u = \dfrac{P}{h^2u^2}e^{2ks}$.

Example 2.8 A particle moves in a resisting medium with a given central acceleration P; the path of the particle being given shows that the resistance is $-\frac{1}{2p^2}\frac{d}{ds}\left(p^3\frac{dr}{dp}P\right)$.

Solution Let Q be the position of the particle at time t. In Fig. 2.16, referred to the center of force O as the pole and a line **OX** as the initial line, let Q be specified by the polar coordinates (r, θ).

The tangential and normal equations of motion of the particle are

$$v\frac{dv}{ds} = -R - P\cos\phi, \tag{2.3.30}$$

$$\frac{v^2}{\rho} = P\sin\phi. \tag{2.3.31}$$

Equation (2.3.31) gives

$$v^2 = \rho P\sin\phi = \rho P\frac{p}{r}. \tag{2.3.32}$$

(2.3.30) yields

$$R = -P\cos\phi - v\frac{dv}{ds} = -P\cos\phi - \frac{d}{ds}\left(\frac{v^2}{2}\right) = -P\frac{dr}{ds} - \frac{d}{ds}\left(\frac{\rho P}{2}\frac{p}{r}\right)$$

$$= -P\frac{dr}{ds} - \frac{1}{2}\frac{d}{ds}\left(\rho P\frac{p}{r}\right) = -P\frac{dr}{ds} - \frac{1}{2}\frac{d}{ds}\left(\frac{\rho Pp}{r}\right) = -\left[P\frac{dr}{ds} + \frac{1}{2}\frac{d}{ds}\left(\frac{\rho Pp}{r}\right)\right]$$

$$= -\frac{1}{2p^2}\left[2p^2 P\frac{dr}{ds} + p^2\frac{d}{ds}\left(\frac{\rho P p}{r}\right)\right] = -\frac{1}{2p^2}\left[2Pp^2\frac{dr}{ds} + p^2\frac{d}{ds}\left(\frac{Pp}{r}\frac{rdr}{dp}\right)\right]$$

$$= -\frac{1}{2p^2}\left[2Pp^2\frac{dr}{ds} + p^2\frac{d}{ds}\left(Pp\frac{dr}{dp}\right)\right]$$

$$= -\frac{1}{2p^2}\left[2Pp^2\frac{dr}{dp}\frac{dp}{ds} + p^2\frac{d}{ds}\left(Pp\frac{dr}{dp}\right)\right]$$

$$= -\frac{1}{2p^2}\left[Pp\ \frac{dr}{dp}2p\frac{dp}{ds} + p^2\frac{d}{ds}\left(Pp\frac{dr}{dp}\right)\right]$$

$$= -\frac{1}{2p^2}\left[Pp\frac{dr}{dp}\frac{d}{ds}(p^2) + p^2\frac{d}{ds}\left(Pp\frac{dr}{dp}\right)\right]$$

$$= -\frac{1}{2p^2}\frac{d}{ds}\left(Pp\frac{dr}{dp}p^2\right) = -\frac{1}{2p^2}\frac{d}{ds}\left(Pp^3\frac{dr}{dp}\right).$$

Example 2.9 A particle moves in an ellipse given by $\frac{l}{r} = 1 + e\cos\theta$ under a force which is directed towards its focus. Find the periodic time.

Solution Let T be the time that the particle takes to describe the whole ellipse. We have

$$\frac{h}{2}T = \text{ area of the ellipse} = \Pi\ ab,$$

$$\text{i.e., } T = \frac{2\pi ab}{h} = \frac{2\pi ab}{\sqrt{\mu l}} = \frac{2\pi ab}{\sqrt{\mu\frac{b^2}{a}}} = \frac{2\pi aba^{\frac{1}{2}}}{\sqrt{\mu}b} = \frac{2\pi a^{\frac{3}{2}}}{\sqrt{\mu}}.$$

Example 2.10 A particle describes a circle under a force P to the pole on the circumference of the circle. Find the law of force.

Referring to a point O on the circle as a pole and the diameter OA as the initial line, the polar equation of the circle (Fig. 2.17) is $r = 2a\cos\theta$, i.e., $u = \dfrac{\sec\theta}{2a}$.

$$\therefore\quad \frac{du}{d\theta} = \frac{1}{2a}\sec\theta\ \tan\theta = u\tan\theta,$$

and so

$$\frac{d^2u}{d\theta^2} = u\sec^2\theta + \tan\theta\ \frac{du}{d\theta} = u\sec^2\theta + u\ \tan^2\theta = u\left(\sec^2\theta + \tan^2\theta\right).$$

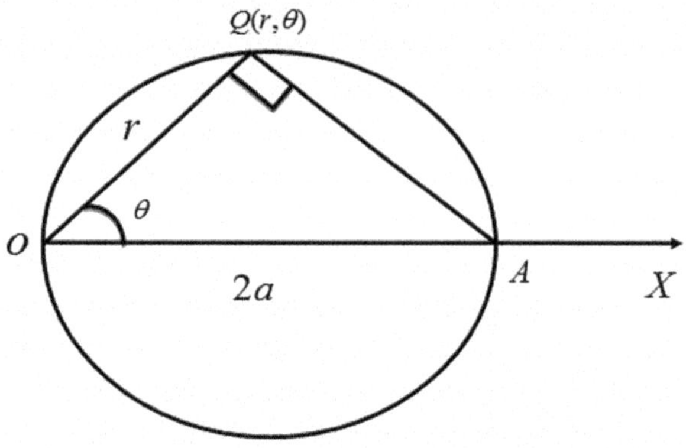

Fig. 2.17 Illustration for Example 2.10

The differential equation of the central orbit is

$$\frac{P}{h^2u^2} = \frac{d^2u}{d\theta^2} + u = u\left(\sec^2\theta + \tan^2\theta\right) + u = 2u\sec^2\theta = 2u(2au)^2 = 8a^2u^3,$$

$$\text{i.e.,}\quad P = 8h^2a^2u^5 = \frac{8h^2a^2}{r^5},$$

$$\therefore\quad P \propto \frac{1}{r^5}.$$

Example 2.11 A particle acted on by a central attractive force $\frac{\mu}{r^3}$ is projected with a velocity $\frac{\sqrt{\mu}}{a}$ at an angle $\frac{\pi}{4}$ with its initial distance a from the center of force; show that its orbit is the equiangular spiral $r = ae^{-\theta}$.

Solution With usual rotations, in case of a central orbit

$$vp = h. \tag{2.3.33}$$

It is given that $v = \frac{\sqrt{\mu}}{a}$ at $t = 0$.

Also from Fig. 2.18, we have

$$p = a\sin\frac{\pi}{4} = \frac{a}{\sqrt{2}} \quad \text{at} \quad t = 0.$$

Subjecting the above condition on Eq. (2.3.33), we find

$$\frac{\sqrt{\mu}}{a}\frac{a}{\sqrt{2}} = h,$$

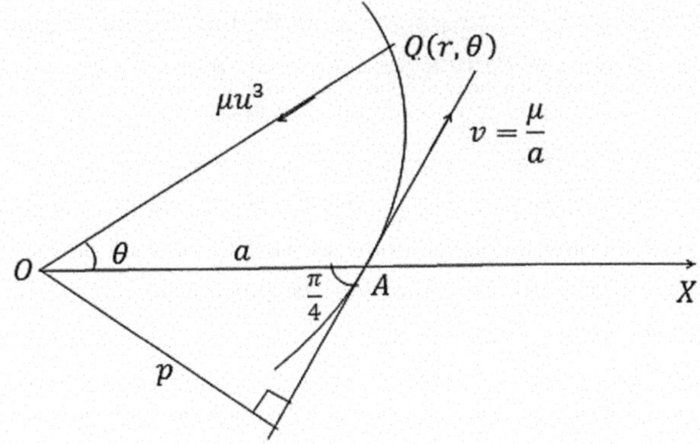

Fig. 2.18 Schematic diagram for Example 2.11

$$\text{i.e.,}\quad h = \sqrt{\frac{\mu}{2}}. \tag{2.3.34}$$

The differential of the central orbit is

$$\frac{d^2u}{d\theta^2} + u = \frac{P}{h^2u^2}. \tag{2.3.35}$$

Given $P = \mu u^3$ and so Eq. (2.3.35) reduces to

$$\frac{d^2u}{d\theta^2} + u = \frac{2\mu\, u^3}{\mu\, u^2} = 2u,$$

$$\text{or}\quad \left(D^2 - 1\right)\, u = 0. \tag{2.3.36}$$

The general solution of (2.3.36) is

$$u = Ae^{\theta} + Be^{-\theta}, \tag{2.3.37}$$

where A and B are two constants to be determined.

Initially, $u = \frac{1}{a}$, $\theta = 0$ when $t = 0$.

From Eq. (2.3.37), we find

$$A + B = \frac{1}{a}. \tag{2.3.38}$$

Further,

$$v^2 = h^2\left[\left(\frac{du}{d\theta}\right)^2 + u^2\right] = \frac{\mu}{2}\left[\left(\frac{du}{d\theta}\right)^2 + u^2\right]. \tag{2.3.39}$$

Initially $v = \frac{\sqrt{\mu}}{a}$, $u = \frac{1}{a}$, $\frac{du}{d\theta} = k$ (say) at time $t = 0$.

Putting these in Eq. (2.3.39), we obtain

$$\frac{\mu}{a^2} = \frac{\mu}{2}\left[k^2 + \frac{1}{a^2}\right],$$

$$\text{or} \quad \frac{1}{a^2} = \frac{1}{2}\left[k^2 + \frac{1}{a^2}\right],$$

$$\text{or} \quad 1 = \frac{1}{2}\left[k^2a^2 + 1\right],$$

$$\text{or} \quad 2 = k^2a^2 + 1,$$

$$\text{or} \quad k^2a^2 = 1,$$

$$\text{or} \quad k = \frac{1}{a},$$

$$\text{or} \quad \frac{du}{d\theta} = \frac{1}{a} \quad \text{at } t = 0.$$

Differentiating (2.3.37) w.r.t. θ, we derive

$$\frac{du}{d\theta} = Ae^{\theta} - Be^{-\theta}. \tag{2.3.40}$$

Initially $\frac{du}{d\theta} = \frac{1}{a}$, $\theta = 0$ at $t = 0$.

Therefore, Eq. (2.3.40) gives

$$A - B = \frac{1}{a}. \tag{2.3.41}$$

Solving (2.3.38) and (2.3.41), we get $A = \frac{1}{a}$, and $B = 0$. Substituting the values of A and B in (2.3.37), we obtain $r = ae^{-\theta}$.

Example 2.12 If a particle is projected from an apse at a distance a with the velocity from infinity under the action of a central force μr^{-2n-3}, prove that the path is $r^n = a^n \cos(n\,\theta)$.

Solution Let V be the velocity of fall from infinity to the point $r = a$.

$$\therefore \quad V^2 = -2\int_{\infty}^{a} \mu r^{-2n-3} dr = -2\mu \left.\frac{r^{-2n-2}}{-2n-2}\right]_{\infty}^{a} = \frac{\mu}{(n+1)\,a^{2n+2}}. \tag{2.3.42}$$

The differential equation of the central orbit is

$$\frac{d^2u}{d\theta^2} + u = \frac{P}{h^2u^2}. \tag{2.3.43}$$

Given

$$P = \mu r^{-2n-3} = \frac{\mu}{r^{2r+3}} = \mu u^{2n+3}. \quad (2.3.44)$$

By the use of (2.3.44), Eq. (2.3.43) takes the form

$$\frac{d^2u}{d\theta^2} + u = \frac{\mu u^{2n+3}}{h^2u^2} = \frac{\mu u^{2n+1}}{h^2},$$

$$\text{or} \quad \frac{du_1}{d\theta} + u = \frac{\mu u^{2n+1}}{h^2}; \quad u_1 = \frac{du}{d\theta},$$

$$\text{or} \quad 2u_1\frac{du_1}{d\theta} + 2uu_1 = \frac{2\mu}{h^2}u^{2n+1}u_1,$$

$$\text{or} \quad 2u_1\frac{du_1}{d\theta} + 2u\frac{du}{d\theta} = \frac{2\mu}{h^2}u^{2n+1}\frac{du}{d\theta},$$

$$\text{or} \quad 2u_1du_1 + 2udu = \frac{2\mu}{h^2}u^{2n+1}du,$$

$$\text{or} \quad u_1^2 + u^2 = \frac{2\mu}{h^2}\frac{u^{2n+1}}{2n+2} + C,$$

$$\text{or} \quad \left(\frac{du}{d\theta}\right)^2 + u^2 = \frac{\mu u^{2n+2}}{(n+1)\,h^2} + C. \quad (2.3.45)$$

Again

$$v^2 = h^2\left[\left(\frac{du}{d\theta}\right)^2 + u^2\right]. \quad (2.3.46)$$

Initially $v = V = \sqrt{\frac{\mu}{n+1}\frac{1}{a^{n+1}}}, \ \frac{du}{d\theta} = 0, \ u = \frac{1}{a}$ at time $t = 0$.
Therefore, Eqs. (2.3.45) and (2.3.46) give

$$\frac{1}{a^2} = \frac{\mu}{h^2\,(n+1)}\,\frac{1}{a^{2n+2}} + C \text{ and } \frac{\mu}{n+1}\,\frac{1}{a^{2n+2}} = \frac{h^2}{a^2}. \quad (2.3.47)$$

Equation (2.3.47) gives

$$h^2 = \frac{\mu}{n+1}\,\frac{1}{a^{2n}}, \quad C = 0.$$

Substituting the values of h^2 and C in (2.3.45), we obtain

$$\left(\frac{du}{d\theta}\right)^2 + u^2 = \frac{\mu u^{2n+2}\,(n+1)\,a^{2n}}{(n+1)\,\mu},$$

$$\text{i.e.,}\quad \left(\frac{du}{d\theta}\right)^2 + u^2 = u^{2n+2}a^{2n},$$

$$\text{i.e.,}\quad \left(\frac{du}{d\theta}\right)^2 = u^2\left\{(au)^{2n} - 1\right\},$$

$$\text{i.e.,}\quad \frac{du}{d\theta} = -u\sqrt{(au)^{2n} - 1},$$

$$\text{i.e.,}\quad \frac{du}{u\sqrt{(au)^{2n} - 1}} = -d\theta,$$

$$\text{i.e.,}\quad I = -\theta + C_1, \tag{2.3.48}$$

where

$$I = \int \frac{du}{u\sqrt{(au)^{2n} - 1}}.$$

Considering the substitution

$$(au)^n = \sec\varphi, \tag{2.3.49}$$

we determine

$$I = \int \frac{\sec\varphi\ \tan\varphi\ d\varphi}{na^n u^n\ \tan\varphi} = \frac{1}{n}\int \frac{\sec\varphi\, d\varphi}{\sec\varphi} = \frac{1}{n}\,\varphi.$$

Initially $u = \frac{1}{a}$, $\varphi = \alpha$ (say) when $t = 0$.
Therefore, Eq. (2.3.49) shows that

$$1 = \sec\alpha,\ \text{i.e.,}\ \alpha = 0,\ \text{i.e.,}\ \varphi = 0\ \text{at}\ t = 0.$$

Equation (2.3.48) finally reduces to

$$\frac{1}{n}\varphi = -\theta + C_1. \tag{2.3.50}$$

Initially $\varphi = 0, \quad \theta = 0 \quad \text{at} \quad t = 0.$

Therefore, Eq. (2.3.50) gives $C_1 = 0$.

Hence $\varphi = -n\theta$ and so Eq. (2.3.49) yields

$$(au)^n = \sec(-n\theta) = \sec(n\theta),$$

$$\text{i.e.,} \quad \left(\frac{a}{r}\right)^n = \sec(n\theta),$$

$$\text{i.e.,} \quad \left(\frac{r}{a}\right)^n = \cos(n\theta),$$

$$\text{i.e.,} \quad r^n = a^n \cos(n\theta).$$

Example 2.13 If the law of force be $2\mu\left(u^3 - a^2u^5\right)$ and the particle be projected from an apse at a distance a with velocity $\frac{\sqrt{\mu}}{a}$, show that it will be at a distance r from the center after a time $\frac{1}{2\sqrt{\mu}}\left\{r\sqrt{r^2 - a^2} + a^2 \ \cos h^{-1}\left(\frac{r}{a}\right)\right\}$.

Solution The differential equation of the central orbit is

$$\frac{d^2u}{d\theta^2} + u = \frac{P}{h^2u^2},$$

$$\text{i.e.,} \quad \frac{d^2u}{d\theta^2} + u = \frac{2\mu\left(u^3 - a^2u^5\right)}{h^2u^2},$$

$$\text{i.e.,} \quad \frac{du_1}{d\theta} + u = \frac{2\mu}{h^2}\left(u - a^2u^3\right),$$

$$\text{i.e.,} \quad 2u_1\frac{du_1}{d\theta} + 2uu_1 = \frac{2\mu}{h^2}\left[2uu_1 - 2a^2u^3u_1\right],$$

$$\text{i.e.,} \quad 2u_1\frac{du_1}{d\theta} + 2u\frac{du}{d\theta} = \frac{2\mu}{h^2}\left[2u\frac{du}{d\theta} - 2a^2u^3\frac{du}{d\theta}\right],$$

$$\text{i.e.,} \quad 2u_1du_1 + 2udu = \frac{2\mu}{h^2}\left[2udu - 2a^2u^3du\right],$$

$$\text{i.e.,} \quad u_1^2 + u^2 = \frac{2\mu}{h^2}\left[u^2 - 2a^2\frac{u^4}{4}\right] + C,$$

$$\left(\frac{du}{d\theta}\right)^2 + u^2 = \frac{2\mu}{h^2}\left[u^2 - \frac{a^2u^4}{2}\right] + C. \tag{2.3.51}$$

Again

$$vp = h. \tag{2.3.52}$$

In (2.3.51) and (2.3.52), C and h are two arbitrary constants to be determined from the initial conditions of the problem.

Initially $v = \frac{\sqrt{\mu}}{a}, \quad p = a$ at $t = 0$.
Equation (2.3.52) gives

$$h = vp = \frac{\sqrt{\mu}}{a}a = \sqrt{\mu}.$$

Equation (2.3.51) determines

$$\frac{1}{a^2} = \frac{2\mu}{\mu}\left[\frac{1}{a^2} - \frac{1}{2a^2}\right] + C,$$

$$\text{i.e.,}\quad \frac{1}{a^2} = 2\left[\frac{1}{a^2} - \frac{1}{2a^2}\right] + C = \frac{2}{a^2} - \frac{1}{a^2} + C = \frac{1}{a^2} + C,$$

$$\therefore \quad C = 0.$$

Hence Eq. (2.3.51) reduces to

$$\left(\frac{du}{d\theta}\right)^2 + u^2 = 2\left[u^2 - \frac{a^2u^4}{2}\right] = 2u^2 - a^2u^4,$$

$$\text{i.e.,}\quad \left(\frac{du}{d\theta}\right)^2 = u^2\left(1 - a^2u^2\right),$$

$$\text{i.e.,}\quad \frac{du}{d\theta} = -u\sqrt{1 - a^2u^2},$$

$$\text{i.e.,}\quad \frac{du}{u\sqrt{1 - a^2u^2}} = -d\theta. \tag{2.3.53}$$

We have $u = \frac{1}{r}$,

$$\therefore \quad du = -\frac{1}{r^2}dr.$$

Equation (2.3.53) reduces to

$$\frac{dr}{\sqrt{r^2 - a^2}} = d\theta. \tag{2.3.54}$$

But

$$r^2\dot{\theta} = h,$$

$$\text{i.e.,}\quad r^2\dot{\theta} = \sqrt{\mu},$$

$$\text{i.e.,}\quad r^2 d\theta = \sqrt{\mu} dt,$$

$$\text{i.e.,}\quad d\theta = \frac{\sqrt{\mu}}{r^2}\, dt. \tag{2.3.55}$$

By eliminating $d\theta$ from (2.3.54) and (2.3.55), we obtain

$$\frac{dr}{\sqrt{r^2 - a^2}} = \frac{\sqrt{\mu}}{r^2}\, dt,$$

$$\text{i.e.,}\quad dt = \frac{1}{\sqrt{\mu}}\frac{r^2 dr}{\sqrt{r^2 - a^2}} = \frac{1}{\sqrt{\mu}}\frac{r^2 - a^2 + a^2}{\sqrt{r^2 - a^2}} dr = \frac{1}{\sqrt{\mu}}\left[\sqrt{r^2 - a^2} + \frac{a^2}{\sqrt{r^2 - a^2}}\right] dr.$$

Integration of the above equation leads

$$t = \frac{1}{\sqrt{\mu}}\left[\frac{1}{2}\left\{r\sqrt{r^2 - a^2} - a^2 \cosh^{-1}\left(\frac{r}{a}\right)\right\} + a^2 \cosh^{-1}\left(\frac{r}{a}\right)\right] + C_1$$

$$= \frac{1}{2\sqrt{\mu}}\left[\left\{r\sqrt{r^2 - a^2} - a^2 \cosh^{-1}\left(\frac{r}{a}\right)\right\} + 2a^2 \cosh^{-1}\left(\frac{r}{a}\right)\right] + C_1,$$

$$\text{i.e.,}\quad t = \frac{1}{2\sqrt{\mu}}\left[\sqrt{r^2 - a^2} + a^2 \cosh^{-1}\left(\frac{r}{a}\right)\right] + C_1. \tag{2.3.56}$$

Initially $r = a$ when $t = 0$.

Equation (2.3.56) gives

$$C_1 = 0.$$

Therefore, Eq. (2.3.56) gets reduced to

$$t = \frac{1}{2\sqrt{\mu}}\left[r\sqrt{r^2 - a^2} + a^2 \cosh^{-1}\left(\frac{r}{a}\right)\right],$$

which is the required time.

Example 2.14 A particle moves with a central acceleration which varies inversely as the cube of the distance, if it is projected from an apse at a distance "a" from the origin with a velocity which is $\sqrt{2}$ times the velocity for a circle of radius a, show that the equation to its path is $r \cos\frac{\theta}{\sqrt{2}} = a$.

Solution Given

$$P \propto \frac{1}{r^3},$$

$$\text{i.e.,} \quad P = \frac{\mu}{r^3} = \mu u^3.$$

Let V_0 be the velocity for a circle of radius a under the same central force.

$$\therefore \quad V_0^2 = \text{Pr}]_{r=a} = \left.\frac{\mu u^3}{u}\right]_{u=\frac{1}{a}} = \frac{\mu}{a^2}.$$

Let V denote the velocity of the projection.

$$\therefore \quad V = \sqrt{2}V_0,$$

$$\text{i.e.,} \quad V^2 = \frac{2\mu}{a^2}.$$

In the case of a central orbit,

$$vp = h. \tag{2.3.57}$$

Initially $v = V, \quad p = a$.

$$\therefore \quad h = Va,$$

$$\text{i.e.,} \quad h^2 = a^2 V^2,$$

$$\text{i.e.,} \quad h^2 = 2\mu. \tag{2.3.58}$$

The d.e. of the central orbit is

$$\frac{d^2u}{d\theta^2} + u = \frac{P}{h^2u^2} = \frac{\mu u^3}{2\mu u^2} = \frac{u}{2},$$

$$\text{i.e.,}\quad \left(D^2 + \frac{1}{2}\right) u = 0. \tag{2.3.59}$$

General solution of (2.3.59) is

$$u = A\cos\left(\frac{\theta}{\sqrt{2}}\right) + B\sin\left(\frac{\theta}{\sqrt{2}}\right). \tag{2.3.60}$$

The values of A and B subject to the condition $u = \frac{1}{a},\ \frac{du}{d\theta} = 0$ at $\theta = 0$ are as follows:
$B = 0$ and $A = \frac{1}{a}$.
Therefore, Eq. (2.3.60) reduces to

$$u = \frac{1}{a}\cos\frac{\theta}{\sqrt{2}},$$

$$\text{i.e.,}\quad r\cos\frac{\theta}{\sqrt{2}} = a.$$

Example 2.15 A particle moves under a central repulsive force $\left\{= \frac{m\mu}{(\text{distance})^3}\right\}$, and is projected from an apse at a distance a with velocity V. Show that the equation to the path is $r\ \cos(p\theta) = a$, and the angle θ described in time t is $\frac{1}{p}\tan^{-1}\left[\frac{pV}{a}t\right]$,where $p^2 = \frac{\mu+a^2V^2}{a^2V^2}$.

Solution The central acceleration is given by $P = -\frac{\mu}{r^3} = -\mu u^3$. If v is the velocity of the particle at the point (r, θ), then

$$vp = h. \tag{2.3.61}$$

Initially $v = V,\quad p = a$.
Equation (2.3.61) gives

$$h = Va. \tag{2.3.62}$$

The differential equation of the central orbit is

$$\frac{d^2u}{d\theta^2} + u = \frac{P}{h^2u^2} = \frac{-\mu u^3}{a^2V^2u^2} = -\frac{\mu u}{a^2V^2},$$

$$\text{i.e.,}\quad \frac{d^2u}{d\theta^2} + \frac{\mu + a^2V^2}{a^2V^2}u = 0,$$

$$\text{i.e.,}\quad \frac{d^2u}{d\theta^2} + p^2u = 0\quad \left[\because p^2 = \frac{\mu + a^2V^2}{a^2V^2}\right],$$

$$\text{i.e.,} \quad \left(D^2 + p^2\right) u = 0. \tag{2.3.63}$$

The general solution of (2.3.63) is

$$u = A\cos(p\theta) + B\sin(p\theta), \tag{2.3.64}$$

and

$$\frac{du}{d\theta} = -Ap\sin(p\theta) + Bp\cos(p\theta). \tag{2.3.65}$$

Initially, $u = \frac{1}{a}, \ \frac{du}{d\theta} = 0, \text{ when } \theta = 0.$
Equations (2.3.64) and (2.3.65) give

$$A = \frac{1}{a} \quad \& \quad B = 0.$$

Thus, Eq. (2.3.64) reduces to

$$u = \frac{1}{a}\cos(p\theta),$$

$$r\cos(p\theta) = a. \tag{2.3.66}$$

Equation (2.3.66) yields

$$r = a\sec(p\theta).$$

For a central orbit,

$$r^2\dot{\theta} = h,$$

$$\text{i.e.,} \quad a^2\sec^2 p\theta\,\dot{\theta} = aV,$$

$$\text{i.e.,} \quad a\sec^2 p\theta\,d\theta = V\,dt,$$

$$\text{i.e.,} \quad dt = \frac{a}{V}\sec^2 p\theta\,d\theta,$$

$$\text{i.e.,} \quad \int_0^t dt = \frac{a}{V}\int_{\theta=0}^{\theta}\sec^2 p\theta\,d\theta,$$

$$\text{i.e.,}\quad t = \frac{a}{V}\left.\frac{\tan p\theta}{p}\right]_0^\theta = \frac{a}{pV}\tan p\theta,$$

$$\text{i.e.,}\quad \tan p\theta = \frac{pVt}{a},$$

$$\text{i.e.,}\quad p\theta = \tan^{-1}\left(\frac{pVt}{a}\right),$$

$$\therefore\quad \theta = \frac{1}{p}\tan^{-1}\left(\frac{pVt}{a}\right).$$

Example 2.16 A particle moves under a central force $\mu\left(r^5 - c^4 r\right)$ per unit mass and is projected from an apse at distance c with a velocity $\sqrt{\frac{2\mu}{3}}c^3$, show that its path is the curve $x^4 + y^4 = c^4$.

Solution The central acceleration is given by

$$P = \mu\left(r^5 - c^4 r\right) = \mu\left(u^{-5} - c^4 u^{-1}\right).$$

The differential equation of the central orbit is

$$\frac{d^2u}{d\theta^2} + u = \frac{\mu}{h^2}\left[u^{-7} - c^4 u^{-3}\right]. \tag{2.3.67}$$

For a central orbit,

$$vp = h. \tag{2.3.68}$$

Initially $v = \sqrt{\frac{2\mu}{3}}c^3,\ \ p = c.$

Equation (2.3.68) gives

$$\sqrt{\frac{2\mu}{3}}c^3 c = h,$$

$$\text{i.e.,}\quad h^2 = \frac{2\mu}{3}c^8.$$

Hence Eq. (2.3.67) reduces to

$$\frac{d^2u}{d\theta^2} + u = \frac{3\mu}{2\mu c^8}\left[u^{-7} - c^4 u^{-3}\right],$$

$$\text{i.e.,}\quad \frac{du_1}{d\theta} + u = \frac{3}{2c^8}\left[u^{-7} - c^4 u^{-3}\right],$$

$$\text{i.e.,}\quad 2u_1\frac{du_1}{d\theta}+2uu_1=\frac{3}{c^8}\left[u^{-7}u_1-c^4u^{-3}u_1\right],$$

$$\text{i.e.,}\quad 2u_1\frac{du_1}{d\theta}+2u\frac{du}{d\theta}=\frac{3}{c^8}\left[u^{-7}\frac{du}{d\theta}-c^4u^{-3}\frac{du}{d\theta}\right],$$

$$\text{i.e.,}\quad 2u_1du_1+2udu=\frac{3}{c^8}\left[u^{-7}du-c^4u^{-3}du\right],$$

$$\text{i.e.,}\quad u_1^2+u^2=\frac{3}{c^8}\left[\frac{u^{-6}}{-6}-c^4\frac{u^{-2}}{-2}\right]+c_1,$$

$$\text{i.e.,}\quad \left(\frac{du}{d\theta}\right)^2+u^2=\frac{3}{c^8}\left[\frac{c^4}{2u^2}-\frac{1}{6u^6}\right]+c_1. \tag{2.3.69}$$

Initially $\frac{du}{d\theta}=0$ when $u=\frac{1}{c}$.

Equation (2.3.69) gives

$$\frac{1}{c^2}=\frac{3}{c^8}\left[\frac{c^6}{2}-\frac{c^6}{6}\right]+c_1,$$

$$\text{i.e.,}\quad \frac{1}{c^2}=\frac{1}{c^2}+c_1,$$

$$\therefore\quad c_1=0.$$

Thus, Eq. (2.3.69) reduces to

$$\left(\frac{du}{d\theta}\right)^2+u^2=\frac{3}{2}\frac{1}{c^4u^2}-\frac{1}{2c^8u^6},$$

$$\text{i.e.,}\quad \left(\frac{du}{d\theta}\right)^2=-\frac{2c^8u^8-3c^4u^4+1}{2c^8u^6}=\frac{\left(1-c^4u^4\right)\left(c^4u^4-\frac{1}{2}\right)}{2c^8u^6},$$

$$\therefore\quad \frac{du}{d\theta}=-\frac{\sqrt{\left(1-c^4u^4\right)\ \left(c^4u^4-{}^1/_2\right)}}{2c^4u^3}. \tag{2.3.70}$$

In consideration of the substitution,

$$c^4u^4=\cos^2\varphi+\frac{1}{2}\sin^2\varphi.$$

Equation (2.3.70) gets transformed to

$$\frac{-\frac{\sin\varphi\cos\varphi\, d\varphi}{4}}{\sqrt{\frac{1}{2}\sin^2\varphi\,\frac{1}{2}\cos^2\varphi}} = -d\theta,$$

$$\text{i.e.,}\quad \frac{\sin\varphi\,\cos\varphi\, d\varphi}{2\sin\varphi\,\cos\varphi} = d\theta,$$

$$\text{i.e.,}\quad d\varphi = 2d\theta. \tag{2.3.71}$$

On the integration of Eq. (2.3.71), we obtain

$$\varphi = 2\theta + c_1. \tag{2.3.72}$$

$$\text{Let}\quad \varphi = \alpha \quad \text{when} \quad u = \frac{1}{c}.$$

Therefore,

$$1 = \cos^2\alpha + \frac{1}{2}\sin^2\alpha,$$

$$\text{i.e.,}\quad \sin^2\alpha = 0,$$

$$\text{i.e.,}\quad \alpha = 0.$$

Therefore,

$$\varphi = 0 \quad \text{when} \quad u = \frac{1}{c},$$

$$\text{i.e.,}\quad \varphi = 0 \quad \text{when} \quad \theta = 0.$$

Hence, Eq. (2.3.72) gives $c_1 = 0$ and Eq. (2.3.72) reduces to $\varphi = 2\theta$, which leads the equation of the path to take the form

$$c^4u^4 = \cos^2 2\theta + \frac{1}{2}\sin^2 2\theta,$$

$$\text{i.e.,}\quad \frac{2c^4}{r^4} = 2\cos^2 2\theta + \sin^2 2\theta = 2\left(\cos^2\theta - \sin^2\theta\right)^2 + 4\sin^2\theta\cos^2\theta,$$

$$\text{i.e.,}\quad \frac{c^4}{r^4} = \left(\cos^2\theta - \sin^2\theta\right)^2 + 2\sin^2\theta\cos^2\theta = \cos^4\theta + \sin^4\theta,$$

$$\text{i.e.,}\quad c^4 = (r\cos\theta)^4 + (r\sin\theta)^4,$$

$$\text{i.e.,}\quad x^4 + y^4 = c^4.$$

Example 2.17 A particle moves in a field of force whose potential is $\mu r^{-2}\cos\theta$ and it is projected at a distance a, $\perp^r$ to the initial line with velocity $\frac{2}{a}\sqrt{\mu}$; show that the orbit described is $r = a\ \sec\left\{\sqrt{2}\ \log\tan\frac{\pi+\theta}{4}\right\}$.

Solution Given $V = -\mu r^{-2}\cos\theta$.

$$\therefore\quad \mathbf{F} = -\nabla V = -\hat{r}\frac{\partial V}{\partial r} - \frac{\hat{\theta}}{r}\frac{\partial V}{\partial \theta} = -\frac{2\mu\cos\theta}{r^3}\hat{r} - \frac{\mu\sin\theta}{r^3}\hat{\theta}. \tag{2.3.73}$$

The equation of the cross-radial motion is

$$\frac{1}{r}\frac{d}{dt}\left(r^2\dot{\theta}\right) = -\frac{\mu\sin\theta}{r^3},$$

$$\text{i.e.,}\quad r^2\frac{d}{dt}\left(r^2\dot{\theta}\right) = -\mu\sin\theta,$$

$$\text{i.e.,}\quad r^2\dot{\theta}\frac{d}{dt}\left(r^2\dot{\theta}\right) = -\mu\sin\theta\ \dot{\theta},$$

$$\text{i.e.,}\quad r^2\dot{\theta}\frac{d}{dt}\left(r^2\dot{\theta}\right) = -\mu\sin\theta\ \frac{d\theta}{dt},$$

$$\text{i.e.,}\quad r^2\dot{\theta}d\left(r^2\dot{\theta}\right) = -\mu\sin\theta\ d\theta,$$

$$\text{i.e.,}\quad \left(r^2\dot{\theta}\right)^2 = 2\mu\cos\theta + c_1. \tag{2.3.74}$$

Initially $a\dot{\theta} = \frac{2}{a}\sqrt{\mu},\quad r = a$ when $\theta = 0$.

$$\text{i.e.,}\quad \dot{\theta} = \frac{2}{a^2}\sqrt{\mu}, r = a\ \text{ when }\ \theta = 0.$$

Equation (2.3.74) gives $C_1 = 2\mu$.

Therefore, Eq. (2.3.74) reduces to

$$\left(r^2\dot{\theta}\right)^2 = 2\mu\left(1 + \cos\theta\right) = 4\mu\cos^2\frac{\theta}{2},$$

$$\therefore\quad r^2\dot{\theta} = 2\sqrt{\mu}\cos\frac{\theta}{2}. \tag{2.3.75}$$

The equation of the radial motion is

$$\ddot{r} - r\dot{\theta}^2 = F_r,$$

$$\text{i.e.,} \quad \ddot{r} - r\dot{\theta}^2 = -\frac{2\mu\cos\theta}{r^3},$$

$$\text{i.e.,} \quad \ddot{r} - \frac{4\mu}{r^3}\cos^2\frac{\theta}{2} = -\frac{2\mu\cos\theta}{r^3},$$

$$\text{i.e.,} \quad \ddot{r} - \frac{2\mu}{r^3}(1+\cos\theta) = -\frac{2\mu\cos\theta}{r^3},$$

$$\text{i.e.,} \quad \ddot{r} - \frac{2\mu}{r^3} = 0,$$

$$\text{i.e.,} \quad \frac{dv_r}{dt} = \frac{2\mu}{r^3},$$

$$\text{i.e.,} \quad \frac{dv_r}{dr}v_r = \frac{2\mu}{r^3},$$

$$\text{i.e.,} \quad v_r dv_r = \frac{2\mu}{r^3}dr,$$

$$\text{i.e.,} \quad \frac{v_r^2}{2} = -\frac{\mu}{r^2} + C_2.$$

Initially $v_r = 0$, when $r = a$.

$$\therefore \quad C_2 = \frac{\mu}{a^2},$$

$$\text{i.e.,} \quad v_r^2 = \frac{2\mu\left(r^2 - a^2\right)}{a^2r^2},$$

$$\text{i.e.,} \quad v_r = \frac{\sqrt{2\mu}}{ar}\sqrt{r^2 - a^2},$$

$$\text{i.e.,} \quad \frac{dr}{d\theta}\dot{\theta} = \frac{\sqrt{2\mu}}{ar}\sqrt{r^2 - a^2},$$

$$\text{i.e.,} \quad \frac{dr}{d\theta}\frac{2\sqrt{\mu}\cos\frac{\theta}{2}}{r^2} = \frac{\sqrt{2\mu}}{ar}\sqrt{r^2 - a^2},$$

$$\text{i.e.,} \quad \frac{dr}{d\theta}\frac{\sqrt{2}\cos\frac{\theta}{2}}{r} = \frac{\sqrt{r^2 - a^2}}{a},$$

$$\text{i.e.,}\quad \frac{\sqrt{2}a\,dr}{r\sqrt{r^2-a^2}} = \sec\frac{\theta}{2}\,d\theta,$$

$$\text{i.e.,}\quad \sqrt{2}a\frac{1}{a}\sec^{-1}\left(\frac{r}{a}\right) = 2\log\left(\sec\frac{\theta}{2}+\tan\frac{\theta}{2}\right)+C_3,$$

$$\text{i.e.,}\quad \sec^{-1}\frac{r}{a} = \sqrt{2}\log\left(\sec\frac{\theta}{2}+\tan\frac{\theta}{2}\right)+C_3.$$

Initially $r = a$ when $t = 0$,
$\therefore\ C_3 = 0$, and so

$$\sec^{-1}\frac{r}{a} = \sqrt{2}\log\left[\frac{1+\sin\frac{\theta}{2}}{\cos\frac{\theta}{2}}\right],$$

$$\text{i.e.,}\quad \sec^{-1}\frac{r}{a} = \sqrt{2}\log\left(\frac{\cos\frac{\theta}{4}+\sin\frac{\theta}{4}}{\cos\frac{\theta}{4}-\sin\frac{\theta}{4}}\right) = \sqrt{2}\log\left(\frac{1+\tan\frac{\theta}{4}}{1-\tan\frac{\theta}{4}}\right) = \sqrt{2}\log\left\{\tan\left(\frac{\pi+\theta}{4}\right)\right\},$$

$\therefore\ r = a\sec\left\{\sqrt{2}\log\tan\frac{\pi+\theta}{4}\right\}$,
which is the required equation for the path.

Example 2.18 Show that the only law for a central attraction, for which the velocity in a circle at any distance is equal to the velocity acquired in falling from infinity to the distance, is that of the inverse cube.

Solution Let P be the central acceleration. The velocity in a circle of radius r under P is given by

$$V^2 = \text{Pr}\,. \tag{2.3.76}$$

The velocity of fall from infinity to a point at a distance r from the center of force is given by

$$\frac{V_0^2}{2} = -\int_{\infty}^{r} P dr,$$

$$\text{i.e.,}\quad V_0^2 = 2\int_{r}^{\infty} P dr = 2\left[\int_{r}^{a} P dr + \int_{a}^{\infty} P dr\right] = -2\int_{a}^{r} P dr + 2A. \tag{2.3.77}$$

Given $V = V_0$

$$\text{i.e.,}\quad V^2 = V_0^2.$$

Equations (2.3.76) and (2.3.77) give

$$Pr = -2\int_a^r P\,dr + 2A,$$

$$\text{i.e.,}\quad \frac{d}{dr}(Pr) = -2\frac{d}{dr}\int_a^r Pdr,$$

$$\text{i.e.,}\quad P + r\frac{dP}{dr} = -2P,$$

$$\text{i.e.,}\quad r\frac{dP}{dr} = -3P,$$

$$\text{i.e.,}\quad \frac{dP}{P} = -3\frac{dr}{r},$$

$$\text{i.e.,}\quad \log P = -3\log r + \log\mu = \log\frac{\mu}{r^3},$$

$$\text{i.e.,}\quad P = \frac{\mu}{r^3},$$

$$\therefore\quad P \propto \frac{1}{r^3},$$

which is the required law of force.

Example 2.19 A particle moves in an orbit under a central acceleration $\frac{\mu}{r^2}$ along the radius vector r; obtain the equation of energy and angular momentum. If the particle is projected with velocity u at right angles to the radius, at distance c from the origin, prove that $\left(\frac{dr}{dt}\right)^2 = \left\{\frac{2\mu}{c} - u^2\left(1 + \frac{c}{r}\right)\right\}\left(\frac{c}{r} - 1\right)$.

Solution The force is given by

$$\mathbf{F} = m\frac{\mu}{r^2}(-\hat{r}) = -\frac{m\mu}{r^2}\frac{\mathbf{r}}{r} = -\frac{m\mu}{r^3}\mathbf{r}.$$

Let V denote the P.E. of the particle.

We have

$$\mathbf{F} = -\nabla V,$$

$$\text{i.e.,}\quad \mathbf{F}\cdot d\mathbf{r} = -\nabla V\cdot d\mathbf{r} = -dV,$$

$$\text{i.e.,}\quad -dV = \mathbf{F}\cdot d\mathbf{r} = -\frac{m\mu}{r^3}\mathbf{r}\cdot d\mathbf{r},$$

$$\text{i.e.,}\quad dV = \frac{m\mu}{r^3}rdr = \frac{m\mu}{r^2}dr,$$

$$\text{i.e.,}\quad V = -m\frac{\mu}{r}.$$

The equation of energy is

$$T + V = E = \text{a constant},$$

$$\text{i.e.,}\quad \frac{1}{2}mv^2 + V = E,$$

$$\text{i.e.,}\quad \frac{1}{2}m\left[\dot{r}^2 + r^2\dot{\theta}^2\right] + V = E,$$

$$\text{i.e.,}\quad m\left(\dot{r}^2 + r^2\dot{\theta}^2\right) - \frac{2m\mu}{r} = 2E. \tag{2.3.78}$$

Let $\mathbf{\Omega}$ stipulate the angular momentum of the particle about O (center of force)

$$\because\quad \frac{d\mathbf{\Omega}}{dt} = \mathbf{r}\times\mathbf{F} = \mathbf{r}\times\left(-\frac{\mu\mathbf{r}}{r^3}\right) = \mathbf{0},$$

$$\therefore\quad \mathbf{\Omega} = \text{a constant vector} = m\mathbf{h}\quad(\text{say}),$$

$$\text{i.e.,}\quad \mathbf{r}\times\mathbf{v} = \mathbf{h},$$

$$\text{i.e.,}\quad r\hat{r}\times\left(\dot{r}\,\hat{r} + r\dot{\theta}\,\hat{\theta}\right) = \mathbf{h},$$

$$\text{i.e.,}\quad r^2\dot{\theta}\,\hat{k} = h\hat{k},$$

$$\text{i.e.,}\quad r^2\dot{\theta} = h. \tag{2.3.79}$$

This equation is termed the equation of the angular momentum equation.

Equation (2.3.79) gives

$$\dot{\theta} = \frac{h}{r^2}. \tag{2.3.80}$$

Equation (2.3.78) reduces to

$$m\left[\dot{r}^2 + \frac{h^2}{r^2}\right] - \frac{2m\mu}{r} = 2E. \tag{2.3.81}$$

Initially $\dot{r} = 0, \quad r\dot{\theta} = u$, when $r = c$,

i.e., $r = c,\ \dot{r} = 0,\ c\dot{\theta} = u$, at $t = 0$,

$$\text{i.e.,}\quad r = c,\ \dot{r} = 0,\ \dot{\theta} = \frac{u}{c}\ \text{ at } t = 0. \tag{2.3.82}$$

By the use of (2.3.82), Eq. (2.3.80) gives $\frac{u}{c} = \frac{h}{c^2}$, i.e., $h = cu$.
Therefore, Eq. (2.3.81) reduces to

$$m\left[\dot{r}^2 + \frac{c^2u^2}{r^2}\right] - \frac{2m\mu}{r} = 2E. \tag{2.3.83}$$

Initially $\dot{r} = 0$ when $r = c$.
Equation (2.3.83) gives

$$mu^2 - \frac{2m\mu}{c} = 2E.$$

Therefore, Eq. (2.3.83) reduces to

$$m\left(\dot{r}^2 + \frac{c^2u^2}{r^2}\right) - \frac{2m\mu}{r} = mu^2 - \frac{2m\mu}{c},$$

$$\text{i.e.,}\quad \dot{r}^2 + \frac{c^2u^2}{r^2} - \frac{2\mu}{r} = u^2 - \frac{2\mu}{c},$$

$$\text{i.e.,}\quad \dot{r}^2 = u^2 - \frac{2\mu}{c} - \frac{c^2u^2}{r^2} + \frac{2\mu}{r},$$

$$\text{i.e.,}\quad \dot{r}^2 = u^2\left(1 - \frac{c^2}{r^2}\right) + 2\mu\left(\frac{1}{r} - \frac{1}{c}\right),$$

$$\text{i.e.,}\quad \dot{r}^2 = -u^2\left(\frac{c^2}{r^2} - 1\right) + \frac{2\mu}{c}\left(\frac{c}{r} - 1\right),$$

$$\text{i.e.,}\quad \dot{r}^2 = \left(\frac{c}{r} - 1\right)\left[\frac{2\mu}{c} - u^2\left(\frac{c}{r} + 1\right)\right],$$

$$\text{i.e.,}\quad \dot{r}^2 = \left[\frac{2\mu}{c} - u^2\left(\frac{c}{r} + 1\right)\right]\left(\frac{c}{r} - 1\right),$$

$$\text{i.e.,}\quad \left(\frac{dr}{dt}\right)^2 = \left\{\frac{2\mu}{c} - u^2\left(1 + \frac{c}{r}\right)\right\}\left(\frac{c}{r} - 1\right).$$

Example 2.20 A particle, moving with a central acceleration $\frac{\mu}{(\text{distance})^2}$, is projected from an apse at a distance a with a velocity V; show that the path is $r\cosh\left[\frac{\sqrt{\mu - a^2V^2}}{aV}\theta\right] = a$, or $r\cos\left[\frac{\sqrt{a^2V^2 - \mu}}{aV}\theta\right] = a$, according as V is $\lesseqgtr$ velocity form infinity.

Solution Given

$$P = \frac{\mu}{r^3} = \mu u^3.$$

Let V_0 be the velocity of fall from infinity to the point at a distance from the center of force under the same force.

$$\frac{V_0^2}{2} = -\int_{\infty}^{a} P\,dr = -\int_{\infty}^{a} \frac{\mu}{r^3}dr = \frac{\mu}{2a^2},$$

$$\text{i.e.,}\quad V_0^2 = \frac{\mu}{a^2}. \tag{2.3.84}$$

The differential equation of the central orbit is

$$\frac{d^2u}{d\theta^2} + u = \frac{\mu u}{h^2}. \tag{2.3.85}$$

The velocity of the particle at a point $Q\ (r,\ \theta)$ is given by

$$v^2 = h^2\left[\left(\frac{du}{d\theta}\right)^2 + u^2\right]. \tag{2.3.86}$$

Initially $v = V,\ \frac{du}{d\theta} = 0$ when $u = \frac{1}{a}$,

Therefore, Eq. (2.3.86) gives

$$h^2 = a^2 V^2.$$

Thus, Eq. (2.3.85) reduces to

$$\frac{d^2u}{d\theta^2} + u = \frac{\mu u}{a^2 V^2},$$

i.e., $\left(D^2 + 1 - \frac{\mu}{a^2 V^2}\right) u = 0,$

i.e., $$\left(D^2 + \frac{a^2 V^2 - \mu}{a^2 V^2}\right) u = 0. \tag{2.3.87}$$

Case I: When $V < V_0$,

i.e., $V^2 < \frac{\mu}{a^2},$

i.e., $a^2 V^2 - \mu < 0,$

i.e., $\frac{\mu - a^2 V^2}{a^2 V^2} > 0.$

Let $\lambda^2 = \frac{\mu - a^2 V^2}{a^2 V^2}$.
Therefore, Eq. (2.3.87) reduces to

$$\left(D^2 - \lambda^2\right) u = 0. \tag{2.3.88}$$

The general solution of (2.3.88) is

$$u = c_1 e^{\lambda\theta} + c_2 e^{-\lambda\theta}, \tag{2.3.89}$$

with

$$\frac{du}{d\theta} = c_1 \lambda e^{\lambda\theta} - c_2 e^{-\lambda\theta} \lambda. \tag{2.3.90}$$

Initially $u = \frac{1}{a}$, $\frac{du}{d\theta} = 0$; when $\theta = 0$,

Equations (2.3.89) and (2.3.90) give

$$c_1 + c_2 = \frac{1}{a}, \tag{2.3.91}$$

$$c_1 - c_2 = 0. \tag{2.3.92}$$

Solving (2.3.91) and (2.3.92), it is found that $c_1 = \frac{1}{2a} = c_2$.
Therefore, Eq. (2.3.89) takes the form

$$u = \frac{1}{2a}\left(e^{\lambda\theta} + e^{-\lambda\theta}\right),$$

$$\text{i.e.,}\quad \frac{1}{r} = \frac{1}{2a} 2\cosh\lambda\theta,$$

$$\text{i.e.,}\quad a = r\cosh\left[\frac{\sqrt{\mu - a^2V^2}}{aV}\right]\theta.$$

Case II: When

$$V > V_0,$$

$$\text{i.e.,}\quad V^2 > \frac{\mu}{a^2},$$

$$\text{i.e.,}\quad \frac{a^2V^2 - \mu}{a^2V^2} > 0.$$

Let $p^2 = \frac{a^2V^2 - \mu}{a^2V^2}$.
Equation (2.3.87) reduces to

$$\left(D^2 + p^2\right) u = 0. \tag{2.3.93}$$

General solution of (2.3.93) is

$$u = A\cos p\theta + B\sin p\theta, \tag{2.3.94}$$

with

$$\frac{du}{d\theta} = -Ap\sin p\theta + Bp\cos p\theta. \tag{2.3.95}$$

Initially, $u = \frac{1}{a},\ \frac{du}{d\theta} = 0$, when $\theta = 0$.

Therefore, Eqs. (2.3.94) and (2.3.95) give $A = \frac{1}{a}$ and $B = 0$.
Equation (2.3.94) transforms to

$$u = \frac{1}{a}\cos p\theta,$$

$$\text{i.e.,}\quad \frac{1}{r} = \frac{1}{a}\cos p\theta,$$

$$\text{i.e.,}\quad a = r\cos\left[\frac{\sqrt{a^2V^2-\mu}}{aV}\right]\theta.$$

Example 2.21 A particle moves in a curve under a central attraction so that its velocity at any point is equal to that in a circle at the same distance and under the same attraction; show that the law of force is that of the inverse cube, and that of the path is an equiangular spiral.

Solution Let v be the velocity at a distance r from the center of force under the central acceleration P.

$$\therefore\quad v^2 = Pr. \tag{2.3.96}$$

Again

$$v^2 = h^2\left[\left(\frac{du}{d\theta}\right)^2 + u^2\right],$$

$$\text{i.e.,}\quad \frac{P}{u} = h^2\left[\left(\frac{du}{d\theta}\right)^2 + u^2\right]. \tag{2.3.97}$$

Differentiation of Eq. (2.3.97) gives

$$\frac{u\frac{dP}{d\theta} - P\frac{du}{d\theta}}{u^2} = h^2\left[2\left(\frac{du}{d\theta}\right)\frac{d^2u}{d\theta^2} + 2u\frac{du}{d\theta}\right],$$

$$\text{i.e.,}\quad \frac{u\frac{dP}{d\theta} - P\frac{du}{d\theta}}{u^2} = 2h^2\frac{du}{d\theta}\left[\frac{d^2u}{d\theta^2} + u\right] = \frac{2P}{u^2}\frac{du}{d\theta},$$

$$\text{i.e.,}\quad u\frac{dP}{d\theta} - P\frac{du}{d\theta} = 2P\frac{du}{d\theta},$$

$$\text{i.e.,}\quad u\frac{dP}{d\theta} = 3P\frac{du}{d\theta},$$

$$\text{i.e.,} \quad u dP = 3P du,$$

$$\text{i.e.,} \quad \frac{dP}{P} = 3\frac{du}{u},$$

$$\text{i.e.,} \quad \log P = 3 \log u + \log \mu = \log \left(\mu u^3\right),$$

$$\therefore \quad P = \frac{\mu}{r^3}.$$

Thus,

$$P \propto \frac{1}{r^3},$$

which is the law of force.

We find from Eq. (2.3.97),

$$h^2 \left[\left(\frac{du}{d\theta} \right)^2 + u^2 \right] = \frac{\mu u^3}{u} = \mu u^2,$$

$$\text{i.e.,} \quad \left(\frac{du}{d\theta} \right)^2 + u^2 = \frac{\mu}{h^2} u^2,$$

$$\text{i.e.,} \quad \left(\frac{du}{d\theta} \right)^2 = \left(\frac{\mu}{h^2} - 1 \right) u^2 = \lambda^2 u^2; \lambda^2 = \frac{\mu}{h^2} - 1,$$

$$\text{i.e.,} \quad \frac{du}{d\theta} = \lambda u,$$

$$\text{i.e.,} \quad \frac{du}{u} = \lambda d\theta,$$

$$\text{i.e.,} \quad \log u = \lambda\theta + \log k,$$

$$\text{i.e.,} \quad \log \frac{u}{k} = \lambda\theta,$$

$$\text{i.e.,} \quad u = k e^{\lambda\theta},$$

$$\text{i.e.,} \quad r = \nu e^{-\lambda\theta},$$

which represents an equiangular spiral.

Example 2.22 A particle of unit mass is describing an elliptic orbit of axes $2a$, $2b$ under an attraction μr to the center C when it is at a distance r from C; prove that its speed v is given by the equation $v^2 = \mu\left(a^2 + b^2 - r^2\right)$, and that the moment of its velocity about C is $\sqrt{\mu}ab$.

Solution The equation of the ellipse is

$$\frac{x^2}{a^2} + \frac{y^2}{b^2} = 1. \tag{2.3.98}$$

Let P (x, y) be the position of the particle at time t. Let (r, θ) be the polar coordinates of P referred to C as a pole and CX (major axis) as the initial line. Join CP.

Now

$$CP = r,$$

$$\text{i.e.,}\quad CP^2 = r^2,$$

$$\text{i.e.,}\quad x^2 + y^2 = r^2. \tag{2.3.99}$$

Let $\mathbf{r}$ be the position vector of the point P relative to C.

$$\therefore\quad \mathbf{r} = \text{P.V. of P relative to O} = x\hat{i} + y\hat{j}. \tag{2.3.100}$$

The equation of the motion of the particle is

$$m\ddot{\mathbf{r}} = \mu r\left(-\hat{r}\right) = -\mu r\frac{\mathbf{r}}{r} = -\mu\mathbf{r},$$

$$\text{i.e.,}\quad \ddot{\mathbf{r}} = -\mu\mathbf{r} \qquad (\because m = 1)\,,$$

$$\text{i.e.,}\quad \ddot{x}\hat{i} + \ddot{y}\hat{j} = -\mu\left(x\hat{i} + y\hat{j}\right). \tag{2.3.101}$$

Equation (2.3.101) gives

$$\ddot{x} = -\mu x, \tag{2.3.102}$$

$$\ddot{y} = -\mu y. \tag{2.3.103}$$

Equation (2.3.102) gives

$$\left(D^2 + \mu\right) x = 0. \tag{2.3.104}$$

General solution of (2.3.104) is

$$x = A \cos \sqrt{\mu} t + B \sin \sqrt{\mu} t, \tag{2.3.105}$$

with

$$\dot{x} = -A\sqrt{\mu} \sin \sqrt{\mu} t + B\sqrt{\mu} \cos \sqrt{\mu} t. \tag{2.3.106}$$

Suppose that the particle is located at A $(a, 0)$ at time $t = 0$,

$$\therefore \quad x = a, \;\; \dot{x} = 0 \;\; at \quad t = 0.$$

Hence Eqs. (2.3.105) and (2.3.106) gives $A = a, \quad B = 0$.
Thus, Eq. (2.3.105) reduces to

$$x = a \cos \sqrt{\mu} t. \tag{2.3.107}$$

On substitution of (2.3.107) in (2.3.98), we get

$$y = b \sin \sqrt{\mu} t. \tag{2.3.108}$$

Let $\mathbf{v}$ be the velocity of the particle at P.

$$\therefore \quad \mathbf{v} = \dot{\mathbf{r}} = \dot{x}\hat{i} + \dot{y}\hat{j} = -a\sqrt{\mu}\hat{i} \sin \sqrt{\mu} t + b\sqrt{\mu}\hat{j} \cos \sqrt{\mu} t,$$

$$\text{i.e.,} \quad v^2 = a^2 \mu \sin^2 \sqrt{\mu} t + b^2 \mu \cos^2 \sqrt{\mu} t,$$

$$\text{i.e.,} \quad v^2 = \mu \left[a^2 \sin^2 \sqrt{\mu t} + b^2 \cos^2 \sqrt{\mu t}\right] = \left[a^2 \left(1 - \cos^2 \sqrt{\mu t}\right) + b^2 \left(1 - \sin^2 \sqrt{\mu t}\right)\right],$$

$$\text{i.e.,} \quad v^2 = \mu \left[a^2 + b^2 - \left(a^2 \cos^2 \sqrt{\mu t} + b^2 \sin^2 \sqrt{\mu t}\right)\right],$$

$$\text{i.e.,} \quad v^2 = \mu \left[a^2 + b^2 - \left(x^2 + y^2\right)\right],$$

$$\text{i.e.,} \quad v^2 = \mu \left(a^2 + b^2 - r^2\right).$$

The moment of the velocity of the particle about C is given by

$$\mathbf{h} = \mathbf{r} \times \mathbf{v} = \begin{vmatrix} \hat{i} & \hat{j} & \hat{k} \\ x & y & 0 \\ \dot{x} & \dot{y} & 0 \end{vmatrix} = \hat{k}\,(x\dot{y} - y\dot{x}),$$

$$\therefore \quad h = |x\dot{y} - y\dot{x}| = \left|a\cos\sqrt{\mu}t\, b\sqrt{\mu}\cos\sqrt{\mu}t + b\sin\sqrt{\mu}t\, a\sqrt{\mu}\sin\sqrt{\mu}t\right| = \sqrt{\mu}ab.$$

Example 2.23 A particle of mass m is moving in a circle of radius a and center O, under a force $\mu m\left(r + \frac{2a^3}{r^2}\right)$ directed towards O. If it is acted on by an impulse tangential to the path and of magnitude $(3\mu m^2a^2)^{\frac{1}{2}}$, show that the velocity of the particle is immediately doubled and that the greatest and least distances from O in the ensuring motion are a and $3a$.

Solution Given

$$P' = \mu\left(r + \frac{2a^3}{r^2}\right). \tag{2.3.109}$$

Let V_0 be the velocity of the particle in the circle just before the impulse.

$$\therefore \quad V_0^2 = P'r\big]_{r=a} = \mu\left(r + \frac{2a^3}{r^2}\right)r\bigg]_{r=a} = 3\mu a^2,$$

$$\therefore \quad V_0 = \sqrt{3\mu}a. \tag{2.3.110}$$

Let V be the velocity just after the impulse.

$$\therefore \quad mV - mV_0 = \sqrt{3\mu m^2a^2},$$

$$\text{i.e.,} \quad V - V_0 = a\sqrt{3\mu},$$

$$\text{i.e.,} \quad V - a\sqrt{3\mu} = a\sqrt{3\mu}$$

$$\text{i.e.,} \quad V = 2a\sqrt{3\mu} = 2V_0. \tag{2.3.111}$$

$\therefore$ Velocity of the particle is doubled.

For a central orbit,

$$vp = h,$$

$$\text{i.e.,} \quad Va = h,$$

i.e., $h = 2a^2\sqrt{3\mu}$,

$$\therefore \quad h^2 = 12a^4\mu. \tag{2.3.112}$$

The differential equation of the central orbit is

$$\frac{d^2u}{d\theta^2} + u = \frac{P'}{h^2u^2} = \frac{\frac{1}{u} + 2a^3u^2}{12a^4u^2} = \frac{1}{12a^4}\left(u^{-3} + 2a^3\right),$$

$$\text{i.e.,} \quad \frac{du_1}{d\theta} + u = \frac{1}{12a^4}\left(u^{-3} + 2a^3\right),$$

$$\text{i.e.,} \quad 2u_1\frac{du_1}{d\theta} + 2u\frac{du}{d\theta} = \frac{1}{6a^4}\left(u^{-3}\frac{du}{d\theta} + 2a^3\frac{du}{d\theta}\right),$$

$$\text{i.e.,} \quad 2u_1du_1 + 2udu = \frac{1}{6a^4}\left(u^{-3}du + 2a^3du\right),$$

$$\text{i.e.,} \quad u_1^2 + u^2 = \frac{1}{6a^4}\left(\frac{u^{-2}}{-2} + 2a^3u\right) + C,$$

$$\text{i.e.,} \quad \left(\frac{du}{d\theta}\right)^2 + u^2 = \frac{1}{6a^4}\left(-\frac{1}{2u^2} + 2a^3u\right) + C. \tag{2.3.113}$$

Initially $\frac{du}{d\theta} = 0$ when $u = \frac{1}{a}$.
Therefore, Eq. (2.3.113) gives

$$\frac{1}{a^2} = \frac{1}{6a^4}\left(-\frac{a^2}{2} + 2a^2\right) + C = \frac{1}{6a^4}\frac{3a^2}{2} + C = \frac{1}{4a^2} + C,$$

$$\text{i.e.,} \quad C = \frac{1}{a^2} - \frac{1}{4a^2} = \frac{3}{4a^2}.$$

Equation (2.3.113) reduces to

$$\left(\frac{du}{d\theta}\right)^2 = \frac{1}{6a^4}\left(-\frac{1}{2u^2} + 2a^3u\right) + \frac{3}{4a^2} - u^2. \tag{2.3.114}$$

The extreme values of u or r are given by

$$\frac{du}{d\theta} = 0,$$

$$\text{i.e.,} \quad \frac{1}{6a^4}\left(-\frac{1}{2u^2} + 2a^3u\right) + \frac{3}{4a^2} - u^2 = 0,$$

$$\text{i.e.,} \quad \frac{1}{6a^4}\frac{-1 + 4a^3u^3}{2u^2} + \frac{3}{4a^2} - u^2 = 0,$$

$$\text{i.e.,} \quad \frac{-1 + 4a^3u^3}{12a^4u^2} + \frac{3}{4a^2} - u^2 = 0,$$

$$\text{i.e.,} \quad -1 + 4a^3u^3 + 9a^2u^2 - 12a^4u^4 = 0,$$

$$\text{i.e.,} \quad 12a^4u^4 - 4a^3u^3 - 9a^2u^2 + 1 = 0,$$

$$\text{i.e.,} \quad \frac{12a^4}{r^4} - \frac{4a^3}{r^3} - \frac{9a^2}{r^2} + 1 = 0,$$

$$\text{i.e.,} \quad 12a^4 - 4a^3r - 9a^2r^2 + r^4 = 0,$$

$$\text{i.e.,} \quad r^4 - 9a^2r^2 - 4a^3r + 12a^4 = 0,$$

$$\text{i.e.,} \quad r^4\left(r^2 - 9a^2\right) - 4a^3\left(r - 3a\right) = 0,$$

$$\text{i.e.,} \quad r^2\left(r + 3a\right)\left(r - 3a\right) - 4a^3\left(r - 3a\right) = 0,$$

$$\text{i.e.,} \quad (r - 3a)\left[r^2\left(r + 3a\right) - 4a^3\right] = 0,$$

$$\text{i.e.,} \quad (r - 3a)\left(r^3 + 3ar^2 - 4a^3\right) = 0,$$

$$\text{i.e.,} \quad (r - 3a)\left\{r^3 + ar^2 + 4ar^2 - 4a^3\right\} = 0,$$

$$\text{i.e.,} \quad (r - 3a)\left\{r^2\left(r - a\right) + 4a\left(r^2 - a^2\right)\right\} = 0,$$

$$\text{i.e.,} \quad (r - 3a)\,r^2\left(r - a\right)\left\{r^2 + 4a\left(r + a\right)\right\} = 0,$$

$$\text{i.e.,} \quad (r - 3a)\left(r - a\right)\left(r^2 + 4ar + 4a^2\right) = 0,$$

$$\text{i.e.,} \quad (r - 3a)\left(r - a\right)\left(r + 2a\right)^2 = 0,$$

$$\therefore \quad r = a \quad or \quad 3a \quad \left[\because \; r \neq -2a\right].$$

Thus, the greatest distance of the particle from O is $3a$ and the least distance of the particle from O is a.

Example 2.24 A particle P, of mass m, is attracted towards the origin of rectangular coordinates $x,\ y$ by a force mn^2OP. Initially, the particle is projected from the point. $(a,\ 0)$ with velocity u parallel to the axis of y. Prove that the particle describes the ellipse $\left(\frac{x}{a}\right)^2 + \left(\frac{ny}{u}\right)^2 = 1$ and the eccentric angle of P on the ellipse increases at a constant rate.

Solution Let P$(x,\ y)$ be the position of the particle at time t. Let $\mathbf{r}$ stipulate the position vector of P relative to O .

The equation of motion of the particle is

$$m\ddot{\mathbf{r}} = mn^2\text{OP}\left(-\hat{r}\right),$$

$$\text{i.e.,}\quad \ddot{\mathbf{r}} = -n^2 r\frac{\mathbf{r}}{r} = -n^2\mathbf{r},$$

$$\text{i.e.,}\quad \ddot{x}\hat{i} + \ddot{y}\hat{j} = -n^2\left(x\hat{i} + y\hat{j}\right). \tag{2.3.115}$$

By equating the coefficients of $\hat{i},\ \hat{j}$ in (2.3.115), we obtain

$$\ddot{x} = -n^2 x, \tag{2.3.116}$$

$$\ddot{y} = -n^2 y. \tag{2.3.117}$$

Equation (2.3.116) can be written as

$$\left(D^2 + n^2\right)x = 0. \tag{2.3.118}$$

General solution of (2.3.118) is

$$x = A\cos nt + B\sin nt, \tag{2.3.119}$$

with

$$\dot{x} = -An\sin nt + Bn\cos nt. \tag{2.3.120}$$

Similarly (2.3.117) gives

$$y = C\cos nt + D\sin nt, \tag{2.3.121}$$

with

$$\dot{y} = -Cn \sin nt + Dn \cos nt. \tag{2.3.122}$$

Initially $x = a, \quad y = 0, \quad \dot{x} = 0, \quad \dot{y} = u \quad \text{at} \quad t = 0.$

Subjecting Eqs. (2.3.119)–(2.3.122) to the above conditions, we derive

$$\begin{aligned} A &= a, \quad B = 0; \\ C &= 0, \quad D = \tfrac{u}{n}. \end{aligned}$$

Thus, we obtain

$$x = a \cos nt, \tag{2.3.123}$$

$$y = \frac{u}{n} \sin nt. \tag{2.3.124}$$

On elimination of t from (2.3.123) and (2.3.124), we get

$$\left(\frac{x}{a}\right)^2 + \left(\frac{ny}{u}\right)^2 = 1,$$

which represents an ellipse.

Let φ be the eccentric angle of P.

$$\therefore \quad x = a \ \cos \varphi,$$

and so

$$\varphi = nt,$$

$$\therefore \quad \frac{d\varphi}{dt} = n = \text{a constant},$$

i.e., the eccentric angle of P increases at a constant rate.

Example 2.25 A particle P moves under a force ω^2**PO** per unit mass towards the fixed point O. If its initial position is A and its initial velocity is represented by ω**OC**, show that its position at time t is given by

$$\mathbf{OP} = \mathbf{OA} \cos \omega t + \mathbf{OC} \sin \omega t,$$

and that the orbit is an ellipse with center O. If the particle passes through a given point B, where **OB** is $\perp^r$ to **OA**, show that the point C must lie on a hyperbola with semi-axes OB, OA.

Solution Let $\mathbf{OP} = \mathbf{r}$, $\mathbf{OA} = \mathbf{a}$ and $\mathbf{OC} = \mathbf{c}$.

The equation of motion of the particle is

$$\ddot{\mathbf{r}} = \omega^2 \mathbf{PO} = -\omega^2 \mathbf{OP} = -\omega^2 \mathbf{r},$$

$$\text{i.e.,} \quad \left(D^2 + \omega^2\right) \mathbf{r} = \mathbf{0}. \tag{2.3.125}$$

General solution of (2.3.125) is

$$\mathbf{r} = \boldsymbol{\Lambda} \cos \omega t + \boldsymbol{\Upsilon} \sin \omega t, \tag{2.3.126}$$

with

$$\dot{\mathbf{r}} = -\boldsymbol{\Lambda} \omega \sin \omega t + \omega \boldsymbol{\Upsilon} \cos \omega t. \tag{2.3.127}$$

Initially

$$\mathbf{r} = \mathbf{OA}, \dot{\mathbf{r}} = \omega \mathbf{OC} \quad \text{at} \quad t = 0,$$

$$\text{i.e.,} \quad \mathbf{r} = \mathbf{a}, \ \dot{\mathbf{r}} = \omega \mathbf{c} \quad \text{at} \quad t = 0.$$

Equations (2.3.126) and (2.3.127) give

$$\mathbf{a} = \boldsymbol{\Lambda} \ \& \ \boldsymbol{\Upsilon} = \mathbf{c}.$$

Equation (2.3.126) gives

$$\mathbf{r} = \mathbf{a} \cos \omega t + \mathbf{c} \sin \omega t,$$

$$\text{i.e.,} \quad \mathbf{OP} = \mathbf{OA} \cos \omega t + \mathbf{OC} \sin \omega t. \tag{2.3.128}$$

2nd Part : Take O as the origin, the X-axis along **OA** and the Y-axis along **OB**.

Let

$$\mathbf{OP} = x\hat{i} + y\hat{j} = (x, \ y),$$

$$\mathbf{OA} = a\hat{i} = (a, \ 0),$$

$$\mathbf{OC} = \alpha\hat{i} + \beta\hat{j} = (\alpha, \ \beta),$$

$$\mathbf{OB} = b\hat{j} = (0, \ b).$$

Therefore, Eq. (2.3.128) gives

$$x\hat{i} + y\hat{j} = a\hat{i}\cos\omega t + \left(\alpha\hat{i} + \beta\hat{j}\right)\sin\omega t. \tag{2.3.129}$$

Equation (2.3.129) yields

$$x = a\cos\omega t + \alpha\sin\omega t, \tag{2.3.130}$$

$$y = \beta\sin\omega t. \tag{2.3.131}$$

On elimination of t from (2.3.130) and (2.3.131), we obtain

$$\text{i.e.,}\quad \frac{y^2}{\beta^2} + \left(\frac{x}{a} - \frac{\alpha y}{a\beta}\right)^2 = 1,$$

$$\text{i.e.,}\quad \beta^2 x^2 - -2\alpha\beta\, xy + \left(a^2 + \alpha^2\right) y^2 - a^2\beta^2 = 0. \tag{2.3.132}$$

In (2.3.132),

$$A = \beta^2,\ \ H = -\alpha\beta,\quad B = \alpha^2 + a^2.$$

Now,

$$AB - H^2 = \beta^2\left(\alpha^2 + a^2\right) - \alpha^2\beta^2 = a^2\beta^2 > 0.$$

$$\therefore\quad AB > H^2.$$

Thus, Eq. (2.3.132) represents an ellipse.

i.e., the orbit of the particle is an ellipse.

Let the ellipse (2.3.132) passes through the point B $(0,\ b)$.

$$\therefore\quad \left(\alpha^2 + a^2\right) b^2 = a^2\beta^2,$$

$$\text{i.e.,}\quad b^2\alpha^2 + a^2 b^2 = a^2\beta^2,$$

$$\text{i.e.,}\quad a^2\beta^2 - b^2\alpha^2 = a^2 b^2,$$

$$\text{i.e.,}\quad \frac{\beta^2}{b^2} - \frac{\alpha^2}{a^2} = 1.$$

Therefore, the point C (α, β) lies on the hyperbola:

$$\frac{y^2}{b^2} - \frac{x^2}{a^2} = 1.$$

Thus, C (α, β) lies on a hyperbola with semi-axes $b =$ OB and $a =$ OA.
Hence proved.

2.4 Exercise-II

1. A particle describes the cardioid $r = a(1 + \cos\theta)$ under a force directed to the pole. Show that the law of force is $P \propto \frac{1}{r^4}$.
2. Find the law of force, when the path described by a particle under a force directed towards the pole, is an equiangular spiral given by $r = \lambda e^{\nu\theta}$.
   (**Answer**: $P \propto \frac{1}{r^3}$)
3. A particle moves under a central acceleration P, and an acceleration R in a direction orthogonal to the direction of P. Under usual notations, show that the path of the particle is specified by $\frac{d^2u}{d\theta^2} + u = \frac{P - \frac{R}{u}\frac{du}{d\theta}}{h^2u^2}$.
4. A particle, of mass m, moves in a resisting medium under a central acceleration P; show that the equation to the orbit is $\frac{d^2u}{d\theta^2} + u = \frac{P}{h^2u^2}$, where $h = h_0 e^{-\int \frac{R}{v}dt}$, and R is the resistance of medium per unit of mass.
5. If the law of force is $5\mu u^3 + 8\mu c^2 u^5$, and the particle is projected from an apse at the distance c with velocity $3\frac{\sqrt{\mu}}{c}$; prove that the orbit is $r = c\cos\frac{2\theta}{3}$.
6. Prove that, if the law of force $3\mu u^3 + 2\mu a^2 u^5$ and a particle is projected in a direction making an angle $\cot^{-1} 2$ with the initial distance a, and with a velocity equal to that in a circle at the same distance, the orbit is $au = \tan\left(\frac{\pi}{4} + \theta\right)$.
7. The law of force is $\frac{\mu}{r^5}$ and a particle is projected from an apse at a distance a from the center of force, with the velocity of projection $\frac{\sqrt{\mu}}{a^2\sqrt{2}}$. Find the orbit.
8. A particle moving under a constant force from a center is projected in a direction $\perp^r$ to the radians vector with the velocity acquired in falling to the point of projection from the center. Show that its path is $\left(\frac{a}{r}\right)^3 = \cos^2\left(\frac{3\theta}{2}\right)$, and the particle will ultimately move in a straight line through the origin in the same way as if its path had always been this line. If the velocity of projection is double that in the previous case, show that its path is $\frac{\theta}{2} = \tan^{-1}\sqrt{\frac{r-a}{a}} - \frac{1}{\sqrt{3}}\tan^{-1}\sqrt{\frac{r-a}{3a}}$.
9. In a central orbit the force is $\mu u^3\left(3 + 2a^2u^2\right)$; if the particle projected at a distance a with a velocity$\sqrt{\frac{5\mu}{a^2}}$ in a direction making $\tan^{-1}\frac{1}{2}$ with the radius, show that the equation to the path is $r = a\tan\theta$.
10. A particle of mass m under a central force $m\mu\left\{3au^4 - 2\left(a^2 - b^2\right)u^5\right\}, a > b$, and is projected from an apse at distance $a + b$, with velocity $\frac{\sqrt{\mu}}{a+b}$; show that its orbit is $r = a + b\cos\theta$.

11. A particle moving under a central force $\mu\left\{2\left(a^2+b^2\right)u^8-3a^2b^2u^7\right\}$, is projected at the distance a with velocity $\frac{\sqrt{\mu}}{a}$ in a direction at right angles to the initial distance, show that the path is the curve $r^2=a^2\cos^2\theta+b^2\sin^2\theta$.
12. A particle of mass m moves under an attractive force to the pole equal to $\frac{m\mu\sin^2\theta}{r^2}$. It is projected with velocity $\sqrt{\frac{2\mu}{3a}}$ from an apse at a distance a. Show that the equation to the orbit is $r\left(1+\cos^2\theta\right)=2a$, and that the time of a complete revolution is $\frac{\pi(3a)^{\frac{3}{2}}}{\sqrt{\mu}}$.
13. A particle describes the curve $r^2=a^2\cos 2\theta$ under the action of a force to the pole. Prove that the square of the force varies as the seventh power of the angular velocity of the radius vector of the particle.
14. A particle moves with a central acceleration $\mu\left(r+\frac{a^4}{r^3}\right)$ being projected from an apse at distance a with a velocity $2\sqrt{\mu a}$; show that it describes the curve $r^2\left[2+\cos\sqrt{3}\theta\right]=3a^2$.
15. A particle is projected at a distance a from a center of force with a velocity $\frac{2\sqrt{\mu}}{a}$, and at an angle $\frac{\pi}{4}$ with the radius vector, the force being $\mu\left(\frac{3}{r^3}+\frac{a^2}{r^5}\right)$; determine the orbit described, and show that the time to the center of force is $\frac{a^2}{\sqrt{2\mu}}\left(2-\frac{\pi}{2}\right)$.
16. A particle, subject to a force producing an acceleration $\mu\frac{r+2a}{r^5}$ towards the origin, is projected from the point $(a, 0)$ with a velocity equal to the velocity form infinity at an angle $\cot^{-1}2$ with the initial line. Show that the equation to the path is $r=a\,(1+2\sin\theta)$, and find the apsidal angle and distance.
17. A particle is projected from an apse at a distance a with the velocity form infinity, the acceleration being μu^7, show that the equation to the path is $r^2=a^2\cos 2\theta$.
Hints: $P=\mu u^7=\frac{\mu}{r^7}$. Let V be the velocity of fall from infinity to the point $(a, 0)$. $\therefore\ V^2=2\int\limits_a^\infty\frac{\mu}{r^7}dr=2\mu\left.\frac{r^{-6}}{-6}\right]_a^\infty=-\frac{\mu}{3}\left.\frac{1}{r^6}\right]_a^\infty=\frac{\mu}{3a^6}$.
18. A particle moves under a central acceleration $\mu\left(\frac{5}{r^3}+\frac{8a^2}{r^5}\right)$, and is projected from an apse at a distance a with velocity $\frac{3\sqrt{\mu}}{a}$, prove that the orbit is $r=a\,\mathrm{Cos}\left(\frac{2\theta}{3}\right)$, and that it will arrive at the origin after a time $\frac{\pi a^2}{8\sqrt{\mu}}$.
19. A particle of mass m moves with a central attractive force $m\mu\left(u^5-\frac{a^2}{8}u^7\right)$; it is projected at a distance a with a velocity $\sqrt{\frac{25}{7}}$ times the velocity for a circle at that distance and at an indirection $\cot^{-1}\frac{3}{4}$ to the radius vector. Show that its path is the curve $4r^2-a^2=\frac{3a^2}{(1-\theta)^2}$.
20. A particle is projected with velocity $\sqrt{\frac{2\mu}{3a^2}}$ at right angles to the radius vector at a distance a from a center of attracting force $\frac{\mu}{r^4}$ per unit mass, r is the distance of the particle for the center of force. Find the path of the particle, and show that the time it takes to reach the center of force is $\frac{3\pi}{8}\sqrt{\frac{3a^5}{2\mu}}$.

21. A particle under that action of a central force $\frac{\mu(r+a)}{r^3}$ is projected from an apse at a distance a with a velocity which is to that in a circle at the same distance as $1:\sqrt{2}$; show that the equation of the orbit is $r\left(2+\theta^2\right)=2a$, and that the particle will arrive at the origin in time $T=\pi\sqrt{\frac{a^3}{8\mu}}$.
22. A particle moves with central acceleration $\mu u^2+\nu u^3$ and the velocity of projection at distance R is V; show that the particle ultimately go off to infinity if $V^2>\frac{2\mu}{R}+\frac{v}{R^2}$.
23. A particle is acted on by a repulsive force $\frac{\mu}{r^3}$ per unit mass from a fixed point, is projected at a distance a from the fixed point perpendicularly to the radius vector with a velocity $\frac{\sqrt{\mu}}{2a\sqrt{2}}$. Prove that the time when it is at a distance $a\sqrt{2}$ from the center of force is $\frac{2\sqrt{2}a^2}{3\sqrt{\mu}}$.
24. A particle moves under a central force μu^5 per unit mass; it is projected at a distance a from the center of force with the velocity $\sqrt{\frac{\mu}{2a^4}}$ at an angle α with the radius vector, show that the path is a circle through the center of force of diameter $a\csc\alpha$
25. A particle moves with a central acceleration $\lambda^2\left(8au^2+a^4u^5\right)$; it is projected with velocity 9λ from an apse at a distance $\frac{a}{3}$ from the origin; show that the equation to its path is

$$\frac{1}{\sqrt{3}}\sqrt{\frac{au+5}{au-3}}=\cot\frac{\theta}{\sqrt{6}}.$$

26. A particle describes an orbit with a control acceleration $\mu u^3-\lambda u^5$, being projected from an apse at distance a with a velocity equal to the from infinity; show that its path is $r=a\cosh\frac{\theta}{n}$, where $n^2+1=\frac{2\mu a^2}{\lambda}$. Prove also that it will be at distance r at end of time $\sqrt{\frac{a^2}{2\lambda}}\left[a^2\log\frac{r+\sqrt{r^2-a^2}}{a}+r\sqrt{r^2-a^2}\right]$.
27. A particle moves under a central force $m\lambda\left(3a^3u^4+8au^2\right)$; it is projected from an apse at a distance a from the center of force with velocity $\sqrt{10\lambda}$; show that the 2nd apsidal distance is half the first and that the equation to the path is $2r=a\left[1+\operatorname{sech}\frac{\theta}{\sqrt{5}}\right]$.
28. The velocity at any point of a central orbit is $\frac{1}{n}$ *th* of what it would be a circular orbit at the same distance; show that the central force varies as $\frac{1}{r^{2n^2+1}}$ and that the equation of the orbit is $r^{n^2-1}=a^{n^2-1}\cos\left\{\left(n^2-1\right)\theta\right\}$.
29. A particle under a central acceleration μr^{-7} is projected from a point at a distance a with a velocity from infinity, the direction of projection making an angle α with the initial line, show that the equation of the path is $r^2=a^2\left(\cos 2\theta-\sin 2\theta\cot\alpha\right)$.
30. A particle is projected at right angles at a distance under a central acceleration μu^n with velocity from infinity; show that the orbit is $r^{\frac{r-3}{2}}=a^{\frac{n-3}{2}}\cos\left\{\left(\frac{n-3}{2}\right)\theta\right\}$.

31. A particle moves with a central acceleration $\frac{\mu}{(\text{distance})^5}$, and is projected from an apse at a distance a with a velocity equal to n times that which would be acquired in falling from infinity; show that the other apsidal distance is $\frac{a}{\sqrt{n^2-1}}$. If $n = 1$, the particle be projected in any direction, show that the path is a circle passing through the center of force.
32. A particle of mass m moving under a central force $m\mu u^4 \left(1 - \frac{10}{9}au\right)$ is projected from an apse at a distance $5a$ from the center of force with the velocity which is equal to $\sqrt{\frac{5}{7}}$ of the velocity in a circle at the same distance; show that the polar equation of the orbit is $r = a\,(3 + 2\cos\theta)$.
33. If $P = \mu\left(u^2 - au^3\right)$, where $a > 0$, and a particle is projected from an apse at a distance a from the center of force with a velocity $\left(\frac{\mu c}{a^2}\right)^{\frac{1}{2}}$, where $a > c$, prove that other apsidal distance of the orbit is $\frac{a(a+c)}{a-c}$.

Chapter 3
Law of Inverse Square and Two-Body Problem

3.1 Basic Concepts

3.1.1 Newton's Law of Gravitation

Let two masses M and m be placed at two points O and P, respectively, at a distance r apart. Then the mass m is attracted towards M with a force **F** given by

$$\mathbf{F} = -\frac{GMm}{r^2}\hat{r}, \ \hat{r} = \frac{\mathbf{r}}{r} = \frac{\mathbf{OP}}{\mathrm{OP}}; \ \mathbf{r} = \mathbf{OP}, \ r = \mathrm{OP}.$$

Here G is defined as the **gravitational constant**.

Theorem 3.1 *Force of gravitation is conservative*

Proof Consider the gravitational force **F** of a big mass M placed at a point O ($\mathbf{0}$) on a mass $m(m < M)$ placed at P(**r**), given by

$$\mathbf{F} = -\frac{GMm}{r^2}\hat{r} = -\frac{\lambda}{r^2}\hat{r} = -\frac{\lambda\mathbf{r}}{r^3}; \quad \lambda = GMm.$$

Now,

$$\nabla \times \mathbf{F} = -\lambda\nabla \times \left(r^{-3}\mathbf{r}\right) = -\lambda\left[\nabla r^{-3} \times \mathbf{r} + r^{-3}\nabla \times (\mathbf{r})\right] = -\lambda\left[-3r^{-4}\frac{\mathbf{r}}{r} \times \mathbf{r}\right] = \mathbf{0}.$$

Therefore, **F** is conservative. □

N. Ahmed et al., *Classical Dynamics*, University Texts in the Mathematical Sciences,
https://doi.org/10.1007/978-981-95-6394-4_3

3.1.2 Gravitational Potential

Derivation: Consider the gravitational force $\mathbf{F} = -\frac{GMm}{r^2}\hat{r}$.

Let V be the force potential corresponding to the force **F**.

By definition of force potential, it follows that

$$\mathbf{F} = -\nabla V,$$

$$\therefore \quad \mathbf{F} \cdot d\mathbf{r} = -\nabla V \cdot d\mathbf{r} = -dV,$$

$$\text{i.e.,} \quad dV = -\mathbf{F} \cdot d\mathbf{r} = \frac{\lambda}{r^3}\mathbf{r} \cdot d\mathbf{r} = \frac{\lambda}{r^2}dr.$$

Integration of the above equation gives

$$V = -\frac{\lambda}{r} = -\frac{GMm}{r},$$

which is the required gravitational potential corresponding to the force **F**.

3.1.3 Orbit Under the Law of Force $P = \frac{\mu}{r^2}$

Derivation: Let Q be the position of the particle at time t, with polar coordinates (r, θ) referred to O as pole.

Here

$$P = \frac{\mu}{r^2} = \mu u^2.$$

The differential equation of the central orbit is

$$\frac{d^2u}{d\theta^2} + u = \frac{P}{h^2u^2} = \frac{\mu u^2}{h^2u^2} = \frac{\mu}{h^2},$$

$$\text{i.e.,} \quad \left(D^2 + 1\right)u = \frac{\mu}{h^2}. \tag{3.1.1}$$

G.S. of (3.1.1) is

$$u = A\cos\theta + B\sin\theta + \frac{\mu}{h^2},$$

i.e., $$\frac{1}{r} = \frac{\mu}{h^2} + A\cos\theta + B\sin\theta,$$

i.e., $$\frac{\frac{h^2}{\mu}}{r} = 1 + \frac{Ah^2}{\mu}\cos\theta + \frac{Bh^2}{\mu}\sin\theta,$$

i.e., $$\frac{l}{r} = 1 + \xi\cos\theta + \eta\sin\theta, \tag{3.1.2}$$

where $l = \frac{h^2}{\mu}$, $\xi = \frac{Ah^2}{\mu}$, $\eta = \frac{Bh^2}{\mu}$.

Let

$$\left.\begin{array}{l} \xi = e\cos\alpha \\ \eta = e\sin\alpha \end{array}\right\}. \tag{3.1.3}$$

$$\therefore \quad \left.\begin{array}{l} e = \sqrt{\xi^2 + \eta^2} \\ \tan\alpha = \frac{\eta}{\xi} \end{array}\right\}. \tag{3.1.4}$$

Substituting ξ and η from (3.1.3) in (3.1.2), we obtain

$$\frac{l}{r} = 1 + e\cos\alpha\cos\theta + e\sin\alpha\sin\theta = 1 + e\cos(\theta - \alpha). \tag{3.1.5}$$

Let us choose the initial line in such a way that $\alpha = 0$.

Therefore, Eq. (3.1.5) reduces to

$$\frac{l}{r} = 1 + e\cos\theta. \tag{3.1.6}$$

Equation (3.1.6) represents a conic section with semi-latus rectum $l = \frac{h^2}{\mu}$, e as eccentricity, and the fixed point O (center of force) as a focus.

Equation (3.1.6) represents

$$\left.\begin{array}{l} \text{an ellipse} \\ \text{a parabola} \\ \text{a hyperbola} \end{array}\right\} \text{according as} \left.\begin{array}{r} e < 1 \\ e = 1 \\ e > 1 \end{array}\right\}.$$

Equation (3.1.6) may be written as

$$lu = 1 + e\cos\theta, \tag{3.1.7}$$

where $u = \frac{1}{r}$.

Equation (3.1.7) gives

$$u = \frac{1 + e\cos\theta}{l}. \tag{3.1.8}$$

Differentiating (3.1.8) w.r.t θ, we derive

$$\frac{du}{d\theta} = -\frac{e\sin\theta}{l}. \tag{3.1.9}$$

Let v be the velocity of the particle at a point (r, θ) .

$$\therefore \quad v^2 = h^2\left[\left(\frac{du}{d\theta}\right)^2 + u^2\right] = \mu l\left[\frac{e^2\sin^2\theta}{l^2} + \frac{(1 + e\cos\theta)^2}{l^2}\right],$$

$$\text{i.e.,} \quad v^2 = \frac{\mu}{l}\left[1 + 2e\cos\theta + e^2\right] = \frac{\mu}{l}\left[1 + e^2 + 2\left(lu - 1\right)\right],$$

(by (3.1.7))

$$\text{i.e.,} \quad v^2 = \frac{\mu}{l}\left[1 + e^2 + \frac{2l}{r} - 2\right],$$

$$\text{i.e.,} \quad v^2 = \frac{\mu}{l}\left[\frac{2l}{r} - 1 + e^2\right],$$

$$\text{i.e.} \quad v^2 = \mu\left[\frac{2}{r} + \frac{e^2 - 1}{l}\right],$$

$$\text{i.e.,} \quad v^2 = \frac{2\mu}{r} + C, \tag{3.1.10}$$

where $C = \frac{\mu(e^2-1)}{l}$.

Thus, the path of the particle is

$$\left.\begin{array}{l}\text{an ellipse}\\ \text{a parabola}\\ \text{a hyperbola}\end{array}\right\} \text{ if } \left\{\begin{array}{r}e < 1\\ e = 1\\ e > 1\end{array}\right\},$$

Thus, the path of the particle is

$$\left.\begin{array}{l}\text{an ellipse}\\ \text{a parabola}\\ \text{a hyperbola}\end{array}\right\} \text{ if } \left\{\begin{array}{r}C < 0\\ C = 0\\ C > 0\end{array}\right\}.$$

Let the particle be projected with velocity V from the point at a distance R from the center of force O. i.e., $v = V$ at $r = R$.

Therefore, Eq. (3.1.10) gives

$$V^2 = \frac{2\mu}{R} + C,$$

$$\therefore \quad C = V^2 - \frac{2\mu}{R}.$$

Thus, the orbit of the particle is an ellipse, parabola, or hyperbola if $V^2 < = > \frac{2\mu}{R}$.

Let V_0 be the velocity of fall from infinity to the point $(R, 0)$.

$$\therefore \quad \frac{V_0^2}{2} = -\int_{\infty}^{R} P dr,$$

$$\text{i.e.,} \quad V_0^2 = 2\int_{R}^{\infty} P dr = 2\int_{R}^{\infty} \frac{\mu}{r^2} dr = 2\mu\left(-\frac{1}{r}\right)\Big]_R^{\infty} = \frac{2\mu}{R}.$$

Thus, we see that the orbit described by the particle is an ellipse, parabola, or hyperbola if $V^2 < = > V_0^2$; V_0 being the velocity of fall form infinity under the acceleration $\frac{\mu}{r^2}$ to the point of projection $(R, 0)$.

If the orbit is an ellipse, the velocity of the particle at any point (r, θ) is given by

$$v^2 = \mu\left[\frac{2}{r} - \frac{1 - e^2}{l}\right]. \tag{3.1.11}$$

For an ellipse, $b^2 = a^2\left(1 - e^2\right)$ and $\frac{b^2}{a} = l$, i.e., $b^2 = al$, i.e., $a^2\left(1 - e^2\right) = al$,

$$\text{i.e.,} \quad 1 - e^2 = \frac{l}{a}. \tag{3.1.12}$$

By use of (3.1.12), Eq. (3.1.11) reduces to

$$v^2 = \mu\left[\frac{2}{r} - \frac{l}{la}\right] = \mu\left[\frac{2}{r} - \frac{1}{a}\right]. \tag{3.1.13}$$

Similarly, for parabolic and hyperbolic orbits, the expressions for the velocity field are respected as follows:

$$v^2 = \frac{2\mu}{r}, \tag{3.1.14}$$

and

$$v^2 = \mu \left(\frac{2}{r} + \frac{1}{a} \right). \tag{3.1.15}$$

If T is the orbital period in case of an elliptic orbit,

$$\frac{h}{2} T = \pi ab,$$

$$\text{i.e.,} \quad T = \frac{2\pi ab}{h} = \frac{2\pi ab}{\sqrt{\mu l}} = \frac{2\pi ab}{\sqrt{\frac{\mu b^2}{a}}} = \frac{2\pi a^{\frac{3}{2}}}{\sqrt{\mu}}.$$

3.1.4 Kepler's Laws of Planetary Motion

1. Each planet describes an ellipse having the sun in one of its foci.
2. The radius vector drawn from the sun to a planet sweeps out areas at a constant rate.
3. The squares of the orbital periods of the various planets are proportional to the cube of the major axes of their orbits.

Sir Isaac Newton later unified these laws through his Law of Gravitation.

3.1.5 Dynamical Significance of Kepler's Laws

- Kepler's 1st law states that the path of a planet around the sun is an ellipse with the sun in one of its foci. Therefore, the law of force on the planet due to the sun is $P \propto \frac{1}{r^2}$ with the usual meaning of the notations.
- From Kepler's 2nd law, it follows that the areal velocity of the radius vector joining the sun and the planet is constant. That is $\frac{h}{2} = \text{Constant}$,

$$\text{or} \quad h = \text{Constant},$$

$$\text{or} \quad r^2\dot{\theta} = \text{Constant},$$

$$\text{or} \quad \frac{m}{r} \frac{d}{dt} \left(r^2 \dot{\theta} \right) = 0, \tag{3.1.16}$$

m being the mass of a planet. Equation (3.1.16) shows that $F_\theta = 0$, where F_θ is the effective force on the planet in the transverse direction. Thus, the force on the planet is entirely directed towards the sun.

- If the central orbit is an ellipse with major axis $2a$, and with T as the orbital period, then

$$T = \frac{2\pi}{\sqrt{\mu}} a^{\frac{3}{2}}. \tag{3.1.17}$$

Kepler's 3rd law says that

$$\begin{aligned} & T^2 \propto (2a)^3, \\ \text{or} \quad & T^2 \propto a^3, \\ \text{or} \quad & T^2 = k^2 a^3, \quad k \text{ is a constant} \\ \text{or} \quad & T = k a^{\frac{3}{2}}. \end{aligned} \tag{3.1.18}$$

From (3.1.17) and (3.1.18), we have

$$\frac{2\pi}{\sqrt{\mu}} a^{\frac{3}{2}} = k a^{\frac{3}{2}},$$

$$\text{or} \quad \frac{2\pi}{\sqrt{\mu}} = k,$$

$$\text{or} \quad \mu = \frac{4\pi^2}{k^2} = \text{a constant}.$$

The force of a planet towards the sun is governed by the law:

$$P = \frac{\mu}{r^2},$$

$$\therefore \quad P]_{r=1} = \mu = \text{a constant}.$$

Thus, the acceleration at a unit distance is constant for all planets.

3.1.6 Kepler's 3rd Law Is Approximately True

Proof Let S be the sun with mass M and P be a planet of mass m ($M >> m$). Let SP $= r$ and $\hat{r}$ be the unit vector along **SP** (see Fig. 3.1). Let **F** be the force of attraction of the sun on the planet.

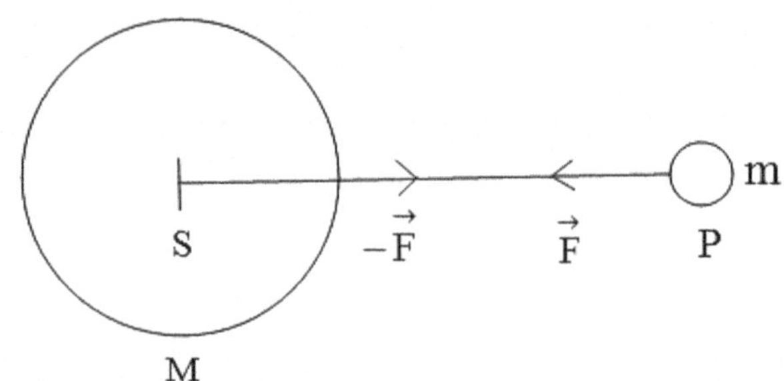

Fig. 3.1 Force of attraction between sun and planet

$$\therefore \quad \mathbf{F} = \frac{GMm}{r^2}(-\hat{r}) = -\frac{GMm}{r^2}\hat{r}.$$

By Newton's 2nd law of motion, we have

$$m\alpha = \mathbf{F} = -\frac{GMm}{r^2}\hat{r},$$

$$\text{i.e.,} \quad \alpha = -\frac{GM}{r^2}\hat{r}.$$

If β is the acceleration of the sun towards the planet then

$$M\beta = -\mathbf{F} = \frac{GMm}{r^2}\hat{r},$$

$$\text{i.e.,} \quad \beta = \frac{Gm}{r^2}\hat{r}.$$

Therefore, acceleration of the planet relative to the sun is

$$\alpha - \beta = -\frac{GM}{r^2}\hat{r} - \frac{Gm}{r^2}\hat{r} = -\frac{G}{r^2}(M+m)\hat{r} = -\frac{\mu}{r^2}\hat{r}; \quad \mu = G(M+m).$$

Therefore, the orbital period T of the planet with usual notations is given by

$$T = \frac{2\pi}{\sqrt{\mu}}a^{\frac{3}{2}} = \frac{2\pi}{\sqrt{G(M+m)}}a^{\frac{3}{2}}. \tag{3.1.19}$$

If T' be the periodic time of another planet with mass m', semi-axis a' then

$$T' = \frac{2\pi}{\sqrt{G(M+m')}}a'^{\frac{3}{2}}. \tag{3.1.20}$$

(3.1.19) $\div$ (3.1.20) gives

$$\frac{T}{T'} = \sqrt{\frac{M+m'}{M+m}}\left(\frac{a}{a'}\right)^{\frac{3}{2}},$$

$$\text{i.e.,}\quad \frac{T^2}{T'^2} = \frac{M+m'}{M+m}\left(\frac{a}{a'}\right)^3. \tag{3.1.21}$$

If the masses of the planets are negligible in comparison to the mass of the sun, then Eq. (3.1.21) reduces to

$$\frac{T^2}{T'^2} = \left(\frac{a}{a'}\right)^3,$$

$$\text{i.e.,}\quad \frac{T^2}{a^3} = \frac{T'^2}{a'^3},$$

$$\text{i.e.,}\quad \frac{T^2}{a^3} = \frac{T_1^2}{a_1^3} = \frac{T_2^2}{a_2^3} = \cdots = \text{a constant} = k\ (\text{say}).$$

Therefore, $T^2 \propto a^3$, which is Kepler's 3rd law for planetary motion. Thus, we see that Kepler's 3rd law is approximately true. □

3.1.7 Solar System: Planets and Satellites

A solar system is composed of a **star** (such as the sun) and objects called **planets** which revolve around it. A star is an object that emits its own light, while planets do not emit light but can reflect it. In addition, there may be objects revolving about planets. These are termed as **satellites**. In our solar system, the moon is a satellite of the Earth, which in turn is a planet revolving about the sun. In addition, there are artificial or man-made satellites that can revolve around the planets or their moons. The path of a planet or satellite is called its orbit. The orbital positions at which a planet reaches its minimum and maximum distances from the Sun are termed **perihelion** and **aphelion**, respectively. The points in a satellite's orbit where it is farthest from and nearest to the Earth are called the **apogee** and **perigee**, respectively. The time for one complete revolution of a body in an orbit is called its **period** or **orbital period**.

3.1.8 Two-Body Problem

The problem of dealing with the motions of two objects under their mutual attractions is often called the two-body problem. The motion of a planet relative to the sun in our solar system is the classical example of the two-body problem. In this example, the gravitational effects arising from the other planets are neglected. These effects depend on their masses and respective distances. It is to be noted that the mass of the sun is about 700 times the total mass of the planets of our solar system, and hence the gravitational effect of the sun is the dominant one on every planet. The motion of a planet relative to the sun is governed by the three Kepler's laws. The idea of the two-body problem is based on the motion of an artificial satellite around the Earth, following the motion of the moon around the Earth.

3.1.9 Motion of the Center of Mass and Motion Relative to the Center of Mass

Derivation: Let M be the mass of the sun and m that of a planet. In Fig. 3.2, let C be the center of mass of M and m, and S and P denote the positions of the sun and the planet, respectively. Let $\mathbf{r}_1$, $\mathbf{r}_2$, $\mathbf{R}$ stipulate the position vectors of S, P, and C, respectively, relative to some fixed origin O. Suppose that $\mathbf{SP} = \mathbf{r}$.

Equation of motion of the sun S is

$$M\ddot{\mathbf{r}}_1 = \frac{GMm}{r^2}\hat{r} = \frac{GMm}{r^3}\mathbf{r},$$

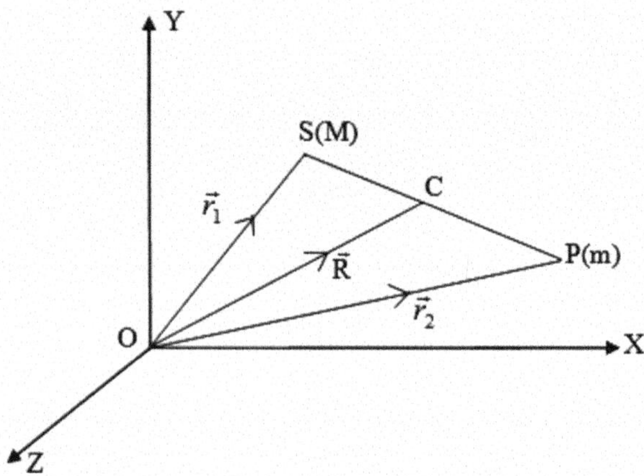

Fig. 3.2 Illustrating the motion between sun and planet

$$\text{i.e.,}\quad \ddot{\mathbf{r}}_1 = \frac{Gm}{r^2}\hat{r} = \frac{Gm}{r^3}\mathbf{r}. \tag{3.1.22}$$

Similarly, the equation of motion of the planet P is

$$\ddot{\mathbf{r}}_2 = -\frac{GM}{r^2}\hat{r} = -\frac{GM}{r^3}\mathbf{r}. \tag{3.1.23}$$

We have

$$\mathbf{r} = \mathbf{SP} = \mathbf{r}_2 - \mathbf{r}_1. \tag{3.1.24}$$

Again

$$\mathbf{R} = \frac{M\mathbf{r}_1 + m\mathbf{r}_2}{M+m},$$

$$\text{i.e.,}\quad (M+m)\,\mathbf{R} = M\mathbf{r}_1 + m\mathbf{r}_2. \tag{3.1.25}$$

Equation (3.1.25) gives

$$(M+m)\,\ddot{\mathbf{R}} = M\ddot{\mathbf{r}}_1 + m\ddot{\mathbf{r}}_2 = \frac{GMm}{r^3}\mathbf{r} - \frac{GMm}{r^3}\mathbf{r} = \mathbf{0},$$

$$\text{i.e.,}\quad \ddot{\mathbf{R}} = \mathbf{0}. \tag{3.1.26}$$

Integration of Eq. (3.1.26) leads to

$$\mathbf{R} = \mathbf{C}_1 t + \mathbf{C}_2. \tag{3.1.27}$$

Equation (3.1.27) shows that the locus of C(**R**) is a straight line. In other words, the center of mass is either at rest or in uniform motion.

Equations (3.1.23)–(3.1.22) give

$$\ddot{\mathbf{r}}_2 - \ddot{\mathbf{r}}_1 = -\frac{G\,(M+m)\,\mathbf{r}}{r^3},$$

$$\ddot{\mathbf{r}} = -\frac{\mu\mathbf{r}}{r^3};\quad \mu = G\,(M+m)\,,$$

$$\text{i.e.,}\quad \ddot{\mathbf{r}} = -\frac{\mu}{r^2}\hat{r}. \tag{3.1.28}$$

Equation (3.1.28) shows that the motion of a planet relative to the sun follows the "Inverse square law" directed towards the sun. Now, closely following Sect. 3.1.4, it is seen that the path of a planet is a conic section with the center of force (sun) at one focus of the orbit. By Kepler's 1st law of planetary motion, it follows that the orbit of a planet around the sun is an ellipse with the sun at one of its two foci.

With usual notations, the equation of the elliptic orbit of a planet with the sun as the pole and the major axis as the initial line is

$$\frac{l}{r} = 1 + e\cos\theta.$$

Hence the expression for the velocity of the planet at distance r from the sun is specified by

$$v^2 = \mu\left(\frac{2}{r} - \frac{1}{a}\right) \text{ and the orbital period is given by } T = \frac{2\pi}{\sqrt{\mu}}a^{\frac{3}{2}}.$$

3.1.10 *Distances of Aphelion and Perihelion from the Sun*

By the principle of energy, we have

$$T + V = E,$$

$$\text{i.e., } \frac{1}{2}mv^2 - \frac{GMm}{r} = E,$$

$$\text{i.e., } \frac{1}{2}v^2 - \frac{GM}{r} = E \quad (\text{taking } m = 1),$$

$$\text{i.e., } v^2 - \frac{2\mu}{r} = 2E \quad (GM = \mu).$$

Thus, the velocity of a planet at a distance r from the sun is given by

$$v^2 = \frac{2\mu}{r} + 2E.$$

In case of a central orbit,

$$v^2 = h^2\left[\left(\frac{du}{d\theta}\right)^2 + u^2\right].$$

$$\therefore \quad \frac{2\mu}{r} + 2E = h^2\left[\left(\frac{du}{d\theta}\right)^2 + u^2\right]. \tag{3.1.29}$$

At aphelion, r is maximum, and at perihelion, r is minimum. As $u = \frac{1}{r}$, therefore u is minimum, at aphelion and u is maximum at perihelion. Thus, $\frac{du}{d\theta} = 0$ for an aphelion as well as for a perihelion.

Hence, the distance of aphelion or perihelion from the sun is given by

$$\frac{2\mu}{r} + 2E = \frac{h^2}{r^2},$$

$$\text{i.e.,}\quad 2Er^2 + 2\mu r - h^2 = 0, \tag{3.1.30}$$

which is a quadratic equation in r determining the distances of aphelion and perihelion from the sun.

3.1.11 True Anomaly, Eccentric Anomaly, and Mean Anomaly

Let the elliptic orbit of a planet around the sun be given by

$$\frac{x^2}{a^2} + \frac{y^2}{b^2} = 1, \tag{3.1.31}$$

where

$$b^2 = a^2\left(1 - e^2\right). \tag{3.1.32}$$

In Fig. 3.3, let C be the center of the orbit, and S denote the position of the sun, which is one of the two foci of the orbit. Let AA′ be the major axis of the orbit. Here, A is the aphelion and A′ is the perihelion of the planet. Let P be the position of the planet at time t with polar coordinates $(r,\ \theta)$ referred to the sun S as the pole and **SA** as the initial line. Draw $\perp^r$ PL from P on AA′ and produce LP to meet the auxiliary circle at Q and join CQ and SP. Suppose that $\angle$ACQ=ϕ and $\angle$ASP=θ. The angle $\angle$ASP=θ is termed as the **true anomaly** and the angle $\angle$ACQ=ϕ is called the **eccentric anomaly** of the planet with respect to the sun S at time t. If T is the orbital period of the planet, then $\frac{2\pi}{T}$ is defined to be the **mean angular velocity** of the planet and it is generally denoted by the symbol n. Suppose that the planet crosses point A, the perihelion at time $t = 0$. Then $m = nt = \frac{2\pi}{T}t$ is defined to be **mean anomaly** of the planet. Thus we have

- $\theta =$ True Anomaly,
- $\phi =$ Eccentric Anomaly,

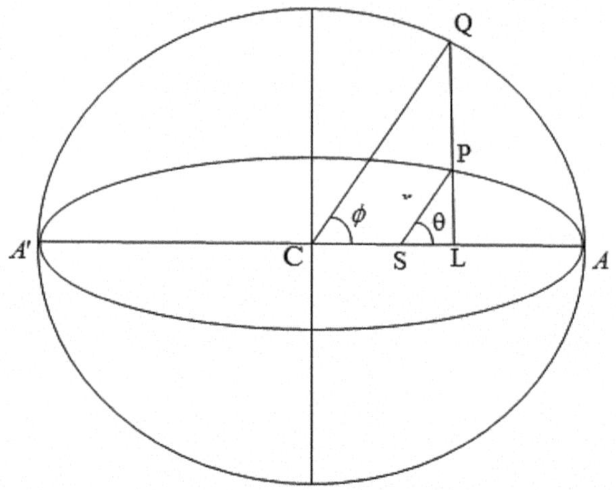

Fig. 3.3 Geometry of true, eccentric, and mean anomaly

- $n = \frac{2\pi}{T}$= Mean Angular Velocity,
- $m = nt = \frac{2\pi}{T}t$ = Mean Anomaly.

3.1.12 *Position of a Planet in Its Orbit*

Let $\theta,\ \phi\, and\, m$ be respectively the true, eccentric, and mean anomalies.

From Fig. 3.4,

$$r\cos\theta = SL = a\cos\phi - ae. \tag{3.1.33}$$

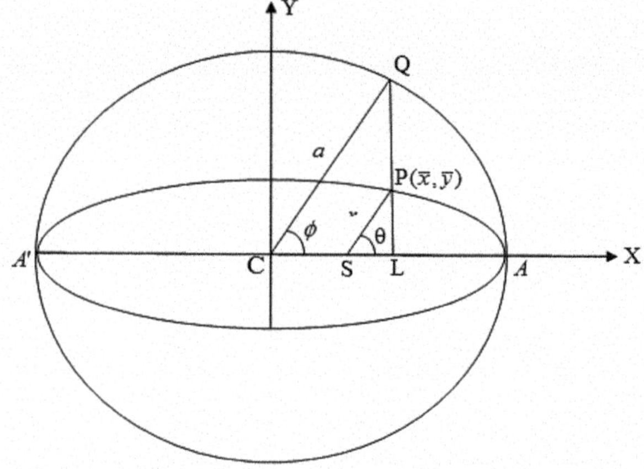

Fig. 3.4 Schematic diagram for anomalies

The equation of the ellipse is

$$\frac{x^2}{a^2} + \frac{y^2}{b^2} = 1. \tag{3.1.34}$$

Let $(\bar{x},\ \bar{y})$ be the coordinates of the planet P.

$$\therefore \quad \bar{x} = a\cos\phi.$$

Since the point P $(\bar{x},\ \bar{y})$ is on (3.1.34),

$$\therefore \quad \frac{\bar{x}^2}{a^2} + \frac{\bar{y}^2}{b^2} = 1,$$

$$\text{i.e.,} \quad \frac{a^2\cos^2\phi}{a^2} + \frac{\mathrm{PL}^2}{b^2} = 1,$$

$$\text{i.e.,} \quad \cos^2\phi + \frac{\mathrm{PL}^2}{b^2} = 1,$$

$$\text{i.e.,} \quad \mathrm{PL}^2 = b^2\left(1 - \cos^2\phi\right) = b^2\sin^2\phi,$$

$$\text{i.e.,} \quad \mathrm{PL} = b\sin\phi,$$

$$\text{i.e.,} \quad r\ \sin\theta = b\sin\phi = a\sqrt{1-e^2}\sin\phi. \tag{3.1.35}$$

On elimination of θ from Eqs. (3.1.33) and (3.1.35), we obtain

$$r^2 = a^2(1 - e\cos\phi)^2,$$

$$\text{i.e.,} \quad r = a\,(1 - e\cos\phi)\,. \tag{3.1.36}$$

Now,

$$2r\sin^2\frac{\theta}{2} = r\,(1 - \cos\theta) = r - r\cos\theta,$$

$$\text{i.e.,} \quad 2r\sin^2\frac{\theta}{2} = a\,(1 - e\cos\phi) - (a\cos\phi - ae)\,, \quad \left[\text{by (3.1.33) and (3.1.36)}\right]$$

$$\text{i.e.,} \quad 2r\sin^2\frac{\theta}{2} = a\,(1+e)\,(1-\cos\phi) = 2a\,(1+e)\sin^2\frac{\phi}{2}. \tag{3.1.37}$$

Again

$$2r\cos^2\frac{\theta}{2} = r\,(1+\cos\theta) = r + r\cos\theta,$$

$$\text{i.e.,}\quad 2r\cos^2\frac{\theta}{2} = a\,(1-e\cos\phi) + (a\cos\phi - ae)\,,$$

$$\text{i.e.,}\quad 2r\cos^2\frac{\theta}{2} = a\,(1-e)\,(1+\cos\phi) = 2a\,(1-e)\cos^2\frac{\phi}{2}. \tag{3.1.38}$$

(3.1.37) ÷ (3.1.38) yields

$$\tan^2\frac{\theta}{2} = \frac{1+e}{1-e}\tan^2\frac{\phi}{2},$$

$$\text{i.e.,}\quad \tan\frac{\theta}{2} = \sqrt{\frac{1+e}{1-e}}\tan\frac{\phi}{2}.$$

Thus, the position $(r,\ \theta)$ of the planet can be determined from the following relations:

$$r = a\,(1-e\cos\phi)\ \text{and}\ \tan\frac{\theta}{2} = \sqrt{\frac{1+e}{1-e}}\tan\frac{\phi}{2},$$

with the known value of ϕ.

3.1.13 Kepler's Equation

Derivation: In the case of the orbit of a planet around the sun, $r,\ \theta,\ \phi,\ m,\ T,\ e,\ \mu,\ t,\ a$ with the usual meaning of the symbols, are connected by the relations:

$$r\cos\theta = a\cos\phi - ae, \tag{3.1.39}$$

$$r\sin\theta = a\sqrt{1-e^2}\sin\phi, \tag{3.1.40}$$

$$r = a\,(1-e\cos\phi)\,, \tag{3.1.41}$$

$$\frac{l}{r} = 1 + e\cos\theta, \tag{3.1.42}$$

$$m = nt = \frac{2\pi}{T}t, \tag{3.1.43}$$

$$T = \frac{2\pi}{\sqrt{\mu}}a^{\frac{3}{2}}, \tag{3.1.44}$$

$$r^2\dot{\theta} = h = \sqrt{\mu l} = \sqrt{\mu a\left(1 - e^2\right)}. \tag{3.1.45}$$

Equation (3.1.41) gives

$$\dot{r} = ae\sin\phi\,\dot{\phi}. \tag{3.1.46}$$

Equation (3.1.42) shows

$$\frac{l}{r^2}\dot{r} = e\sin\theta\,\dot{\theta},$$

$$\text{i.e.,}\quad \dot{r} = \frac{e\sin\theta}{l}\left(r^2\dot{\theta}\right) = \frac{eh\sin\theta}{l}. \tag{3.1.47}$$

Equations (3.1.46) and (3.1.47) give

$$ae\sin\phi\,\dot{\phi} = \frac{eh\sin\theta}{l},$$

$$\text{i.e.,}\quad a\sin\phi\,\dot{\phi} = \frac{\sin\theta\sqrt{\mu l}}{l} = \sin\theta\sqrt{\frac{\mu}{l}} = \frac{\sqrt{\mu}\sin\theta}{\sqrt{a\left(1 - e^2\right)}},$$

$$\text{i.e.,}\quad \dot{\phi} = \frac{\sqrt{\mu}\sin\theta}{\sqrt{a\left(1 - e^2\right)}}\frac{1}{a\sin\phi} = \frac{\sqrt{\mu}\sin\theta}{\sqrt{a}}\frac{1}{a\sqrt{1 - e^2}\sin\phi} = \sqrt{\frac{\mu}{a}}\frac{\sin\theta}{r\sin\theta},$$

$$\text{i.e.,}\quad \dot{\phi} == \sqrt{\frac{\mu}{a}}\frac{1}{r} = \sqrt{\frac{\mu}{a}}\frac{1}{a\left(1 - e\cos\phi\right)},$$

$$\text{i.e.,}\quad \left(1 - e\cos\phi\right)d\phi = \sqrt{\frac{\mu}{a^3}}\,dt. \tag{3.1.48}$$

On integration of Eq. (3.1.48), we find

$$\phi - e\sin\phi = \sqrt{\frac{\mu}{a^3}}\,t + C. \tag{3.1.49}$$

At the perihelion A, $\phi = 0$, $t = 0$, $\therefore\ C = 0$.

Hence, Eq. (3.1.49) reduces to

$$\phi - e \sin \phi = \sqrt{\frac{\mu}{a^3}}\, t. \tag{3.1.50}$$

Equation (3.1.44) shows that

$$\sqrt{\mu} = \frac{2\pi}{T} a^{\frac{3}{2}} = n a^{\frac{3}{2}} = n\sqrt{a^3},$$

$$\text{i.e.,} \quad \sqrt{\frac{\mu}{a^3}} = n. \tag{3.1.51}$$

Equations (3.1.50) and (3.1.51) leads to

$$\phi - e \sin \phi = nt = m,$$

$$\text{i.e.,} \quad \phi - e \sin \phi = m.$$

This equation is termed as Kepler's Equation.

3.2 Illustrations

Example 3.1 Show that the velocity of a particle moving in an ellipse about a center of force in focus is the compound of two constant velocities, $\frac{\mu}{h}$ perpendicular to the radius and $\frac{\mu e}{h}$ perpendicular to the major axis.

Solution Let the equation of the elliptic path of the particle with the focus S as the pole and the major axis as the initial line be

$$\frac{l}{r} = 1 + e \cos \theta. \tag{3.2.1}$$

Here, S is the center of force. Let P be the position of the particle at time t, with polar coordinates $(r,\ \theta)$. Let $\hat{i},\ \hat{j}$ stipulate the unit vectors along **OX** and **OY** respectively as shown in Fig. 3.5. Let $\hat{r},\ \hat{\theta}$ denote the unit vectors along the increasing directions of r and θ respectively.

$\left(\hat{r},\ \hat{\theta}\right)$ and $\left(\hat{i},\ \hat{j}\right)$ are connected by the relations:

$$\hat{r} = \hat{i} \cos \theta + \hat{j} \sin \theta, \tag{3.2.2}$$

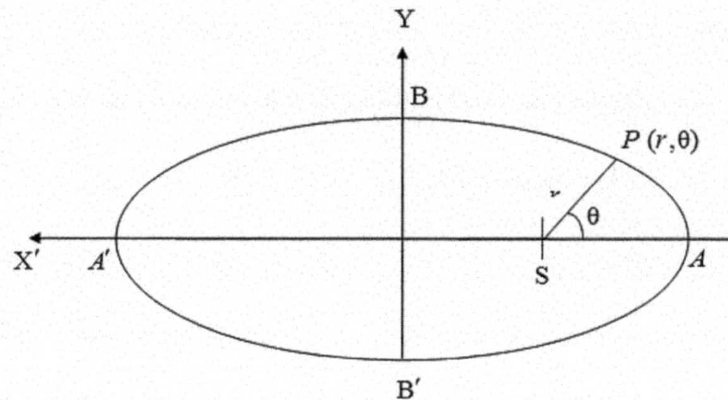

Fig. 3.5 Illustrating diagram for Example 3.1

$$\hat{\theta} = -\hat{i}\sin\theta + \hat{j}\cos\theta. \tag{3.2.3}$$

The velocity of the particle at $Q\,(r, \theta)$ is given by

$$\mathbf{v} = v_r\hat{r} + v_\theta\hat{\theta} = v_r\hat{i}\cos\theta + v_r\hat{j}\sin\theta + v_\theta\hat{\theta}. \tag{3.2.4}$$

Equation (3.2.3) gives

$$\hat{i} = \frac{\hat{j}\cos\theta - \hat{\theta}}{\sin\theta} = \hat{j}\cot\theta - \hat{\theta}\csc\theta. \tag{3.2.5}$$

Eliminating $\hat{i}$ from (3.2.4) and (3.2.5) we get

$$\mathbf{v} = v_r\cos\theta\left(\hat{j}\cot\theta - \hat{\theta}\csc\theta\right) + v_r\hat{j}\sin\theta + v_\theta\hat{\theta},$$

i.e., $\mathbf{v} = (v_\theta - v_r\cos\theta\csc\theta)\,\hat{\theta} + (v_r\cos\theta\cot\theta + v_r\sin\theta)\,\hat{j}$,

i.e., $\mathbf{v} = u_\theta\hat{\theta} + u_j\hat{j}$, (3.2.6)

where

$$u_\theta = v_\theta - v_r\cos\theta\csc\theta, \tag{3.2.7}$$

and

$$u_j = v_r\left(\cos\theta\cot\theta + \sin\theta\right). \tag{3.2.8}$$

Equation (3.2.7) simplifies to

$$u_\theta = r\dot{\theta} - \dot{r}\cot\theta, \tag{3.2.9}$$

and Equation (3.2.8) reduces to

$$u_j = v_r \left(\frac{\cos^2\theta}{\sin\theta} + \sin\theta \right) = \frac{\dot r}{\sin\theta}. \tag{3.2.10}$$

For a central orbit, we have

$$r^2\dot\theta = h,$$

$$\text{i.e.,}\quad \dot\theta = \frac{h}{r^2}. \tag{3.2.11}$$

Equation (3.2.1) gives

$$-\frac{l}{r^2}\dot r = -e\sin\theta\,\dot\theta,$$

$$\text{i.e.,}\quad \dot r = e\sin\theta\,\frac{r^2\dot\theta}{l} = \frac{eh}{l}\sin\theta. \tag{3.2.12}$$

Therefore, Eq. (3.2.9) reduces to

$$u_\theta = r\frac{h}{r^2} - \frac{eh\sin\theta\cot\theta}{l} = \frac{h}{r} - \frac{eh\cos\theta}{l} = h\left(\frac{1}{r} - \frac{e\cos\theta}{l}\right),$$

$$\text{i.e.,}\quad u_\theta = \frac{h}{l}\left[\frac{l}{r} - e\cos\theta\right] = \frac{h}{l}\left[1 + e\cos\theta - e\cos\theta\right],$$

$$\text{i.e.,}\quad u_\theta = \frac{h}{l} = \frac{h^2}{lh} = \frac{\mu l}{lh} = \frac{\mu}{h} = \text{a constant.}$$

We find from (3.2.10),

$$u_j = \frac{1}{\sin\theta}\frac{eh\sin\theta}{l} = \frac{eh}{l} = \frac{eh^2}{lh} = \frac{e\mu l}{lh} = \frac{e\mu}{h} = \text{a constant.}$$

Thus, Eq. (3.2.6) reduces to $\mathbf{v} = \frac{\mu}{h}\hat\theta + \frac{\mu e}{h}\hat j$, which establishes the required result.

Example 3.2 A planet is describing an ellipse about the sun as focus; show that its velocity away from the sun is greatest when the radius vector to the planet is at right angles to the major axis of the path, and then it is $\dfrac{2\pi ae}{T\sqrt{1-e^2}}$, where $2a$ is the major axis, e the eccentricity, and T the periodic time.

Solution Let the equation of the elliptic orbit of the planet with the sun at a focus, as the pole and the major axis as the initial line be

$$\frac{l}{r} = 1 + e\cos\theta. \tag{3.2.13}$$

Because T is the periodic time,

$$\therefore \quad T = \frac{2\pi}{\sqrt{\mu}} a^{\frac{3}{2}}. \tag{3.2.14}$$

Differentiating (3.2.13), w.r.t t, we get

$$-\frac{l}{r^2}\dot{r} = e\,(-\sin\theta)\,\dot{\theta},$$

$$\text{i.e.,} \quad \dot{r} = \frac{e \sin\theta\, r^2\dot{\theta}}{l} = \frac{e \sin\theta h}{l},$$

$$\text{i.e.,} \quad v_r = \frac{eh \sin\theta}{l}. \tag{3.2.15}$$

From (3.2.15) it follows that
v_r is greatest when $\sin\theta$ is greatest,
i.e., v_r is greatest when $\sin\theta = 1$,
i.e., v_r is greatest when $\theta = \frac{\pi}{2}$,
i.e., v_r is greatest when the radius vector SP is $\perp^r$ to the major axes.
The maximum value of v_r is given by

$$v_r]_{\max} = \frac{eh}{l} = \frac{e\sqrt{\mu l}}{l} = e\sqrt{\frac{\mu}{l}} = e\sqrt{\frac{a\mu}{b^2}} = \frac{\sqrt{a}e\sqrt{\mu}}{b} = \frac{e\sqrt{\mu}}{\sqrt{a\left(1-e^2\right)}}. \tag{3.2.16}$$

Equation (3.2.14) gives

$$\sqrt{\mu} = \frac{2\pi a^{\frac{3}{2}}}{T}. \tag{3.2.17}$$

Eliminating $\sqrt{\mu}$ from (3.2.16) and (3.2.17), we get

$$v_r]_{\max} = \frac{e}{\sqrt{a}\sqrt{\left(1-e^2\right)}} \frac{2\pi a^{\frac{3}{2}}}{T} = \frac{2\pi a e}{T\sqrt{\left(1-e^2\right)}}.$$

Example 3.3 If a planet were suddenly stopped in its orbit, suppose circular, show that it would fall into the sun in a time which is $\frac{\sqrt{2}}{8}$ times the period of the planet's revolution.

Solution Let a be the radius of the circular orbit with the sun S as the center. Let A be the position of the planet when it is suddenly stopped. Let P be the position of the planet at time t and suppose that SP $= x$ (Fig. 3.6).

The equation of motion of the planet is

$$\ddot{x} = -\frac{\mu}{x^2},$$

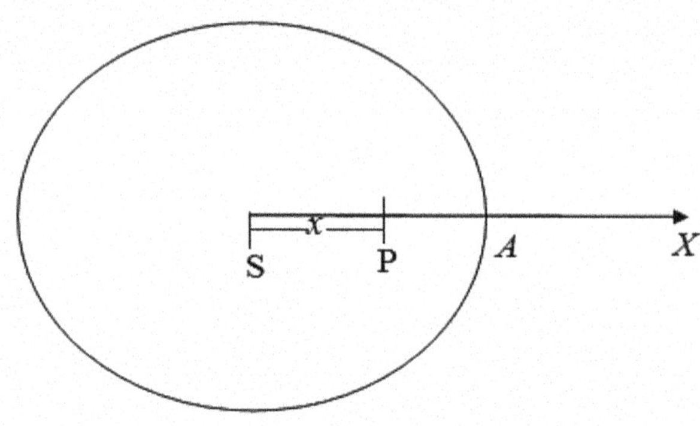

Fig. 3.6 Diagram depicting Example 3.3

$$\text{i.e.,}\quad vdv = -\frac{\mu}{x^2}dx,$$

$$\text{i.e.,}\quad \frac{v^2}{2} = \frac{\mu}{x} + C. \tag{3.2.18}$$

Initially $v = 0$, when $x = a$.

$$\therefore\quad 0 = \frac{\mu}{a} + C,$$

$$\text{i.e.,}\quad C = -\frac{\mu}{a}.$$

Therefore, Eq. (3.2.18) reduces to

$$\frac{v^2}{2} = \mu\left(\frac{1}{x} - \frac{1}{a}\right),$$

$$\text{i.e.,}\quad v^2 = 2\mu\frac{a-x}{ax},$$

$$\text{i.e.,}\quad v = -\sqrt{2\mu}\sqrt{\frac{a-x}{ax}},$$

$$\text{i.e.,}\quad \frac{dx}{dt} = -\sqrt{2\mu}\sqrt{\frac{a-x}{ax}},$$

$$\text{i.e.,}\quad \sqrt{\frac{ax}{a-x}}dx = -\sqrt{2\mu}dt,$$

$$\text{i.e.,}\quad \int_a^0 \sqrt{\frac{ax}{a-x}}dx = -\sqrt{2\mu}\int_0^{T_0} dt = -\sqrt{2\mu}\,T_0,$$

$$\text{i.e.,}\quad \sqrt{2\mu}\,T_0 = \int_0^a \sqrt{\frac{ax}{a-x}}dx. \tag{3.2.19}$$

Let $x = a\sin^2\theta$,
$\therefore \quad dx = 2a\sin\theta\cos\theta d\theta.$
$x = 0 \Rightarrow \theta = 0;\ x = a \Rightarrow \theta = \frac{\pi}{2}.$
Equation (3.2.19) gives

$$\sqrt{2\mu}T_0 = \int_0^{\frac{\pi}{2}} \sqrt{\frac{a^2\sin^2\theta}{a\cos^2\theta}} 2a\sin\theta\cos\theta d\theta = 2a^{\frac{3}{2}}\int_0^{\frac{\pi}{2}} \sin^2\theta\, d\theta = \frac{\pi a^{\frac{3}{2}}}{2},$$

$$\therefore \quad T_0 = \frac{\pi a^{\frac{3}{2}}}{2\sqrt{2\mu}}. \tag{3.2.20}$$

The periodic time of the planet is given by

$$T = \frac{2\pi}{\sqrt{\mu}}a^{\frac{3}{2}}. \tag{3.2.21}$$

Now (3.2.20) $\div$ (3.2.21) gives

$$\frac{T_0}{T} = \frac{\pi a^{\frac{3}{2}}}{2\sqrt{2\mu}} \times \frac{\sqrt{\mu}}{2\pi a^{\frac{3}{2}}} = \frac{1}{4\sqrt{2}} = \frac{\sqrt{2}}{8},$$

$$\therefore \quad T_0 = \frac{\sqrt{2}}{8}T.$$

Example 3.4 A particle moves with a central acceleration $\left[= \frac{\mu}{(\text{distance})^2}\right]$; it is projected with velocity Vat a distance R. Show that its path is a rectangular hyperbola if the angle of projection is $\frac{\mu}{VR(V^2-\frac{2\mu}{R})^{\frac{1}{2}}}$.

Solution Let v be the velocity of the particle at the point (r, θ), and let $P_0 = p$ be the perpendicular distance of the tangent at A from the center of force O as seen in Fig. 3.7.

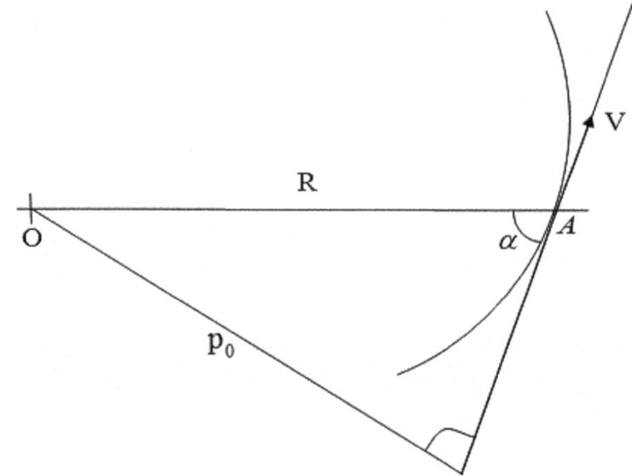

Fig. 3.7 Diagram depicting Example 3.4

$$\therefore \quad v^2 = \mu\left(\frac{2}{r} + \frac{1}{a}\right). \tag{3.2.22}$$

Given that $v = V$ when $r = R$.
Equation (3.2.22) gives

$$V^2 = \mu\left(\frac{2}{R} + \frac{1}{a}\right). \tag{3.2.23}$$

In case of a central orbit,

$$vp = h. \tag{3.2.24}$$

Initially $v = V, \;\; p = R \sin\alpha$ when $t = 0$.
Equation (3.2.24) gives

$$h = VR \sin\alpha. \tag{3.2.25}$$

Given

$$\alpha = \sin^{-1} \frac{\mu}{VR\left(V^2 - \frac{2\mu}{R}\right)^{1/2}},$$

$$\therefore \quad \sin\alpha = \frac{\mu}{VR\left(V^2 - \frac{2\mu}{R}\right)^{1/2}}. \tag{3.2.26}$$

Education (3.2.25) gives

$$h^2 = V^2 R^2 \sin^2 \alpha,$$

$$\text{i.e.,}\quad \mu l = V^2 R^2 \frac{\mu^2}{V^2 R^2 \left(V^2 - \frac{2\mu}{R}\right)} = \frac{\mu^2}{V^2 - \frac{2\mu}{R}},$$

$$\text{i.e.,}\quad l = \frac{\mu}{V^2 - \frac{2\mu}{R}}. \tag{3.2.27}$$

Education (3.2.23) gives

$$V^2 = \frac{2\mu}{R} + \frac{\mu}{a},$$

$$\text{i.e.,}\quad V^2 - \frac{2\mu}{R} = \frac{\mu}{a}. \tag{3.2.28}$$

Using (3.2.28) in (3.2.27), we get

$$l = \frac{\mu a}{\mu},$$

$$\text{i.e.,}\quad l = a,$$

$$\text{i.e.,}\quad \frac{b^2}{a} = a,$$

$$\text{i.e.,}\quad b^2 = a^2,$$

$$\text{i.e.,}\quad b = a.$$

Therefore, the path of the particle is a rectangular hyperbola.

Example 3.5 A satellite has its largest and smallest orbital speeds given by $v_{\max}$ and $v_{\min}$, respectively. Prove that the eccentricity of the orbit in which the satellite moves is equal to $\frac{v_{\max} - v_{\min}}{v_{\max} + v_{\min}}$.

Solution With usual rotations, let the equation of the elliptic orbit (Fig. 3.8) of the satellite be

$$\frac{l}{r} = 1 + e \cos \theta. \tag{3.2.29}$$

Let v be the speed of the satellite at the (r, θ).

$$\therefore \quad v^2 = \mu \left(\frac{2}{r} - \frac{1}{a}\right). \tag{3.2.30}$$

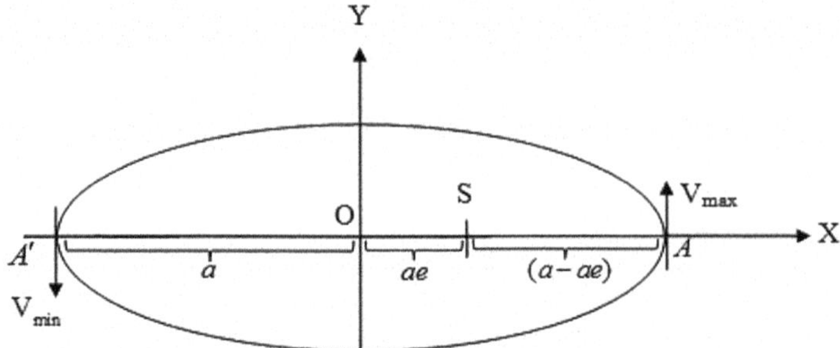

Fig. 3.8 Diagram depicting Example 3.5

It is known that v is maximum at A, where $r = a - ae$, and v is minimum at A′ where $r = a + ae$.
Therefore Eq. (3.2.30) gives

$$v_{\max}^2 = \mu\left[\frac{2}{a(1-e)} - \frac{1}{a}\right] = \frac{\mu(1+e)}{a(1-e)}, \tag{3.2.31}$$

and

$$v_{\min}^2 = \mu\left[\frac{2}{a(1+e)} - \frac{1}{a}\right] = \frac{\mu(1-e)}{a(1+e)}. \tag{3.2.32}$$

(3.2.31) ÷ (3.2.32) gives

$$\frac{v_{\max}^2}{v_{\min}^2} = \frac{\mu(1+e)}{a(1-e)} \times \frac{a(1+e)}{\mu(1-e)} = \frac{(1+e)^2}{(1-e)^2}.$$

$$\therefore \quad \frac{v_{\max}}{v_{\min}} = \frac{1+e}{1-e},$$

$$\text{i.e.,} \quad \frac{v_{\max} - v_{\min}}{v_{\max} + v_{\min}} = \frac{2e}{2} = e,$$

$$\text{i.e., eccentricity} = \frac{v_{\max} - v_{\min}}{v_{\max} + v_{\min}}.$$

Example 3.6 A satellite has its largest and smallest orbital speeds given by $v_{\max}$ and $v_{\min}$, respectively. If τ is the periodic time of the satellite, show that the length of the semi-major axis of the elliptic orbit is $\frac{\tau}{2\pi}\sqrt{v_{\max} v_{\min}}$.

Solution Let v be the speed of the satellite at a distance r from the Earth.

$$\therefore \quad v^2 = \mu\left(\frac{2}{r} - \frac{1}{a}\right). \tag{3.2.33}$$

Proceeding as in the previous example, we obtain

$$v_{\max}^2 = \mu\left[\frac{2}{a(1-e)} - \frac{1}{a}\right] = \mu\frac{2-(1-e)}{a(1-e)} = \frac{\mu(1+e)}{a(1-e)}, \tag{3.2.34}$$

and

$$v_{\min}^2 = \mu\left[\frac{2}{a(1+e)} - \frac{1}{a}\right] = \mu\frac{2-(1+e)}{a(1+e)} = \frac{\mu(1-e)}{a(1+e)}. \tag{3.2.35}$$

Now (3.2.34) × (3.2.35) gives

$$v_{\max}^2 v_{\min}^2 = \frac{\mu(1+e)}{a(1-e)}\frac{\mu(1-e)}{a(1+e)} = \frac{\mu^2}{a^2},$$

$$\therefore \quad v_{\max} v_{\min} = \frac{\mu}{a}. \tag{3.2.36}$$

Because τ is the orbital period,

$$\therefore \quad \tau = \frac{2\pi}{\sqrt{\mu}} a^{\frac{3}{2}},$$

$$\text{i.e.,} \quad \tau^2 = \frac{4\pi^2}{\mu} a^3,$$

$$\text{i.e.,} \quad \mu = \frac{4\pi^2}{\tau^2} a^3. \tag{3.2.37}$$

Equation (3.2.36) gives

$$v_{\max} v_{\min} = \frac{4\pi^2 a^3}{\tau^2 a} = \frac{4\pi^2 a^2}{\tau^2},$$

$$\therefore \quad a^2 = \frac{v_{\max} v_{\min} \tau^2}{4\pi^2},$$

$$\text{i.e.,} \quad a = \frac{\tau}{2\pi}\sqrt{v_{\max} v_{\min}},$$

which is the semi-major axis of the orbit.

Example 3.7 A particle describes an ellipse as a central orbit about the focus. Prove that the velocity at the end of the minor axis is the geometric mean between the velocities at the ends of any diameter.

Solution Let $2a$ be the major and $2b$ be the minor axis of the ellipse given by

$$\frac{l}{r} = 1 + e\cos\theta \tag{3.2.38}$$

referred to the focus S $(ae,\ \theta)$ as the pole and the major axis A′A as the initial line.

Let O be the center of the ellipse and PQ be any diameter of the ellipse. Take O as the origin, the X-axis along **OA**, and the Y-axis along **OB**. Let ϕ be the eccentric angle of P.

$$\therefore \quad \mathrm{P} = (a\cos\phi,\ b\sin\phi)\,.$$

Let Q $= (\alpha,\ \beta)$. Therefore, O(0, 0) is the midpoint of PQ.

$$\therefore \quad 0 = \frac{a\cos\phi + \alpha}{2}, \tag{3.2.39}$$

&

$$0 = \frac{b\sin\phi + \beta}{2}. \tag{3.2.40}$$

(3.2.39) and (3.2.40) gives

$$\alpha = -a\cos\phi, \quad \beta = -b\sin\phi.$$

From Fig. 3.9, we have

$$\mathrm{SB}^2 = a^2e^2 + b^2 = a^2e^2 + a^2\left(1 - e^2\right) = a^2\left(e^2 + 1 - e^2\right) = a^2,$$

$$\therefore \quad \mathrm{SB} = a. \tag{3.2.41}$$

Let SP $= r_1$ and SQ $= r_2$.

Now

$$r_1^2 = \mathrm{SP}^2 = (a\cos\phi - ae)^2 + (b\sin\phi)^2 = a^2(\cos\phi - e)^2 + b^2\sin^2\phi.$$

$$\text{i.e.,}\quad r_1^2 = a^2\left[(\cos\phi - e)^2 + \left(1 - e^2\right)\sin^2\phi\right],$$

$$\text{i.e.,}\quad r_1^2 = a^2\left[1 - 2e\cos\phi + e^2\cos^2\phi\right] = a^2(1 - e\cos\phi)^2,$$

$$\therefore \quad r_1 = a\,(1 - e\cos\phi)\,. \tag{3.2.42}$$

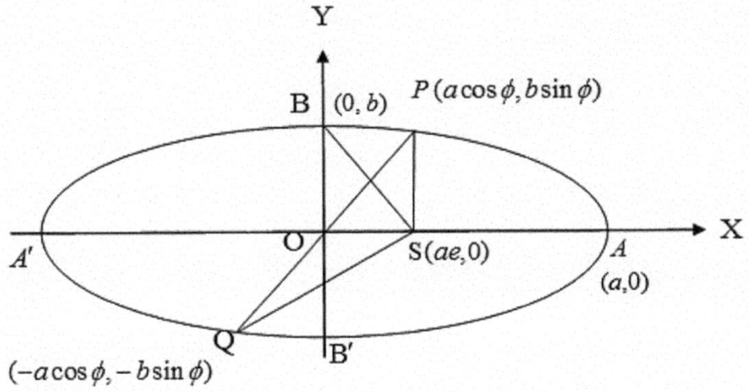

Fig. 3.9 Diagram depicting Example 3.7

Similarly

$$r_2 = \mathrm{SQ} = a\,(1 + e\cos\phi)\,. \tag{3.2.43}$$

The velocity of the particle at a point R $(r,\ \theta)$ on the ellipse is given by

$$v^2 = \mu\left(\frac{2}{r} - \frac{1}{a}\right). \tag{3.2.44}$$

Let $v = v_0$ at B, where $r = a$.
Equation (3.2.44) gives

$$v_0^2 = \mu\left(\frac{2}{a} - \frac{1}{a}\right) = \frac{\mu}{a}. \tag{3.2.45}$$

Let $v = v_1$ at P, where $r = r_1$ and $v = v_2$ at Q, where $r = r_2$.
Equation (3.2.44) gives

$$v_1^2 = \mu\left(\frac{2}{r_1} - \frac{1}{a}\right) = \mu\left[\frac{2}{a\,(1 - e\cos\phi)} - \frac{1}{a}\right] = \frac{\mu}{a}\frac{1 + e\cos\phi}{1 - e\cos\phi}. \tag{3.2.46}$$

Similarly

$$v_2^2 = \frac{\mu}{a}\cdot\frac{1 - e\cos\phi}{1 + e\cos\phi}. \tag{3.2.47}$$

(3.2.46) × (3.2.47) gives

$$v_1^2 v_2^2 = \frac{\mu^2}{a^2},$$

$$\text{i.e.,} \quad v_1 v_2 = \frac{\mu}{a},$$

$$\text{i.e.,} \quad v_1 v_2 = v_0^2,$$

$$\text{i.e.,} \quad v_0 = \sqrt{v_1 v_2},$$

$$\text{i.e.,} \quad v_0 = \text{Geometric mean of } v_1,\ v_2.$$

Hence proved.

Example 3.8 If ω be the angular velocity of a planet at the nearer end of the major axis, prove that its period is $\frac{2\pi}{\omega}\sqrt{\frac{1+e}{(1-e)^3}}$.

Solution For a central orbit, we have

$$r^2\dot{\theta} = h = \sqrt{\mu l}. \tag{3.2.48}$$

Given $\dot{\theta} = \omega$ when $r = a\,(1-e)$.

Therefore, Eq. (3.2.48) gives

$$a^2(1-e)^2\omega = \sqrt{\mu l},$$

$$\text{i.e.,} \quad a^4(1-e)^4\omega^2 = \mu l = \mu\frac{b^2}{a} = \mu\frac{a^2\left(1-e^2\right)}{a} = \mu a\left(1-e^2\right),$$

$$\text{i.e.,} \quad a^3(1-e)^3\omega^2 = \mu\,(1+e)\,.$$

$$\therefore \quad \mu = \frac{a^3(1-e)^3\omega^2}{1+e}.$$

Therefore, the Orbital period is given by

$$T = \frac{2\pi}{\sqrt{\mu}}a^{\frac{3}{2}} = 2\pi a^{\frac{3}{2}}\frac{\sqrt{1+e}}{\sqrt{a^3(1-e)^3\omega^2}} = \frac{2\pi}{\omega}\sqrt{\frac{1+e}{(1-e)^3}}.$$

Example 3.9 A particle describes an ellipse under a force $\frac{\mu}{(\text{distance})^2}$ towards the focus; if it was projected with velocity V from a point at a distance r from the center of force, show that its periodic time is $\frac{2\pi}{\sqrt{\mu}}\left[\frac{2}{r} - \frac{V^2}{\mu}\right]^{-\frac{3}{2}}$.

Solution Let v be the velocity of the particle at a distance r' from the center of force.

$$\therefore \quad v^2 = \mu\left(\frac{2}{r'} - \frac{1}{a}\right), \tag{3.2.49}$$

where $2a$ is the major axis of the orbit.

Given $v = V$ when $r' = r$.

$$\therefore \quad V^2 = \mu\left(\frac{2}{r} - \frac{1}{a}\right),$$

$$\text{i.e.,} \quad \frac{V^2}{\mu} = \frac{2}{r} - \frac{1}{a},$$

$$\text{i.e.,} \quad \frac{1}{a} = \frac{2}{r} - \frac{V^2}{\mu},$$

$$\text{i.e.,} \quad a = \left(\frac{2}{r} - \frac{V^2}{\mu}\right)^{-1}. \tag{3.2.50}$$

Let T be the periodic time.

$$\therefore \quad T = \frac{2\pi}{\sqrt{\mu}} a^{\frac{3}{2}} = \frac{2\pi}{\sqrt{\mu}}\left(\frac{2}{r} - \frac{V^2}{\mu}\right)^{-\frac{3}{2}}.$$

Example 3.10 Show that, in elliptic motion about a focus under attraction μr^{-2}, the radial velocity is given by the equation $r^2\left(\frac{dr}{dt}\right)^2 = \frac{\mu}{a}\left\{a(1+e) - r\right\}\left\{r - a(1-e)\right\}$.

Solution Let the equation of the elliptic orbit be

$$\frac{l}{r} = 1 + e\cos\theta. \tag{3.2.51}$$

Differentiating (3.2.51) w.r.t. t, we get

$$\frac{l}{r^2}\dot{r} = e\sin\theta\,\dot{\theta},$$

$$\text{i.e.,} \quad \dot{r} = \frac{e\sin\theta}{l} r^2\dot{\theta} = \frac{e\sin\theta}{l} h,$$

$$\text{i.e.,} \quad \dot{r}^2 = \frac{e^2\sin^2\theta}{l^2} h^2 = \frac{e^2\sin^2\theta}{l^2}\mu l = \frac{e^2\sin^2\theta\,\mu}{l}\dot{r}^2 = \frac{e^2\sin^2\theta}{l^2} h^2,$$

$$\text{i.e.,}\quad \dot{r}^2 = \frac{e^2\mu\left(1-\cos^2\theta\right)}{l} = \frac{e^2\mu}{l}\left[1-\left(\frac{\left(\frac{l}{r}-1\right)^2}{e^2}\right)\right],$$

$$\text{i.e.,}\quad \dot{r}^2 = \frac{\mu}{l}\left[e^2-\left(\frac{l}{r}-1\right)^2\right] = \frac{\mu}{l}\left[\left(e+\frac{l}{r}-1\right)\left(e-\frac{l}{r}+1\right)\right],$$

$$\therefore\quad r^2\dot{r}^2 = \frac{\mu}{l}\left\{l-r\left(1-e\right)\right\}\left\{\left(e+1\right)r-l\right\},$$

$$\text{i.e.,}\quad r^2\dot{r}^2 = \frac{\mu}{b^2a}\left\{b^2-ar\left(1-e\right)\right\}\left\{a\left(1+e\right)r-b^2\right\},$$

$$\text{i.e.,}\quad r^2\dot{r}^2 = \frac{\mu}{ab^2}\left\{a^2\left(1-e^2\right)-ar\left(1-e\right)\right\}\left\{a\left(1+e\right)r-a^2\left(1-e^2\right)\right\},$$

$$\text{i.e.,}\quad r^2\dot{r}^2 = \frac{\mu}{ab^2}a\left(1-e\right)\left\{a\left(1+e\right)-r\right\}a\left(1+e\right)\left\{r-a\left(1-e\right)\right\},$$

$$\text{i.e.,}\quad r^2\dot{r}^2 = \frac{\mu}{ab^2}a^2\left(1-e^2\right)\left\{a\left(1+e\right)-r\right\}\left\{r-a\left(1-e\right)\right\},$$

$$\text{i.e.,}\quad r^2\dot{r}^2 = \frac{\mu b^2}{ab^2}\left\{a\left(1+e\right)-r\right\}\left\{r-a\left(1-e\right)\right\} = \frac{\mu}{a}\left\{a\left(1+e\right)-r\right\}\left\{r-a\left(1-e\right)\right\}.$$

Example 3.11 A particle is describing an ellipse of eccentricity e about a center of force at a focus. Prove, with the usual rotation $v^2 = \mu\left(\frac{2}{r}-\frac{1}{a}\right),\ \ h^2 = \mu a\left(1-e^2\right).$

Solution We have

$$h^2 = \mu l = \mu\frac{b^2}{a} = \mu\frac{a^2\left(1-e^2\right)}{a} = \mu a\left(1-e^2\right).$$

Let the equation of the ellipse with the focus (center of force) as the pole and the major axis as the initial line be

$$\frac{l}{r} = 1+e\cos\theta. \tag{3.2.52}$$

Equation (3.2.52) can be written as

$$lu = 1+e\cos\theta,\quad u=\frac{1}{r},$$

$$\text{i.e.,}\quad u = \frac{1+e\cos\theta}{l}. \tag{3.2.53}$$

Differentiating (3.2.53) w.r.t. θ, we get

$$\frac{du}{d\theta} = \frac{-e \sin \theta}{l}. \tag{3.2.54}$$

The velocity of the particle at a point (r, θ) is given by

$$v^2 = h^2 \left[\left(\frac{du}{d\theta} \right)^2 + u^2 \right] = \mu a \left(1 - e^2\right) \left[\frac{e^2 \sin^2 \theta}{l^2} + \frac{(1 + e \cos \theta)^2}{l^2} \right]. \tag{3.2.55}$$

We have for an ellipse,

$$b^2 = a^2 \left(1 - e^2\right),$$

$$\text{i.e.,} \quad 1 - e^2 = \frac{b^2}{a^2} = \frac{al}{a^2} = \frac{l}{a}. \tag{3.2.56}$$

Therefore, Eq. (3.2.55) reduces to

$$v^2 = \mu a \frac{l}{a} \frac{1}{l^2} \left[1 + 2e \cos \theta + e^2\right] = \frac{\mu}{l} \left[1 + e^2 + 2e \cos \theta\right],$$

$$\text{i.e.,} \quad v^2 = \frac{\mu}{l} \left[1 + e^2 + 2 \left(\frac{l}{r} - 1 \right) \right] = \frac{\mu}{l} \left[\frac{2l}{r} - 1 + e^2 \right] = \frac{\mu}{l} \left[\frac{2l}{r} - \left(1 - e^2\right) \right],$$

$$\text{i.e.,} \quad v^2 = \frac{\mu}{l} \left[\frac{2l}{r} - \frac{l}{a} \right] = \mu \left[\frac{2}{r} - \frac{1}{a} \right], \quad \text{with } h^2 = \mu a \left(1 - e^2\right).$$

This is the required result.

Example 3.12 Let v_0 be the speed at $r = a$ of a particle subject to a central force $f(r)$ per unit mass and let v be the velocity at a distance r from the center of attraction. Show that $v^2 = v_0^2 - V^2 + 2 \int_r^\infty f(r)\, dr$, where $V^2 - 2 \int_a^\infty f(r)\, dr$.

Solution The central force acting on the particle of mass is given by

$$\mathbf{F} = mf(r)\left(-\hat{r}\right) = -mf(r) \frac{\mathbf{r}}{r}. \tag{3.2.57}$$

Let Ω be the potential energy at a distance r from the center of force.

$$\therefore \quad \mathbf{F} = -\nabla \Omega. \tag{3.2.58}$$

Therefore, Eq. (3.2.57) gives

$$-\nabla\Omega = -mf(r)\frac{\mathbf{r}}{r},$$

$$\text{i.e.,}\quad \nabla\Omega\cdot d\mathbf{r} = mf(r)\frac{\mathbf{r}\cdot d\mathbf{r}}{r},$$

$$\text{i.e.,}\quad d\Omega = mf(r)\,dr,$$

$$\text{i.e.,}\quad \int_0^\Omega d\Omega = m\int_\infty^r f(r)\,dr,$$

$$\text{i.e.,}\quad \Omega = m\int_\infty^r f(r)\,dr. \tag{3.2.59}$$

We note that a central force is conservative and hence under a central force field,

K.E. + P.E. = a constant = E = Total Energy

$$\text{i.e.,}\quad \frac{1}{2}mv^2 + m\int_\infty^r f(r)\,dr = \text{E}. \tag{3.2.60}$$

But $v = v_0$ *at* $r = a$ and hence Eq. (3.2.60) gives

$$\frac{1}{2}mv_0^2 + m\int_\infty^a f(r)\,dr = \text{E}. \tag{3.2.61}$$

Now (3.2.60)–(3.2.61) gives

$$\frac{1}{2}mv^2 - \frac{1}{2}mv_0^2 + m\int_\infty^r f(r)\,dr - m\int_\infty^a f(r)\,dr = 0,$$

$$\text{i.e.,}\quad v^2 - v_0^2 + 2\int_\infty^r f(r)\,dr - 2\int_\infty^a f(r)\,dr = 0,$$

$$\text{i.e.,}\quad v^2 = v_0^2 - 2\int_\infty^r f(r)\,dr + 2\int_\infty^a f(r)\,dr,$$

$$\text{i.e.,}\quad v^2 = v_0^2 + 2\int_r^\infty f(r)\,dr - 2\int_a^\infty f(r)\,dr,$$

$$\text{i.e.,}\quad v^2 = v_0^2 - V^2 + 2\int_r^\infty f(r)\,dr,$$

where $V^2 = 2\int_r^\infty f(r)\,dr$.

Example 3.13 A particle describes a rectangular hyperbola under the action of a force to the center; prove that if the radius vector, during the time t after leaving the apse passes over an angle θ, $\tan\theta = \tan h\sqrt{\mu t}$, μr being the force at distance r on unit mass.

Solution Let the equation of the hyperbola be

$$x^2 - y^2 = a^2. \tag{3.2.62}$$

Let P be the position of the particle at time t with position vector **r** relative to O, (r, θ) the polar coordinates of P relative to O as pole and **OX** as initial line and (x, y) the Cartesian coordinates of P (Fig. 3.10).

$$\therefore \left.\begin{array}{l} x = r\cos\theta \\ y = r\sin\theta \end{array}\right\}. \tag{3.2.63}$$

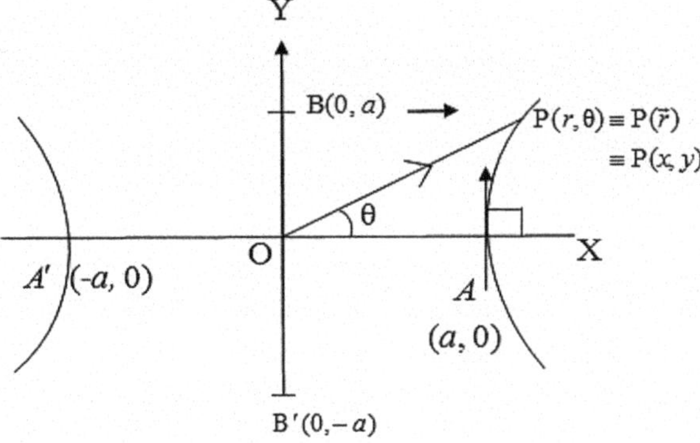

Fig. 3.10 Illustrating diagram for Example 3.13

The equation of motion of the particle is

$$
\begin{aligned}
& m\ddot{\mathbf{r}} = \mathbf{F}, \\
\text{i.e.,}\quad & m\ddot{\mathbf{r}} = m\mu r\,\hat{r}, \\
\text{i.e.,}\quad & \ddot{\mathbf{r}} = \mu\mathbf{r}.
\end{aligned} \tag{3.2.64}
$$

We have

$$
\left.\begin{aligned}
\mathbf{r} &= x\hat{i} + y\hat{j} \\
\ddot{\mathbf{r}} &= \ddot{x}\hat{i} + \ddot{y}\hat{j}
\end{aligned}\right\}. \tag{3.2.65}
$$

Substituting (3.2.65) in (3.2.64), we obtain

$$
\ddot{x}\hat{i} + \ddot{y}\hat{j} = \mu\left(x\hat{i} + y\hat{j}\right). \tag{3.2.66}
$$

By equating coefficient of $\hat{i}$, in (3.2.66) we get

$$
\ddot{x} = \mu x,
$$

$$
\text{i.e.,}\quad \frac{d^2x}{dt^2} - \mu x = 0,
$$

$$
\text{i.e.,}\quad \left(D^2 - \mu\right)x = 0. \tag{3.2.67}
$$

The general solution of (3.2.67) is

$$
x = C_1 e^{\sqrt{\mu}t} + C_2 e^{-\sqrt{\mu}t}, \tag{3.2.68}
$$

$$
\therefore\quad \dot{x} = C_1\sqrt{\mu}e^{\sqrt{\mu}t} - C_2\sqrt{\mu}e^{-\sqrt{\mu}t}. \tag{3.2.69}
$$

At A, $x = a$, $\dot{x} = 0$ at $t = 0$.
Therefore, Eqs. (3.2.68) and (3.2.69) give

$$
C_1 = C_2 = \frac{a}{2}. \tag{3.2.70}
$$

Thus, (3.2.68) gives

$$
x = a\frac{e^{\sqrt{\mu}t} + e^{-\sqrt{\mu}t}}{2} = a\cosh\sqrt{\mu}t.
$$

$$
\because\quad \mathrm{P}(x, y) \text{ is on } x^2 - y^2 = a^2,
$$

$$\therefore \quad a^2 \cosh \sqrt{\mu}t - y^2 = a^2,$$

$$\therefore \quad y^2 = a^2 \sinh^2 \sqrt{\mu}t,$$

$$\therefore \quad y = a \sinh \sqrt{\mu}t.$$

Equation (3.2.63) gives

$$\frac{y}{x} = \tan\theta.$$

$$\text{i.e.,} \quad \tan\theta = \frac{a \sinh \sqrt{\mu}t}{a \cosh \sqrt{\mu}t} = \tanh\left(\sqrt{\mu}t\right).$$

Example 3.14 A particle is describing an ellipse under a force to a focus S. Prove that the product of speeds when the particle is at the ends of any diameter of the ellipse, is constant and equal to $\frac{4\pi^2 a^2}{T^2}$ where T is the periodic time.

Solution Let the equation of the ellipse referred to the focus S as pole and SX (major axis) as initial line be

$$\frac{l}{r} = 1 + e\cos\theta. \tag{3.2.71}$$

Let PP′ be a diameter of the ellipse (3.2.71) joining the points P $(a\cos\varphi,\ b\sin\varphi)$ and P′ $(-a\cos\varphi, -b\sin\varphi)$. In Fig. 3.11, let $\mathrm{SP} = r_1$ *and* $\mathrm{SP}' = r_2$.

The velocity of the particle at the point Q (r, θ) is given by

$$v^2 = \mu\left(\frac{2}{r} - \frac{1}{a}\right). \tag{3.2.72}$$

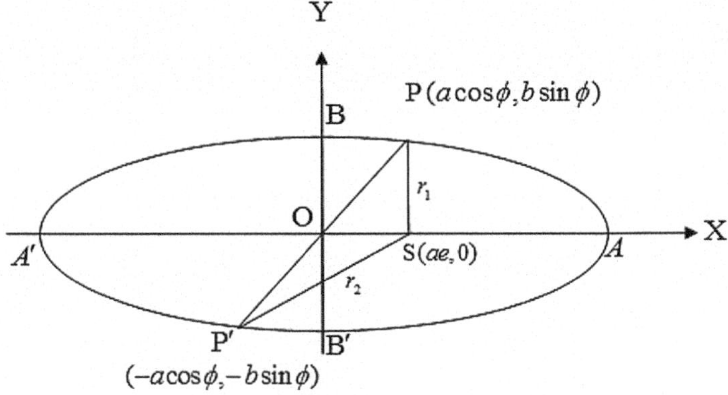

Fig. 3.11 Illustrating diagram for Example 3.14

Now

$$
\begin{aligned}
r_1^2 = \mathrm{SP}^2 &= (a\cos\varphi - ae)^2 + (b\sin\varphi)^2,\\
&= a^2(\cos\varphi - e)^2 + a^2\left(1 - e^2\right)\sin^2\varphi,\\
&= a^2\left[(\cos\varphi - e)^2 + \left(1 - e^2\right)\sin^2\varphi\right],\\
&= a^2\left[1 - 2e\cos\varphi + e^2 - e^2\sin^2\varphi\right],\\
&= a^2\left[1 - 2e\cos\varphi + e^2\cos^2\varphi\right],\\
&= a^2(1 - e\cos\varphi)^2,
\end{aligned}
$$

$$
\therefore \quad r_1 = a\,(1 - e\cos\varphi)\,.
$$

Similarly

$$
r_2 = \mathrm{SP}' = a\,(1 + e\cos\varphi)\,.
$$

Let $v = v_1$ at P and $v = v_2$ at P′.
i.e., $v = v_1$ for $r = r_1$ and $v = v_2$ for $r = r_2$.
Thus, Eq. (3.2.72) gives

$$
v_1^2 = \mu\left(\frac{2}{r_1} - \frac{1}{a}\right) = \mu\left[\frac{2}{a\,(1 - e\cos\varphi)} - \frac{1}{a}\right] = \frac{\mu}{a}\frac{1 + e\cos\varphi}{1 - e\cos\varphi}. \tag{3.2.73}
$$

Similarly,

$$
v_2^2 = \frac{\mu}{a}\frac{1 - e\cos\varphi}{1 + e\cos\varphi}. \tag{3.2.74}
$$

(3.2.73) × (3.2.74) gives

$$
v_1^2\,v_2^2 = \frac{\mu^2}{a^2},
$$

$$
\therefore \; v_1\,v_2 = \frac{\mu}{a}. \tag{3.2.75}
$$

Given that T is the orbital period.

$$
\therefore \quad T = \frac{2\pi}{\sqrt{\mu}}a^{\frac{3}{2}},
$$

$$
\text{i.e.,} \quad \sqrt{\mu} = \frac{2\pi}{T}a^{\frac{3}{2}},
$$

$$\text{i.e.,}\quad \mu = \frac{4\pi^2}{T^2}a^3.$$

Therefore, Eq. (3.2.75) gives

$$v_1 v_2 = \frac{1}{a}\frac{4\pi^2 a^3}{T^2} = \frac{4\pi^2 a^2}{T^2}.$$

Example 3.15 A planet is describing an elliptical orbit about the sun. When it is at the end of the minor axis, suddenly its velocity is increased by half of its original velocity. Show that its orbit will be a hyperbola of eccentricity $\frac{1}{4}\sqrt{25 - 9e^2}$, where e is the eccentricity of the original orbit.

Solution Let the equation of the original elliptic orbit referred to the sun (focus) as the pole and the positive major axis as the initial line be

$$\frac{l}{r} = 1 + e\cos\theta. \tag{3.2.76}$$

The velocity of the planet at a distance r from the sun is given by

$$v^2 = \mu\left(\frac{2}{r} - \frac{1}{a}\right). \tag{3.2.77}$$

Let v_0 be the velocity of the planet at B, the end of the minor axis. Let $r_0 = \text{SB}$. From Fig. 3.12, we have

$$r_0^2 = \text{SB}^2 = a^2e^2 + b^2 = a^2e^2 + a^2\left(1 - e^2\right) = a^2,$$

$$\therefore\quad r_0 = a.$$

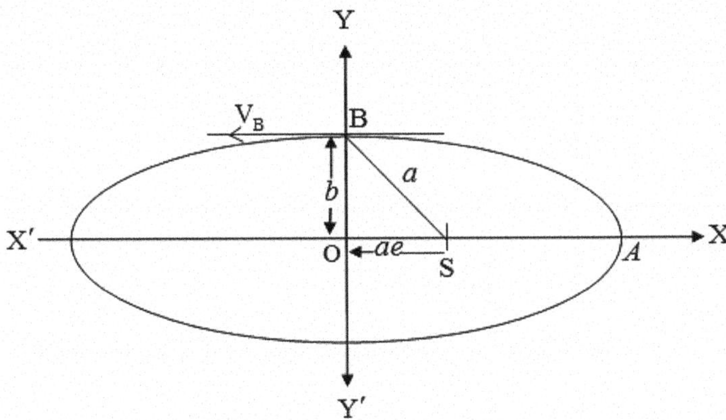

Fig. 3.12 Illustrating diagram for Example 3.15

Thus, $v = v_0$ at $r = a$.
Equation (3.2.77) gives

$$v_0{}^2 = \mu\left(\frac{2}{a} - \frac{1}{a}\right) = \frac{\mu}{a},$$

$$\therefore \quad v_0 = \sqrt{\frac{\mu}{a}}. \tag{3.2.78}$$

In the 2nd case, as per question,

$$v_B = v_0 + \frac{1}{2}v_0 = \frac{3}{2}v_0.$$

We have

$$vp = h',$$

$$\text{i.e.,} \quad v_B b = \sqrt{\mu l'},$$

$$\text{i.e.,} \quad \frac{3}{2}v_0 b = \sqrt{\mu l'},$$

$$\text{i.e.,} \quad \frac{3}{2}\sqrt{\frac{\mu}{a}}b = \sqrt{\mu l'},$$

$$\text{i.e.,} \quad l' = \frac{9b^2}{4a}. \tag{3.2.79}$$

The velocity v' for the 2nd orbit at a distance r from the sun is given by

$$v'^2 = \frac{2\mu}{r} + C. \tag{3.2.80}$$

But $v' = v_B$ when $r = a$.
Equation (3.2.80) gives

$$v_B^2 = \frac{2\mu}{a} + C,$$

$$\text{i.e.,} \quad \frac{9}{4}v_0^2 = \frac{2\mu}{a} + C,$$

$$\text{i.e.,} \quad \frac{9\mu}{4a} = \frac{2\mu}{a} + C,$$

$$\text{i.e.,}\quad C = \frac{\mu}{4a} > 0.$$

Therefore, the new orbit is a hyperbola of transverse axis a' (say) . Thus, the velocity field for the new orbit is given by

$$v'^2 = \mu\left(\frac{2}{r} + \frac{1}{a'}\right). \tag{3.2.81}$$

But $v' = v_B = \frac{3}{2}v_0 = \frac{3}{2}\sqrt{\frac{\mu}{a}}$, when $r = a$.

Therefore, Eq. (3.2.81) gives

$$\frac{9\mu}{4a} = \mu\left(\frac{2}{a} + \frac{1}{a'}\right),$$

$$\text{i.e.,}\quad \frac{9}{4a} = \frac{2}{a} + \frac{1}{a'},$$

$$\text{i.e.,}\quad a' = 4a.$$

But $l' = \frac{9b^2}{4a}$, (by (3.2.79))

$$\text{i.e.,}\quad \frac{b'^2}{a'} = \frac{9b^2}{4a},$$

$$\text{i.e.,}\quad \frac{b'^2}{a'} = \frac{9a^2\left(1 - e^2\right)}{4a} = \frac{9a\left(1 - e^2\right)}{4},$$

$$\text{i.e.,}\quad \frac{a'^2\left(e'^2 - 1\right)}{a'} = \frac{9a\left(1 - e^2\right)}{4},$$

$$\text{i.e.,}\quad a'\left(e'^2 - 1\right) = \frac{9a\left(1 - e^2\right)}{4},$$

$$\text{i.e.,}\quad 4a\left(e'^2 - 1\right) = \frac{9a\left(1 - e^2\right)}{4},$$

$$\text{i.e.,}\quad 16\left(e'^2 - 1\right) = 9\left(1 - e^2\right),$$

$$\text{i.e.,}\quad e'^2 = \frac{25 - 9e^2}{16},$$

$$\text{i.e.,}\quad e' = \frac{\sqrt{25 - 9e^2}}{4}.$$

Example 3.16 A particle of unit mass describes an ellipse under the action of a central force μr. Show that the normal component of the acceleration at any instant is $\dfrac{ab\mu^{\frac{3}{2}}}{v}$, where v is the velocity at that instant and a, b are the semi-axes of the ellipse.

Solution The equation of the ellipse with the center O as origin, X-axis along the major axis and Y-axis along the minor axis is

$$\frac{x^2}{a^2} + \frac{y^2}{b^2} = 1. \tag{3.2.82}$$

In Fig. 3.13, let P (x, y) be the position of the particle at time t with position vector $\mathbf{r}$ relative to O.

$$\therefore \quad \mathbf{r} = \mathbf{OP} = \text{P.V. of P}\,(x, y) = x\hat{i} + y\hat{j}.$$

Equation of motion of the particle is

$$m\ddot{\mathbf{r}} = m\mu r\left(-\hat{r}\right),$$

$$\text{i.e.,} \quad \ddot{\mathbf{r}} = -\mu\mathbf{r},$$

$$\text{i.e.,} \quad \ddot{x}\hat{i} + \ddot{y}\hat{j} = -\mu\left(x\hat{i} + y\hat{j}\right). \tag{3.2.83}$$

Equation (3.2.83) gives

$$\left.\begin{aligned} \ddot{x} &= -\mu x \\ \ddot{y} &= -\mu y \end{aligned}\right\}. \tag{3.2.84}$$

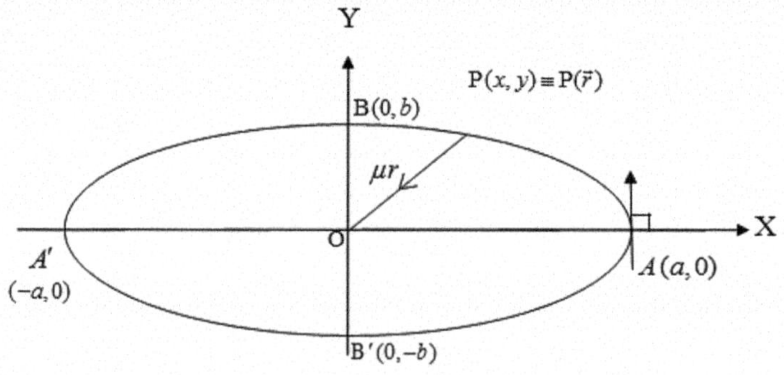

Fig. 3.13 Illustration of Example 3.16

1st equation of (3.2.84) can be written as

$$\frac{d^2x}{dt^2} + \mu x = 0,$$

$$\text{i.e.,} \quad \left(D^2 + \mu\right) x = 0. \tag{3.2.85}$$

G.S. of (3.2.85) is

$$x = A \cos \sqrt{\mu t} + \sin \sqrt{\mu t}, \tag{3.2.86}$$

with

$$\dot{x} = -A\sqrt{\mu} \sin \sqrt{\mu t} + B\sqrt{\mu} \cos \sqrt{\mu t}. \tag{3.2.87}$$

At A, $x = a, \quad \dot{x} = 0, \quad t = 0.$
Therefore, Eqs. (3.2.86) and (3.2.87) give

$$A = a, \quad B = 0.$$

$$\therefore \quad x = a \cos \sqrt{\mu t}. \tag{3.2.88}$$

Similarly,

$$y = b \sin \sqrt{\mu t}. \tag{3.2.89}$$

$$\therefore \quad \mathbf{r} = x\hat{i} + y\hat{j} = a \cos \sqrt{\mu t}\, \hat{i} + b \sin \sqrt{\mu t}\, \hat{j}.$$

$$\therefore \quad \mathbf{v} = \dot{\mathbf{r}} = -a\sqrt{\mu} \sin \sqrt{\mu t}\, \hat{i} + b\sqrt{\mu} \cos \sqrt{\mu t}\, \hat{j}, \tag{3.2.90}$$

and

$$\mathbf{f} = \dot{\mathbf{v}} = -a\mu \cos \sqrt{\mu t}\, \hat{i} - b\mu \sin \sqrt{\mu t}\, \hat{j}, \tag{3.2.91}$$

$$\hat{t} = \frac{\mathbf{v}}{|\mathbf{v}|} = \frac{-a \sin \sqrt{\mu t}\, \hat{i} + b \cos \sqrt{\mu t}\, \hat{j}}{\sqrt{a^2 \sin^2 \sqrt{\mu t} + b^2 \cos^2 \sqrt{\mu t}}} = \frac{-a \sin \sqrt{\mu t}\, \hat{i} + b \cos \sqrt{\mu t}\, \hat{j}}{\lambda}; \tag{3.2.92}$$

$$\lambda = \sqrt{a^2 \sin^2 \sqrt{\mu t} + b^2 \cos^2 \sqrt{\mu t}}.$$

Now

$$\mathbf{f} \times \hat{t} = \frac{1}{\lambda} \begin{vmatrix} \hat{i} & \hat{j} & \hat{k} \\ -a\mu \cos \sqrt{\mu}t & -b\mu \sin \sqrt{\mu}t & 0 \\ -a \sin \sqrt{\mu}t & b \cos \sqrt{\mu}t & 0 \end{vmatrix} = -\frac{ab\mu}{\lambda}\hat{k}.$$

$$\therefore \quad f_n = \text{Normal acceleration at P}\,(x,\ y) = \left|\mathbf{f} \times \hat{t}\right| = \frac{ab\mu}{\lambda}. \tag{3.2.93}$$

Equation (3.2.90) yields

$$v = |\mathbf{v}| = \sqrt{a^2\mu \sin^2 \sqrt{\mu}t + b^2\mu \cos^2 \sqrt{\mu}t} = \sqrt{\mu}\sqrt{a^2 \sin^2 \sqrt{\mu}t + b^2 \cos^2 \sqrt{\mu}t} = \sqrt{\mu}\,\lambda,$$

$$\therefore \quad \lambda = \frac{v}{\sqrt{\mu}}.$$

Equation (3.2.93) reduces to $f_n = \frac{ab\mu\sqrt{\mu}}{v} = \frac{ab\mu^{\frac{3}{2}}}{v}$. Hence the required result.

Example 3.17 A particle describes an ellipse about a center of force at the center. Show that the law of force is $P = \mu r$.

Solution Let O be the center of the ellipse with $2a$ and $2b$ as axes ($a > b$). Take O as the origin, X-axis along **OA**, and Y-axis along **OB**.

The equation of the ellipse is

$$\frac{x^2}{a^2} + \frac{y^2}{b^2} = 1. \tag{3.2.94}$$

In Fig. 3.14, let $(r,\ \theta)$ be the polar coordinates of $(x,\ y)$ referred to O as a pole and **OX** as the initial line. $(x,\ y)$ and $(r,\ \theta)$ are corrected by the relation

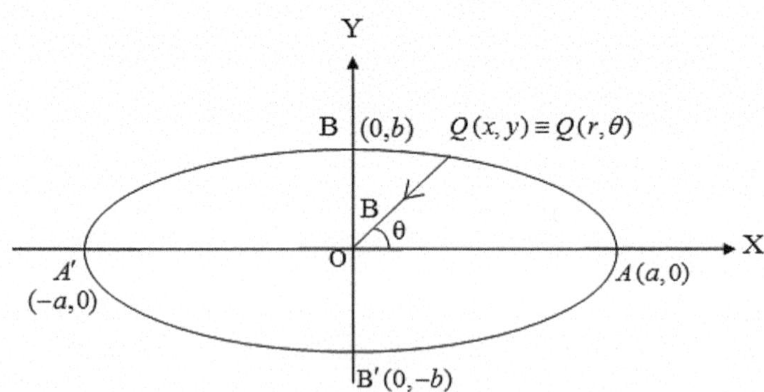

Fig. 3.14 Schematic diagram for Example 3.17

$$\left.\begin{array}{l} x = r\cos\theta \\ y = r\sin\theta \end{array}\right\}. \tag{3.2.95}$$

Therefore, the polar equation of the ellipse, referred to the center of the ellipse as the pole and the +ve major axis as the initial line, is

$$\frac{r^2\cos^2\theta}{a^2} + \frac{r^2\sin^2\theta}{b^2} = 1,$$

$$\text{i.e.,}\quad b^2r^2\cos^2\theta + a^2r^2\sin^2\theta = a^2b^2,$$

$$\text{i.e.,}\quad r^2 = \frac{a^2b^2}{b^2\cos^2\theta + a^2\sin^2\theta},$$

$$\text{i.e.,}\quad u^2 = \frac{b^2\cos^2\theta + a^2\sin^2\theta}{a^2b^2}, \tag{3.2.96}$$

where $u = \frac{1}{r}$. Equation (3.2.96) gives

$$2u\frac{du}{d\theta} = \frac{1}{a^2b^2}\left[-2b^2\cos\theta\sin\theta + 2a^2\sin\theta\cos\theta\right] = \frac{1}{a^2b^2}\left(a^2 - b^2\right)\sin 2\theta.$$

$$\therefore\quad \frac{du}{d\theta} = \frac{\left(a^2 - b^2\right)\sin 2\theta}{2a^2b^2u}. \tag{3.2.97}$$

Differentiating (3.2.97) w.r.t., θ we obtain

$$\begin{aligned}
\frac{d^2u}{d\theta^2} &= \frac{a^2 - b^2}{2a^2b^2u^2}\left[2u\cos 2\theta - \sin 2\theta\frac{du}{d\theta}\right], \\
&= \frac{a^2 - b^2}{2a^2b^2u^2}\left[2u\cos 2\theta - \sin 2\theta\,\frac{\left(a^2 - b^2\right)\sin 2\theta}{2a^2b^2u}\right], \\
&= \frac{\left(a^2 - b^2\right)\left[4a^2b^2u^2\cos 2\theta - \left(a^2 - b^2\right)\sin^2 2\theta\right]}{4a^4b^4u^3}, \\
&= \frac{a^2 - b^2}{4a^4b^4u^3}\left[4\left(b^2\cos^2\theta + a^2\sin^2\theta\right)\cos 2\theta - \left(a^2 - b^2\right)\sin^2 2\theta\right], \\
&= \frac{a^2 - b^2}{4a^4b^4u^3}\left[b^2\left(\sin^2 2\theta + 4\cos^2\theta\cos 2\theta\right) + a^2\left(4\sin^2\theta\cos 2\theta - \sin^2 2\theta\right)\right], \\
&= \frac{a^2 - b^2}{4a^4b^4u^3}\left[4b^2\cos^2\theta\left(\sin^2\theta + \cos 2\theta\right) + 4a^2\sin^2\theta\left(\cos 2\theta - \cos^2\theta\right)\right], \\
&= \frac{4\left(a^2 - b^2\right)}{4a^4b^4u^3}\left[b^2\cos^2\theta\cos^2\theta + a^2\sin^2\theta\left(-\sin^2\theta\right)\right],
\end{aligned}$$

$$= \frac{a^2 - b^2}{a^4 b^4 u^3}\left[b^2 \cos^4\theta - a^2 \sin^4\theta\right].$$

The d.e. of the central orbit is

$$\begin{aligned}
\frac{P}{h^2u^2} &= \frac{d^2u}{d\theta^2} + u, \\
&= \frac{a^2 - b^2}{a^4 b^4 u^3}\left[b^2 \cos^4\theta - a^2 \sin^4\theta\right] + u, \\
&= \frac{\left(a^2 - b^2\right)\left(b^2 \cos^4\theta - a^2 \sin^4\theta\right) + a^4 b^4 u^3}{a^4 b^4 u^3}, \\
&= \frac{\left(a^2 - b^2\right)\left(b^2 \cos^4\theta - a^2 \sin^4\theta\right) + \left(a^2 b^2 u^2\right)^2}{a^4 b^4 u^3}, \\
&= \frac{\left(a^2 - b^2\right)\left(b^2 \cos^4\theta - a^2 \sin^4\theta\right) + \left(b^2\cos^2\theta + a^2\sin^2\theta\right)^2}{a^4 b^4 u^3}, \\
&= \frac{1}{a^4 b^4 u^3}\left[a^2b^2 \cos^4\theta + a^2b^2 \sin^4\theta + 2a^2b^2\cos^2\theta\sin^2\theta\right], \\
&= \frac{a^2b^2}{a^4 b^4 u^3} = \frac{1}{a^2b^2u^3}, \\
\therefore \quad P &= \frac{h^2u^2}{a^2b^2u^3} = \frac{\frac{h^2}{a^2b^2}}{u} = \frac{\mu}{u} = \mu r.
\end{aligned}$$

Which implies that $P \propto r$.

Example 3.18 A particle describes an ellipse under a central force directed towards the center. Prove that the eccentric angle of the particle increases at a constant rate.

Solution Given that the particle moves on an ellipse about a center of force at the center of the ellipse. In Fig. 3.15, take the center O of the ellipse as the origin, the X-axis along **OA**, and the Y-axis along **OB**. Let $2a$ be the major axis, and $2b$ be the minor axis of an ellipse.

The equation of the ellipse is

$$\frac{x^2}{a^2} + \frac{y^2}{b^2} = 1. \tag{3.2.98}$$

Since the particle describes the ellipse (3.2.98) about the center of force in O.

Therefore, the law of force is

$$P = \mu r. \tag{3.2.99}$$

Let $Q\,(x,\ y)$ be the position of the particle on the ellipse at time t. Let $\mathbf{r}$ stipulate the position vector of Q relative to O.

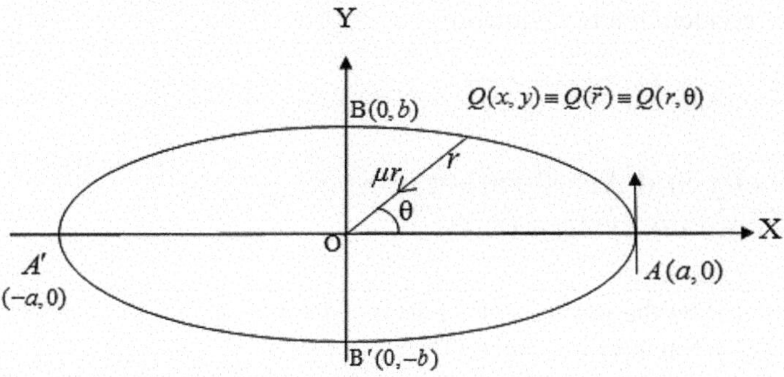

Fig. 3.15 Shematic diagram for Example 3.18

The equation of motion of the particle is

$$m\ddot{\mathbf{r}} = m\mu r\left(-\hat{r}\right),$$

$$\text{i.e.,}\quad \ddot{\mathbf{r}} = -\mu\mathbf{r}. \tag{3.2.100}$$

We have

$$\mathbf{r} = \text{P.V. of P relative to O} = x\hat{i} + y\hat{j}. \tag{3.2.101}$$

Equation (3.2.100) reduces to

$$\ddot{x}\hat{i} + y\hat{j} = -\mu\left(x\hat{i} + y\hat{j}\right). \tag{3.2.102}$$

By equating the coefficients of $\hat{i}$ and $\hat{j}$ in (3.2.102), we obtain

$$\left.\begin{array}{l}\ddot{x} = -\mu x\\ \ddot{y} = -\mu y\end{array}\right\}. \tag{3.2.103}$$

Equation (3.2.103) gives

$$\frac{d^2x}{dt} + \mu x = 0,$$

$$\text{i.e.,}\quad \left(D^2 + \mu\right)x = 0. \tag{3.2.104}$$

The general solution of (3.2.104) is

$$x = A\cos\sqrt{\mu}t + B\sin\sqrt{\mu}t, \tag{3.2.105}$$

with

$$\dot{x} = -A\sqrt{\mu}\sin\sqrt{\mu}t + B\sqrt{\mu}\cos\sqrt{\mu}t. \tag{3.2.106}$$

Initially $x = a, \quad \dot{x} = 0$ when $t = 0$.

Equations (3.2.105) and (3.2.106) give $A = a,\ B = 0$.

$$\therefore \quad x = a\cos\sqrt{\mu}t. \tag{3.2.107}$$

$\because$ Q(x, y) is on (3.2.98),

$$\therefore \quad \frac{a^2\cos^2\sqrt{\mu}t}{a^2} + \frac{y^2}{b^2} = 1,$$

$$\text{i.e.,} \quad y^2 = b^2\sin^2\sqrt{\mu}t,$$

$$\text{i.e.,} \quad y = b\sin\sqrt{\mu}t. \tag{3.2.108}$$

Let ϕ be the eccentric angle of the point Q $(x,\ y)$.

$$\therefore \quad \left.\begin{array}{l} x = a\cos\phi \\ y = b\sin\phi \end{array}\right\},$$

$$\text{i.e.,} \quad \left.\begin{array}{l} \cos\phi = \cos\sqrt{\mu}t \\ \sin\phi = \sin\sqrt{\mu}t \end{array}\right\}.$$

$$\therefore \quad \phi = \sqrt{\mu}t \; and \; \dot{\phi} = \sqrt{\mu} = \text{Constant}.$$

Example 3.19 If a particle be describing an ellipse about a center of force in the center, show that the sum of the reciprocals of its angular velocities about the foci is constant

Solution Let O be the center, $2a$ the major axis, and $2b$ the minor axis of the ellipse. The (x, y) equation of the ellipse is

$$\frac{x^2}{a^2} + \frac{y^2}{b^2} = 1 \tag{3.2.109}$$

referred to O as the origin, the X-axis along the +ve major axis, and the Y-axis along the +ve minor axis. Let P be the position of the particle at time t.

Let (r, θ) be the polar coordinates of P referred to the focus S as a pole and **SX** as the initial line. Let ϕ denote the eccentric angle of P.

$$\therefore \quad \left.\begin{array}{l} x = a\cos\phi \\ y = b\sin\phi \end{array}\right\}. \tag{3.2.110}$$

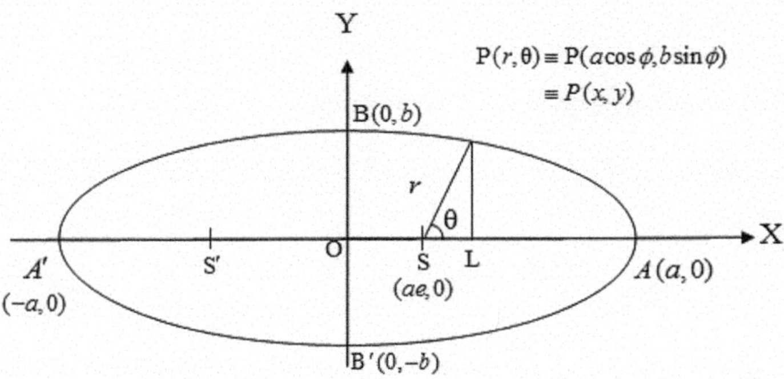

Fig. 3.16 Shematic diagram for Example 3.19

From Fig. 3.16, we have

$$\left.\begin{aligned} x &= \mathrm{OL} = \mathrm{OS} + \mathrm{SL} = ae + r\cos\theta \\ y &= \mathrm{PL} = \mathrm{SP}\,\sin\theta = r\sin\theta \end{aligned}\right\}. \tag{3.2.111}$$

(3.2.111) gives

$$r\cos\theta = a\left(\cos\phi - e\right), \tag{3.2.112}$$

and

$$r\sin\theta = b\sin\phi. \tag{3.2.113}$$

Now

$$\begin{aligned} r^2 = \mathrm{SP}^2 &= (a\cos\phi - ae)^2 + (b\sin\phi)^2, \\ &= a^2(\cos\phi - e)^2 + a^2\left(1 - e^2\right)\sin^2\phi, \\ &= a^2\left[(\cos\phi - e)^2 + \left(1 - e^2\right)\sin^2\phi\right], \\ &= a^2\left[1 - 2e\,\cos\phi + e^2\cos^2\phi\right], \\ &= a^2(1 - e\,\cos\phi)^2, \\ \therefore\quad r &= a\left(1 - e\cos\phi\right). \end{aligned} \tag{3.2.114}$$

Therefore, (3.2.112) and (3.2.114) give

$$\cos\theta = \frac{\cos\phi - e}{1 - e\cos\phi}. \tag{3.2.115}$$

As the particle describes an ellipse about a center of force in the center of the ellipse, Therefore, the law of force is

$$P = \mu r, \tag{3.2.116}$$

and the eccentric angle of the point P is given by

$$\phi = \sqrt{\mu} t, \tag{3.2.117}$$

$$\therefore \ \dot{\phi} = \sqrt{\mu}. \tag{3.2.118}$$

Equation (3.2.115) gives

$$-\sin\theta\dot{\theta} = \frac{d}{d\phi}\frac{\cos\phi - e}{1 - e\cos\phi}\dot{\phi} = \dot{\phi}\frac{(1 - e\cos\phi)(-\sin\phi) - (\cos\phi - e)(e\sin\phi)}{(1 - e\cos\phi)^2},$$

$$\text{i.e.,} \quad -\sin\theta\dot{\theta} = \dot{\phi}\frac{-\sin\phi + e\sin\phi\cos\phi - e\sin\phi\cos\phi + e^2\sin\phi}{(1 - e\cos\phi)^2},$$

$$\text{i.e.,} \quad -\sin\theta\dot{\theta} = \frac{\dot{\phi}\left(e^2 - 1\right)\sin\phi}{(1 - e\cos\phi)^2},$$

$$\text{i.e.,} \quad \sin\theta\dot{\theta} = \frac{\left(1 - e^2\right)\dot{\phi}\sin\phi}{(1 - e\cos\phi)^2} = \frac{\left(1 - e^2\right)\sqrt{\mu}\sin\phi}{(1 - e\cos\phi)^2},$$

$$\text{i.e.,} \quad \frac{b\sin\phi}{r}\dot{\theta} = \frac{\left(1 - e^2\right)\sqrt{\mu}\sin\phi}{(1 - e\cos\phi)^2},$$

$$\text{i.e.,} \quad \frac{b\sin\phi\dot{\theta}}{a(1 - e\cos\phi)} = \frac{\left(1 - e^2\right)\sqrt{\mu}\sin\phi}{(1 - e\cos\phi)^2},$$

$$\text{i.e.,} \quad \frac{b\dot{\theta}}{a} = \frac{\left(1 - e^2\right)\sqrt{\mu}}{1 - e\cos\phi},$$

$$\text{i.e.,} \quad \dot{\theta} = \frac{a\left(1 - e^2\right)\sqrt{\mu}}{b(1 - e\cos\phi)},$$

$$\text{i.e.,} \quad \dot{\theta} = \frac{b^2\sqrt{\mu}}{ab(1 - e\cos\phi)} = \frac{b\sqrt{\mu}}{a(1 - e\cos\phi)}.$$

$\therefore \ \omega = \dot{\theta} =$ angular velocity of the particle about the focus S $= \frac{b\sqrt{\mu}}{a(1-e\cos\phi)}$.
Similarly
$\omega' =$ angular velocity of the particle about the other focus $= \frac{b\sqrt{\mu}}{a(1+e\cos\phi)}$.

$$\therefore \quad \frac{1}{\omega} + \frac{1}{\omega'} = \frac{a\,(1 - e\cos\phi)}{b\sqrt{\mu}} + \frac{a\,(1 + e\cos\phi)}{b\sqrt{\mu}} = \frac{2a}{b\sqrt{\mu}} = \text{a constant.}$$

Hence proved.

Example 3.20 With usual notations, prove that in case of a two-body problem $\mathrm{GM}=\mathrm{gR}^2$.

Solution In Fig. 3.17, let O be the center of the Earth with mass M, and let a mass m be placed at a point A on the Earth's surface. If R is the radius of the Earth, then the force of attraction of the Earth on the object is given by

$$F = \frac{\mathrm{GMm}}{\mathrm{R}^2}. \tag{3.2.119}$$

The weight W of the mass m is given by

$$\mathrm{W} = mg. \tag{3.2.120}$$

We have

$$F = W,$$

$$\text{i.e.,} \quad \frac{\mathrm{GMm}}{\mathrm{R}^2} = \mathrm{mg},$$

$$\text{i.e.,} \quad \mathrm{GM} = g\mathrm{R}^2.$$

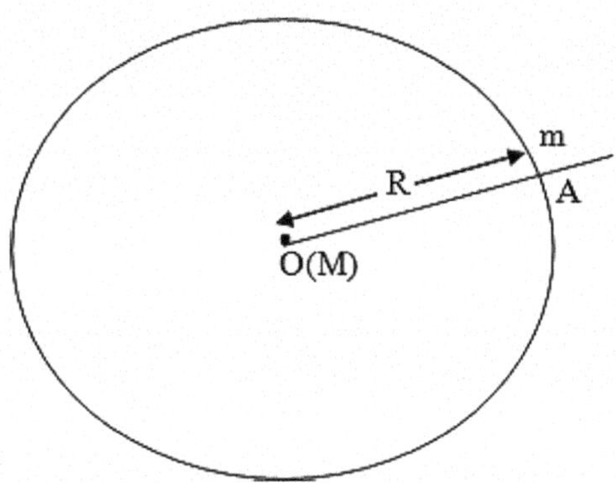

Fig. 3.17 Illustrating diagram for Example 3.20

Example 3.21 Calculate the mass of the Earth.

Solution We note that

$$GM = gR^2,$$

$$\text{i.e.,}\quad M = \frac{gR^2}{G}. \tag{3.2.121}$$

Given

$$R = 6.38 \times 10^6\,\text{m},\quad g = 9.80\,\text{ms}^{-2},\quad G = 6.67 \times 10^{-11}\frac{\text{m}^3}{\text{kg s}^2}.$$

Therefore, Eq. (3.2.121) gives

$$M = \frac{9.80 \times (6.38 \times 10^6)^2}{6.67 \times 10^{-11}}\,\text{kg} = \frac{9.80 \times (6.38)^2}{6.67} \times 10^{23}\,\text{kg} = 5.98 \times 10^{24}\,\text{kg}.$$

Example 3.22 The periodic time of Saturn is 29.5 years. Find its mean distance from the Sun.

Solution By Kepler's 3rd law of planetary motion,

$$D^3 \propto T^2,$$

$$\text{i.e.,}\quad D^3 = kT^2. \tag{3.2.122}$$

Suppose that

$$D = D_1 \text{ at } T = T_1;\quad D = D_2 \text{ at } T = T_2.$$

$$\therefore\quad {D_1}^3 = k{T_1}^2 \text{ and } {D_2}^3 = k{T_2}^2,$$

$$\frac{{D_1}^3}{{T_1}^2} = \frac{{D_2}^3}{{T_2}^2}. \tag{3.2.123}$$

Here

$$D_1 = D;\quad D_2 = 1\text{AU};\quad T_1 = 29.5\,\text{years},\quad T_2 = 1\,\text{year}.$$

Thus, Eq. (3.2.123) gives

$$\frac{D^3}{(29.5)^2} = \frac{1}{1},$$

$$\text{i.e.,}\quad D^3 = (29.5)^2,$$

$$\text{i.e.,}\quad D = (29.5)^{\frac{2}{3}} = 9.5\ \text{AU (Approx.)}.$$

Example 3.23 An object is projected vertically upward from the Earth's surface with initial speed V_0. Neglecting air resistance, (**A**) find the speed at a distance H above the Earth's surface and (**B**) the smallest velocity of projection needed in order that the object never return.

Solution (A)

Refer to Fig. 3.18, let O be the center of the Earth, M the mass of the Earth, R the radius of the Earth, m the mass of the planet, A the point of projection on the Earth's surface, OX the upward vertical line through O, and P the position of the particle at time t. Suppose that OP $= x$.

The equation of motion of the particle is

$$m\ddot{x} = -\frac{GMm}{x^2},$$

$$\text{i.e.,}\quad \ddot{x} = -\frac{GM}{x^2},$$

$$\text{i.e.,}\quad v\frac{dv}{dx} = -\frac{GM}{x^2},$$

$$\text{i.e.,}\quad vdv = -\frac{GM}{x^2}dx,$$

$$\text{i.e.,}\quad \frac{v^2}{2} = \frac{\mu}{x} + C;\quad \mu = GM. \tag{3.2.124}$$

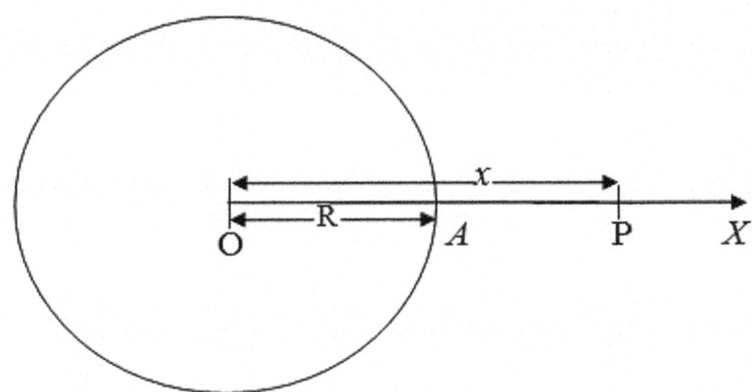

Fig. 3.18 Illustration for Example 3.23

Initially $v = V_0$ at $x = \text{OA} = R$.
Therefore, Eq. (3.2.124) gives

$$\frac{{V_0}^2}{2} = \frac{\mu}{R} + C,$$

$$\text{i.e.,}\quad C = \frac{{V_0}^2}{2} - \frac{\mu}{R},$$

$$\therefore\quad \frac{v^2}{2} = \frac{\mu}{x} + \frac{V_0^2}{2} - \frac{\mu}{R},$$

$$\text{i.e.,}\quad v^2 = 2\mu\left(\frac{1}{x} - \frac{1}{R}\right) + V_0^2. \tag{3.2.125}$$

Let u be the velocity of the object at a height H above the Earth's surface.

$$\therefore\quad v = u \text{ when } x = R + H.$$

Therefore, Eq. (3.2.125) gives

$$u^2 = 2\mu\left(\frac{1}{R+H} - \frac{1}{R}\right) + V_0^2 = V_0^2 - \frac{2\mu H}{R\,(R+H)},$$

$$u^2 = V_0^2 - \frac{2GMH}{R\,(R+H)} = V_0^2 - \frac{2gR^2H}{R\,(R+H)} = V_0^2 - \frac{2gRH}{R+H}.$$

$$u = \sqrt{V_0^2 - \frac{2gRH}{R+H}}.$$

(B) Let us choose V_0 in such a way that $v \to 0$ *as* $x \to \infty$.
Therefore, Eq. (3.2.125) gives

$$0 = 2\mu\left(-\frac{1}{R}\right) + V_0^2,$$

$$\text{i.e.,}\quad V_0^2 = \frac{2\mu}{R} = \frac{2GM}{R} = \frac{2gR^2}{R} = 2gR,$$

$$\text{i.e.,}\quad V_0 = \sqrt{2gR},$$

which is the minimum speed of projection needed in order that the object never return. Here $V_0 = \sqrt{2gR}$ is called the escape velocity.

Example 3.24 Show that the magnitude of the escape velocity of an object from the Earth's surface is about 11 km/s.

Solution Let V_0 denote the escape velocity and R be the radius of the Earth. Then V_0 and R are related by the relation

$$V_0 = \sqrt{2gR}. \tag{3.2.126}$$

Now $g = 9.8\,\mathrm{m/s^2},\ R = 6.38 \times 10^6\,\mathrm{m}$

Therefore, Eq. (3.2.126) gives

$$V_0 = \sqrt{2 \times 9.8 \times 6.38 \times 10^6}\,\mathrm{m/s},$$

$$\text{i.e.,}\ \ V_0 = \sqrt{4 \times 4.9 \times 6.38}\,10^3\,\mathrm{m/s},$$

$$\text{i.e.,}\ \ V_0 = \sqrt{4 \times 49 \times 0.638}\,\mathrm{km/s},$$

$$\text{i.e.,}\ \ V_0 = 14\sqrt{0.638}\,\mathrm{km/s},$$

$$\text{i.e.,}\ \ V_0 \approx 14 \times 0.8\,\mathrm{km/s} = 11.2\,\mathrm{km/s},$$

$$\text{i.e.,}\ \ V_0 \approx 11\,\mathrm{km/s}.$$

Example 3.25 An artificial satellite revolves about the Earth at a height H above the surface. Determine (a) Orbital speed and (b) Orbital period so that a man in the satellite will be in a state of weightless.

Solution (a) R be radius and M the mass of the Earth. Let V_0 be the orbital speed and m the mass of the satellite.

Centrifugal force on the satellite at height H above the Earth's surface $= \frac{mV_0^2}{R+H}$.

Force of gravitation on the satellite due to the Earth $= \frac{GMm}{(R+H)^2}$.

The man in the satellite will feel weightless for

$$\frac{mV_0^2}{R+H} = \frac{GMm}{(R+H)^2},$$

$$\text{i.e.,}\ \ V_0^2 = \frac{GM}{R+H} = \frac{gR^2}{R+H},$$

$$\text{i.e.,}\ \ V_0 = R\sqrt{\frac{g}{R+H}} = R\sqrt{\frac{\frac{g}{R}}{1+\frac{H}{R}}}.$$

If $H \ll R$, then $\frac{H}{R} \ll 1$; then $V_0 = R\sqrt{\frac{g}{R}} = \sqrt{gR}$ (Approx.) (b)

$$\text{Orbital speed} = \frac{\text{distance covered in 1 revolution}}{\text{Period}} = \frac{2\pi(R+H)}{P}$$

$$\text{i.e.,}\quad P = \frac{2\pi\,(R+H)}{V_0} = \frac{2\pi\,(R+H)}{R\sqrt{\frac{g}{R+H}}} = 2\pi\left(1+\frac{H}{R}\right)\sqrt{\frac{1+\frac{H}{R}}{\frac{g}{R}}}.$$

But $\frac{H}{R} \to 0$.

$$\therefore\quad P = 2\pi\sqrt{\frac{R}{g}}\ \text{(Approx.)}.$$

Example 3.26 Express the true anomaly θ in terms of the eccentricity e and eccentric anomaly ϕ in a series: $\theta = \phi + \left(e + \frac{1}{4}e^3\right)\sin\phi + \frac{1}{4}e^2\sin 2\phi + \frac{1}{12}e^3\sin 3\phi$.

Solution The true anomaly θ and eccentric anomaly ϕ are connected by the relation

$$\tan\frac{\theta}{2} = \sqrt{\frac{1+e}{1-e}}\tan\frac{\phi}{2}. \tag{3.2.127}$$

Introduce a function ψ defined by

$$\sin\psi = e\left(0 < \psi < \frac{\pi}{2}\right). \tag{3.2.128}$$

Therefore, Eq. (3.2.127) gives

$$\tan\frac{\theta}{2} = \sqrt{\frac{1+\sin\psi}{1-\sin\psi}}\tan\frac{\phi}{2},$$

$$\text{i.e.,}\quad \tan\frac{\theta}{2} = \frac{\cos\frac{\psi}{2}+\sin\frac{\psi}{2}}{\cos\frac{\psi}{2}-\sin\frac{\psi}{2}}\tan\frac{\phi}{2},$$

$$\text{i.e.,}\quad \tan\frac{\theta}{2} = \frac{1+\tan\frac{\psi}{2}}{1-\tan\frac{\psi}{2}}\tan\frac{\phi}{2},$$

$$\text{i.e.,}\quad \frac{2i\sin\frac{\theta}{2}}{2\cos\frac{\theta}{2}} = \frac{1+x}{1-x}\,\frac{2i\sin\frac{\phi}{2}}{2\cos\frac{\phi}{2}};\quad x = \tan\frac{\psi}{2},$$

$$\text{i.e.,}\quad \frac{e^{\frac{i\theta}{2}}-e^{-\frac{i\theta}{2}}}{e^{\frac{i\theta}{2}}+e^{-\frac{i\theta}{2}}} = \frac{1+x}{1-x}\,\frac{e^{\frac{i\phi}{2}}-e^{-\frac{i\phi}{2}}}{e^{\frac{i\phi}{2}}+e^{-\frac{i\phi}{2}}},$$

$$\text{i.e.,}\quad \frac{2e^{\frac{i\theta}{2}}}{2e^{-\frac{i\theta}{2}}} = \frac{2e^{\frac{i\phi}{2}} - 2xe^{-\frac{i\phi}{2}}}{2e^{-\frac{i\varphi}{2}} - 2xe^{\frac{i\varphi}{2}}},$$

$$\text{i.e.,}\quad e^{i\theta} = \frac{e^{\frac{i\phi}{2}}\left(1 - xe^{-i\phi}\right)}{e^{-\frac{i\phi}{2}}\left(1 - xe^{i\phi}\right)},$$

$$\text{i.e.,}\quad e^{i\theta} = e^{i\phi}\frac{1 - xe^{-i\phi}}{1 - xe^{i\phi}},$$

$$\text{i.e.,}\quad \log e^{i\theta} = \log e^{i\phi} + \log\frac{1 - xe^{-i\phi}}{1 - xe^{i\phi}},$$

$$\text{i.e.,}\quad \log e^{i\theta} = \log e^{i\phi} + \log\left(1 - xe^{-i\phi}\right) - \log\left(1 - xe^{i\phi}\right),$$

$$\text{i.e.,}\quad i\theta = i\phi + \left[-xe^{-i\phi} - \frac{x^2e^{-2i\phi}}{2} - \frac{x^3e^{-3i\phi}}{3} - \cdots\right]$$

$$-\left[-xe^{i\phi} - \frac{x^2e^{2i\phi}}{2} - \frac{x^3e^{3i\phi}}{3} - \cdots\right],$$

$$\text{i.e.,}\quad i\theta = i\phi + x\,2i\sin\phi + \frac{x^2}{2}2i\sin 2\phi + \frac{x^3}{3}2i\sin 3\phi + \cdots$$

$$\text{i.e.,}\quad \theta = \phi + 2x\sin\phi + \frac{2x^2}{2}\sin 2\phi + \frac{2x^3}{3}\sin 3\phi + \cdots \tag{3.2.129}$$

Now

$$x = \tan\frac{\psi}{2} = \frac{1 - \cos\psi}{\sin\psi} = \frac{1 - \sqrt{1 - e^2}}{e},$$

$$\text{i.e.,}\quad x = \frac{1}{e}\left[1 - \left(1 - e^2\right)^{\frac{1}{2}}\right],$$

$$\text{i.e.,}\quad x = \frac{1}{e}\left[1 - \left\{1 + \frac{1}{2}\left(-e^2\right) + \frac{\frac{1}{2}\left(-\frac{1}{2}\right)}{2!}\left(-e^2\right)^2 + \right\}\right],$$

$$\text{i.e.,}\quad x = \frac{1}{e}\left(1 - 1 + \frac{1}{2}e^2 + \frac{1}{8}e^4 + \cdots\right),$$

$$\text{i.e.,}\quad x = \frac{1}{2}e + \frac{1}{8}e^3 + \cdots,$$

$$\text{i.e.,}\quad x = \frac{1}{2}\left(e + \frac{1}{4}e^3\right)\ \text{(upto 3rd degree)},$$

$$\therefore\quad \theta = \phi + 2\frac{1}{2}\left(e + \frac{1}{4}e^3\right)\sin\phi + \frac{\left(e+\frac{1}{4}e^3\right)^2}{4}\sin 2\phi + \frac{2}{3}\frac{\left(e+\frac{1}{4}e^3\right)^3}{8}\sin 3\phi,$$

$$\text{i.e.,}\quad \theta = \phi + \left(e + \frac{1}{4}e^3\right)\sin\phi + \frac{1}{4}e^2\sin 2\phi + \frac{1}{12}e^3\sin 3\phi.$$

Example 3.27 Show that $\frac{m}{2} = (1-e)\,k - \frac{1-3e}{3}k^3 + \frac{1-5e}{5}k^5 - \cdots$; $k = \sqrt{\frac{1-e}{1+e}}\tan\frac{\theta}{2}$.

Solution Given

$$k = \sqrt{\frac{1-e}{1+e}}\tan\frac{\theta}{2} = \tan\frac{\phi}{2},$$

$$\text{i.e.,}\quad k = \tan\frac{\phi}{2},$$

$$\text{i.e.,}\quad \tan^{-1}k = \frac{\phi}{2}.$$

Now

$$\begin{aligned}
\text{RHS} &= (1-e)\,k - \frac{1-3e}{3}k^3 + \frac{1-5e}{5}k^5 - \cdots \\
&= \left(k - \frac{k^3}{3} + \frac{k^5}{5} - \cdots\right) - e\left(k - k^3 + k^5 - \cdots\right) \\
&= \tan^{-1}k - ek\left(1 - k^2 + k^4 - \cdot\right) \\
&= \frac{\phi}{2} - ek\left(1 - x + x^2 - \cdot\right);\quad x = k^2 \\
&= \frac{\phi}{2} - ek(1+x)^{-1} \\
&= \frac{\phi}{2} - ek\left(1+k^2\right)^{-1} \\
&= \frac{\phi}{2} - ek\left(1 + \tan^2\frac{\phi}{2}\right)^{-1} \\
&= \frac{\phi}{2} - ek\left(s^2\frac{\phi}{2}\right)^{-1} \\
&= \frac{\phi}{2} - ek\cos^2\frac{\phi}{2} \\
&= \frac{\phi}{2} - e\tan\frac{\phi}{2}\cos^2\frac{\phi}{2}
\end{aligned}$$

$$
\begin{aligned}
&= \frac{\phi}{2} - e \sin\frac{\phi}{2}\cos\frac{\phi}{2} \\
&= \frac{\phi}{2} - \frac{e}{2}\sin\phi \\
&= \frac{1}{2}(\phi - e\sin\phi) = \frac{m}{2} = \text{LHS.}
\end{aligned}
$$

Example 3.28 An artificial satellite is released at an altitude of 400 km with speed 8.85 km/s in a horizontal direction. Determine the energy per unit mass of this satellite, taking the center of the Earth as the center of force, the radius of the Earth as 6378 km, and $GM = 398603.6\,\text{km}^3/\text{s}^2$. Find also the Perigee, Apogee, and Orbital Period.

Solution Let O be the center of the Earth and R be the radius of the Earth.
Therefore, from Fig. 3.19, we have

$$p = r = \text{radius of the orbit of the satellite} = (R + 400)\ \text{km} = 6778\,\text{km},$$

$$Vp = h,$$

$$h = 8.85 \times 6778\,\text{km}^2/\text{s} = 59985.3\,\text{km}^2/\text{s}.$$

Energy E per unit mass of the satellite is given by

$$E = \text{K.E.} + \text{P.E.} = \frac{1}{2}mV^2 - \frac{GMm}{r} = \frac{1}{2}V^2 - \frac{GM}{r},$$

$$\text{i.e.,}\quad E = \left[\frac{1}{2}(8.85)^2 - \frac{398603.6}{6778}\right] J = -19.6472J < 0.$$

This shows that the orbit is an ellipse.

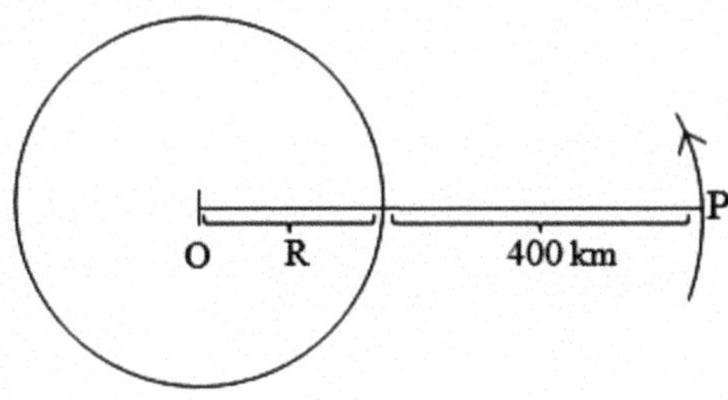

Fig. 3.19 Illustrating diagram for Example 3.28

The positions of perigee and apogee are given by

$$r^2 + \frac{\mu}{E}r - \frac{h^2}{2E} = 0,$$

$$\text{i.e.,}\quad r^2 + \frac{398603.6}{-19.6472}r + \frac{(59985.3)^2}{2 \times 19.6472} = 0,$$

$$\text{i.e.,}\quad r^2 - 20288r + 91572154 = 0,$$

$$\text{i.e.,}\quad r = \frac{20288 \pm \sqrt{(20288)^2 - 4 \times 91572154}}{2},$$

$$\text{i.e.,}\quad r = \frac{20288 \pm \sqrt{411602944 - 366288616}}{2},$$

$$\text{i.e.,}\quad r = \frac{20288 \pm \sqrt{45314328}}{2},$$

$$\text{i.e.,}\quad r = \frac{20288 \pm 6731.59}{2}.$$

$$r_1 = 13509.8 \text{ and } r_2 = 6778.21.$$

$\therefore$ The distance of perigee from center = 6778 km (approx.).
The distance of apogee from center = 13510 km (approx.).
But

$$2a = r_1 + r_2 = 20288\,\text{km},$$

$$\text{i.e.,}\quad a = 10144\,\text{km}.$$

Hence

$$T = \frac{2\pi}{\sqrt{\mu}}a^{\frac{3}{2}} = \frac{2 \times 3.14}{\sqrt{398603.6}} \times (10144)^{\frac{3}{2}}\,\text{s},$$

$$\text{i.e.,}\quad T = \frac{6.28}{631.35} \times 1021147\,\text{s},$$

$$\text{i.e.,}\quad T = 6.28 \times 1617.4\,\text{s},$$

$$\text{i.e.,}\quad T = 10157.3\,\text{s} = 169.29\,\text{min} = 2.82\,\text{h}.$$

Example 3.29 Neglecting the fourth and higher power of e, prove that

$$\varphi = m + \frac{e \sin m}{1 - e \cos m} - \frac{1}{2}\left(\frac{e \sin m}{1 - e \cos m}\right)^3$$

is a solution of Kepler's equation.

Solution By Kepler's equation,

$$m = \phi - e \sin \phi,$$

$$\text{i.e.,} \quad \phi - e \sin \phi - m = 0,$$

$$\text{i.e.,} \quad f(\phi) = 0, \tag{3.2.130}$$

where $f(\phi) = \phi - e \sin \phi - m$.

Let $\phi = m + x$; $x = e \sin \phi$

Equation (3.2.130) gives

$$f(m + x) = 0,$$

$$\text{i.e.,} \quad f(m) + x f'(m) + \frac{x^2}{2!} f''(m) + \cdots = 0,$$

$$\text{i.e.,} \quad x = -\frac{f(m)}{f'(m)} - \frac{x^2}{2!}\frac{f''(m)}{f'(m)} - \cdots \tag{3.2.131}$$

We have

$$f(\phi) = \phi - e \sin \phi - m,$$

$$f'(\phi) = 1 - e \cos \phi,$$

$$f''(\phi) = e \sin \phi,$$

$$\therefore \quad f(m) = m - e \sin m - m = -e \sin m,$$

$$f'(m) = -e \cos m,$$

$$f''(m) = e \sin m.$$

Thus, Eq. (3.2.131) gives

$$x = \frac{e \sin m}{1 - e \cos m} - \frac{x^2}{2} \frac{e \sin m}{(1 - e \cos m)}. \tag{3.2.132}$$

For first approximation, (3.2.132) gives

$$x = \frac{e \sin m}{1 - e \cos m}.$$

Equation (3.2.132) yields

$$x = \frac{e \sin m}{1 - e \cos m} - \frac{1}{2}\left(\frac{e \sin m}{1 - e \cos m}\right)^2 \frac{e \sin m}{1 - e \cos m},$$

$$\text{i.e.,} \quad x = \frac{e \sin m}{1 - e \cos m} - \frac{1}{2}\left(\frac{e \sin m}{1 - e \cos m}\right)^3,$$

$$\text{i.e.,} \quad \phi - m = \frac{e \sin m}{1 - e \cos m} - \frac{1}{2}\left(\frac{e \sin m}{1 - e \cos m}\right)^3,$$

$$\text{i.e.,} \quad \phi = m + \frac{e \sin m}{1 - e \cos m} - \frac{1}{2}\left(\frac{e \sin m}{1 - e \cos m}\right)^3,$$

which is the required result.

Example 3.30 Communication satellites are launched in circular orbit in the equatorial plane of the Earth so that they remain stationary relative to the surface of the Earth. What is the minimum number of satellites so that every point (or plane) may view at least one satellite?

Solution Let S be the position of a communication satellite, C be the center of the Earth, R be the radius of the Earth, and x be the height of S above the Earth's surface.

Draw two tangents from S to the Earth to touch the surface of the Earth at two points P and Q such that the points C, P, S, and Q are co-planar.

Let m∠SCP = m∠SCQ = α.

From Fig. 3.20, we have

$$\cos \alpha = \frac{\text{CP}}{\text{CS}} = \frac{R}{R + x}. \tag{3.2.133}$$

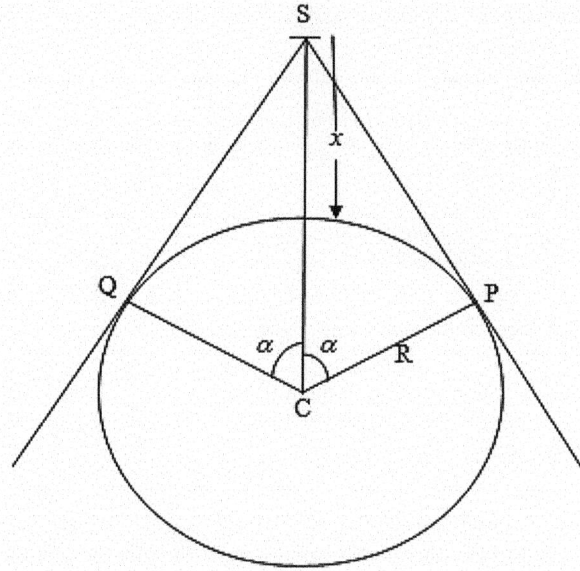

Fig. 3.20 Illustration for Example 3.30

The orbital period of the satellite is given by

$$T = \frac{2\pi}{\sqrt{\mu}} a^{\frac{3}{2}},$$

$$\text{i.e.,}\quad T^2 = \frac{4\pi^2}{\mu} a^3 = \frac{4\pi^2}{GM} (R + x)^3,$$

$$\text{i.e.,}\quad (R + x)^3 = \frac{T^2 GM}{4\pi^2}. \tag{3.2.134}$$

Here

$$T = 1\,\text{day} = 24\,\text{h} = 24 \times 60 \times 60\,\text{s} = 8.64 \times 10^4\,\text{s},$$

$$GM = 398603.6\,\text{km}^3/\text{s}^2.$$

Equation (3.2.134) gives

$$(R + x)^3 = \frac{(8.64 \times 10^4)^2 \times 398603.6}{4 \times (3.1415)^2}\,\text{km}^3 = 75376.83 \times 10^9\,\text{km}^3,$$

$$\text{i.e.,}\quad R + x = 40.690 \times 10^3\,\text{km} = 40690\,\text{km}.$$

Equation (3.2.133) gives

$$\cos\alpha = \frac{R}{R+x} = \frac{6378}{40690} = 0.157,$$

$$\text{i.e.,}\quad \alpha = \cos^{-1}(0.157) = 80.967^\circ,$$

$$\text{i.e.,}\quad 2\alpha = 162^\circ\ (\text{approx.})\,.$$

$\therefore$ The two satellite can cover maximum $= 162^\circ \times 2 = 324^\circ < 360^\circ$.
Hence the minimum number of satellites required $= 2 + 1 = 3$.

3.3 Exercise-III

1. A particle is projected from the Earth's surface with velocity v; show that, if the diminution of gravity be taken into account, but the resistance of the air neglected, the path is an ellipse of major axis $\frac{2ga^2}{2ga-v^2}$, where a is the Earth's radius.
2. Show that, if a particle describes an ellipse under a force to a focus, the velocity at the mean distance from the center of force is a mean proportional between the velocities at the ends of any diameter.
3. A particle is repelled from a center of force O with a force μr per unit mass, where r is the distance of the particle from O. Show that if the particle is projected from a point P in any direction with velocity OP$\sqrt{\mu}$, its path is a rectangular hyperbola with O as the center.
4. If ω be the angular velocity of a planet at the far end of the major axis, prove that its period is $\frac{2\pi}{\omega}\sqrt{\frac{1-e}{(1+e)^3}}$.
5. A particle is describing a circle of radius under an inverse square law of force $\frac{\mu}{r^2}$ per unit mass directed towards the center of the circle. An outward radial velocity $\sqrt{\frac{\mu}{5a}}$ is suddenly impressed on it in such a way as to leave the transverse velocity unaltered; prove that it will describe an ellipse and find the new periodic time.
6. A particle P of unit mass is moving under a central force $\frac{\mu}{r^2}$ directed towards a fixed point S. Establish the relations $V^2 = \mu\left(\frac{2}{r} - \frac{1}{a}\right), h^2 = \mu a\left(1 - e^2\right), (e < 1)$,where the initial conditions are such that $V^2 < \frac{2\mu}{r}$ and the symbols have their usual meanings. If when the particle is at P its radial component of velocity is suddenly destroyed, show that the focal choral PSP' of the original elliptical orbit becomes the major axis of the new elliptical orbit.
7. A particle is describing an ellipse of eccentricity e about the center of force at a focus. Prove with usual rotation $v^2 = \mu\left(\frac{2}{r} - \frac{1}{a}\right), h^2 = \mu a\left(1 - e^2\right)$ .When the particle is at one end of a minor axis, its velocity is doubled. Prove that th e new path is a hyperbola of eccentricity $\left(9 - 8e^2\right)^{\frac{1}{2}}$.

8. A particle describes an ellipse of major axis $2a$ and eccentricity$e$ under a force directed to a focus S. Find an expression for the velocity of the particle at any point. When the particle is at the end of the latus rectum through S its velocity is suddenly turned through a right angle without change of magnitude. Prove that the particle will now describe an ellipse whose major axis is still of length $2a$ but is inclined at an angle $\tan^{-1}\left\{\frac{1-e^2}{e}\right\}$ to the major axis of the original orbit. Find the eccentricity of the new orbit.
9. A particle of mass m moving in a circle of radius c under an attractive force $\frac{\mu}{r^2}$ per unit mass towards the center, collides and coalesces with a particle of mass λm which is at rest. Show that the orbit of compound mass is an ellipse with major axis $c\csc^2\alpha$, latus rectum $4c\cos^2\alpha$, and eccentricity—$\cos 2\alpha$, where $s^2\alpha = 2(1+\lambda)^2$.
10. If the path of a particle P moving under inverse square law about the center of focus S is an ellipse, ω_1, ω_2 are the greatest and least angular velocities of SP, show the mean angular velocity of SP is $\frac{2(\omega_1\omega_2)^{\frac{3}{4}}}{\sqrt{\omega_1}+\sqrt{\omega_2}}$.
11. A particle is describing a parabola of latus rectum $4a$, under a force to the focus. When it is at the end of the latus rectum, its velocity is suddenly halved. Show that it now proceeds to describe an ellipse of major axis $\frac{8}{3}a$. What is the eccentricity of the ellipse?
12. A missile is projected from the Earth's surface at A with speed $(2gd)^{1/2}$, the Earth being supposed spherical, of radius and center O, and $d < c$. The missile is assumed to be attracted by a force $\frac{gc^2}{r^2}$ per unit mass towards O, where OP=r. Show that the path of P is an ellipse having O as focus, and that the speed v of P is given by $v^2 = 2g\left(d - c + \frac{c_2}{r}\right)$.
13. A particle which is describing a circle under an acceleration$\frac{\mu}{r^2}$ directed towards the center of the circle, collides and coalesces with an equal particle which is at rest. Show that, if the composite particle moves under the law of force, it will describe an ellipse of eccentricity $\frac{3}{4}$ and also show that the periods of the description of the circle and the ellipse are in the ratio $7\sqrt{7} : 8$.
14. A particle is projected from a given point under the action of forces whose accelerating effects are μPS, μPS$'$ directed towards fined points S, S$'$. Find the magnitude and direction of the velocity of projection in order that the orbit may be an ellipse with S, S$'$ as foci.
15. Two particles are describing the same ellipse about a center of force in the center in opposite directions, the mass of one being double the other. If the particles meet and coalesce at the end of the minor axis, show that the new orbit trisects the major axis of the old.
16. A particle is moving in an ellipse under a force to the focus. At any point in its path, its direction of motion is suddenly turned through a right angle, the magnitude of the velocity being unaltered. Prove that its subsequent orbit is an ellipse whose eccentricity varies as the distance of P from the center of the original orbit.

17. Evaluate the orbital sped of a satellite if R = 6380 Km, g = 9.8 m/s^2.
Hints: $V_0 = \sqrt{gR}$.
Ans.7.91 km/s (Approx.)
18. A satellite of radius a revolves in a circular orbit about a planet of radius b with period P. If the shortest distance between their surfaces is c, prove that the mass of the planet is $\frac{4\pi^2(a+b+c)^3}{GP^2}$.
Hints: Centrifugal force on the satellite $F = \frac{mV_0^2}{a+b+c}$, Gravitational force on the satellite $W = \frac{GMm}{(a+b+c)^2}$, $F = W$
19. If V_1 and V_2 are the linear velocities of a planet at perihelion and aphelion respectively, prove that the eccentricity e and the semi-major axis a of the orbit are given by $e = \frac{V_1 - V_2}{V_1 + V_2}$ and $a = \frac{\mathcal{T}}{2\pi}\sqrt{V_1 V_2}$, τ being the orbital period.
20. If ψ is the angle between the direction of a planet's motion and the direction $\perp^r$ to the radius vector, prove that $\tan\psi = \frac{e\sin\phi}{\sqrt{1-e^2}}$.
21. Show that the time taken by the Earth to describe a vectorial angle θ from perihelion is $\frac{T}{2\pi}\left(\theta - 2e\sin\theta + \frac{3}{4}e^2\sin 2\theta\right)$, T being the orbital period and cubes and higher powers of e being neglected.
22. With usual notations, prove that $\cos\theta = \frac{\cos\phi - e}{1 - e\cos\phi}$ and $\sin\theta = \frac{\sqrt{1-e^2}\sin\phi}{1 - e\cos\varphi}$.

Chapter 4
Motion of a Particle in Three Dimensions

4.1 General Orthogonal Curvilinear Coordinates

Consider three functions defined by

$$\left.\begin{aligned} \lambda &= \lambda\,(x, y, z) \\ \phi &= \phi\,(x, y, z) \\ \upsilon &= \upsilon\,(x, y, z) \end{aligned}\right\}. \tag{4.1.1}$$

Consider the surfaces represented by

$$\left.\begin{aligned} \lambda\,(x, y, z) &= c_1 \\ \phi\,(x, y, z) &= c_2 \\ \upsilon\,(x, y, z) &= c_3 \end{aligned}\right\}. \tag{4.1.2}$$

Let P (x, y, z) be the point of intersection of the surfaces given by (4.1.2).

From Fig. 4.1, Let PA, PB, PC be the curves specified by

$$\left.\begin{aligned} \mu &= c_2 \\ \nu &= c_3 \end{aligned}\right\}, \quad \left.\begin{aligned} \nu &= c_3 \\ \lambda &= c_1 \end{aligned}\right\}, \quad \left.\begin{aligned} \lambda &= c_1 \\ \mu &= c_2 \end{aligned}\right\} \tag{4.1.3}$$

respectively.

Let the functions λ, μ, ν be such that the curves PA, PB, PC are orthogonal, i.e., the tangents to PA, PB, PC at P are mutually $\perp^r$.

The set of Eq. (4.1.1) depicts that, for known values of λ, μ, ν; x, y, z can be evaluated and conversely. If follows that (λ, μ, ν) is an alternative and equivalent coordinate system that defines the point P in space. The system (x, y, z) is termed as the Cartesian system. The other system of coordinates (λ, μ, ν) is called a general orthogonal curvilinear coordinate system because in general the three surfaces specified by Eq. (4.1.2) intersect along the curved lines which are mutually orthogonal at each point P.

N. Ahmed et al., *Classical Dynamics*, University Texts in the Mathematical Sciences,
https://doi.org/10.1007/978-981-95-6394-4_4

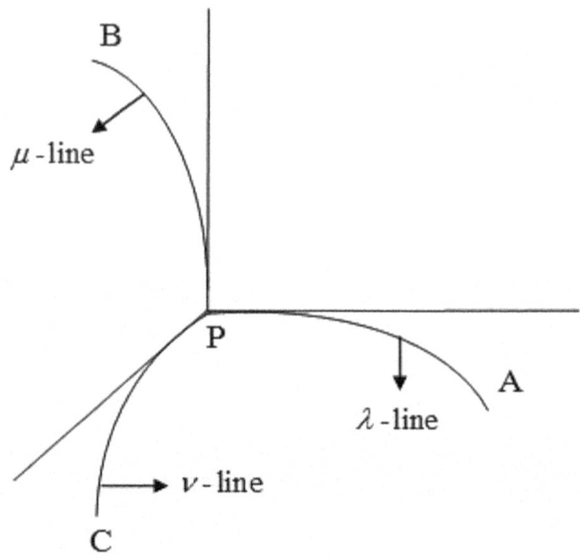

Fig. 4.1 Schematic diagram for general orthogonal curvilinear system

The simplest of all orthogonal coordinate is the Cartesian coordinate system (x, y, z), because the three planes given by

$$\left.\begin{array}{l} x = \text{constant} \\ y = \text{constant} \\ z = \text{constant} \end{array}\right\} \tag{4.1.4}$$

through the point P intersect along three lines which are mutually orthogonal in many problems involved in vector field. It is however sometimes convenient to work with other systems of orthogonal curvilinear coordinates. The most common such systems are cylindrical polar coordinate and spherical polar coordinates.

Figure 4.2 exhibits a point P in a Cartesian tri-rectangular frame specified by the axes **OX**, **OY**, **OZ**. A line segment PM is drawn perpendicular to the coordinate plane XOY. If $\angle XOM = \varphi$, $OM = \rho$, $PM = z$, then (ρ, φ, z) are called the cylindrical polar coordinates of P. It is recalled that $\rho =$ constant represents a right circular cylinder through P with OZ as axis and radius ρ. $\varphi =$ Constant defines a plane through P that makes a constant angle φ with the coordinate plane ZOX and z = constant specifies a plane parallel to XOY plane at distance z from it. Clearly these three surfaces given by

$$\left.\begin{array}{l} \rho = \text{constant} \\ \varphi = \text{constant} \\ z = \text{constant} \end{array}\right\} \tag{4.1.5}$$

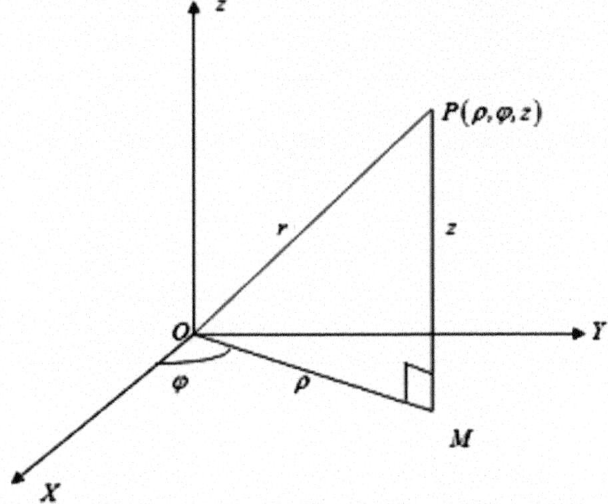

Fig. 4.2 Representation of a point P in cylindrical polar coordinates

intersect orthogonally at P, so the system of coordinates $(\rho,\ \varphi,\ z)$ is an orthogonal one. Suppose that $(x,\ y,\ z)$ are the Cartesian coordinates of P. It follows from Fig. 4.1 that

$$\left.\begin{array}{l} x = \rho\cos\varphi \\ y = \rho\sin\ \varphi \\ z = z \end{array}\right\}. \tag{4.1.6}$$

The set of Eq. (4.1.6) gives

$$\left.\begin{array}{l} \rho = \sqrt{x^2 + y^2} \\ \varphi = \tan^{-1}\frac{y}{x} \\ z = z \end{array}\right\}. \tag{4.1.7}$$

The system of Eq. (4.1.6) expresses the Cartesian coordinates $(x,\ y,\ z)$ in terms of the cylindrical polar coordinates $(\rho,\ \varphi,\ z)$. The other system (4.1.7) gives the cylindrical polar coordinates $(\rho,\ \varphi,\ z)$ in terms of the cartesian $(x,\ y,\ z)$

In Fig. 4.3, P is the point in a cartesian tri-rectangular frame of reference **OX**, **OY**, **OZ**. We draw a perpendicular PM from P on the plane XOY. Suppose that $OP = r,\ \angle ZOP = \theta$ and $\angle XOM = \varphi$. Then $(r,\ \theta,\ \varphi)$ are called the spherical polar coordinates of P. We note that r = constant represents a sphere through P with center O and radius r; $\theta =$ Constant gives a right circular cone with OZ as axis, θ as the semi-vertical angle and O is the vertex; φ constant specifies a plane through OZ making an angle φ with the coordinate plane ZOX. If $(x,\ y,\ z)$ are the Cartesian coordinates of P, then $(x,\ y,\ z)$ and (r, θ, φ) are connected as follows:

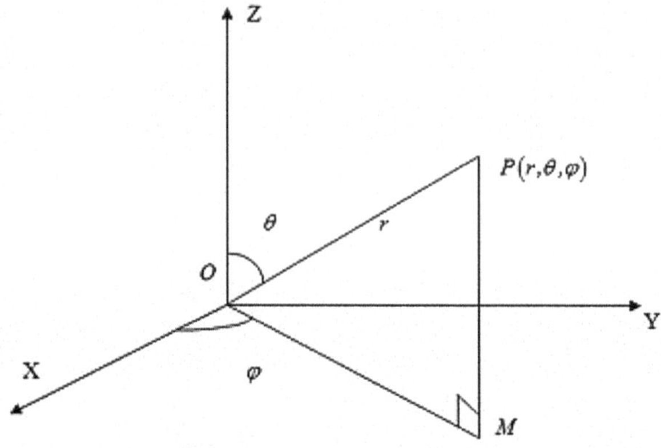

Fig. 4.3 Representation of a point P in spherical polar coordinates

$$\left.\begin{array}{l} x = r\sin\theta\cos\varphi \\ y = r\sin\theta\cos\varphi \\ z = r\cos\theta \end{array}\right\}. \tag{4.1.8}$$

From (4.1.8), we have

$$\left.\begin{array}{l} r = \sqrt{x^2+y^2+z^2} \\ \theta = \cos^{-1}\frac{z}{\sqrt{x^2+y^2+z^2}} \\ \varphi = \tan^{-1}\frac{y}{x} \end{array}\right\}. \tag{4.1.9}$$

4.2 Unit Vectors in Terms of Coordinates

Let $(\lambda,\ \mu,\ \nu)$ represent the orthogonal curvilinear coordinates describing the position of P of a particle at time t. If $\hat{\lambda},\ \hat{\mu},\ \hat{\nu}$ denote the unit vectors in the increasing direction of $\lambda,\ \mu,\ \nu$, then

$$\hat{\lambda} = \frac{\frac{\partial \mathbf{r}}{\partial \lambda}}{\left|\frac{\partial \mathbf{r}}{\partial \lambda}\right|},\quad \hat{\mu} = \frac{\frac{\partial \mathbf{r}}{\partial \mu}}{\left|\frac{\partial \mathbf{r}}{\partial \mu}\right|},\quad \hat{\nu} = \frac{\frac{\partial \mathbf{r}}{\partial \nu}}{\left|\frac{\partial \mathbf{r}}{\partial \nu}\right|}, \tag{4.2.1}$$

where $\mathbf{r}$ is the position vector of the particle relative to a fixed origin O.

Let the particle move to the point $Q\ (\lambda+\delta\lambda,\ \mu+\delta\mu,\ \nu+\delta\nu)$ at time $t+\delta t$ and suppose that $\mathbf{r}+\delta\mathbf{r}$ is the position vector of the point Q relative to O.

We have

$$\mathbf{r} = \mathbf{r}(\lambda,\ \mu,\ \nu), \tag{4.2.2}$$

$$\mathbf{r} + \delta\mathbf{r} = \mathbf{r}\,(\lambda + \delta\lambda,\ \mu + \delta\mu,\ \nu + \delta\nu)\,. \tag{4.2.3}$$

By expanding $\mathbf{r}\,(\lambda + \delta\lambda,\ \mu + \delta\mu,\ \nu + \delta\nu)$ in Taylor's series for three variables and neglecting the higher powers of $\delta\lambda,\ \delta\mu,\ \delta\nu$; we derive

$$\mathbf{r}\,(\lambda + \delta\lambda,\ \mu + \delta\mu,\ \nu + \delta\nu) = \mathbf{r}\,(\lambda,\ \mu,\ \nu) + \delta\lambda\frac{\partial \mathbf{r}}{\partial \lambda} + \delta\mu\frac{\partial \mathbf{r}}{\partial \mu} + \delta\nu\frac{\partial \mathbf{r}}{\partial \nu}. \tag{4.2.4}$$

By the use of the Eqs. (4.2.4) and (4.2.3) yields

$$\delta\mathbf{r} = \delta\lambda\frac{\partial \mathbf{r}}{\partial \lambda} + \delta\mu\frac{\partial \mathbf{r}}{\partial \mu} + \delta\nu\frac{\partial \mathbf{r}}{\partial \nu}. \tag{4.2.5}$$

Letting $\delta t \to 0$ in (4.2.5), we obtain

$$d\mathbf{r} = d\lambda\frac{\partial \mathbf{r}}{\partial \lambda} + d\mu\frac{\partial \mathbf{r}}{\partial \mu} + d\nu\frac{\partial \mathbf{r}}{\partial \nu}. \tag{4.2.6}$$

We note that $d\mathbf{r}$ is a vector tangent to any surface passing through the point $P\,(\mathbf{r})$.

Let C_1 be the curve of intersection of the surfaces $\mu =$ constant and $\nu =$ constant. For the curve C_1, Eq. (4.2.6) transforms to

$$d\mathbf{r} = d\lambda\frac{\partial \mathbf{r}}{\partial \lambda}. \tag{4.2.7}$$

Equation (4.2.7) yields that $\frac{\partial \mathbf{r}}{\partial \lambda}$ is a vector tangent to the curve C_1 for which $\mu =$ constant and $\nu =$ constant. That is $\frac{\partial \mathbf{r}}{\partial \lambda}$ is a vector in λ-direction and hence $\hat{\lambda}$ = a unit vector in λ-direction.

Hence $\hat{\lambda} = \frac{\frac{\partial \mathbf{r}}{\partial \lambda}}{\left|\frac{\partial \mathbf{r}}{\partial \lambda}\right|}$ and similarly $\hat{\mu} = \frac{\frac{\partial \mathbf{r}}{\partial \mu}}{\left|\frac{\partial \mathbf{r}}{\partial \mu}\right|}$ and $\hat{\nu} = \frac{\frac{\partial \mathbf{r}}{\partial \nu}}{\left|\frac{\partial \mathbf{r}}{\partial \nu}\right|}$.

- Let $(x,\ y,\ z)$ denote the Cartesian coordinates of the point P. If $(\rho, \varphi,\ z)$ are the cylindrical polar of P, then $x = \rho\,\cos\varphi,\ \ y = \rho\,\sin\varphi,\ \ z = z$. The position vector $\mathbf{r}$ of the point P is given by $\mathbf{r} = x\hat{i} + y\hat{j} + z\hat{k} = \rho\cos\varphi\hat{i} + \rho\sin\varphi\hat{j} + z\hat{k}$

  $\hat{\rho}$ = A unit vector in ρ-direction=$\frac{\frac{\partial \mathbf{r}}{\partial \rho}}{\left|\frac{\partial \mathbf{r}}{\partial \rho}\right|} = \hat{i}\,\cos\varphi + \hat{j}\,\sin\varphi$,

  $\hat{\varphi}$ = A unit vector in φ-direction=$\frac{\frac{\partial \mathbf{r}}{\partial \varphi}}{\left|\frac{\partial \mathbf{r}}{\partial \varphi}\right|} = -\hat{i}\,\sin\varphi + \hat{j}\,\cos\varphi$

  and $\hat{z}$ = A unit vector in z-direction= $\frac{\frac{\partial \mathbf{r}}{\partial z}}{\left|\frac{\partial \mathbf{r}}{\partial z}\right|} = \hat{k}$.
- If $(r,\ \theta,\ \varphi)$ are the spherical polar coordinates of the point P, then $x = r\sin\theta\cos\varphi$, $y = r\sin\theta\sin\varphi,\ z = r\cos\theta$ and
  $\mathbf{r}$=position vector of P=$x\hat{i} + y\hat{j} + z\hat{k} = r\sin\theta\cos\varphi\hat{i} + r\sin\theta\sin\varphi\hat{j} + r\cos\theta\hat{k}$.
  Then

$$\hat{r} = \frac{\frac{\partial \mathrm{r}}{\partial r}}{\left|\frac{\partial \mathrm{r}}{\partial r}\right|} = \hat{i}\ \sin\theta\cos\varphi + \hat{j}\ \sin\theta\sin\varphi + \hat{k}\ \cos\theta,$$

$$\hat{\theta} = \frac{\frac{\partial \mathrm{r}}{\partial \theta}}{\left|\frac{\partial \mathrm{r}}{\partial \theta}\right|} = \hat{i}\ \cos\theta\cos\varphi + \hat{j}\ \cos\theta\sin\varphi - \hat{k}\ \sin\theta.$$

4.3 Velocity and Acceleration in Cartesian Coordinate System (x, y, z)

Let a particle move along a path C in three dimensions as shown in Fig. 4.4. Let P and Q be the positions of the particle at times t and $t + \delta t$ respectively, where P and Q have position vectors $\mathbf{OP} = \mathbf{r}$, $\mathbf{OQ} = \mathbf{r} + \delta\mathbf{r}$ referred to a fixed origin O. It is noted that $\mathbf{r} = \mathbf{r}(t)$ and $\mathbf{r} + \delta\mathbf{r} = \mathbf{r}(t + \delta t)$. Then the velocity $\mathbf{q}$ of the particle at P is given by

$$\mathbf{q} = \frac{d\mathbf{r}}{dt} = \lim_{\delta t \to 0} \frac{\delta\mathbf{r}}{\delta t} = \lim_{\delta t \to 0} \frac{\mathbf{r}(t + \delta t) - \mathbf{r}(t)}{\delta t}, \tag{4.3.1}$$

and it is a vector tangent to C at P.

We define the acceleration of the particle at P as

$$\mathbf{a} = \frac{d\mathbf{q}}{dt} = \lim_{\delta t \to 0} \frac{\mathbf{q}(t + \delta t) - \mathbf{q}(t)}{\delta t}. \tag{4.3.2}$$

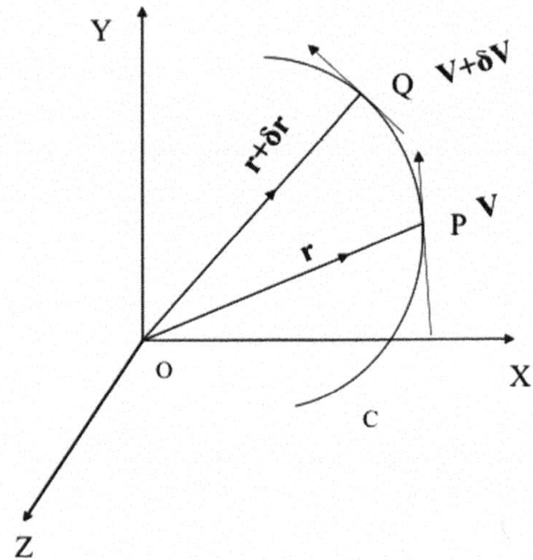

Fig. 4.4 Illustrating the velocity and acceleration of a particle in (x, y, z) system

Introduce a Cartesian coordinate system (x, y, z) with O are the origin and suppose that $\hat{i}, \hat{j}, \hat{k}$ denote unit vectors along **OX**, **OY** and **OZ**, respectively. We assume that

$$\mathbf{r} = \mathbf{r}(t) = x\hat{i} + y\hat{j} + z\hat{k} = x(t)\hat{i} + y(t)\hat{j} + z(t)\hat{k}. \tag{4.3.3}$$

Then we can write $\mathbf{q} = \frac{d\mathbf{r}}{dt} = \frac{dx}{dt}\hat{i} + \frac{dy}{dt}\hat{j} + \frac{dz}{dt}\hat{k} = u\hat{i} + v\hat{j} + w\hat{k}$ where

$$u = \frac{dx}{dt}, \quad v = \frac{dy}{dt}, \quad w = \frac{dz}{dt}, \tag{4.3.4}$$

are the components of the velocity vector **q** along X-axis, Y-axis, and Z-axis respectively. The magnitude of the velocity **q** called the speed is given by

$$q = |\mathbf{q}| = \left|u\hat{i} + v\hat{j} + w\hat{k}\right| = \sqrt{u^2 + v^2 + w^2} = \sqrt{\left(\frac{dx}{dt}\right)^2 + \left(\frac{dy}{dt}\right)^2 + \left(\frac{dz}{dt}\right)^2} = \frac{ds}{dt}, \tag{4.3.5}$$

where s is the length along measured from some initial point to P.

Equation (4.3.2) yields

$$\mathbf{a} = \frac{d\mathbf{q}}{dt} = \frac{d}{dt}\left(u\hat{i} + v\hat{j} + w\hat{k}\right) = \frac{du}{dt}\hat{i} + \frac{dv}{dt}\hat{j} + \frac{dw}{dt}\hat{k} = \frac{d^2x}{dt^2}\hat{i} + \frac{d^2y}{dt^2}\hat{j} + \frac{d^2z}{dt^2}\hat{k}. \tag{4.3.6}$$

The components of acceleration **a** in the direction of +ve X-axis, Y-axis, and Z-axis are as follows:

$$\left.\begin{aligned} a_x &= \frac{du}{dt} = \frac{d^2x}{dt} = \ddot{x} \\ a_y &= \frac{dv}{dt} = \frac{d^2y}{dt^2} = \ddot{y} \\ a_z &= \frac{dw}{dt} = \frac{d^2z}{dt^2} = \ddot{z} \end{aligned}\right\}, \tag{4.3.7}$$

where $\cdot$ denotes differentiation w.r.t. time t.

4.4 Velocity and Acceleration in Cylindrical Polar Coordinate System (ρ, φ, z)

Let a particle be moving along a space curve C and P be the positions of the particle at time t with **OP** $=$ **r** as shown in Fig. 4.5. Introduce a cylindrical polar coordinate system (ρ, φ, z) with O as the origin. If (x, y, z) is the Cartesian coordinate of P, then (x, y, z) and (ρ, φ, z) are connected by the relations:

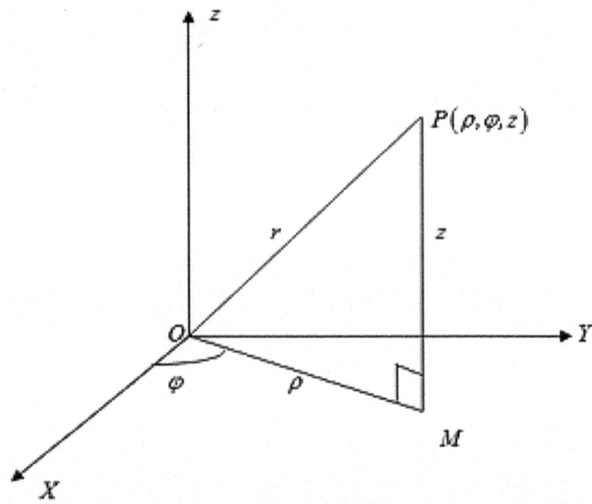

Fig. 4.5 Illustrating the velocity and acceleration of a particle in (r, φ, z) system

$$\left.\begin{aligned} x &= \rho\ \cos\varphi \\ y &= \rho\ \sin\varphi \\ z &= z \end{aligned}\right\}. \tag{4.4.1}$$

Let $\hat{\rho},\ \hat{\varphi},\ \hat{z}$ denote unit vectors along the increasing directions of $\rho,\ \varphi,\ z$ respectively.

We note that

$$\mathbf{r} = x\hat{i} + y\hat{j} + z\hat{k} = \rho\cos\varphi\,\hat{i} + \rho\sin\varphi\,\hat{j} + z\hat{k}. \tag{4.4.2}$$

Now

$$\hat{\rho} = \frac{\frac{\partial \mathbf{r}}{\partial \rho}}{\left|\frac{\partial \mathbf{r}}{\partial \rho}\right|} = \hat{i}\cos\varphi + \hat{j}\sin\varphi, \tag{4.4.3}$$

$$\hat{\varphi} = \frac{\frac{\partial \mathbf{r}}{\partial \varphi}}{\left|\frac{\partial \mathbf{r}}{\partial \varphi}\right|} = \hat{i}\sin\varphi + \hat{j}\cos\varphi, \tag{4.4.4}$$

$$\hat{z} = \frac{\frac{\partial \mathbf{r}}{\partial z}}{\left|\frac{\partial \mathbf{r}}{\partial z}\right|} = \hat{k}, \tag{4.4.5}$$

$$\dot{\hat{\rho}} = -\hat{\imath}\sin\varphi\,\dot{\varphi} + \hat{\jmath}\cos\varphi\,\dot{\varphi} = \left(-\hat{\imath}\sin\varphi + \hat{\jmath}\cos\varphi\right)\dot{\varphi} = \dot{\varphi}\hat{\varphi}, \tag{4.4.6}$$

$$\dot{\hat{\varphi}} = -\hat{\imath}\cos\varphi\,\dot{\varphi} + \hat{\jmath}\sin\varphi\,\dot{\varphi} = \left(-\hat{\imath}\cos\varphi + \hat{\jmath}\sin\varphi\right)\dot{\varphi} = -\hat{\rho}\dot{\varphi}. \tag{4.4.7}$$

Now

$$\mathbf{r} = \rho\cos\varphi i + \rho\sin\varphi\hat{\jmath} + z\hat{z} = \rho\left(\hat{\imath}\cos\varphi + \hat{\jmath}\sin\varphi\right) + z\hat{z} = \rho\hat{\rho} + z\hat{z}.$$

The velocity of the particle at time t is given by

$$\mathbf{q} = \frac{d\mathbf{r}}{dt} = \rho\dot{\hat{\rho}} + \dot{\rho}\hat{\rho} + \dot{z}\hat{z} = \rho\dot{\varphi}\hat{\varphi} + \dot{\rho}\hat{\rho} + \dot{z}\hat{z} = \dot{\rho}\hat{\rho} + \rho\dot{\varphi}\hat{\varphi} + \dot{z}\hat{z}. \tag{4.4.8}$$

The components of the velocity vector $\mathbf{q}$ in (ρ, φ, z) system are as follows:

$q_\rho = \dot{\rho}$ in ρ-direction,

$q_\varphi = \rho\dot{\varphi}$ in φ-direction,

$q_z = \dot{z}$ in z-direction.

If $\mathbf{a}$ is the acceleration of the particle at P, then

$$\mathbf{a} = \frac{d\mathbf{q}}{dt} = \ddot{\rho}\hat{\rho} + \dot{\rho}\dot{\hat{\rho}} + \dot{\rho}\dot{\varphi}\hat{\varphi} + \rho\ddot{\varphi}\hat{\varphi} + \rho\dot{\varphi}\dot{\hat{\varphi}} + \ddot{z}z,$$

i.e., $\mathbf{a} = \ddot{\rho}\hat{\rho} + \dot{\rho}\dot{\varphi}\hat{\varphi} + \dot{\rho}\dot{\varphi}\hat{\varphi} + \rho\ddot{\varphi}\hat{\varphi} + \rho\dot{\varphi}\left(-\hat{\rho}\dot{\varphi}\right) + \ddot{z}\hat{z},$

i.e., $\mathbf{a} = \left(\ddot{\rho} - \rho\dot{\varphi}^2\right)\hat{\rho} + \left(2\dot{\rho}\ddot{\varphi} + \rho\ddot{\varphi}\right)\hat{\varphi} + \ddot{z}\,\hat{z},$

i.e., $\mathbf{a} = \left(\ddot{\rho} - \rho\dot{\varphi}^2\right)\hat{\rho} + \frac{1}{\rho}\left(2\rho\dot{\rho}\dot{\varphi} + \rho^2\ddot{\varphi}\right)\hat{\varphi} + \ddot{z}\hat{z},$

i.e., $\mathbf{a} = \left(\ddot{\rho} - \rho\dot{\varphi}^2\right)\hat{\rho} + \frac{1}{\rho}\left\{\dot{\varphi}\frac{d}{dt}\rho^2 + \rho^2\frac{d}{dt}\dot{\varphi}\right\}\hat{\varphi} + \ddot{z}\,\hat{z},$

i.e., $\mathbf{a} = \left(\ddot{\rho} - \rho\dot{\varphi}^2\right)\hat{\rho} + \frac{1}{\rho}\frac{d}{dt}\left(\rho^2\dot{\varphi}\right)\hat{\varphi} + \ddot{z}\hat{z}.$ (4.4.9)

It follows from Eq. (4.4.9) that

$$a_\rho = \ddot{\rho} - \rho\dot{\varphi}^2 \quad \text{acceleration in } \rho\text{-direction},$$
$$a_\varphi = \frac{1}{\rho}\frac{d}{dt}\left(\rho^2\dot{\varphi}\right) \quad \text{acceleration in } \varphi\text{-direction},$$
$$a_z = \ddot{z} \quad \text{acceleration in } z\text{-direction}.$$

4.5 Velocity and Acceleration in Spherical Polar Coordinate System (r, θ, φ)

In Fig. 4.6, let P be the position of a particle at time t, moving in space. Consider a tri-rectangular frame of reference O-XYZ fixed at a point O. Let $(r,\ \theta,\ \varphi)$ be the spherical polar coordinates of the point P with Cartesian coordinates $(x,\ y,\ z)$. Then

$$x = r\sin\theta\cos\varphi, \quad y = r\sin\theta\sin\varphi, \quad z = r\cos\theta. \tag{4.5.1}$$

If $\mathbf{OP} = \mathbf{r}$ is the position vector of P relative to the point O, then

$$\mathbf{r} = x\hat{i} + y\hat{j} + z\hat{k} = r\left(\sin\theta\cos\varphi\,\hat{i} + \sin\theta\sin\varphi\,\hat{j} + \cos\theta\,\hat{k}\right). \tag{4.5.2}$$

We have

$$\hat{r} == \frac{\frac{\partial \mathbf{r}}{\partial r}}{\left|\frac{\partial \mathbf{r}}{\partial r}\right|} = \hat{i}\sin\theta\cos\varphi + \hat{j}\sin\theta\sin\varphi + \hat{k}\cos\theta, \tag{4.5.3}$$

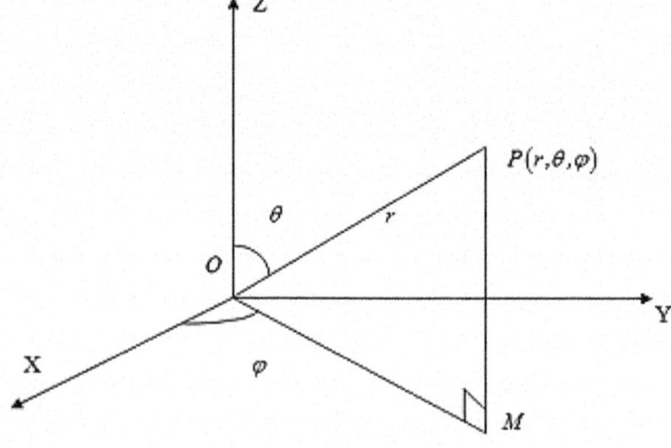

Fig. 4.6 Illustrating the velocity and acceleration of a particle in (r, θ, φ) system

$$\hat{\theta} = \frac{\frac{\partial \mathbf{r}}{\partial \theta}}{\left|\frac{\partial \mathbf{r}}{\partial \theta}\right|} = \hat{i}\cos\theta\cos\varphi + \hat{j}\cos\theta\sin\varphi - \hat{k}\sin\theta, \tag{4.5.4}$$

$$\hat{\varphi} = \frac{\frac{\partial \mathbf{r}}{\partial \varphi}}{\left|\frac{\partial \mathbf{r}}{\partial \varphi}\right|} = -\hat{i}\sin\varphi + \hat{j}\cos\varphi. \tag{4.5.5}$$

Equation (4.5.2) gives

$$\mathbf{r} = r\hat{r}. \tag{4.5.6}$$

Now

$$\dot{\hat{r}} = \hat{i}\left(-\sin\theta\sin\varphi\dot{\varphi} + \cos\varphi\cos\theta\dot{\theta}\right) + \hat{j}\left(\sin\theta\cos\varphi\dot{\varphi} + \sin\varphi\cos\theta\dot{\theta}\right) - \hat{k}\sin\theta\dot{\theta},$$

$$\text{i.e.,}\quad \dot{\hat{r}} = \left[\hat{i}\cos\theta\cos\varphi + \hat{j}\cos\theta\sin\varphi - \hat{k}\sin\theta\right]\dot{\theta} + \left[-\hat{i}\sin\theta\sin\varphi + \hat{j}\sin\theta\cos\varphi\right]\dot{\varphi},$$

$$\text{i.e.,}\quad \dot{\hat{r}} = \hat{\theta}\dot{\theta} + \hat{\varphi}\sin\theta\,\dot{\varphi}. \tag{4.5.7}$$

$$\dot{\hat{\theta}} = \hat{i}\left(-\cos\theta\sin\varphi\dot{\varphi} - \cos\varphi\sin\theta\dot{\theta}\right) + \hat{j}\left(\cos\theta\cos\varphi\dot{\varphi} - \sin\varphi\sin\theta\dot{\theta}\right) - \hat{k}\cos\theta\dot{\theta},$$

$$\text{i.e.,}\quad \dot{\hat{\theta}} = -\left(\hat{i}\sin\theta\cos\varphi + \hat{j}\sin\theta\sin\varphi + \hat{k}\cos\theta\right)\dot{\theta} + \dot{\varphi}\left(-\hat{i}\cos\theta\sin\varphi + \hat{j}\cos\theta\cos\varphi\right),$$

$$\dot{\hat{\theta}} = -\hat{r}\dot{\theta} + \left(-\hat{i}\sin\varphi + \hat{j}\cos\varphi\right)\cos\theta,$$

$$\text{i.e.,}\quad \dot{\hat{\theta}} = -\hat{r}\dot{\theta} + \hat{\varphi}\cos\theta\dot{\varphi}. \tag{4.5.8}$$

$$\dot{\hat{\varphi}} = -\hat{i}\cos\varphi\dot{\varphi} - \hat{j}\sin\varphi\dot{\varphi} = -\left(\hat{i}\cos\varphi + \hat{j}\sin\varphi\right)\dot{\varphi}. \tag{4.5.9}$$

Equations (4.5.3) and (4.5.6) yield

$$\hat{r}\sin\theta + \hat{\theta}\cos\theta = \hat{i}\cos\varphi + \hat{j}\sin\varphi$$

and hence (4.5.9) takes the form

$$\dot{\hat{\varphi}} = -\left(\hat{r}\sin\theta + \hat{\theta}\cos\theta\right)\dot{\varphi}. \tag{4.5.10}$$

The velocity of the particle at time t is given by

$$\mathbf{q} = \frac{d\mathbf{r}}{dt} = \dot{r}\hat{r} + r\dot{\hat{r}} = \dot{r}\hat{r} + r\left(\dot{\theta}\hat{\theta} + \dot{\varphi}\sin\theta\hat{\varphi}\right) = \dot{r}\hat{r} + r\dot{\theta}\hat{\theta} + r\sin\theta\,\dot{\varphi}\,\hat{\varphi}. \tag{4.5.11}$$

The velocity components in $r,\ \theta,\ \varphi$ directions are as follows:

$$\left.\begin{aligned} q_r &= \dot{r} \\ q_\theta &= r\dot{\theta} \\ q_\varphi &= r\sin\theta\dot{\varphi} \end{aligned}\right\}. \tag{4.5.12}$$

Now, the acceleration of the particle at time t is specified by $\mathbf{a} = \frac{d\mathbf{q}}{dt}$, which gives

$$\mathbf{a} = \ddot{r}\hat{r} + \dot{r}\dot{\hat{r}} + \dot{r}\dot{\theta}\hat{\theta} + r\ddot{\theta}\hat{\theta} + r\dot{\theta}\dot{\hat{\theta}} + \dot{r}\sin\theta\dot{\varphi}\hat{\varphi} + r\cos\theta\,\dot{\theta}\dot{\varphi}\hat{\varphi} + r\sin\theta\ddot{\varphi}\hat{\varphi} + r\sin\theta\dot{\varphi}\dot{\hat{\varphi}}. \tag{4.5.13}$$

By the use of (4.5.7), (4.5.8), (4.5.9) in Eq. (4.5.13), it is obtained that $\mathbf{a} = a_r\hat{r} + a_\theta\hat{\theta} + a_\varphi\hat{\varphi}$, where

$$\left.\begin{aligned} a_r &= \ddot{r} - r\dot{\theta}^2 - r\sin^2\theta\dot{\varphi}^2 \\ a_\theta &= 2\dot{r}\dot{\theta} + r\ddot{\theta} - r\sin\theta\cos\theta\dot{\varphi}^2 = \tfrac{1}{r}\tfrac{d}{dt}\left(r^2\dot{\theta}\right) - r\sin\theta\cos\theta\dot{\varphi}^2 \\ a_\varphi &= r\sin\theta\ddot{\varphi} + 2\dot{r}\dot{\varphi}\sin\theta + 2r\dot{\theta}\dot{\varphi}\cos\theta = \tfrac{1}{r\sin\theta}\tfrac{d}{dt}\left(r^2\sin^2\theta\dot{\varphi}\right) \end{aligned}\right\}. \tag{4.5.14}$$

4.6 Illustrations

Example 4.1 A particle moves on the inner surface of a smooth cone, of vertical angle 2α, being acted on by a force towards the vertex of the cone, and its direction of motion cuts the generator at a constant angle β. Find the motion and the law of force.

Solution Figure 4.7 depicts the situation at time t. Let mF denote the force, where m is the mass of the particle. Let R be the reaction of the cone on the particle. Let $\mathrm{P}(r,\ \theta,\ \phi)$ be the position of the particle at time t.

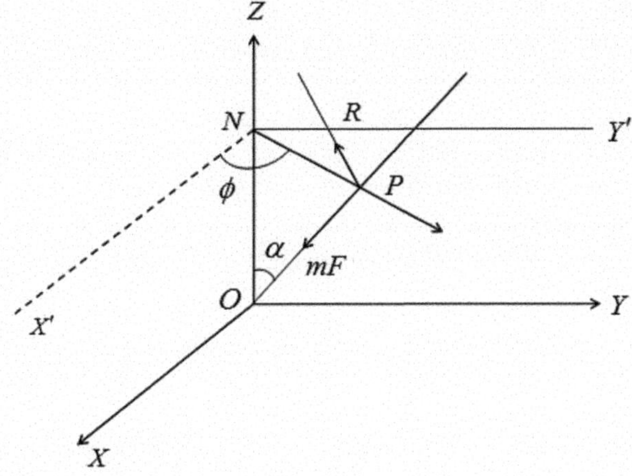

Fig. 4.7 Configuration of a particle moving on the smooth inner surface of a cone of vertical angle 2α

The equations of motion of the particle are

$$m\left(\ddot{r} - r\dot{\theta}^2 - r\sin^2\theta\dot{\phi}^2\right) = -mF, \tag{4.6.1}$$

$$m\left\{\frac{1}{2}\frac{d}{dt}\left(r^2\dot{\theta}\right) - r\sin\theta\cos\theta\dot{\phi}^2\right\} = -R, \tag{4.6.2}$$

$$m\,\frac{1}{r\sin\theta}\,\frac{d}{dt}\left(r^2\sin^2\theta\dot{\phi}\right) = 0. \tag{4.6.3}$$

In the present case, $\theta = \alpha =$ a constant and hence

$$\dot{\theta} = 0. \tag{4.6.4}$$

Further, the direction of motion cuts **OP** at a constant angle β as visualized in Fig. 4.8. Resolving the velocity vector **V** at P in the direction of $\hat{\theta}$ $\hat{\varphi}$, we obtain

$$V_r = \dot{r} = V\cos\beta, \tag{4.6.5}$$

$$V_\theta = r\sin\theta\dot{\phi} = V\sin\beta. \tag{4.6.6}$$

(4.6.5) $\div$ (4.6.6) gives

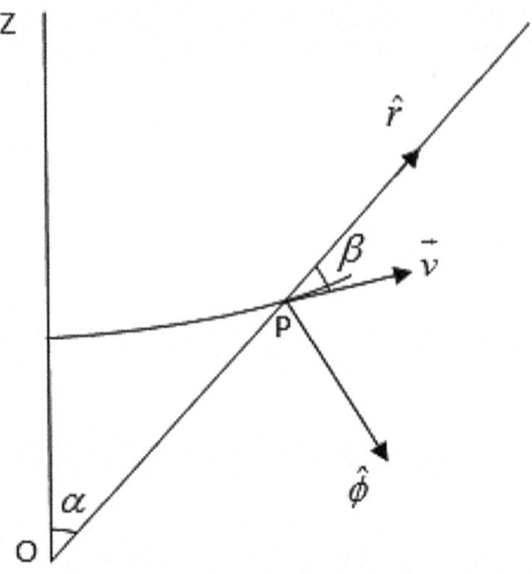

Fig. 4.8 Resolution of the particle's velocity at point P

$$\frac{\dot{r}}{r\sin\theta\dot{\phi}} = \cot\beta,$$

$$\text{i.e.,}\quad \dot{r} = r\sin\alpha\cot\beta\dot{\phi}. \tag{4.6.7}$$

By the use of the relation (4.6.4), Eqs. (4.6.1)–(4.6.3) become respectively,

$$\ddot{r} - r\sin^2\alpha\dot{\phi} = -F, \tag{4.6.8}$$

$$r\sin\alpha\cos\alpha\dot{\phi}^2 = \frac{R}{m}, \tag{4.6.9}$$

$$\frac{d}{dt}\left(r^2\dot{\phi}\right) = 0. \tag{4.6.10}$$

Equation (4.6.10) yields

$$r^2\dot{\phi} = C = a\ \text{ constant.} \tag{4.6.11}$$

On the use of (4.6.11), (4.6.7) reduces to

$$\dot{r} = r\sin\alpha\cot\beta\frac{C}{r^2} = \frac{C}{r}\sin\alpha\cot\beta. \tag{4.6.12}$$

Differentiation of Eq. (4.6.12) w.r.t. t, we derive

$$\ddot{r} = -\frac{C}{r^2} \sin\alpha \cot\beta \, \dot{r} = -\frac{C^2}{r^3} \sin^2\alpha \cot^2\beta. \tag{4.6.13}$$

Equation (4.6.8) gives

$$F = \frac{C^2}{r^3} \sin^2\alpha \cot^2\beta + r \sin^2\alpha \frac{C^2}{r^4} = \frac{C^2 \sin^2\alpha \csc^2\beta}{r^3} = \frac{\mu}{r^3}, \tag{4.6.14}$$

where $\mu = C^2 \sin^2\alpha \csc^2\beta$.

Thus, we see that $F \propto \frac{1}{r^3}$ which is the law of force.

The velocity $\mathbf{V}$ of the particle is given by

$$\mathbf{V} = \dot{r}\hat{r} + r \sin\alpha \dot{\phi}\hat{\phi}. \tag{4.6.15}$$

Equation (4.6.15) gives

$$V^2 = \dot{r}^2 + r^2\dot{\phi}^2 \sin^2\alpha = \left(\frac{C}{r} \sin\alpha \; \cot\beta\right)^2 + r^2\frac{C^2}{r^4} \sin^2\alpha = \frac{C^2}{r^2} \sin^2\alpha \csc^2\beta. \tag{4.6.16}$$

Thus,

$$V = |\mathbf{V}| = \frac{C \sin\alpha \; \csc\beta}{r} = \frac{\sqrt{\mu}}{r}. \tag{4.6.17}$$

Therefore, $V \propto \frac{1}{r}$.

Equation (4.6.9) yields

$$R = m\, r \sin\alpha \cos\alpha \, \dot{\phi}^2 = m\, r \sin\alpha \cos\alpha \frac{C^2}{r^4} = \frac{\nu}{r^3}, \tag{4.6.18}$$

where $\nu = mC^2 \sin\alpha \cos\alpha$.

It follows from Eq. (4.6.18) that $R \propto \frac{1}{r^3}$.

Equations (4.6.14), (4.6.17) and (4.6.18) define the motion of the particle.

Equation (4.6.7) can be expressed as

$$\frac{\dot{r}}{r} = \sin\alpha \cot\beta \dot{\phi},$$

which may be simplified to

$$\frac{dr}{r} = \sin\alpha \cot\beta \, d\phi,$$

and integration of this differential equation leads to the relation

$$r = r_0\, e^{\sin\alpha \cot\beta\, \phi},$$

which represents the path of the particle.

Example 4.2 A particle is projected horizontally along the interior surface of a smooth hemisphere whose axis is vertical and whose vertex is downwards; the point of projection being at an angular distance β from the lowest point, show that the initial velocity of the particle may just ascend to the rim of the hemisphere is $\sqrt{2ag \sec\beta}$.

Solution In Fig. 4.9, let P(r, θ, ϕ) be the position of the particle at time t, and R denote the reaction of the interior surface on the particle. The velocity of the particle at P is given by

$$\mathbf{v} = \dot{r}\hat{r} + r\dot{\theta}\hat{\theta} + r\sin\theta\dot{\phi}\,\hat{\phi}. \tag{4.6.19}$$

Here $r = a =$ radius of the hemisphere, which is a constant and hence

$$\dot{r} = 0. \tag{4.6.20}$$

By the use of (4.6.20), (4.6.19) reduces to

$$\mathbf{v} = a\dot{\theta}\hat{\theta} + a\sin\theta\dot{\phi}\hat{\phi}$$

and it gives

$$v^2 = a^2\dot{\theta}^2 + a^2\sin^2\theta\,\dot{\phi}^2. \tag{4.6.21}$$

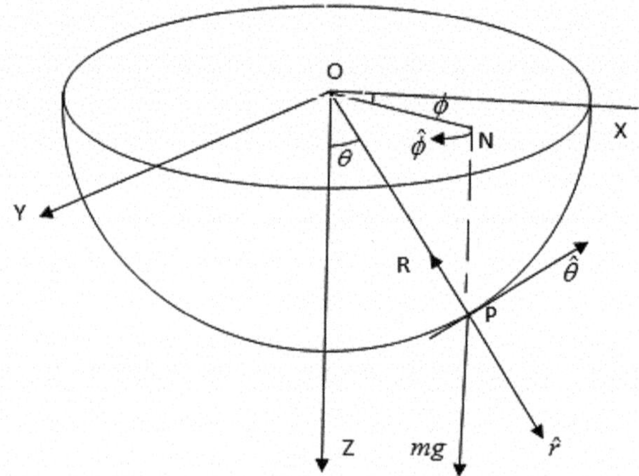

Fig. 4.9 Illustration for Example 4.2

The equation of motion of the particle in φ direction is

$$\frac{m}{a \sin\theta} \frac{d}{dt}\left(a^2 \sin^2\theta\, \dot{\varphi}\right) = 0,$$

which gives

$$\sin^2\theta\, \dot{\phi} = h \ \text{(say)}. \tag{4.6.22}$$

The energy equation yields $\frac{1}{2}mv^2 - mga\cos\theta = C$, C being a constant.
That is

$$\frac{1}{2}m\left(a^2\dot{\theta}^2 + a^2\sin^2\theta\dot{\phi}^2\right) - mga\cos\theta = C. \tag{4.6.23}$$

Initially

$$v_\theta = 0, \quad v_\varphi = u \ \text{(say)} \ \text{when} \ \theta = \beta,$$

$$\text{i.e.,} \quad \dot{\theta} = 0, \quad a\ \sin\beta\dot{\phi} = u \ \text{when} \ \theta = \beta,$$

$$\text{i.e.,} \quad \dot{\theta} = 0, \quad \dot{\phi} = \frac{u}{a \sin\beta} \ \text{at} \ \theta = \beta.$$

Subjecting this condition in (4.6.22), we obtain

$$h = \sin^2\beta \frac{u}{a\sin\beta} = \frac{u\sin\beta}{a}. \tag{4.6.24}$$

Eliminating h from (4.6.22) and (4.6.24), we get

$$\dot{\phi} = \frac{u\sin\beta}{a\sin^2\theta}.$$

Eq. (4.6.23) yields

$$C = \frac{1}{2}mu^2 - mga\cos\beta$$

and so Eq. (4.6.23) takes the from

$$\frac{m}{2}\left(a^2\dot{\theta}^2 + a^2\sin^2\theta\frac{u^2\sin^2\beta}{a^2\sin^4\theta}\right) - mga\cos\theta = \frac{1}{2}mu^2 - mga\cos\beta,$$

$$\text{i.e.,}\quad a^2\dot{\theta}^2 + \frac{u^2 \sin^2\beta}{\sin^2\theta} - 2ga\cos\theta = u^2 - 2ga\cos\beta. \tag{4.6.25}$$

We choose u in such a way that $\dot{\theta} = 0$ at $\theta = \frac{\pi}{2}$.
Equation (4.6.25) gives

$$u^2 \sin^2\beta = u^2 - 2ga\cos\beta.$$

On simplification, we obtain

$$u^2 = 2ga\sec\beta,$$

$$\text{i.e.,}\quad u = \sqrt{2ga\sec\beta}.$$

It establishes the required result.

Example 4.3 A heavy particle of mass m moves on the smooth inner surface of a sphere of radius a and its greatest and least depths below the center are $\frac{a}{2}$ and $\frac{a}{4}$. Show that when the depth below the center is z, the normal reaction is $3mg\frac{z+\frac{a}{2}}{a}$.

Solution Let O be the center of the sphere and introduce a Cartesian coordinate system (x, y, z) with O as the origin, and the Z-axis vertically downwards as depicted in Fig. 4.10. Let P denote the position of the particle with spherical polar coordinates (r, θ, ϕ) at time t. Let R denote the normal reaction of the sphere on the particle at P.

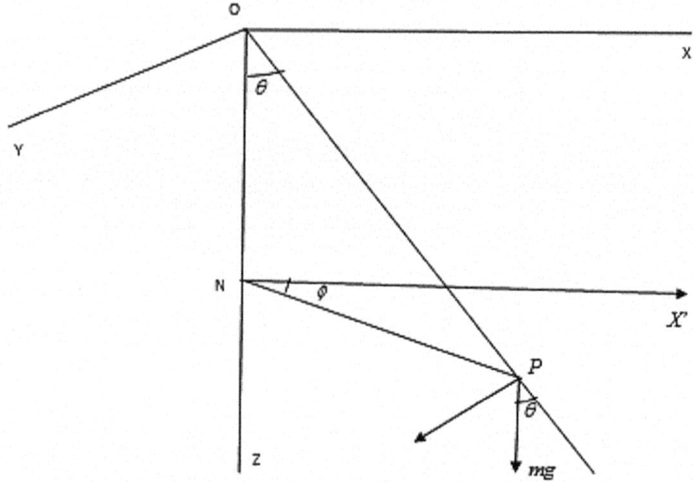

Fig. 4.10 Illustrated diagram for Example 4.3

The equations of motion of the particle are

$$m\left(\ddot{r} - r\dot{\theta}^2 - r\sin^2\theta\,\dot{\phi}^2\right) = -R + mg\,\cos\theta, \tag{4.6.26}$$

$$m\left\{\frac{1}{r}\frac{d}{dt}\left(r^2\dot{\theta}\right) - r\sin\theta\cos\theta\,\dot{\phi}^2\right\} = -mg\sin\theta, \tag{4.6.27}$$

$$m\frac{1}{r\sin\theta}\frac{d}{dt}\left(r^2\sin^2\theta\dot{\phi}\right) = 0. \tag{4.6.28}$$

In the present case $r = a =$ radius of the sphere for which $\dot{r} = \ddot{r} = 0$. Hence the Eqs. (4.6.26)–(4.6.28) respectively take the forms

$$-a\dot{\theta}^2 - a\sin^2\theta\,\dot{\phi}^2 = -\frac{R}{m} + g\cos\theta, \tag{4.6.29}$$

$$a\ddot{\theta} - a\sin\theta\cos\theta\,\dot{\phi}^2 = -g\sin\theta, \tag{4.6.30}$$

$$\frac{d}{dt}\left(\sin^2\theta\,\dot{\phi}\right) = 0. \tag{4.6.31}$$

Equation (4.6.31) gives

$$\sin^2\theta\,\dot{\phi} = \text{a constant} = h\ \ \text{(say)},$$

$$\text{i.e.,}\quad \dot{\phi} = \frac{h}{\sin^2\theta}. \tag{4.6.32}$$

By the use of (4.6.32), Eq. (4.6.30) reduces to

$$\ddot{\theta} = \frac{h^2\cos\theta}{\sin^3\theta} - \frac{g}{a}\sin\theta,$$

$$\text{i.e.,}\quad 2\dot{\theta}\frac{d\dot{\theta}}{dt} = \frac{2h^2\cos\theta\,d\theta}{\sin^3\theta\ \ dt} - \frac{2g}{a}\sin\theta\frac{d\theta}{dt},$$

$$\text{i.e.,}\quad 2\dot{\theta}d\dot{\theta} = \frac{2h^2 d\,(\sin\theta)}{\sin^3\theta} - \frac{2g}{a}\sin\theta\,d\theta.$$

Integrating the above differential equation, we obtain

$$\dot{\theta}^2 = -\frac{h^2}{\sin^2\theta} + \frac{2g}{a}\cos\theta + C, \tag{4.6.33}$$

where C is the constant of integration.

Equation (4.6.33) can be simplified to

$$\dot{\theta}^2 = -\frac{h^2}{1-\cos^2\theta} + \frac{2g}{a}\cos\theta + C,$$

$$\text{i.e.,}\quad \dot{\theta}^2 = -\frac{h^2}{1-\frac{z^2}{a^2}} + \frac{2g}{a}\frac{z}{a} + C,$$

$$\text{i.e.,}\quad \dot{\theta}^2 = -\frac{a^2h^2}{a^2-z^2} + \frac{2gz}{a^2} + C. \tag{4.6.34}$$

In Eq. (4.6.34), C and h are two arbitrary constants to be determined from the given conditions:

$$\dot{\theta} = 0 \quad \text{at} \quad z = \frac{a}{2}, \tag{4.6.35}$$

and

$$\dot{\theta} = 0 \quad \text{at} \quad z = \frac{a}{4}. \tag{4.6.36}$$

Subjecting Eq. (4.6.34) to the conditions (4.6.35) and (4.6.36), we obtain

$$-\frac{4h^2}{3} + \frac{ga}{a} + C = 0, \tag{4.6.37}$$

and

$$-\frac{16h^2}{15} + \frac{ga}{2a} + C = 0. \tag{4.6.38}$$

Solving (4.6.37) and (4.6.38), we derive

$$h^2 = \frac{15g}{8a} \text{ and } C = \frac{3g}{2a}.$$

On substitution of these expressions, Eq. (4.6.29) yields

$$
\begin{aligned}
\frac{R}{m} &= g\cos\theta + a\dot{\theta}^2 + a\sin^2\theta\dot{\phi}^2,\\
&= g\cos\theta + a\left(-\frac{h^2}{\sin^2\theta} + \frac{2g}{a}\cos\theta + C\right) + a\sin^2\theta\frac{h^2}{\sin^4\theta},\\
&= g\cos\theta - \frac{ah^2}{\sin^2\theta} + 2g\cos\theta + aC + \frac{ah^2}{\sin^2\theta},\\
&= 3g\cos\theta + aC,\\
&= 3g\cos\theta + a\frac{3g}{2a},\\
&= 3g\left(\cos\theta + \frac{1}{2}\right),\\
&= 3g\left(\frac{z}{a} + \frac{1}{2}\right),\\
&= \frac{3g}{a}\left(z + \frac{a}{2}\right)
\end{aligned}
$$

which is the required result.

Example 4.4 A particle moves on a smooth sphere under no force except the pressure of the surface; show that its path is given by the equation $\cot\theta = \cot\beta\cos\phi$, where θ and ϕ are its angular coordinates.

Solution Let O be the center of the sphere with a as the radius. In Fig. 4.11, introduce a tri-rectangular Cartesian coordinate system $(x,\ y,\ z)$ with the point O as the origin. Let R denote the normal reaction of the sphere on the particle at the point P on the sphere where P is the position of the particle at time t with spherical polar coordinates $(r,\ \theta,\ \varphi)$.

Equations of motion of the particle are

$$m\left(\ddot{r} - r\dot{\theta}^2 - r\sin^2\theta\dot{\phi}^2\right) = R, \tag{4.6.39}$$

$$m\left\{\frac{1}{r}\frac{d}{dt}\left(r^2\dot{\theta}\right) - r\sin\theta\cos\theta\dot{\phi}^2\right\} = 0, \tag{4.6.40}$$

$$m\frac{1}{r\sin\theta}\frac{d}{dt}\left(r^2\sin^2\theta\dot{\phi}\right) = 0. \tag{4.6.41}$$

In the present case

$$\left.\begin{array}{l} r = a \\ \dot{r} = \ddot{r} = 0 \end{array}\right\}. \tag{4.6.42}$$

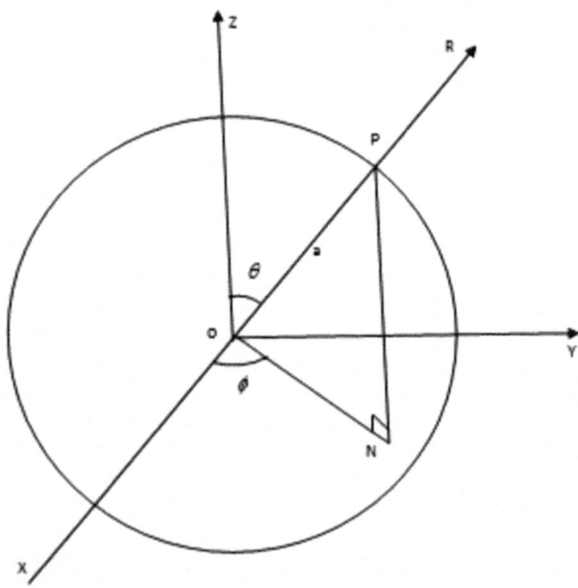

Fig. 4.11 Illustrated diagram for Example 4.4

On substitution of (4.6.42), in (4.6.39)–(4.6.41), we have

$$a\dot{\theta}^2 + a\sin^2\theta\dot{\phi}^2 = \frac{R}{m}, \tag{4.6.43}$$

$$a\ddot{\theta} + a\sin\theta\cos\theta\dot{\phi}^2 = 0, \tag{4.6.44}$$

$$\frac{d}{dt}\left(\sin^2\theta\dot{\phi}\right) = 0. \tag{4.6.45}$$

Equation (4.6.45) gives

$$\sin^2\theta\dot{\phi} = \text{a constant } = h \ \ (\text{say}),$$

$$\text{i.e.,} \quad \dot{\phi} = \frac{h^2}{\sin^2\theta}. \tag{4.6.46}$$

Equation (4.6.44) yields

$$\ddot{\theta} = \frac{h^2}{\sin^3\theta}\cos\theta,$$

$$\text{i.e.,}\quad \frac{d\dot{\theta}}{dt} = \frac{h^2}{\sin^3\theta}\cos\theta,$$

$$\text{i.e.,}\quad \frac{d\dot{\theta}}{d\theta}\dot{\theta} = \frac{h^2}{\sin^3\theta}\cos\theta,$$

$$\text{i.e.,}\quad \dot{\theta}\,d\dot{\theta} = \frac{h^2}{\sin^3\theta}\cos\theta\;d\theta.$$

Integrating the above equation, we obtain

$$\dot{\theta}^2 = C - \frac{h^2}{\sin^2\theta}, \tag{4.6.47}$$

where C is the constant of integration.

Initially

$$\dot{\theta} = 0,\quad \theta = \beta \text{ when } t = 0.$$

Hence (4.6.47) gives

$$C = \frac{h^2}{\sin^2\beta}.$$

Thus Eq. (4.6.47) reduces to

$$\dot{\theta}^2 = h^2(\csc^2\beta - \csc^2\theta),$$

$$\text{i.e.,}\quad \dot{\theta} = h\sqrt{\csc^2\beta - \csc^2\theta}. \tag{4.6.48}$$

Now (4.6.48) ÷ (4.6.46) gives

$$\frac{\dot{\theta}}{\dot{\phi}} = \frac{\sqrt{\csc^2 \beta - \csc^2 \theta}}{\csc^2 \theta},$$

$$\text{i.e.,} \quad \frac{d\theta}{d\phi} = \frac{\sqrt{\csc^2 \beta - \csc^2 \theta}}{\csc^2 \theta},$$

$$\text{i.e.,} \quad \frac{\csc^2 \theta \, d\theta}{\sqrt{\csc^2 \beta - \csc^2 \theta}} = d\phi,$$

$$\text{i.e.,} \quad \frac{\csc^2 \theta \, d\theta}{\sqrt{\cot^2 \beta - \cot^2 \theta}} = d\phi,$$

$$\text{i.e.,} \quad I = \phi + C_2, \tag{4.6.49}$$

where

$$I = \int \frac{\csc^2 \theta \, d\theta}{\sqrt{\cot^2 \beta - \cot^2 \theta}}.$$

Consider the transformation

$$\xi = \cot \theta,$$

$$\text{i.e.,} \quad d\xi = -\csc^2 \theta d\theta,$$

$$\therefore \quad I = \int \frac{-d\xi}{\sqrt{\lambda^2 - \xi^2}}; \quad \lambda = \cot \beta,$$

$$\text{i.e.,} \quad I = \cos^{-1} \frac{\xi}{\lambda} = \cos^{-1} \left(\frac{\cot \theta}{\cot \beta} \right).$$

Hence (4.6.49) gives

$$\cos^{-1} \left(\frac{\cot \theta}{\cot \beta} \right) = \phi + C_2. \tag{4.6.50}$$

In (4.6.50), C_2 is the constant of integration to be determined from the condition:

$$\theta = \beta \text{ at } \phi = 0. \tag{4.6.51}$$

Subjecting (4.6.50) in (4.6.51), we have $C_2 = 0$.

Thus Eq. (4.6.50) reduces to

$$\cos^{-1} \left(\frac{\cot \theta}{\cot \beta} \right) = \phi,$$

$$\text{i.e.,} \quad \cot \theta = cot \beta \cos \phi,$$

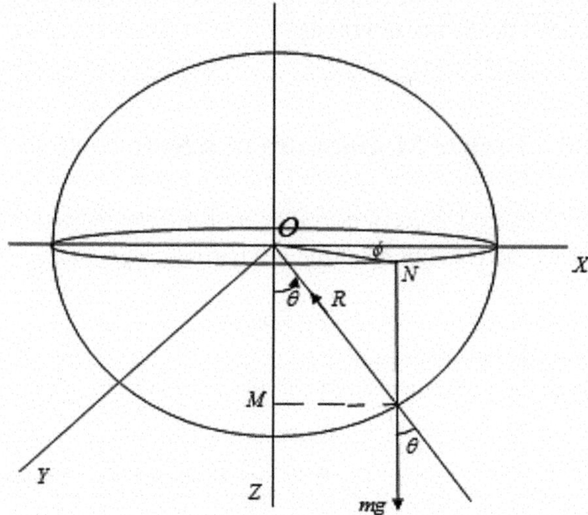

Fig. 4.12 Schematic diagram for Example for 4.5

which is the required result.

Example 4.5 A particle is attached to one end of a string, of length l, the other end of which is tied to a fixed point O. When the string is inclined at an acute angle α to the downward-drawn vertical the particle is projected horizontally and perpendicular to the string with a velocity V; find the tension of the string.

Solution In Fig. 4.12, let OX, OY, and OZ be three mutually perpendicular axes at O where OZ is downward vertical. Let P denote the position of the particle at time t with spherical polar coordinates $(r,\ \theta,\ \varphi)$.

The forces acting on the particle are
(I) The tension T of the string,
(II) The weight mg of the particle. Equations of the motion of the particle are

$$m\left(\ddot{r} - r\dot{\theta}^2 - r\sin^2\theta\,\dot{\varphi}^2\right) = -T + mg\cos\theta, \tag{4.6.52}$$

$$m\left\{\frac{1}{r}\frac{d}{dt}\left(r^2\dot{\theta}\right) - r\sin\theta\cos\theta\dot{\varphi}^2\right\} = -mg\cos\theta, \tag{4.6.53}$$

$$m\frac{1}{r\sin\theta}\frac{d}{dt}\left(r^2\sin^2\theta\,\dot{\varphi}\right) = 0. \tag{4.6.54}$$

For the present problem under consideration,

$$r = l,\ \ \dot{r} = \ddot{r} = 0. \tag{4.6.55}$$

On substitution of (4.6.55) in (4.6.52)–(4.6.54), we obtain

$$m\left(-l\dot{\theta}^2 - l\sin^2\theta\,\dot{\varphi}^2\right) = -T + mg\cos\theta, \tag{4.6.56}$$

$$m\left\{\frac{1}{l}\frac{d}{dt}\left(l^2\dot{\theta}\right) - l\sin\theta\cos\theta\,\dot{\varphi}^2\right\} = -mg\sin\theta, \tag{4.6.57}$$

$$\frac{d}{dt}\left(l^2\sin^2\theta\,\dot{\varphi}\right) = 0. \tag{4.6.58}$$

(4.6.58) gives

$$\sin^2\theta\,\dot{\varphi} = C = \text{a constant}. \tag{4.6.59}$$

Initially,

$$\begin{aligned} & l\sin\theta\,\dot{\varphi} = V, \quad \theta = \alpha \ \text{ at } \ t = 0, \\ \text{i.e.,}\ & l\sin\alpha\,\dot{\varphi} = V \ \text{ at } \ t = 0, \\ \text{i.e.,}\ & \dot{\varphi} = \frac{V}{l\sin\alpha} \ \text{ at } \ t = -0. \end{aligned}$$

Equation (4.6.59) yields

$$C = \sin^2\alpha\,\frac{V}{l\sin\alpha} = \frac{V\sin\alpha}{l}.$$

Hence Eq. (4.6.59) takes the form

$$\dot{\varphi} = \frac{V\sin\alpha}{l\sin^2\theta}. \tag{4.6.60}$$

Equation (4.6.57) can be simplified to

$$\begin{aligned} & \ddot{\theta} - \sin\theta\cos\theta\,\dot{\varphi}^2 = -\frac{g}{l}\sin\theta, \\ \text{i.e.,}\ & \ddot{\theta} - \frac{V^2\sin^2\alpha}{l^2}\frac{\cos\theta}{\sin^3\theta} = -\frac{g}{l}\sin\theta, \\ \text{i.e.,}\ & 2\dot{\theta}\ddot{\theta} - \frac{2V^2\sin^2\alpha\cos\theta}{l^2\sin^3\theta}\dot{\theta} = -\frac{2g}{l}\sin\theta\dot{\theta}, \end{aligned}$$

$$\text{i.e.,}\quad 2\dot{\theta}\frac{d\dot{\theta}}{dt} - \frac{2V^2\sin^2\alpha\cos\theta}{l^2\sin^3\theta}\frac{d\theta}{dt} = -\frac{2g}{l}\sin\theta\frac{d\theta}{dt},$$

$$\text{i.e.,}\quad 2\dot{\theta}d\dot{\theta} - \frac{2V^2\sin^2\alpha}{l^2\sin^3\theta}\cos\theta d\theta = -\frac{2g}{l}\sin\theta d\theta,$$

$$\text{i.e.,}\quad \dot{\theta}^2 + \frac{V^2\sin^2\alpha}{l^2}\frac{1}{\sin^2\theta} = \frac{2g}{l}\cos\theta + C_1. \tag{4.6.61}$$

Initially $\dot{\theta} = 0,\ \ \theta = \alpha$ when $t = 0$.
Equation (4.6.61) gives

$$\frac{V^2}{l^2} = \frac{2g}{l}\cos\alpha + C_1,$$

$$C_1 = \frac{V^2}{l^2} - \frac{2g}{l}\cos\alpha. \tag{4.6.62}$$

Substituting the value of C_1 from (4.6.62) in (4.6.61), we have

$$\dot{\theta}^2 + \frac{V^2}{l^2}\left(\frac{\sin^2\alpha}{\sin^2\theta} - 1\right) = \frac{2g}{l}\left(\cos\theta - \cos\alpha\right),$$

$$\text{i.e.,}\quad \dot{\theta}^2 = \frac{V^2\sin^2\alpha}{l^2}\left(\frac{1}{\sin^2\alpha} - \frac{1}{\sin^2\theta}\right) - \frac{2g}{l}\left(\cos\alpha - \cos\theta\right). \tag{4.6.63}$$

Equation (4.6.56) gives

$$T = mg\,\cos\theta + m\left(l\dot{\theta}^2 + l\sin^2\theta\,\dot{\varphi}^2\right). \tag{4.6.64}$$

Eliminating $\dot{\theta}^2$, $\dot{\varphi}^2$ from (4.6.60), (4.6.63) and (4.6.64), we obtain

$$T = m\left[g\cos\theta + l\left\{\frac{V^2\sin^2\alpha}{l^2}\left(\frac{1}{\sin^2\alpha} - \frac{1}{\sin^2\theta}\right) - \frac{2g}{l}\left(\cos\alpha - \cos\theta\right)\right\} + l\sin^2\theta\frac{V^2\sin^2\alpha}{l^2\sin^4\theta}\right],$$

$$\text{i.e.,}\quad T = m\left[g\cos\theta + \frac{V^2\sin^2\alpha}{l}\left(\frac{1}{\sin^2\alpha} - \frac{1}{\sin^2\theta}\right) - 2g\left(\cos\alpha - \cos\theta\right) + \frac{V^2\sin^2\alpha}{l\sin^2\theta}\right],$$

$$\text{i.e.,}\quad T = m\left[g\cos\theta + \frac{V^2}{l} - \frac{V^2\sin^2\alpha}{l\sin^2\theta} - 2g\left(\cos\alpha - \cos\theta\right) + \frac{V^2\sin^2\alpha}{l\sin^2\theta}\right],$$

$$\text{i.e.,}\quad T = m\left[\frac{V^2}{l} + g\left(3\cos\theta - 2\cos\alpha\right)\right].$$

Example 4.6 A heavy particle moves on a rough vertical circular cylinder of radius a and is projected horizontally with velocity V. Prove that at the point where the path cuts the generator at an angle φ, the velocity v is given by

$$\frac{ag}{v^2\sin^2\varphi} = \frac{ag}{V^2} + 2\mu\log\left(\cot\varphi + \csc\varphi\right),$$

where μ is the coefficient of friction.

Solution Let A be the point of projection and P be the position of the particle at time t. Let the path of the particle at P cut the generator at an angle φ. Let O be the center of the normal section of the cylinder through O. From Fig. 4.13, take O as the origin, X-axis along **OA**, Y-axis along **OB**, and Z-axis along the downward vertical through O.

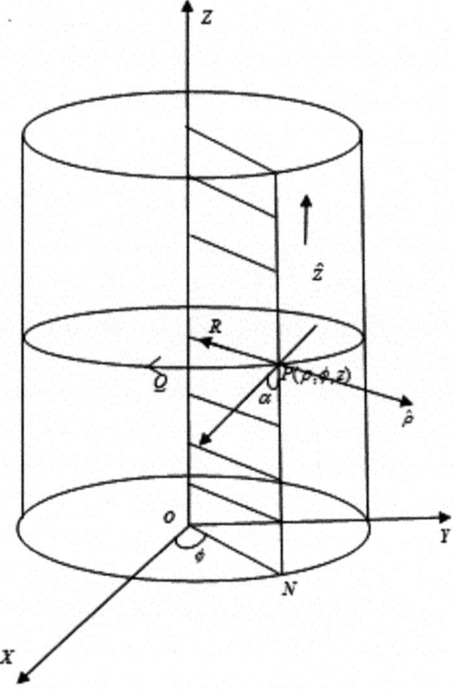

Fig. 4.13 Geometry for Example 4.6

Let the generator through P meet the XY plane at Q.

Let OQ $= r$, $\angle$AOQ $= \theta$, QP $= z$. Then, (r, θ, z) may be regarded as the cylindrical polar coordinates of P.

The forces acting on the particle at P are

(i) the normal reaction R of the surface at P $\parallel$ to **QO**,

(ii) the friction μR in the direction opposite to **v**,

(iii) the weight mg acting vertically downward along Z-axis.

Equations of motion of the particle are

$$m\left(\ddot{r} - r\dot{\theta}^2\right) = -R, \tag{4.6.65}$$

$$m\frac{1}{r}\frac{d}{dt}\left(r^2\dot{\theta}\right) = -\mu R \sin\varphi, \tag{4.6.66}$$

$$m\ddot{z} = mg - \mu R \cos\varphi. \tag{4.6.67}$$

From the geometry of the problem, it follows that

$$r = a. \tag{4.6.68}$$

By the use of (4.6.68) in (4.6.65) and (4.6.66), we obtain

$$ma\dot{\theta}^2 = R, \tag{4.6.69}$$

$$ma\ddot{\theta} = -\mu R \sin\varphi. \tag{4.6.70}$$

Eliminating R from (4.6.69) and (4.6.70), we have

$$\frac{\ddot{\theta}}{\dot{\theta}^2} = -\mu \sin\varphi,$$

$$\text{i.e.,} \quad \ddot{\theta} = -\mu\dot{\theta}^2 \sin\varphi. \tag{4.6.71}$$

Further, we have

$$v\cos\varphi = \dot{z}, \tag{4.6.72}$$

$$v\sin\varphi = r\dot{\theta} = a\dot{\theta}. \tag{4.6.73}$$

(4.6.71) and (4.6.73) yield

$$\ddot{\theta} = -\mu \frac{v^2 \sin^2 \varphi}{a^2} \sin \varphi = -\frac{\mu v^2}{a^2} \sin^3 \varphi,$$

i.e., $$\frac{d\dot{\theta}}{dt} = -\frac{\mu v^2}{a^2} \sin^3 \varphi,$$

i.e., $$\frac{d\dot{\theta}}{d\varphi} \dot{\varphi} = -\frac{\mu v^2}{a^2} \sin^3 \varphi. \tag{4.6.74}$$

Now (4.6.72) ÷ (4.6.73) gives

$$\cot \varphi = \frac{\dot{z}}{a\dot{\theta}}. \tag{4.6.75}$$

Differentiating (4.6.75) w.r.t. t, and using (4.6.67), (4.6.70), (4.6.72) and (4.6.73) we obtain

$$-\csc^2 \varphi \frac{d\varphi}{dt} = \frac{1}{a} \frac{\dot{\theta}\ddot{z} - \dot{z}\ddot{\theta}}{\dot{\theta}^2},$$

i.e., $$-\csc^2 \varphi \frac{d\varphi}{dt} = \frac{av \sin \varphi \left(g - \dfrac{\mu R \cos \varphi}{m} \right) + av \cos \varphi \dfrac{\mu R \sin \varphi}{m}}{av^2 \sin^2 \varphi},$$

i.e., $$-\csc^2 \varphi \frac{d\varphi}{dt} = \frac{gv \sin \varphi - \dfrac{\mu R v \sin \varphi \cos \varphi}{m} + \dfrac{\mu R v \sin \varphi \cos \varphi}{m}}{v^2 \sin^2 \varphi},$$

i.e., $$-\csc^2 \varphi \frac{d\varphi}{dt} = \frac{gv \sin \varphi}{v^2 \sin^2 \varphi} = \frac{g}{v \sin \varphi},$$

i.e., $$-\csc^2 \varphi \dot{\varphi} = \frac{g}{v \sin \varphi},$$

i.e., $$\dot{\varphi} = -\frac{g \sin \varphi}{v}. \tag{4.6.76}$$

Using (4.6.76) in (4.6.74), we get

$$\frac{d\dot{\theta}}{d\varphi} \left(-\frac{g}{v} \sin \varphi \right) = -\frac{\mu v^2 \sin^3 \varphi}{a^2},$$

$$\text{i.e.,}\quad \frac{d\dot{\theta}}{d\varphi} = \frac{\mu v^3 \sin^2\varphi}{ga^2},$$

$$\text{i.e.,}\quad \frac{d}{d\varphi}\left(\frac{v\sin\varphi}{a}\right) = \frac{\mu v^3 \sin^2\varphi}{ga^2},$$

$$\text{i.e.,}\quad d\,(v\sin\varphi) = \frac{\mu v^3 \sin^2\varphi}{ga} d\varphi,$$

$$\text{i.e.,}\quad \frac{d\,(v\sin\varphi)}{v^3\sin^3\varphi} = \frac{\mu}{ga}\csc\varphi\, d\varphi.$$

Integrating the above equation, we derive

$$\frac{(v\sin\varphi)^{-2}}{-2} = -\frac{\mu}{ga}\log(\csc\varphi + \cot\varphi) - C$$

$$\text{i.e.,}\quad \frac{1}{2v^2\sin^2\varphi} = \frac{\mu}{ga}\log(\csc\varphi + \cot\varphi) + C. \tag{4.6.77}$$

In (4.6.77), C is the constant of integration to be determined from the initial condition:

$$v = V,\quad \varphi = \frac{\pi}{2};\quad t = 0.$$

Equation (4.6.77) gives

$$\frac{1}{2\,V^2} = C,$$

and hence Eq. (4.6.77) takes the form

$$\frac{1}{2v^2\sin^2\varphi} = \frac{\mu}{ga}\log(\csc\varphi + \cot\varphi) + \frac{1}{2V^2},$$

$$\text{i.e.,}\quad \frac{ag}{v^2\sin^2\varphi} = 2\mu\log(\csc\varphi + \cot\varphi) + \frac{ag}{V^2},$$

which is the required result.

Example 4.7 A heavy particle is projected horizontally along the inner surface of a smooth spherical shell of radius $\frac{a}{\sqrt{2}}$ with velocity $\sqrt{\frac{7ag}{3}}$ at a depth $\frac{2a}{3}$ below the center. Show that it will rise to a height $\frac{a}{3}$ above the center, and the pressure on the sphere just vanishes at the highest point of the path.

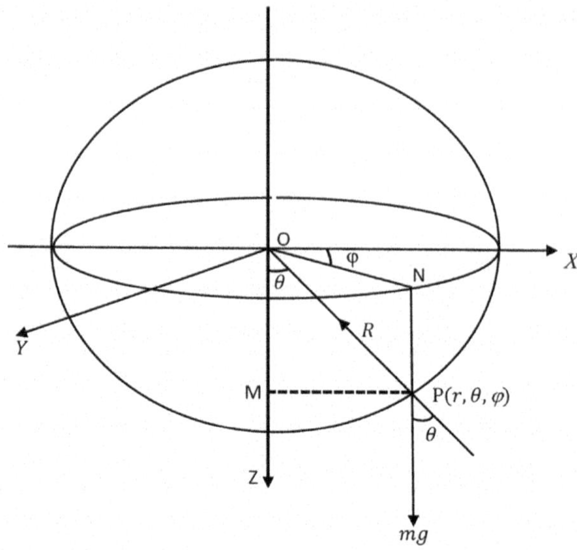

Fig. 4.14 Motion of a particle projected inside a smooth spherical shell

Solution In Fig. 4.14, take the center O of the sphere as the origin, the X-axis and Y-axis horizontally, and the Z-axis vertically downward. Let P denote the position of the particle at time t with spherical polar coordinates (r, θ, φ).

The forces acting on the particle are

(I) the weight mg acting $\|$ to Z-axis,

(II) the normal reaction R acting along **PO**.

The equations of motion of the particle in r and φ directions are

$$m\left(\ddot{r} - r\dot{\theta}^2 - r\sin^2\theta\dot{\varphi}^2\right) = -R + mg\cos\theta, \tag{4.6.78}$$

$$\frac{m}{r\sin\theta}\frac{d}{dt}\left(r^2\sin^2\theta\dot{\varphi}\right) = 0. \tag{4.6.79}$$

From geometry

$$r = b = \frac{a}{\sqrt{2}}. \tag{4.6.80}$$

By the use of (4.6.80) in (4.6.78) we obtain

$$-m\left(b\dot{\theta}^2 + b\sin^2\theta\dot{\varphi}^2\right) = -R + mg\cos\theta. \tag{4.6.81}$$

Equation (4.6.79) gives

$$\sin^2\theta\,\dot{\varphi} = h = \text{a constant}. \tag{4.6.82}$$

Initially

$$v_\varphi = r\sin\theta\dot{\varphi} = \sqrt{\frac{7ag}{3}}, \quad z = \frac{2a}{3} \quad \text{at} \quad t = 0,$$

$$\text{i.e.,} \quad \frac{a}{\sqrt{2}}\sin\theta\dot{\varphi} = \sqrt{\frac{7ag}{3}}, \quad \frac{a}{\sqrt{2}}\cos\theta = \frac{2a}{3} \quad \text{at} \quad t = 0,$$

$$\text{i.e.,} \quad \sin\theta\dot{\varphi} = \frac{\sqrt{2}}{a}\sqrt{\frac{7ag}{3}}, \quad \cos\theta = \frac{2\sqrt{2}}{3} = \cos\beta \quad \text{at} \quad t = 0,$$

$$\text{i.e.,} \quad \sin\beta\,\dot{\varphi} = \sqrt{\frac{14g}{3a}}, \quad \theta = \beta \ \text{at} \ t = 0,$$

$$\text{i.e.,} \quad \sqrt{1-\cos^2\beta}\,\dot{\varphi} = \sqrt{\frac{14g}{3a}}, \quad \theta = \beta \ \text{at} \ t = 0,$$

$$\text{i.e.,} \quad \sqrt{1-\frac{8}{9}}\,\dot{\varphi} = \sqrt{\frac{14g}{3a}}, \quad \theta = \beta \ \text{at} \ t = 0,$$

$$\text{i.e.,} \quad \frac{1}{3}\dot{\varphi} = \sqrt{\frac{14g}{3a}}, \quad \theta = \beta \ \text{at} \ t = 0,$$

$$\text{i.e.,} \quad \dot{\varphi} = 3\sqrt{\frac{14g}{3a}} = \sqrt{\frac{42g}{a}}, \quad \theta = \beta \ \text{at} \ t = 0.$$

Equation (4.6.82) gives

$$h = \sin^2\beta\sqrt{\frac{42g}{a}} = (1-\cos^2\beta)\sqrt{\frac{42g}{a}} = \left(1-\frac{8}{9}\right)\sqrt{\frac{42g}{a}} = \sqrt{\frac{14g}{27a}}. \tag{4.6.83}$$

Thus Eq. (4.6.82) gives

$$\dot{\varphi} = h\csc^2\theta. \tag{4.6.84}$$

Law of conservation of energy yields K.E + P.E = Constant,

$$\text{i.e.,} \quad \frac{1}{2}mv^2 - mgz = Cm,$$

$$\text{i.e.,} \quad \frac{1}{2}v^2 - gz = C, \tag{4.6.85}$$

C is the constant of integration. Initially

$$v = v_\varphi = \sqrt{\frac{7ag}{3}}, \quad z = \frac{2a}{3} \text{ at } t = 0.$$

Equation (4.6.85) gives

$$C = \frac{1}{2}\frac{7ag}{3} - g\frac{2a}{3} = \frac{ag}{2}.$$

Hence Eq. (4.6.85) reduces to

$$\frac{1}{2}v^2 - gz = \frac{ag}{2},$$

$$\text{i.e.,} \quad v^2 - 2gz = ag. \tag{4.6.86}$$

The velocity of the particle is given by

$$\mathbf{v} = \dot{r}\hat{r} + r\dot{\theta}\hat{\theta} + r\sin\theta\dot{\varphi}\hat{\varphi} = \frac{a}{\sqrt{2}}\dot{\theta}\,\hat{\theta} + \frac{a}{\sqrt{2}}\sin\theta\,\dot{\varphi}\,\hat{\varphi}.$$

$$\therefore \quad v^2 = \frac{a^2\dot{\theta}^2}{2} + \frac{a^2\sin^2\theta\dot{\varphi}^2}{2}. \tag{4.6.87}$$

Substituting (4.6.87) in (4.6.86), we have

$$\frac{a^2\dot{\theta}^2}{2} + \frac{a^2\sin^2\theta\dot{\varphi}^2}{2} - 2g\frac{a}{\sqrt{2}}\cos\theta = ag,$$

$$\text{i.e.,} \quad a^2\dot{\theta}^2 + a^2\sin^2\theta\dot{\varphi}^2 - \frac{4ag}{\sqrt{2}}\cos\theta = 2ag,$$

$$\text{i.e.,} \quad \dot{\theta}^2 + \sin^2\theta\dot{\varphi}^2 - \frac{4g}{\sqrt{2}a}\cos\theta = \frac{2g}{a},$$

$$\text{i.e.,} \quad \dot{\theta}^2 + \sin^2\theta\,h^2\csc^4\theta = \frac{4g}{\sqrt{2}a}\cos\theta + \frac{2g}{a},$$

$$\text{i.e.,} \quad \dot{\theta}^2 + \frac{14g}{27a}\csc^2\theta = \frac{4g}{\sqrt{2}a}\cos\theta + \frac{2g}{a},$$

$$\text{i.e.,}\quad \dot{\theta}^2 = -\frac{14g}{27a}\csc^2\theta + \frac{4g}{\sqrt{2}a}\cos\theta + \frac{2g}{a}. \tag{4.6.88}$$

We note that

$$z = r\cos\theta = \frac{a}{\sqrt{2}}\cos\theta,$$

$$\therefore\quad \cos\theta = \frac{\sqrt{2}z}{a},\quad \sin^2\theta = 1-\cos^2\theta = 1-\frac{2z^2}{a^2} = \frac{a^2-2z^2}{a^2},$$

$$\text{i.e.,}\quad \csc^2\theta = \frac{a^2}{a^2-2z^2}.$$

Equation (4.6.88) gives

$$\dot{\theta}^2 = -\frac{14g}{27a}\frac{a^2}{a^2-2z^2} + \frac{4g}{\sqrt{2}a}\frac{\sqrt{2}z}{a} + \frac{2g}{a},$$

$$\text{i.e.,}\quad \dot{\theta}^2 = -\frac{14g}{27a}\frac{a^2}{a^2-2z^2} + \frac{4gz}{a^2} + \frac{2g}{a}. \tag{4.6.89}$$

Now

$$\dot{\theta}^2\Big]_{z=-\frac{a}{3}} = -\frac{14\,g}{27a}\frac{a^2}{a^2-2\frac{a^2}{9}} - \frac{4g}{a^2}\frac{a}{3} + \frac{2g}{a} = -\frac{2}{3a} - \frac{4g}{3a} + \frac{2g}{a} = 0.$$

$\therefore\quad \dot{\theta} = 0$ at $z = -\frac{a}{3}$,

i.e., $v_\theta = 0$ at the height $\frac{a}{3}$ above the center.

Hence the particle will rise to the height $\frac{a}{3}$ above the center. Equation (4.6.78) gives

$$R = mg\cos\theta - m\left(\ddot{r} - r\dot{\theta}^2 - r\sin^2\theta\,\dot{\varphi}^2\right) = mg\cos\theta + \frac{ma}{\sqrt{2}}\left(\dot{\theta}^2 + \sin^2\theta\,\dot{\varphi}^2\right). \tag{4.6.90}$$

Substituting the expression for $\dot{\theta}^2$ from (4.6.88) and $\dot{\varphi}$ from (4.6.84) in (4.6.90), we obtain

$$R = mg\cos\theta + \frac{ma}{\sqrt{2}}\left[\frac{-14g}{27a}\csc^2\theta + \frac{4g}{\sqrt{2}a}\cos\theta + \frac{2g}{a} + \sin^2\theta\,h^2\csc^2\theta\right],$$

$$\text{i.e.,}\quad R = mg\cos\theta + \frac{ma}{\sqrt{2}}\left[\frac{-14g}{27a}\csc^2\theta + \frac{4g}{\sqrt{2}a}\cos\theta + \frac{2g}{a} + h^2\csc^2\theta\right],$$

$$\text{i.e.,}\quad R = mg\cos\theta + \frac{ma}{\sqrt{2}}\left[\frac{-14g}{27a}\csc^2\theta + \frac{4g}{\sqrt{2}a}\cos\theta + \frac{2g}{a} + \frac{14g}{27a}\csc^2\theta\right],$$

$$\text{i.e.,}\quad R = mg\cos\theta + \frac{ma}{\sqrt{2}}\left[\frac{2g}{a} + \frac{4g}{\sqrt{2}a}\cos\theta\right],$$

$$\text{i.e.,}\quad R = mg\cos\theta + m\sqrt{2}g + 2gm\cos\theta = 3mg\cos\theta + \sqrt{2}mg. \tag{4.6.91}$$

We note that for the highest position of the particle

$$z = -\frac{a}{3},$$

$$\text{i.e.,}\quad r\cos\theta = -\frac{a}{3},$$

$$\text{i.e.,}\quad \frac{a}{\sqrt{2}}\cos\theta = -\frac{a}{3},$$

$$\text{i.e.,}\quad \cos\theta = -\frac{\sqrt{2}}{3} = \cos\gamma\ \text{(say)},$$

$$\text{i.e.,}\quad \theta = \gamma,$$

where γ is given by $\cos\gamma = -\frac{\sqrt{2}}{3}$. Now (4.6.91) gives

$$R\Big]_{\theta=\gamma} = 3mg\cos\gamma + \sqrt{2}mg = 3mg\left(-\frac{\sqrt{2}}{3}\right) + \sqrt{2}mg = 0.$$

Thus, we see that the pressure of the particle on the sphere vanishes at the highest point attained by the particle.

Example 4.8 A particle moves on the surface of a rough circular cylinder under no external force. Prove that its path makes a constant angle with the transverse plane of the cylinder.

Solution Let O be the center of a normal section of the cylinder. Take the X-axis and Y-axis horizontally, and the Z-axis vertically upward along the axis of the cylinder. Let P be the position of the particle at time t with cylindrical polar coordinates (ρ, φ, z).

In Fig. 4.15a, draw $\perp^r$ PQ from P on OZ.

The forces acting on the particle at P are

(I) the normal reaction R acting along **PQ**,

(II) The force of friction μR acting along the $-$ve direction of the velocity of the particle.

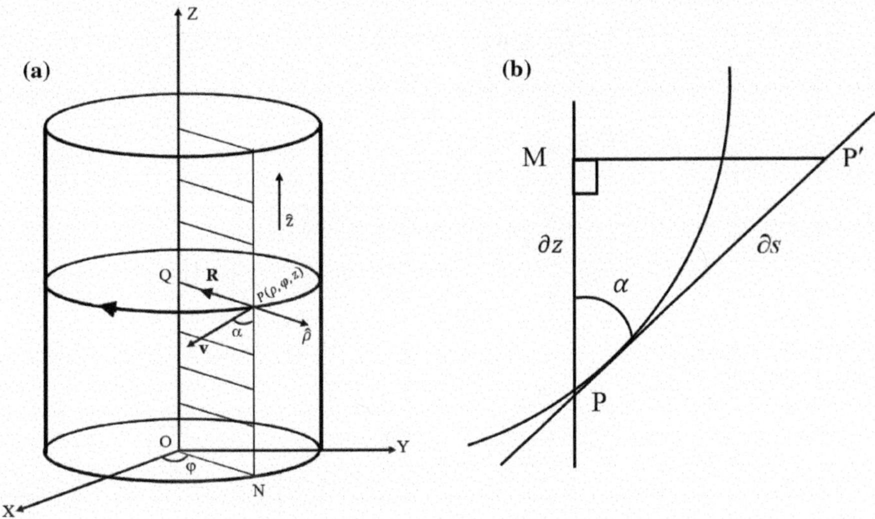

Fig. 4.15 **a** Forces acting on particle on rough cylinder. **b** Geometry showing constant angle of motion

Equations of motion of the particle are

$$m\left(\ddot{\rho} - \rho\dot{\varphi}^2\right) = -R, \tag{4.6.92}$$

$$m\frac{1}{\rho}\frac{d}{dt}\left(\rho^2\dot{\varphi}\right) = -\mu R\sin\alpha, \tag{4.6.93}$$

$$m\ddot{z} = -\mu R\cos\alpha, \tag{4.6.94}$$

where α is the angle between $\mathbf{v}$ and $\hat{z}$. The equation of the cylinder is $\rho = a$, a being the radius of the cylinder. Substituting $\rho = a$ in (4.6.92) and (4.6.93), we obtain

$$m\left(-a\dot{\varphi}^2\right) = -R, \tag{4.6.95}$$

$$ma\ddot{\varphi} = -\mu R\sin\alpha. \tag{4.6.96}$$

(4.6.94) ÷ (4.6.96) gives

$$\frac{\ddot{z}}{a\ddot{\varphi}} = \cot\alpha = \frac{\cos\alpha}{\sin\alpha},$$

$$\text{i.e.,}\quad a\ddot{\varphi}\cos\alpha = \ddot{z}\sin\alpha. \tag{4.6.97}$$

The elementary are length in $(\rho,\ \varphi,\ z)$ system is given by

$$ds^2 = d\rho^2 + \rho^2 d\varphi^2 + dz^2. \tag{4.6.98}$$

In the present case, $\rho = a$,
(4.6.96) reduces to

$$ds^2 = a^2 d\varphi^2 + dz^2,$$

$$\text{i.e.,}\quad 1 = a^2\frac{d\varphi^2}{ds^2} + \left(\frac{dz}{ds}\right)^2. \tag{4.6.99}$$

Fig. 4.15b shows that

$$\cos\alpha = \frac{\text{PM}}{\text{PP}'} = \frac{\delta z}{\delta s},$$

$$\text{i.e.,}\quad \lim_{\delta s\to 0}\cos\alpha = \lim_{\delta s\to 0}\frac{\delta z}{\delta s},$$

$$\text{i.e.,}\quad \cos\alpha = \frac{dz}{ds}. \tag{4.6.100}$$

Hence, Eq. (4.6.99) gives

$$1 = a^2\left(\frac{d\varphi}{ds}\right)^2 + \cos^2\alpha,$$

$$\text{i.e.,}\quad a^2\left(\frac{d\varphi}{ds}\right)^2 = \sin^2\alpha,$$

$$\text{i.e.,}\quad a\frac{d\varphi}{ds} = \sin\alpha. \tag{4.6.101}$$

By virtue of (4.6.100), (4.6.101), Eq. (4.6.97) can be expressed as

$$a\ddot{\varphi}\frac{dz}{ds} = \ddot{z}a\frac{d\varphi}{ds},$$

i.e., $\ddot{\varphi}\frac{dz}{dt}\frac{dt}{ds} = \ddot{z}\frac{d\varphi}{dt}\frac{dt}{ds},$

i.e., $\ddot{\varphi}\dot{z} = \ddot{z}\dot{\varphi},$

i.e., $\frac{d\dot{\varphi}}{dt}\dot{z} = \frac{d\dot{z}}{dt}\dot{\varphi},$

i.e., $\dot{z}d\dot{\varphi} = \dot{\varphi}d\dot{z},$

i.e., $\frac{d\dot{\varphi}}{\dot{\varphi}} = \frac{d\dot{z}}{\dot{z}},$

i.e., $\log\dot{\varphi} = \log\dot{z} + \log c,$

i.e., $\dot{\varphi} = c\dot{z},$

i.e., $\frac{d\varphi}{dt} = c\frac{dz}{dt},$

i.e., $\frac{d\varphi}{ds}\frac{ds}{dt} = c\frac{dz}{ds}\frac{ds}{dt}$

i.e., $\frac{\sin\alpha}{a}\dot{s} = c\dot{s}\cos\alpha,$

i.e., $\sin\alpha = ac\cos\alpha,$

i.e., $\tan\alpha = ac = \text{a constant},$

i.e., $\alpha = \text{a constant},$

i.e., $\frac{\pi}{2} - \alpha = \text{Constant}.$

Hence the path of the particle makes a constant angle with the transverse plane of the cylinder.

Example 4.9 A heavy particle moves on the surface of a smooth sphere under gravity. Initially it was projected with horizontal velocity u from a position $\theta = \alpha$. Show that during the motion θ lies between 0 and α or α and π according as $ag\sin\alpha > u^2\cot\alpha$ or $ag\sin\alpha < u^2\cot\alpha$, where a is the radius of the sphere.

Fig. 4.16 Illustration for Example 4.9

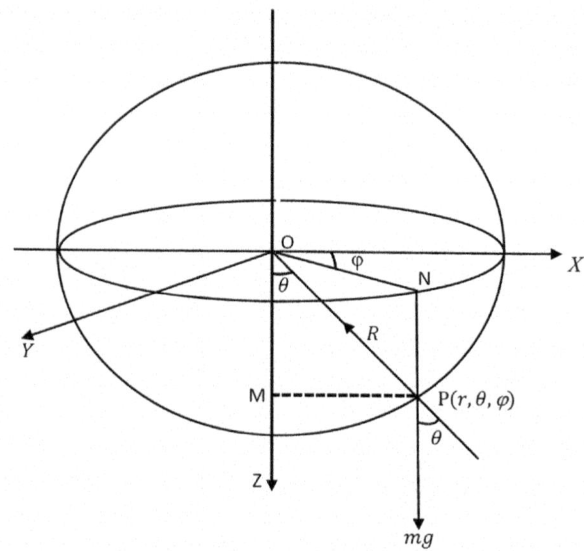

Solution In Fig. 4.16, take the center O of the sphere as the origin, X-axis and Y-axis horizontally, and Z-axis along the downward vertical through O. Let P be the position of the particle at time t with spherical polar coordinates $(r,\ \theta,\ \varphi)$.

Equation of motion in φ-direction is

$$\frac{m}{r\sin\theta}\frac{d}{dt}\left(r^2\sin^2\theta\dot{\varphi}\right) = 0. \tag{4.6.102}$$

From geometry, $r = a =$ a constant.

Equation (4.6.102) reduces to

$$\sin^2\theta\,\dot{\varphi} = \text{a constant}. \tag{4.6.103}$$

Initially

$$v_\varphi = u,\ \ \theta = \alpha \ \text{at}\ \ t = 0,$$

$$\text{i.e.,}\ \ a\sin\theta\,\dot{\varphi} = u,\ \ \theta = \alpha\ \text{at}\ \ t = 0,$$

$$\therefore\ \ a\sin\alpha\,\dot{\varphi} = u\ \ \text{at}\ \ t = 0,$$

which implies

$$\dot{\varphi} = \frac{u}{a\sin\alpha}\ \ \text{at}\ \ t = 0. \tag{4.6.104}$$

Subjecting the condition (4.6.104) in (4.6.103) we have

$$h = \sin^2 \alpha \frac{u}{a \sin \alpha} = \frac{u \sin \alpha}{a}.$$

Hence (4.6.103) gives

$$\sin^2 \theta \, \dot{\varphi} = \frac{u \sin \alpha}{a}. \tag{4.6.105}$$

The velocity of the particle at P is given by

$$\mathbf{v} = \dot{r}\hat{r} + r\dot{\theta}\hat{\theta} + r \sin \theta \dot{\varphi}\hat{\varphi} = a\dot{\theta}\hat{\theta} + a \sin \theta \dot{\varphi}\hat{\varphi},$$

which gives

$$v^2 = a^2\dot{\theta}^2 + a^2 \sin^2 \theta \dot{\varphi}^2 = a^2\dot{\theta}^2 + a^2 \sin^2 \theta \frac{h^2}{\sin^4 \theta} = a^2\dot{\theta}^2 + a^2h^2 \csc^2 \theta. \tag{4.6.106}$$

The energy equation gives

$$\text{K.E.} + \text{P.E.} = \text{Constant},$$

$$\text{i.e.,} \quad \frac{1}{2}mv^2 + mg\,(-z) = \text{a constant},$$

$$\text{i.e.,} \quad \frac{1}{2}mv^2 - mga \cos \theta = \text{a constant},$$

$$\text{i.e.,} \quad v^2 - 2ag \cos \theta = C, \tag{4.6.107}$$

C being a constant.

Initially

$$\mathbf{v} = u\hat{\varphi}, \quad \theta = \alpha \text{ at } \quad t = 0,$$

$$\text{i.e.,} \quad v = u, \quad \theta = \alpha \text{ at } t = 0. \tag{4.6.108}$$

On the use of the condition (4.6.108) in (4.6.107), we obtain

$$C = u^2 - 2ag\cos\alpha.$$

Hence, Eq. (4.6.107) yields

$$v^2 - 2ag\cos\theta = u^2 - 2ag\cos\alpha,$$

$$\text{i.e.,}\quad v^2 = u^2 + 2ag\left(\cos\theta - \cos\alpha\right),$$

$$\text{i.e.,}\quad a^2\dot{\theta}^2 + a^2h^2\csc^2\theta = u^2 + 2ag\left(\cos\theta - \cos\alpha\right),$$

$$\text{i.e.,}\quad a^2\dot{\theta}^2 + u^2\sin^2\alpha\csc^2\theta = u^2 + 2ag\left(\cos\theta - \cos\alpha\right),$$

$$\text{i.e.,}\quad a^2\dot{\theta}^2 + u^2\left(1 - \frac{\sin^2\alpha}{\sin^2\theta}\right) + 2ag\left(\cos\theta - \cos\alpha\right),$$

$$\text{i.e.,}\quad a^2\dot{\theta}^2\sin^2\theta = u^2\left(\sin^2\theta - \sin^2\alpha\right) + 2ag\sin^2\theta\left(\cos\theta - \cos\alpha\right). \quad (4.6.109)$$

The points of rest are given by

$$\dot{\theta} = 0. \quad (4.6.110)$$

Suppose that (4.6.110) gives

$$\theta = \beta, \quad (4.6.111)$$

i.e., $\dot{\theta} = 0$ at $\theta = \beta$.

Substituting $\dot{\theta} = 0, \quad \theta = \beta$, in (4.6.109), we obtain

$$u^2\left(\sin^2\alpha - \sin^2\beta\right) = 2ga\sin^2\beta\left(\cos\beta - \cos\alpha\right),$$

$$\text{i.e.,}\quad u^2\left(\cos^2\beta - \cos^2\alpha\right) = 2ga\ \sin^2\beta\left(\cos\beta - \cos\alpha\right),$$

$$\text{i.e.,}\quad \left(\cos\beta - \cos\alpha\right)\left[u^2\left(\cos\beta + \cos\alpha\right) - 2ga\sin^2\beta\right] = 0,$$

$$\text{i.e.,}\quad u^2\left(cos\beta + cos\alpha\right) - 2ga\ \sin^2\beta = 0 \quad \left(\because\quad \beta \neq \alpha\right). \quad (4.6.112)$$

Now consider a function f defined by

$$f\left(\cos\theta\right) = u^2\left(\cos\theta + \cos\alpha\right) - 2ga\left(1 - \cos^2\theta\right). \quad (4.6.113)$$

$$\therefore\quad f\left(\cos\beta\right) = u^2\left(\cos\beta + \cos\alpha\right) - 2ga\sin^2\beta = 0.$$

Now

$$f(\cos 0) = u^2(1+\cos\alpha) = 2u^2\cos^2\frac{\alpha}{2} > 0,$$

$$f(\cos\alpha) = 2u^2\cos\alpha - 2ga\ \left(1-\cos^2\alpha\right) = 2\left[u^2\cos\alpha - ga\sin^2\alpha\right],$$

$$f(\cos\pi) = u^2(-1+\cos\alpha) = -u^2(1-\cos\alpha) = -2u^2\sin^2\frac{\alpha}{2} < 0.$$

Case I: Suppose that

$$ga\ \sin\alpha > u^2\cot\alpha,$$

i.e., $ga\sin^2\alpha > u^2\cot\alpha,$

i.e., $u^2\cos\alpha < ga\sin^2\alpha$

i.e., $2\left(u^2\cos\alpha - ga\ \sin^2\alpha\right) < 0,$

i.e., $f(\cos\alpha) < 0.$

Case II: Suppose that

$$ga\sin\alpha < u^2\cot\alpha,$$

i.e., $ga\sin^2\alpha < u^2\cos\alpha,$

i.e., $u^2\cos\alpha > ga\sin^2\alpha,$

i.e., $2\left(u^2\cos\alpha - ga\sin^2\alpha\right) > 0,$

i.e., $f(\cos\alpha) > 0.$

Thus for $ga\sin^2\alpha > u^2\cot\alpha$, we have

$$f(\cos 0) > 0 \text{ and } f(\cos\alpha) < 0.$$

Further

$$f(\cos\beta) = 0,$$

i.e., β lies between 0 and α.

Again for

$$ga\sin\alpha < u^2\cot\alpha,\quad f(\cos\alpha) > 0 \text{ and } f(\cos\pi) < 0,\quad f(\cos\beta) = 0.$$

Therefore β lies between α and π.

Thus, during the motion of the particle, θ lies between 0 and α if $ga \sin\alpha > u^2 \cot\alpha$, and α and π if $ga \sin\alpha < u^2 \cot\alpha$.

Example 4.10 A smooth circular cone, of angle 2α, has its axis vertical and its vertex, which is pierced with a small hole, downwards. A mass M hangs at rest by a string which passes through the vertex, and a mass m attached to the upper end describes a horizontal circle on the inner surface of the cone. Find the time T of a complete revolution, and show that small oscillations about the steady take place in the time $T \csc\alpha\sqrt{\frac{M+m}{3m}}$

Solution Let O be the vertex of the cone with 2α as the vertical angle. Let P be the position of the mass m at time t. Let (r, α, φ) be the spherical polar coordinates of P. Let Q be the position of mass M at time t.

In Fig. 4.17, $OP = r$

therefore, $OQ = l - r$, where l is the length of the string.

Let T_1 be the tension of the string.

Equation of motion of the mass m in r-direction and φ-direction are

$$m\left(\ddot{r} - r\sin^2\alpha\,\dot{\varphi}^2\right) = -m\,g\cos\alpha - T_1 \tag{4.6.114}$$

$$\frac{m}{r\sin\theta}\frac{d}{dt}\left(r^2\sin^2\alpha\,\dot{\varphi}\right) = 0. \tag{4.6.115}$$

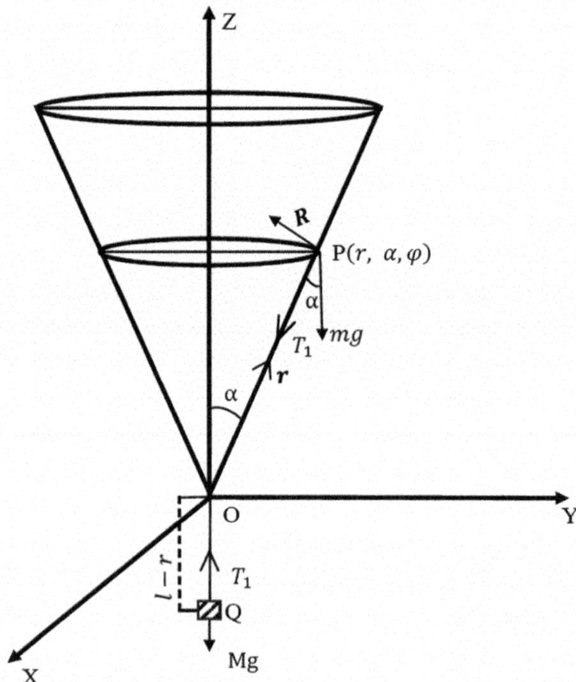

Fig. 4.17 Right circular cone with vertex at origin and semi-vertical angle α

The equation of motion of M is

$$M\frac{d^2}{dt^2}(l-r) = Mg - T_1. \tag{4.6.116}$$

Equation (4.6.116) gives

$$\begin{aligned} -M\ddot{r} &= Mg - T_1 \\ \therefore\quad T_1 &= Mg + M\ddot{r}. \end{aligned} \tag{4.6.117}$$

Equation (4.6.114) gives

$$m\left(\ddot{r} - r\sin^2\alpha\,\dot{\varphi}^2\right) = -mg\cos\alpha - Mg - M\ddot{r}$$

$$\text{i.e.,}\quad (M+m)\ddot{r} - m\,r\sin^2\alpha\,\dot{\varphi}^2 = -A,\quad \text{where } A = Mg + mg\cos\alpha. \tag{4.6.118}$$

Equation (4.6.115) gives

$$\begin{aligned} &r^2\sin^2\alpha\,\dot{\varphi} = \text{constant} \\ &\text{i.e.,}\quad r^2\dot{\varphi} = C_1. \end{aligned}$$

Initially $r = l$, $\dot{\varphi} = \omega$ (say) at $t = 0$.

$$\therefore\quad l^2\omega = C_1$$

which gives

$$\begin{aligned} r^2\,\dot{\varphi} &= l^2\omega \\ \therefore\quad \dot{\varphi} &= \frac{l^2\omega}{r^2}. \end{aligned} \tag{4.6.119}$$

Therefore, Eq. (4.6.118) reduces to

$$(M+m)\ddot{r} - mr\sin^2\alpha\,\frac{l^4\omega^2}{r^4} = -A. \tag{4.6.120}$$

In the steady case, $r = l$, $\ddot{r} = 0$ and hence (4.6.120) reduces to

$$A = m\,l\,\sin^2\alpha\,\frac{l^4\omega^2}{l^4} = m\,l\,\sin^2\alpha\,\omega^2. \tag{4.6.121}$$

Let a disturbance be given to the steady motion. Let $r = l + x$, where x is small.

$$\therefore\quad \ddot{r} = \ddot{x}.$$

Therefore, Eq. (4.6.120) reduces to

$$(M+m)\ddot{x} - m(l+x)\sin^2\alpha\,\frac{l^4\omega^2}{(l+x)^4} = -A,$$

$$\text{i.e.,}\quad (M+m)\ddot{x} - m\,\omega^2 l^4 \sin^2\alpha\,\frac{1}{(l+x)^3} = -A,$$

$$\text{i.e.,}\quad (M+m)\ddot{x} - m\,\omega^2 l^4 \sin^2\alpha\, l^{-3}\left(1+\frac{x}{l}\right)^{-3} = -A,$$

$$\text{i.e.,}\quad A = -(M+m)\ddot{x} + m\,\omega^2 l \sin^2\alpha\left(1+\frac{x}{l}\right)^{-3},$$
$$= -(M+m)\ddot{x} + A\left(1-3\frac{x}{l}\right),$$

$$\text{i.e.,}\quad (M+m)\ddot{x} = -3A\frac{x}{l},$$
$$= -3\,m\,l\,\omega^2 \sin^2\alpha\,\frac{x}{l},$$
$$= -3\,m\,\omega^2 \sin^2\alpha\, x.$$

which implies

$$\therefore \quad \ddot{x} = -\frac{3\,m\,\omega^2\sin^2\alpha}{M+m}\,x,$$
$$= -\mu\,x;\quad \mu = \frac{3\,m\,\omega^2\sin^2\alpha}{M+m}. \tag{4.6.122}$$

Equation (4.6.122) represented an SHM of a time period

$$T = \frac{2\pi}{\sqrt{\mu}}$$
$$= \frac{2\pi\sqrt{M+m}}{\sqrt{3\,m\,\omega^2\sin^2\alpha}}$$
$$= \frac{2\pi\sqrt{M+m}}{\sqrt{3m}\;\omega\,\sin\alpha}. \tag{4.6.123}$$

Given that T is the period of revolution.

$$\therefore \quad T = \frac{2\pi}{\omega}. \tag{4.6.124}$$

Therefore, Eq. (4.6.123) reduces to

$$\mathcal{T} = T\,\frac{\sqrt{M+m}}{\sqrt{3m}}\,\csc\alpha = T\,\csc\alpha\,\frac{\sqrt{M+m}}{\sqrt{3m}}.$$

Example 4.11 A smooth hollow right circular cone is placed with its vertex downward and axis vertical, and at a point on its interior surface at a height h above the vertex, a particle is projected horizontally along the surface with a velocity $\sqrt{\frac{2gh}{n^2+n}}$. Show that the lowest point of its path will be at a height $\frac{h}{n}$ above the vertex.

Solution It is assumed that O is the vertex of the cone, α is the semi-vertical angle, and OZ is the axis of the cone as displayed in Fig. 4.18.

The equation of motion of the particle in the φ-direction is

$$\frac{m}{r\sin\alpha}\frac{d}{dt}\left(r^2\sin^2\alpha\,\dot{\varphi}\right) = 0$$

$$\therefore\ r^2\dot{\varphi} = C. \tag{4.6.125}$$

Initially $v_\varphi = V$, $z = h$ at $t = 0$.

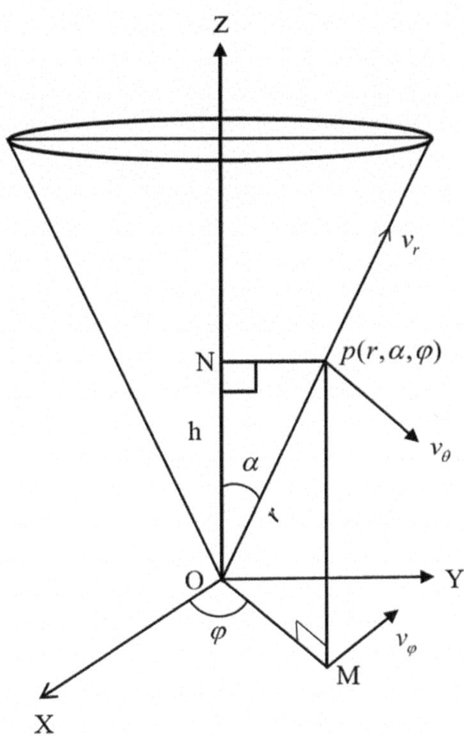

Fig. 4.18 Schematic diagram of right circular cone for Example 4.11

$$\text{i.e.,}\quad r\sin\alpha\,\dot{\varphi} = V,\quad r\cos\alpha = h,\quad \text{at } t = 0.$$
$$\text{i.e.,}\quad h\sec\alpha\,\sin\alpha\,\dot{\varphi} = V,\quad r = h\sec\alpha \quad \text{at } t = 0.$$
$$\text{i.e.,}\quad h\tan\alpha\,\dot{\varphi} = V,\quad r = h\sec\alpha \quad \text{at } t = 0.$$
$$\text{i.e.,}\quad \dot{\varphi} = \frac{V}{h}\cot\alpha,\quad r = h\sec\alpha \quad \text{at } t = 0.$$

Equation (4.6.125) gives

$$\begin{aligned} C &= h^2\sec^2\alpha\,\frac{V\,\cot\alpha}{h} \\ &= Vh\sec^2\alpha\,\cot\alpha \\ &= \frac{Vh}{\sin\alpha\,\cos\alpha} = Vh\csc\alpha\,\sec\alpha. \end{aligned}$$

Therefore, Eq. (4.6.125) gives

$$\begin{aligned} r^2\dot{\varphi} &= Vh\csc\alpha\,\sec\alpha. \\ \therefore\quad \dot{\varphi} &= \frac{Vh\csc\alpha\,\sec\alpha}{r^2}. \end{aligned} \tag{4.6.126}$$

Let $\mathbf{v} = (v_r,\ v_\theta,\ v_\varphi)$.

$$\begin{aligned} \therefore\quad v^2 &= v_r^2 + v_\theta^2 + v_\varphi^2 \\ &= \dot{r}^2 + r^2\dot{\theta}^2 + (r\sin\alpha\,\dot{\varphi})^2 \\ &= \dot{r}^2 + r^2\sin^2\alpha\cdot\frac{V^2h^2\csc^2\alpha\,\sec^2\alpha}{r^4} \\ &= \dot{r}^2 + \frac{V^2h^2\sec^2\alpha}{r^2}. \end{aligned} \tag{4.6.127}$$

The energy equation gives

$$\frac{1}{2}mv^2 + mgz = E. \tag{4.6.128}$$

Initially, $v = V$, $z = h$, when $t = 0$.

$$E = \frac{1}{2}mV^2 + mgh.$$

Therefore, Eq. (4.6.128) reduces to

$$\begin{aligned} \frac{1}{2}mv^2 + mgz &= \frac{1}{2}mV^2 + mgh \\ \text{i.e.,}\quad v^2 + 2gz &= V^2 + 2gh \end{aligned}$$

$$\text{i.e.,}\quad \dot{r}^2 + \frac{V^2 h^2 \sec^2\alpha}{r^2} + 2gz = V^2 + 2gh$$

$$\text{i.e.,}\quad \dot{r}^2 + \frac{V^2 h^2 \sec^2\alpha}{z^2 \sec^2\alpha} + 2gz = V^2 + 2gh$$

$$\text{i.e.,}\quad \dot{r}^2 + \frac{V^2 h^2}{z^2} + 2gz = V^2 + 2gh. \tag{4.6.129}$$

At the lowest position, $\dot{r} = 0$, $z = k$ (say).

Therefore, Eq. (4.6.129) gives

$$\frac{V^2 h^2}{k^2} + 2gz = V^2 + 2gh$$

$$\text{i.e.,}\quad V^2\left(\frac{h^2}{k^2} - 1\right) = 2g(h-k)$$

$$\text{i.e.,}\quad V^2 \frac{h^2 - k^2}{k^2} = 2g(h-k)$$

$$\text{i.e.,}\quad V^2 \frac{h+k}{k^2} = 2g$$

$$\text{i.e.,}\quad V^2(h+k) = 2gk^2. \tag{4.6.130}$$

But $V = \sqrt{\dfrac{2gh}{n^2+n}}$.

Therefore, Eq. (4.6.130) gives

$$\frac{2gh}{n^2+n}(h+k) = 2gk^2$$

$$\text{i.e.,}\quad \frac{h}{n^2+n}(h+k) = k^2$$

$$\text{i.e.,}\quad h^2 + hk = k^2(n^2+n)$$

$$\text{i.e.,}\quad k^2(n^2+n) - hk - h^2 = 0$$

$$\therefore\quad k = \frac{h \pm \sqrt{h^2 + 4(n^2+n)h^2}}{2(n^2+n)}$$

$$= \frac{h \pm h\sqrt{1 + 4n^2 + 4n}}{2(n^2+n)}$$

$$= \frac{h + h\sqrt{1 + 4n^2 + 4n}}{2(n^2+n)}$$

$$= \frac{h + h\sqrt{(2n+1)^2}}{2(n^2+n)}$$

$$= \frac{h + h(2n+1)}{2n(n+1)}$$

$$= \frac{2h(n+1)}{2n(n+1)}$$
$$= \frac{h}{n}.$$

4.7 Exercise-IV

1. A heavy particle is projected with velocity V from the end of a horizontal diameter of a sphere of radius along the inner surface, the direction of projection making an angle β with the equator. If the particle never leaves the surface, prove that $3\sin^2\beta < 2 + \left(\frac{V^2}{3ga}\right)^2$.
2. A heavy particle moves in a smooth sphere; show that, if the velocity is that due to the level of the center, the reaction of the surface will vary as the depth below the center.
3. A particle describes a Rhumb line on a sphere in such a way that its longitude increases uniformly. Show that the resultant acceleration varies as the cosine of the latitude and that its direction makes with the normal an angle equal to the latitude [N.B. A Rhumb line of a sphere is a curve on it which cuts all the meridians at constant angle constant angle α; its equation is $\frac{\dot{\varphi}\sin\theta}{\dot{\theta}} = \tan\alpha$].
4. A particle moves on a rough circular cylinder under the action of no external forces. Initially, the particle has a velocity V in a direction making an angle α with the transverse plane of the cylinder; show that the space described in time t is $\frac{a\sec^2\alpha}{\mu}\log\left[1 + \frac{\mu V\cos^2\alpha}{a}t\right]$.
5. A particle is moving under gravity in contact with the inside of a rough circular cylinder with a vertical generator. At time t, the particle is moving with velocity v in a direction that makes an acute angle φ with the downward drawn vertical. Establish the equation $\frac{1}{v}\frac{dv}{d\varphi} = \frac{\mu v^2}{ga}\sin\varphi - \cot\varphi$, where a is the radius of the cylinder and is the coefficient of friction.
6. A smooth conical surface is fixed with its axis vertical and vertex downwards. A particle is in steady motion on its concave side in a horizontal circle and is slightly disturbed. Show that the time of oscillation about this state of steady motion is $2\pi\sqrt{\frac{l}{3g\cos\alpha}}$, where α is the semi-vertical angle of the cone and l is the length of the generator to the circle of steady motion.
7. A particle is attached to one end of an inextensible string of length l, the other end of which is tied to a fixed point O. If the particle be started horizontally with velocity V at the level of O. Prove that $(l^2 - z^2)\dot{\varphi} = Vl$ and $l^2\dot{z}^2 = z\left\{2g\left(l^2 - z^2\right) - V^2 z\right\}$, where z is the depth of the particle from O. Also show that the tension of the string is given by $\frac{T}{m} = g\cos\theta - l\left\{\frac{z^4}{\left(l^2 - z^2\right)^2} - \frac{V^2}{l^2\sin^2\theta}\right\}$.
8. A heavy particle is projected horizontally along the inner surface of a smooth spherical shell with velocity v from a point where the radius makes an angle

α with the downward drawn vertical. Show that in the subsequent motion it will remain always above or below the starting point according to v^2 is $<$ or $>$ $ga \sin \alpha \tan \alpha$, where a is the radius of the sphere.

9. Obtain the expressions for velocity and acceleration in (ρ, φ, z) coordinate system, in case of a particle moving in space.
10. Derive the velocity and acceleration in (r, θ, φ) system for particle moving in three-dimensions.

Chapter 5
Motion of a System of Particles

5.1 Center of Mass

Let $m_1, m_2, \ldots, m_N$ be the masses of a system of N particles having position vector $\mathbf{r}_1, \mathbf{r}_2, \ldots, \mathbf{r}_N$ relative to a fixed origin O, as shown in Fig. 5.1.

The center of mass or centroid of the system of particles is defined to be the point C having position vector $\mathbf{r_c}$ given by

$$\mathbf{r}_c = \frac{\sum_{\nu=1}^{N} m_\nu \mathbf{r}_\nu}{\sum_{\nu=1}^{N} m_\nu} = \frac{\sum_{\nu=1}^{N} m_\nu \mathbf{r}_\nu}{M}, \tag{5.1.1}$$

where $M = \sum_{\nu=1}^{N} m_\nu =$ total mass of the system. Equation (5.1.1) gives

$$M\mathbf{r}_c = \sum_{\nu=1}^{N} m_\nu \mathbf{r}_\nu. \tag{5.1.2}$$

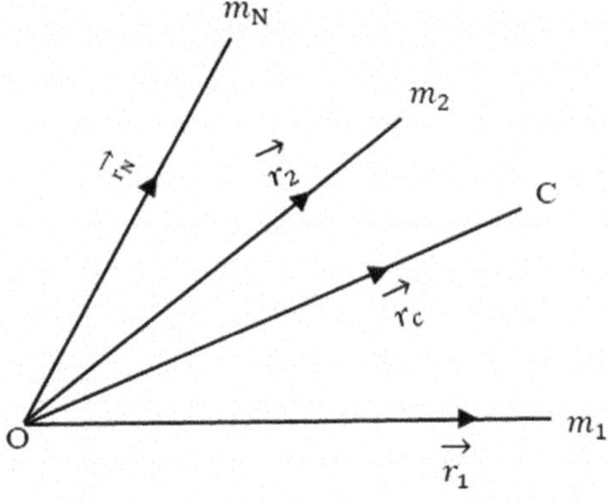

Fig. 5.1 Center of mass of a system of particles

N. Ahmed et al., *Classical Dynamics*, University Texts in the Mathematical Sciences,
https://doi.org/10.1007/978-981-95-6394-4_5

5.2 Center of Gravity

If the system of particles is in a uniform gravitational field, the center of mass is also termed the center of gravity.

5.3 Linear Momentum of a System of Particles

Let m_ν be the mass of the νth particle of the system having position vector $\mathbf{r}_\nu$ relative to O. The velocity $\mathbf{v}_\nu$ of the νth particle is given by

$$\mathbf{v}_\nu = \dot{\mathbf{r}}_\nu.$$

Then the total linear momentum of the system of particles is defined as

$$\mathbf{P} = \sum_{\nu=1}^{N} m_\nu \mathbf{v}_\nu. \tag{5.3.1}$$

Equation (5.3.1) gives

$$\begin{aligned}
\mathbf{P} = \sum_{\nu=1}^{N} m_\nu \dot{\mathbf{r}}_\nu = \sum_{\nu=1}^{N} m_\nu \frac{d\mathbf{r}_\nu}{dt}, \\
= \frac{d}{dt} \sum_{\nu=1}^{N} m_\nu \mathbf{r}_\nu, \\
= \frac{d}{dt} M\mathbf{r}_c, \\
= M\dot{\mathbf{r}}_c = M\mathbf{V_0}
\end{aligned}$$

where $\mathbf{V_0} = \dot{\mathbf{r}}_c =$ Velocity of the centroid.

5.4 Motion of the Center of Mass

Let m_ν, m_λ denote the masses of the νth and λth particle respectively of a system of particles. Let the νth particle be acted on by an external force $\mathbf{F}_\nu$ and an internal force $\mathbf{I}_{\nu\lambda}$ due to the λth particle. Let $\mathbf{r}_\nu$ be the position vector of the νth particle and $\mathbf{r}_c$ that of the center of mass as shown in Fig. 5.2.

By Newton's 2nd law of motion

$$m_\nu \ddot{\mathbf{r}}_\nu = \mathbf{F}_\nu + \sum_{\lambda=1}^{N} \mathbf{I}_{\nu\lambda}. \tag{5.4.1}$$

Equation (5.4.1) gives

$$\sum_{\nu=1}^{N} m_\nu \ddot{\mathbf{r}}_\nu = \sum_{\nu=1}^{N} \mathbf{F}_\nu + \sum_{\substack{\nu=1 \\ \nu\neq\lambda}}^{N} \sum_{\lambda=1}^{N} \mathbf{I}_{\nu\lambda}. \tag{5.4.2}$$

In Eq. (5.4.2),

$$\sum_{\nu=1}^{N} \sum_{\lambda=1}^{N} \mathbf{I}_{\nu\lambda} = \mathbf{0}, \quad \text{as consequence of Newton's 3rd law.}$$

Hence Eq. (5.4.2) reduces to

$$\sum_{\nu=1}^{N} m_\nu \ddot{\mathbf{r}}_\nu = \sum_{\nu=1}^{N} \mathbf{F}_\nu$$

$$\text{i.e.,} \quad \frac{d^2}{dt^2} \sum_{\nu=1}^{N} m_\nu \mathbf{r}_\nu = \sum_{\nu=1}^{N} \mathbf{F}_\nu$$

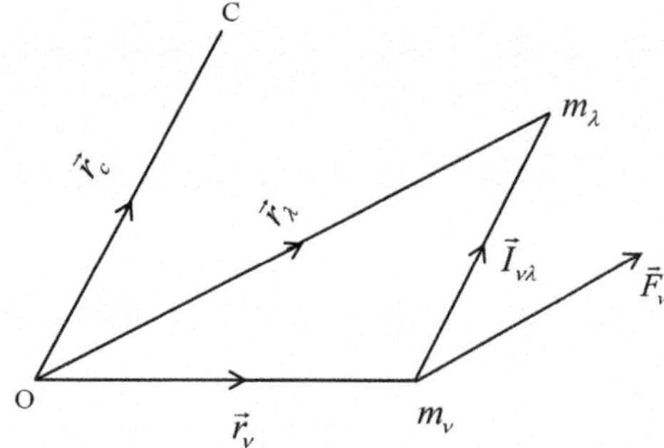

Fig. 5.2 Illustrating the motion of the center of mass of a system of particles

$$\text{i.e.,} \qquad \frac{d^2}{dt^2} M\mathbf{r}_c = \sum_{\nu=1}^{N} \mathbf{F}_\nu$$

$$\text{i.e.,} \qquad M\ddot{\mathbf{r}}_c = \sum_{\nu=1}^{N} \mathbf{F}_\nu. \tag{5.4.3}$$

Thus, the center of mass of a system of particles moves as if the total mass and the resultant impressed force were applied at this point.

5.5 Conservation of Linear Momentum

If $\sum_{\nu=1}^{N} \mathbf{F}_\nu = \mathbf{0}$, then Eq. (5.4.3) gives

$$M\ddot{\mathbf{r}}_c = \mathbf{0},$$

$$\text{i.e.,} \qquad \frac{d}{dt} M\dot{\mathbf{r}}_c = \mathbf{0},$$

$$\therefore \qquad M\dot{\mathbf{r}}_c = \text{a constant vector} = \mathbf{A},$$

$$\text{i.e.,} \qquad \frac{d}{dt} M\mathbf{r}_c = \mathbf{A},$$

$$\text{i.e.,} \qquad \frac{d}{dt} \sum_{\nu=1}^{N} m_\nu \mathbf{r}_\nu = \mathbf{A},$$

$$\text{i.e.,} \qquad \sum_{\nu=1}^{N} m_\nu \dot{\mathbf{r}}_\nu = \mathbf{A},$$

$$\text{i.e.,} \qquad \sum_{\nu=1}^{N} m_\nu \mathbf{v}_\nu = \mathbf{A},$$

$$\text{i.e.,} \qquad \mathbf{P} = \mathbf{A},$$

i.e., the linear momentum is conserved.

Thus, if the resultant impressed force acting on a system of particles is zero, then the total linear momentum is conserved.

5.6 Angular Momentum of a System of Particles

Let m_ν be the mass of the νth particle with position vector $\mathbf{r}_\nu$ relative to a fixed origin. The velocity $\mathbf{v}_\nu$ of the νth particle is given by $\mathbf{v}_\nu = \dot{r}_\nu$.

Then

$$\boldsymbol{\Omega} = \sum_{\nu=1}^{N} \mathbf{r}_\nu \times m_\nu \mathbf{v}_\nu$$

is defined to be the angular momentum of the system of particles about O.

5.7 Total External Torque Applied to a System of Particles

Let the νth particle with mass m_ν be acted on by an external force $\mathbf{F}_\nu$. Let $\mathbf{r}_\nu$ denote the position vector of the νth particle relative to the fixed origin.

Then

$$\boldsymbol{\Lambda} = \sum_{\nu=1}^{N} \mathbf{r}_\nu \times \mathbf{F}_\nu$$

is defined as the external torque applied to the system of particles about the origin O.

5.8 Relation Between External Torque and Angular Momentum

Let m_ν be the mass of the νth particle of the system. Let P_ν, P_λ be respectively the position vector of the νth and λth particles of the system.

From Fig. 5.3, let $\mathbf{r}_\nu$ and $\mathbf{r}_\lambda$ denote the position vector of the points P_ν and P_λ respectively relative to a fixed origin O. Let the νth particle be acted on by an external force $\mathbf{F}_\nu$ and an internal force $\mathbf{I}_{\nu\lambda}$ due to the λth particle.

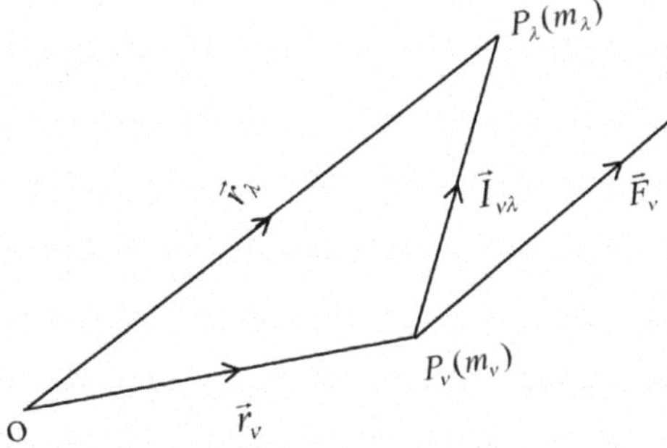

Fig. 5.3 Angular momentum and external torque of a system of particles

The angular momentum of the system of particles about O is given by

$$\mathbf{\Omega} = \sum_{\nu=1}^{N} \mathbf{r}_\nu \times m_\nu \mathbf{v}_\nu; \qquad \mathbf{v}_\nu = \dot{r}_\nu. \tag{5.8.1}$$

The total external torque applied to the system about O is given by

$$\mathbf{\Lambda} = \sum_{\nu=1}^{N} \mathbf{r}_\nu \times \mathbf{F}_\nu. \tag{5.8.2}$$

By Newton's 2nd law of motion

$$m_\nu \ddot{r}_\nu = \mathbf{F}_\nu + \sum_{\lambda=1}^{N} \mathbf{I}_{\nu\lambda}. \tag{5.8.3}$$

Equation (5.8.1) gives

$$\begin{aligned}
\frac{d\mathbf{\Omega}}{dt} &= \sum_{\nu=1}^{N} \left[\mathbf{r}_\nu \times m_\nu \dot{\mathbf{v}}_\nu + \dot{\mathbf{r}}_\nu \times m_\nu \mathbf{v}_\nu \right] \\
&= \sum_{\nu=1}^{N} \left[\mathbf{r}_\nu \times m_\nu \ddot{\mathbf{r}}_\nu + \mathbf{v}_\nu \times m_\nu \mathbf{v}_\nu \right] \\
&= \sum_{\nu=1}^{N} \mathbf{r}_\nu \times m_\nu \ddot{\mathbf{r}}_\nu \\
&= \sum_{\nu=1}^{N} \mathbf{r}_\nu \times \left[\mathbf{F}_\nu + \sum_{\lambda=1}^{N} \mathbf{I}_{\nu\lambda} \right] \\
&= \sum_{\nu=1}^{N} \mathbf{r}_\nu \times \mathbf{F}_\nu + \sum_{\nu=1}^{N} \sum_{\substack{\lambda=1 \\ \nu\neq\lambda}}^{N} \mathbf{r}_\nu \times \mathbf{I}_{\nu\lambda} \\
&= \mathbf{\Lambda} + \sum_{\nu=1}^{N} \sum_{\substack{\lambda=1 \\ \nu\neq\lambda}}^{N} \mathbf{r}_\nu \times \mathbf{I}_{\nu\lambda}.
\end{aligned} \tag{5.8.4}$$

Now

$$
\begin{aligned}
\mathbf{r}_\nu \times \mathbf{I}_{\nu\lambda} + \mathbf{r}_\lambda \times \mathbf{I}_{\lambda\nu} &= \mathbf{r}_\nu \times \mathbf{I}_{\nu\lambda} - \mathbf{r}_\lambda \times \mathbf{I}_{\nu\lambda} \\
&= \left(\mathbf{r}_\nu - \mathbf{r}_\lambda\right) \times \mathbf{I}_{\nu\lambda} \\
&= \mathbf{P}_\lambda \mathbf{P}_\nu \times \mathbf{I}_{\nu\lambda} \\
&= -\mathbf{P}_\nu \mathbf{P}_\lambda \times \mathbf{I}_{\nu\lambda}.
\end{aligned} \tag{5.8.5}
$$

But $\mathbf{P}_\nu \mathbf{P}_\lambda \parallel \mathbf{I}_{\nu\lambda}$

$$\therefore \quad \mathbf{P}_\nu \mathbf{P}_\lambda \times \mathbf{I}_{\nu\lambda} = \mathbf{0}.$$

Hence Eq. (5.8.5) gives

$$\sum_{\nu=1}^{N} \sum_{\substack{\lambda=1 \\ \nu \neq \lambda}}^{N} \mathbf{r}_\nu \times \mathbf{I}_{\nu\lambda} = \mathbf{0}. \tag{5.8.6}$$

Therefore, Eq. (5.8.4) reduces to

$$\frac{d\boldsymbol{\Omega}}{dt} = \boldsymbol{\Lambda}. \tag{5.8.7}$$

This relation is known as the principle of angular momentum.

5.9 Conservation of Angular Momentum

If $\boldsymbol{\Lambda} = \mathbf{0}$, then Eq. (5.8.7) gives

$$\frac{d\boldsymbol{\Omega}}{dt} = \mathbf{0}.$$

i.e., $\boldsymbol{\Omega}$ = a constant vector.

Thus, if the total external torque on a system of particles is zero, then the total angular momentum of the system of particles is conserved.

5.10 Kinetic Energy of a System of Particles

With usual notations, the total kinetic energy of a system of particles is defined by

$$T = \frac{1}{2} \sum_{\nu=1}^{N} m_\nu \dot{\mathbf{r}}_\nu^2 = \frac{1}{2} \sum_{\nu=1}^{N} m_\nu \mathbf{v}_\nu^2.$$

5.11 Work

Let $\mathbf{R}_\nu$ be the resultant force acting on the νth particle having mass m_ν. Then the total work done in moving the system of particles from one state to another is given by

$$W_{12} = \sum_{\nu=1}^{N} \int_1^2 \mathbf{R}_\nu \cdot d\mathbf{r}_\nu. \tag{5.11.1}$$

Let $\mathbf{F}_\nu$ be the external force acting on m_ν and $\mathbf{I}_{\nu\lambda}$ be the internal force acting on the νth particle due to λth particle. Let $\mathbf{r}_\nu$ be the position vector of the νth particle relative to some origin O.

By Newton's 2nd law of motion,

$$\begin{aligned} m_\nu \ddot{\mathbf{r}}_\nu &= \mathbf{F}_\nu + \sum_{\substack{\lambda=1 \\ \lambda\neq\nu}}^{N} \mathbf{I}_{\nu\lambda} \\ &= \mathbf{R}_\nu, \end{aligned} \tag{5.11.2}$$

where $\mathbf{R}_\nu = \mathbf{F}_\nu + \sum_{\substack{\lambda=1 \\ \lambda\neq\nu}}^{N} \mathbf{I}_{\nu\lambda}.$

Therefore, Eq. (5.11.1) gives

$$\begin{aligned} W_{12} &= \sum_{\nu=1}^{N} \int_1^2 \mathbf{R}_\nu \cdot d\mathbf{r}_\nu \\ &= \sum_{\nu=1}^{N} \int_1^2 m_\nu \ddot{\mathbf{r}}_\nu \cdot d\mathbf{r}_\nu \\ &= \sum_{\nu=1}^{N} \int_1^2 m_\nu \frac{d\mathbf{v}_\nu}{dt} \cdot \frac{d\mathbf{r}_\nu}{dt} dt \\ &= \sum_{\nu=1}^{N} \int_1^2 m_\nu \frac{d\mathbf{v}_\nu}{dt} \cdot \mathbf{v}_\nu dt \\ &= \sum_{\nu=1}^{N} \int_1^2 m_\nu v_\nu \frac{dv_\nu}{dt} dt \\ &= \sum_{\nu=1}^{N} \int_1^2 m_\nu v_\nu dv_\nu \end{aligned}$$

$$
\begin{aligned}
&= \sum_{\nu=1}^{N} \int_{1}^{2} m_\nu d\left(\frac{v_\nu^2}{2}\right) \\
&= \sum_{\nu=1}^{N} \int_{1}^{2} d\left(\frac{1}{2} m_\nu v_\nu^2\right) \\
&= \int_{1}^{2} d \sum_{\nu=1}^{N} \left(\frac{1}{2} m_\nu v_\nu^2\right) \\
&= \int_{1}^{2} dT \\
&= T_2 - T_1 .
\end{aligned}
$$

Thus, the total work done in moving a system of particles from one to another is given by

$$W_{12} = T_2 - T_1 .$$

5.12 Principle of Conservation of Energy for a System of Particles

Let the νth particle of the system be acted upon by the external force $\mathbf{F}_\nu$. Let $\mathbf{I}_{\nu\lambda}$ be the internal force on the νth particle due to the λth particle. Let $\mathbf{r}_\nu$ be the position vector of the νth particle relative to some origin O.

Therefore, the resultant impressed force on the νth particle is given by

$$\mathbf{R}_\nu = \mathbf{F}_\nu + \sum_{\substack{\lambda=1 \\ \lambda \neq \nu}}^{N} \mathbf{I}_{\nu\lambda} . \tag{5.12.1}$$

Let $\mathbf{F}_\nu$ be conservative with force potential V_ν.

$$\therefore \qquad \mathbf{F}_\nu = -\nabla V_\nu . \tag{5.12.2}$$

The internal force $\mathbf{I}_{\nu\lambda}$ being a central force is conservative with the force potential $V_{\nu\lambda}$ (say).

$$\therefore \qquad \mathbf{I}_{\nu\lambda} = -\nabla V_{\nu\lambda} . \tag{5.12.3}$$

The elementary work done on the system of particles is given by

$$
\begin{aligned}
dW &= \sum_{\nu=1}^{N} \mathbf{R}_{\nu} \cdot d\mathbf{r}_{\nu} \\
&= \sum_{\nu=1}^{N} \Big\{ \mathbf{F}_{\nu} + \sum_{\substack{\lambda=1 \\ \lambda\neq\nu}}^{N} \mathbf{I}_{\nu\lambda} \Big\} \cdot d\mathbf{r}_{\nu} \\
&= \sum_{\nu=1}^{N} \mathbf{F}_{\nu} \cdot d\mathbf{r}_{\nu} + \sum_{\nu=1}^{N} \sum_{\substack{\lambda=1 \\ \lambda\neq\nu}}^{N} \mathbf{I}_{\nu\lambda} \cdot d\mathbf{r}_{\nu}.
\end{aligned} \tag{5.12.4}
$$

We have

$$
\begin{aligned}
\sum_{\nu=1}^{N} \mathbf{F}_{\nu} \cdot d\mathbf{r}_{\nu} &= -\sum_{\nu=1}^{N} \nabla V_{\nu} \cdot d\mathbf{r}_{\nu} \\
&= -\sum_{\nu=1}^{N} dV_{\nu} \\
&= -d \sum_{\nu=1}^{N} V_{\nu} \\
&= -dV_E
\end{aligned} \tag{5.12.5}
$$

where

$$
V_E = \sum\nolimits_{\nu=1}^{N} V_{\nu} = \text{total external potential energy.}
$$

Now

$$
\begin{aligned}
\mathbf{I}_{\nu\lambda} \cdot d\mathbf{r}_{\nu} + \mathbf{I}_{\lambda\nu} \cdot d\mathbf{r}_{\lambda} &= -\nabla V_{\nu\lambda} \cdot d\mathbf{r}_{\nu} - \nabla V_{\lambda\nu} \cdot d\mathbf{r}_{\lambda} \\
&= -\nabla V_{\nu\lambda} \cdot d\mathbf{r}_{\nu} - \nabla V_{\nu\lambda} \cdot d\mathbf{r}_{\lambda} \\
&= -dV_{\nu}^{\nu\lambda} - dV_{\lambda}^{\nu\lambda} \\
&= -\left[\frac{\partial V_{\nu\lambda}}{\partial x_{\nu}} dx_{\nu} + \frac{\partial V_{\nu\lambda}}{\partial y_{\nu}} dy_{\nu} + \frac{\partial V_{\nu\lambda}}{\partial z_{\nu}} dz_{\nu} + \frac{\partial V_{\nu\lambda}}{\partial x_{\lambda}} dx_{\lambda} + \frac{\partial V_{\nu\lambda}}{\partial y_{\lambda}} dy_{\lambda} + \frac{\partial V_{\nu\lambda}}{\partial z_{\lambda}} dz_{\lambda} \right] \\
&= -dV_{\nu\lambda}.
\end{aligned}
$$

$$\therefore \quad \sum_{\nu=1}^{N}\sum_{\substack{\lambda=1\\ \lambda\neq\nu}}^{N} \mathbf{I}_{\nu\lambda}\cdot d\mathbf{r}_\nu = -\sum_{\nu=1}^{N}\sum_{\substack{\lambda=1\\ \lambda\neq\nu}}^{N} dV_{\nu\lambda}$$

$$= -d\sum_{\nu=1}^{N}\sum_{\substack{\lambda=1\\ \lambda\neq\nu}}^{N} V_{\nu\lambda}$$

$$= -dV_I. \tag{5.12.6}$$

where

$$V_I = \sum_{\nu=1}^{N}\sum_{\substack{\lambda=1\\ \lambda\neq\nu}}^{N} V_{\nu\lambda}$$

$=$ total internal potential energy of the system.

Equation (5.12.4) gets reduced to

$$\begin{aligned} dW &= -dV_E - dV_I \\ &= -d(V_E - V_I) \\ &= -dV \end{aligned} \tag{5.12.7}$$

where

$$V = V_E + V_I$$

$=$ total PE of the system.

Equation (5.12.7) gives

$$W_{12} = -\int_1^2 dV = V_1 - V_2. \tag{5.12.8}$$

In Sect. 5.9, it is established that

$$W_{12} = T_2 - T_1. \tag{5.12.9}$$

Equations (5.12.8) and (5.12.9) give

$$T_2 - T_1 = V_1 - V_2$$

i.e., $\quad T_1 + V_1 = T_2 + V_2$

i.e., $\quad T + V = \text{Constant}.$

Hence the total energy is conserved. This is known as the principle of conservation of energy.

Theorem 5.1 *The total angular momentum of a system of particles about any point O equals the angular momentum of the total mass concentrated at the center of mass plus the angular momentum about the center of mass.*

Proof Let G be the center of mass with position vector r_c relative to O. Let P be the position of the νth particle having position vector $\mathbf{r}_{\nu'}$ relative to G and suppose that $\mathbf{OP} = \mathbf{r}_\nu$ as shown in Fig. 5.4.

$$\mathbf{r}_\nu = \mathbf{r}_c + \mathbf{r}_{\nu'}. \tag{5.12.10}$$

As $\mathbf{r}_c$ is the position vector of G relative to O.

$$\therefore \quad \mathbf{r}_c = \frac{\sum_{\nu=1}^{N} m_\nu \mathbf{r}_\nu}{\sum_{\nu=1}^{N} m_\nu} = \frac{\sum_{\nu=1}^{N} m_\nu \mathbf{r}_\nu}{M} \tag{5.12.11}$$

where $M = \sum\limits_{\nu=1}^{N} m_\nu =$ total mass of the system of particles.

Equation (5.12.11) gives

$$\sum_{\nu=1}^{N} m_\nu \mathbf{r}_\nu = M\mathbf{r}_c. \tag{5.12.12}$$

Now angular momentum of the system of particles about O is given by

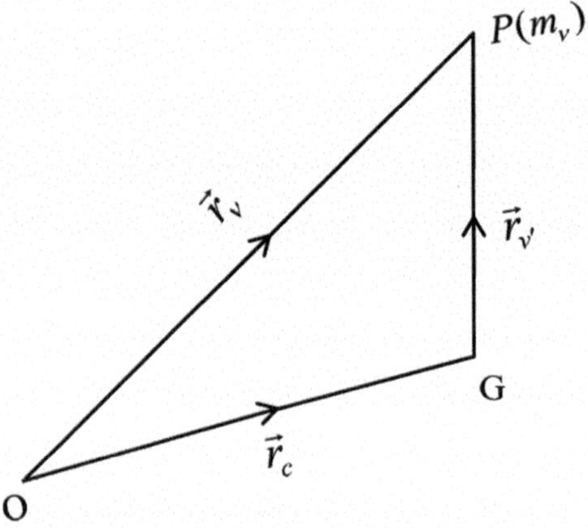

Fig. 5.4 Geometrical illustration for Theorem 5.1

$$
\begin{aligned}
\boldsymbol{\Omega} &= \sum_{\nu=1}^{N} \mathbf{r}_\nu \times m_\nu \dot{\mathbf{r}}_\nu \\
&= \sum_{\nu=1}^{N} \left(\mathbf{r}_c + \mathbf{r}_{\nu'}\right) \times m_\nu \left(\dot{\mathbf{r}}_c + \dot{\mathbf{r}}_{\nu'}\right) \\
&= \sum_{\nu=1}^{N} \mathbf{r}_c \times m_\nu \dot{\mathbf{r}}_c + \sum_{\nu=1}^{N} \mathbf{r}_c \times m_\nu \dot{\mathbf{r}}_{\nu'} + \sum_{\nu=1}^{N} \mathbf{r}_{\nu'} \times m_\nu \dot{\mathbf{r}}_c + \sum_{\nu=1}^{N} \mathbf{r}_{\nu'} \times m_\nu \dot{\mathbf{r}}_{\nu'} \\
&= \boldsymbol{\Omega}_1 + \boldsymbol{\Omega}_2 + \boldsymbol{\Omega}_3 + \boldsymbol{\Omega}_4.
\end{aligned}
\tag{5.12.13}
$$

Now

$$
\begin{aligned}
\boldsymbol{\Omega}_1 &= \sum_{\nu=1}^{N} \mathbf{r}_c \times m_\nu \dot{\mathbf{r}}_c \\
&= \mathbf{r}_c \times \dot{\mathbf{r}}_c \sum_{\nu=1}^{N} m_\nu \\
&= \mathbf{r}_c \times \dot{\mathbf{r}}_c M \\
&= \mathbf{r}_c \times M \dot{\mathbf{r}}_c \\
&= \mathbf{r}_c \times M \mathbf{V}_0; \qquad \mathbf{V}_0 = \dot{\mathbf{r}}_c = \text{velocity of the center of mass.}
\end{aligned}
$$

$$
\begin{aligned}
\boldsymbol{\Omega}_2 &= \sum_{\nu=1}^{N} \mathbf{r}_c \times m_\nu \dot{\mathbf{r}}_{\nu'} \\
&= \mathbf{r}_c \times \sum_{\nu=1}^{N} m_\nu \dot{\mathbf{r}}_{\nu'} \\
&= \mathbf{r}_c \times \sum_{\nu=1}^{N} m_\nu \frac{d\mathbf{r}_{\nu'}}{dt} \\
&= \mathbf{r}_c \times \sum_{\nu=1}^{N} \frac{d}{dt}\left(m_\nu \mathbf{r}_{\nu'}\right) \\
&= \mathbf{r}_c \times \frac{d}{dt} \sum_{\nu=1}^{N} m_\nu \mathbf{r}_{\nu'}.
\end{aligned}
$$

Now

$$
\frac{\sum_{\nu=1}^{N} m_\nu \mathbf{r}_{\nu'}}{\sum_{\nu=1}^{N} m_\nu} = \frac{\sum_{\nu=1}^{N} m_\nu \mathbf{r}_{\nu'}}{M} = \text{P.V. of G relative to G} = \mathbf{0}.
$$

$$\therefore \qquad \sum_{\nu=1}^{N} m_\nu \mathbf{r}_{\nu'} = \mathbf{0}. \tag{5.12.14}$$

Hence
$$\boldsymbol{\Omega}_2 = \mathbf{0}. \qquad \text{[by (5.12.14)]}$$

$$\begin{aligned}
\boldsymbol{\Omega}_3 &= \sum_{\nu=1}^{N} \mathbf{r}_{\nu'} \times m_\nu \dot{\mathbf{r}}_c \\
&= \sum_{\nu=1}^{N} m_\nu \mathbf{r}_{\nu'} \times \dot{\mathbf{r}}_c \\
&= \mathbf{0}. \qquad \left[\text{by } (5.12.14)\right]
\end{aligned}$$

$$\begin{aligned}
\boldsymbol{\Omega}_4 &= \sum_{\nu=1}^{N} \mathbf{r}_{\nu'} \times m_\nu \dot{\mathbf{r}}_{\nu'} \\
&= \sum_{\nu=1}^{N} \mathbf{r}_{\nu'} \times m_\nu \mathbf{V}_{\nu'},
\end{aligned}$$

where $\mathbf{V}_{\nu'} = \dot{\mathbf{r}}_{\nu'}$ = velocity of the νth particle relative to G.
Hence (5.12.13) reduces to

$$\boldsymbol{\Omega} = \mathbf{r}_c \times M\mathbf{V}_0 + \sum_{\nu=1}^{N} \mathbf{r}_{\nu'} \times m_\nu \mathbf{V}_{\nu'}.$$

Thus, the result is established. □

Theorem 5.2 *The total kinetic energy of a system of particles about any point O equals the kinetic energy of translation of the center of mass as if all the mass is located at the center of mass plus the kinetic energy of the motion about the center of mass.*

Proof Closely following Sect. 5.11, The kinetic energy of the system of particles is given by

$$\begin{aligned}
T &= \frac{1}{2}\sum_{\nu=1}^{N} m_\nu \dot{\mathbf{r}}_\nu^2 \\
&= \frac{1}{2}\sum_{\nu=1}^{N} m_\nu \left(\dot{\mathbf{r}}_c + \dot{\mathbf{r}}_{\nu'}\right)^2 \\
&= \frac{1}{2}\sum_{\nu=1}^{N} m_\nu \left(\dot{\mathbf{r}}_c^2 + 2\dot{\mathbf{r}}_c \cdot \dot{\mathbf{r}}_{\nu'} + \dot{\mathbf{r}}_{\nu'}^2\right)
\end{aligned}$$

$$= \frac{1}{2}\sum_{\nu=1}^{N} m_\nu \dot{\mathbf{r}}_c^2 + \sum_{\nu=1}^{N} m_\nu \dot{\mathbf{r}}_c \cdot \dot{\mathbf{r}}_{\nu'} + \frac{1}{2}\sum_{\nu=1}^{N} m_\nu \dot{\mathbf{r}}_{\nu'}^2$$

$$= T_1 + T_2 + T_3. \tag{5.12.15}$$

Now,

$$T_1 = \frac{1}{2}\sum_{\nu=1}^{N} m_\nu \dot{\mathbf{r}}_c^2$$

$$= \frac{1}{2}\Big(\sum_{\nu=1}^{N} m_\nu\Big)\dot{\mathbf{r}}_c^2 = \frac{1}{2} M V_0^2,$$

$$T_2 = \sum_{\nu=1}^{N} m_\nu \dot{\mathbf{r}}_c \cdot \dot{\mathbf{r}}_{\nu'}$$

$$= \dot{\mathbf{r}}_c \cdot \sum_{\nu=1}^{N} m_\nu \dot{\mathbf{r}}_{\nu'}$$

$$= \dot{\mathbf{r}}_c \cdot \frac{d}{dt}\sum_{\nu=1}^{N} m_\nu \mathbf{r}_{\nu'} = 0 \quad \Big[\text{by } (5.12.14)\Big],$$

$$T_3 = \frac{1}{2}\sum_{\nu=1}^{N} m_\nu \dot{\mathbf{r}}_{\nu'}^2$$

$$= \frac{1}{2}\sum_{\nu=1}^{N} m_\nu \mathbf{V}_{\nu'}^2.$$

Therefore (5.12.15) gives

$$T = \frac{1}{2} M V_0^2 + \frac{1}{2}\sum_{\nu=1}^{N} m_\nu \mathbf{V}_{\nu'}^2.$$

Hence the theorem. □

5.13 Impulse

If $\mathbf{R}$ is the total external force acting on a system of particles, then the time integral

$$\mathbf{I} = \int_{t_1}^{t_2} \mathbf{R} dt$$

is called the total linear impulse applied to the system.

Theorem 5.3 *The total linear impulse equals the change in linear momentum.*

Proof Let the νth particle of the system be acted upon by the force $\mathbf{R}_\nu$ (external+internal). Let m_ν be the mass of νth particle with position vector $\mathbf{r}_\nu$ relative to some origin. Let $\mathbf{R}$ be the resultant of the forces acting on the system.

$$\mathbf{R} = \sum_{\nu=1}^{N} \mathbf{R}_\nu.$$

By Newton's 2nd law of motion,

$$m_\nu \ddot{\mathbf{r}}_\nu = \mathbf{R}_\nu,$$

$$\therefore \sum_{\nu=1}^{N} m_\nu \ddot{\mathbf{r}}_\nu = \sum_{\nu=1}^{N} \mathbf{R}_\nu = \mathbf{R}.$$

Now,

$$\begin{aligned}
\mathbf{I} &= \int_{t_1}^{t_2} \mathbf{R} dt \\
&= \int_{t_1}^{t_2} \sum_{\nu=1}^{N} m_\nu \ddot{\mathbf{r}}_\nu \, dt \\
&= \sum_{\nu=1}^{N} \int_{t_1}^{t_2} m_\nu \ddot{\mathbf{r}}_\nu \, dt \\
&= \sum_{\nu=1}^{N} \int_{t_1}^{t_2} \frac{d}{dt}(m_\nu \dot{\mathbf{r}}_\nu) dt \\
&= \sum_{\nu=1}^{N} \int_{t_1}^{t_2} d(m_\nu \mathbf{V}_\nu) \\
&= \int_{t_1}^{t_2} \sum_{\nu=1}^{N} d(m_\nu \mathbf{V}_\nu) \\
&= \int_{t_1}^{t_2} d(\sum_{\nu=1}^{N} m_\nu \mathbf{V}_\nu) \\
&= \int_{t_1}^{t_2} d\mathbf{P} \\
&= \mathbf{P}_2 - \mathbf{P}_1 \\
&= \text{Change in linear momentum.}
\end{aligned}$$

Hence the theorem. □

5.14 Angular Impulse

If $\mathbf{\Lambda}$ is the total external torque applied to a system of particles about the origin O, then

$$\mathbf{J} = \int_{t_1}^{t_2} \mathbf{\Lambda}\, dt$$

is called the total angular impulse applied to the system.

Theorem 5.4 $\mathbf{J} = \mathbf{\Omega}_2 - \mathbf{\Omega}_1$.

Proof By the principle of angular momentum,

$$\frac{d\mathbf{\Omega}}{dt} = \mathbf{\Lambda}. \tag{5.14.1}$$

Now,

$$\begin{aligned} \mathbf{J} &= \int_{t_1}^{t_2} \mathbf{\Lambda}\, dt \\ \text{i.e.,}\quad \mathbf{J} &= \int_{t_1}^{t_2} \frac{d\mathbf{\Omega}}{dt}\, dt \\ &= \mathbf{\Omega}\Big]_{t_1}^{t_2} \\ &= \mathbf{\Omega}(t_2) - \mathbf{\Omega}(t_1) \\ &= \mathbf{\Omega}_2 - \mathbf{\Omega}_1. \end{aligned}$$

□

5.15 Constraints

The limitations imposed on the motion of a particle or a system of particles are often termed as constraints. In the case of the motion of a system of particles consisting of N number of particles, the constraints imposed on the system are said to be holonomic, if it can be expressed as an equation $f(\mathbf{r}_1, \mathbf{r}_2, \ldots, \mathbf{r}_N, t) = 0$. If the constraint condition cannot be expressed as in the above equation, the constraint is called non-holonomic.

5.16 Virtual Displacement and Virtual Work

If a system of particles undergoes some displacement under the action of forces, then some amount of actual work gets done. In case, when the system of particles is in equilibrium under the action of the given forces, then there is no displacement, and hence no work is done. But in some problems of statics, it is often convenient to assume that the system undergoes some imaginary displacement consistent with the constraints or configuration of the system. Such a displacement is said to be virtual because of its hypothetical nature. In such cases, the amount of work done by the forces is called virtual work.

Consider two possible configurations of a system of particles at a particular instant consistent with the forces and constraints. To go from one configuration to another, let the νth particle be displaced through a distance $\delta \mathbf{r}_\nu$, $\mathbf{r}_\nu$ being the position vector of the νth particle. Here $\delta \mathbf{r}_\nu$ is the virtual displacement of the νth particle. In fact, the actual displacement of the νth particle is denoted by $d\,\mathbf{r}_\nu$ to distinguish it from virtual displacement.

5.17 Principle of Virtual Work

Consider a system of N particles with masses m_ν $(\nu = 1, 2, \ldots, N)$. Now, in order to keep the system in equilibrium, the resultant force acting on each particle must be zero, i.e., $\mathbf{F}_\nu = \mathbf{0}$. This leads $\mathbf{F}_\nu \cdot \delta \mathbf{r}_\nu = 0$, where $\mathbf{F}_\nu \cdot \delta \mathbf{r}_\nu$ is called the virtual work. Adding these, we obtain

$$\sum_{\nu=1}^{N} \mathbf{F}_\nu \cdot \delta \mathbf{r}_\nu = 0. \tag{5.17.1}$$

Equation (5.17.1) leads to conclude that a system of particles is in equilibrium if and only if $\sum_{\nu=1}^{N} \mathbf{F}_\nu \cdot \delta \mathbf{r}_\nu = 0$. This is often termed as the principle of virtual work.

5.18 D′Alembert's Principle

Let the νth particle be acted on by an external force $\mathbf{F}_\nu$. Let m_ν be the mass of the νth particle with position vector r_ν relative to some origin O. Let $\mathbf{I}_{\nu\lambda}$ be the internal force on the νth particle due to λth particle. By Newton's 2nd law of motion

$$m_\nu \ddot{\mathbf{r}}_\nu = \mathbf{F}_\nu + \sum_{\substack{\lambda=1 \\ \lambda \neq \nu}}^{N} \mathbf{I}_{\nu\lambda}. \tag{5.18.1}$$

Equation (5.18.1) gives

$$\sum_{\nu=1}^{N} m_\nu \ddot{\mathbf{r}}_\nu = \sum_{\nu=1}^{N} \mathbf{F}_\nu + \sum_{\nu=1}^{N} \sum_{\substack{\lambda=1 \\ \lambda\neq\nu}}^{N} \mathbf{I}_{\nu\lambda}. \tag{5.18.2}$$

By Newton's 3rd law,

$$\sum_{\nu=1}^{N} \sum_{\substack{\lambda=1 \\ \lambda\neq\nu}}^{N} \mathbf{I}_{\nu\lambda} = \mathbf{0}. \tag{5.18.3}$$

Therefore Eq. (5.18.2) reduces to

$$\sum_{\nu=1}^{N} m_\nu \ddot{\mathbf{r}}_\nu = \sum_{\nu=1}^{N} \mathbf{F}_\nu,$$

$$\text{i.e.} \quad -\sum_{\nu=1}^{N} m_\nu \ddot{\mathbf{r}}_\nu + \sum_{\nu=1}^{N} \mathbf{F}_\nu = \mathbf{0}. \tag{5.18.4}$$

Equation (5.18.4) conclude that the forces

$$-m_\nu \ddot{\mathbf{r}}_\nu + \mathbf{F}_\nu \quad (\nu = 1, 2, \ldots, N)$$

are in equilibrium. The force $-m_\nu \ddot{\mathbf{r}}_\nu$ is called the reversal effective force on the νth particle.

Equation (5.18.4) leads to conclude that a moving system of particles can be considered to be in equilibrium under a system of forces $\mathbf{F}_\nu - m_\nu \ddot{\mathbf{r}}_\nu \;\; (\nu = 1, 2, \ldots, N)$; i.e., the actual force together with the reversal effective force.

Now, by using the principle of virtual work, we arrive

$$\sum_{\nu=1}^{N} \left(\mathbf{F}_\nu - m_\nu \ddot{\mathbf{r}}_\nu\right) \cdot \delta\mathbf{r}_\nu = 0.$$

That is a system of particles that move in such a way that the total virtual work $\sum_{\nu=1}^{N} \left(\mathbf{F}_\nu - m_\nu \ddot{\mathbf{r}}_\nu\right) \cdot \delta\mathbf{r}_\nu = 0$. This is often called D'Alembert's principle.

Example 5.1 Two particles having masses m_1 and m_2 move so that their relative velocity is $\mathbf{V}$ and the velocity of their center of mass is $\mathbf{V_0}$. If $M = m_1 + m_2$ is the total mass and $\mu = \frac{m_1 m_2}{m_1 + m_2}$ is the reduced mass of the system, prove that the total kinetic energy is $\frac{1}{2} M \mathbf{V_0^2} + \frac{1}{2} \mu \mathbf{V^2}$.

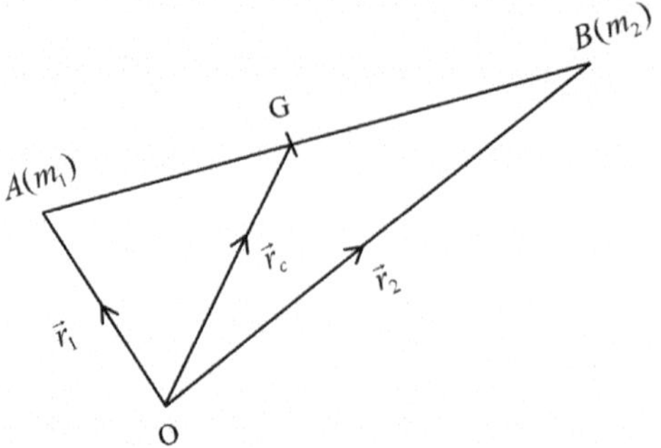

Fig. 5.5 Illustration for Example 5.1

Solution Let A and B be the positions of two masses m_1 and m_2 respectively. Let $\mathbf{r}_1$ and $\mathbf{r}_2$ be the position vectors of A and B, respectively, relative to some origin O. Let $\mathbf{r}_c$ be the position vector of the center of mass G as shown in Fig. 5.5.

Let $$\mathbf{r}_1 = \mathbf{V}_1, \quad \dot{\mathbf{r}}_2 = \mathbf{V}_2$$

$$\therefore \quad \mathbf{r}_c = \frac{m_1\mathbf{r}_1 + m_2\mathbf{r}_2}{m_1 + m_2}$$
$$= \frac{m_1\mathbf{r}_1 + m_2\mathbf{r}_2}{M}$$
$$\therefore \quad M\mathbf{r}_c = m_1\mathbf{r}_1 + m_2\mathbf{r}_2. \tag{5.18.5}$$

Equation (5.18.5) gives

$$M\dot{\mathbf{r}}_c = m_1\dot{\mathbf{r}}_1 + m_2\dot{\mathbf{r}}_2$$
$$\text{i.e.,} \quad M\mathbf{V}_0 = m_1\mathbf{V}_1 + m_2\mathbf{V}_2. \tag{5.18.6}$$

Given

$$m_1 + m_2 = M. \tag{5.18.7}$$

According to the question

$$\mathbf{V}_2 - \mathbf{V}_1 = \mathbf{V},$$
$$\text{i.e.,} \quad \mathbf{V}_2 = \mathbf{V} + \mathbf{V}_1. \tag{5.18.8}$$

Using (5.18.8) in (5.18.6), we obtain

$$
\begin{aligned}
M\mathbf{V}_0 &= m_1\mathbf{V}_1 + m_2(\mathbf{V} + \mathbf{V}_1) \\
&= m_1\mathbf{V}_1 + m_2\mathbf{V} + m_2\mathbf{V}_1 \\
&= (m_1 + m_2)\mathbf{V}_1 + m_2\mathbf{V} \\
&= M\mathbf{V}_1 + m_2\mathbf{V}, \\
\therefore \quad M\mathbf{V}_1 &= M\mathbf{V}_0 - m_2\mathbf{V}, \\
\therefore \quad \mathbf{V}_1 &= \mathbf{V}_0 - \frac{m_2}{M}\mathbf{V}.
\end{aligned} \tag{5.18.9}
$$

Therefore Eq. (5.18.8) gives

$$
\begin{aligned}
\mathbf{V}_2 &= \mathbf{V} + \mathbf{V}_0 - \frac{m_2}{M}\mathbf{V}, \\
\text{i.e.,} \quad \mathbf{V}_2 &= \mathbf{V}_0 + \left(1 - \frac{m_2}{M}\right)\mathbf{V} \\
&= \mathbf{V}_0 + \frac{m_1 + m_2 - m_2}{M}\mathbf{V} \\
&= \mathbf{V}_0 + \frac{m_1}{M}\mathbf{V}.
\end{aligned} \tag{5.18.10}
$$

Therefore, the total kinetic energy of the system of particles is given by

$$
\begin{aligned}
T &= \frac{1}{2}m_1\mathbf{V}_1^2 + \frac{1}{2}m_2\mathbf{V}_2^2 \\
&= \frac{1}{2}m_1\left(\mathbf{V}_0 - \frac{m_2}{M}\mathbf{V}\right)^2 + \frac{1}{2}m_2\left(\mathbf{V}_0 + \frac{m_1}{M}\mathbf{V}\right)^2 \\
&= \frac{1}{2}\left[(m_1 + m_2)\mathbf{V}_0^2 + \left(m_1\frac{m_2^2}{M^2} + m_2\frac{m_1^2}{M^2}\right)\mathbf{V}^2\right], \\
\text{i.e.,} \quad T &= \frac{1}{2}\left[M\mathbf{V}_0^2 + \frac{m_1 m_2}{M^2}(m_2 + m_1)\mathbf{V}^2\right] \\
&= \frac{1}{2}\left[M\mathbf{V}_0^2 + \frac{m_1 m_2}{M}\mathbf{V}^2\right] \\
&= \frac{1}{2}\left[M\mathbf{V}_0^2 + \frac{m_1 m_2}{m_1 + m_2}\mathbf{V}^2\right] \\
&= \frac{1}{2}\left[M\mathbf{V}_0^2 + \mu\mathbf{V}^2\right] \\
&= \frac{1}{2}M\mathbf{V}_0^2 + \frac{1}{2}\mu\mathbf{V}^2.
\end{aligned}
$$

Hence proved.

Example 5.2 Two particles of masses m_1 and m_2 are located on a frictionless double incline and connected by an in-extensible massless string passing over a smooth peg. α_1 and α_2 are angles of the incline. Use D′Alembert's principle to describe the masses.

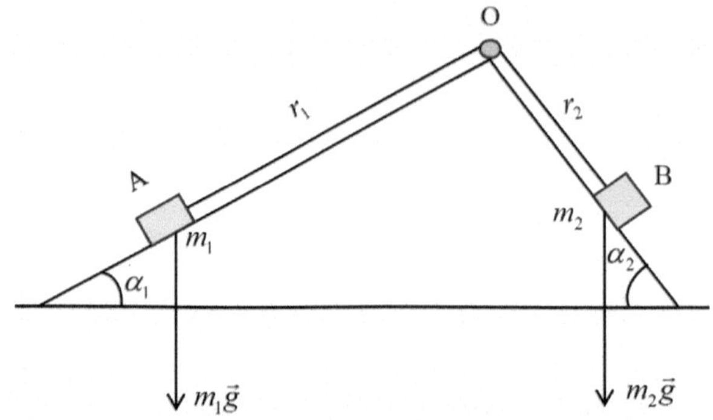

Fig. 5.6 Illustrated diagram for Example 5.2

Solution Let O be the peg and A and B be the positions of the masses m_1 and m_2 respectively. From Fig. 5.6, let OA $= r_1$ and OB $= r_2$ and $\mathbf{OA} = \mathbf{r}_1$ and $\mathbf{OB} = \mathbf{r}_2$.

Because the string is inextensible,

$$\therefore \quad \text{OA} + \text{OB} = \text{a constant} = L \quad (say)$$

$$\text{i.e.,} \quad r_1 + r_2 = L. \tag{5.18.11}$$

Equation (5.18.11) gives

$$\left.\begin{aligned} \delta r_1 + \delta r_2 &= 0 \\ \text{and} \quad \ddot{r}_1 + \ddot{r}_2 &= 0 \end{aligned}\right\} \tag{5.18.12}$$

Let the reversed effective forces $-m_1\ddot{r}_1$ and $-m_2\ddot{r}_2$ be introduced to the masses m_1 and m_2 respectively.

By D′Alembert's principle, it follows that

$$(m_1\mathbf{g} - m_1\ddot{\mathbf{r}}_1) \cdot \delta\mathbf{r}_1 + (m_2\mathbf{g} - m_2\ddot{\mathbf{r}}_2) \cdot \delta\mathbf{r}_2 = 0,$$

$$\text{i.e.,} \quad m_1\mathbf{g} \cdot \delta\mathbf{r}_1 - m_1\ddot{\mathbf{r}}_1 \cdot \delta\mathbf{r}_1 + m_2\mathbf{g} \cdot \delta\mathbf{r}_2 - m_2\ddot{\mathbf{r}}_2 \cdot \delta\mathbf{r}_2 = 0,$$

$$\text{i.e.,} \quad m_1\, g\, \delta r_1 \cos(90^\circ - \alpha_1) - m_1\, \ddot{r}_1\, \delta r_1 \cos 0 + m_2\, g\, \delta r_2 \cos(90^\circ - \alpha_2) - m_2\, \ddot{r}_2\, \delta r_2 \cos 0 = 0,$$

$$\text{i.e.,} \quad m_1\, g\, \delta r_1 \sin\alpha_1 - m_1\, \ddot{r}_1\, \delta r_1 + m_2\, g\, \delta r_2 \sin\alpha_2 - m_2\, \ddot{r}_2\, \delta r_2 = 0. \tag{5.18.13}$$

Equation (5.18.12) gives

$$\left.\begin{aligned} \delta r_2 &= -\delta r_1 \\ \text{and} \quad \ddot{r}_2 &= -\ddot{r}_1 \end{aligned}\right\}. \tag{5.18.14}$$

Using (5.18.14) in (5.18.13), we obtain

$$m_1\, g\, \delta r_1 \sin\alpha_1 - m_1\, \ddot{r}_1\, \delta r_1 - m_2\, g\, \delta r_1 \sin\alpha_2 - m_2\, \ddot{r}_1\, \delta r_1 = 0,$$

$$\text{i.e.,}\quad m_1\, g\, \sin\alpha_1 - m_1\, \ddot{r}_1 - m_2\, g\, \sin\alpha_2 - m_2\, \ddot{r}_1 = 0,$$

$$\text{i.e.,}\quad m_1\ddot{r}_1 + m_2\ddot{r}_1 = m_1\, g\, \sin\alpha_1 - m_2\, g\, \sin\alpha_2,$$

$$\text{i.e.,}\quad (m_1 + m_2)\ddot{r}_1 = g\,(m_1\, \sin\alpha_1 - m_2\, \sin\alpha_2),$$

$$\text{i.e.}\quad \ddot{r}_1 = \frac{g\,(m_1\, \sin\alpha_1 - m_2\, \sin\alpha_2)}{m_1 + m_2}.$$

$$\text{Similarly,}\quad \ddot{r}_2 = \frac{g\,(m_2\, \sin\alpha_2 - m_1\, \sin\alpha_1)}{m_1 + m_2}.$$

These two equations describe the motions of the masses.

Example 5.3 Prove that if the total momentum of a system is conserved, then the center of mass is either at rest or in motion with constant velocity.

Solution With usual notations, given that

$$\sum_{\nu=1}^{N} m_\nu \mathbf{V}_\nu = \text{a constant vector} = \mathbf{P}\,(\text{say}),$$

$$\text{i.e.,}\quad \sum_{\nu=1}^{N} m_\nu \frac{d\mathbf{r}_\nu}{dt} = \mathbf{P},$$

$$\text{i.e.,}\quad \frac{d}{dt}\sum_{\nu=1}^{N} m_\nu \mathbf{r}_\nu = \mathbf{P}. \tag{5.18.15}$$

Now,

$$\frac{\sum_{\nu=1}^{N} m_\nu \mathbf{r}_\nu}{M} = \mathbf{r}_c = \text{P.V. of the center of mass.}$$

$$\therefore\quad \sum_{\nu=1}^{N} m_\nu \mathbf{r}_\nu = M\mathbf{r}_c. \tag{5.18.16}$$

Using (5.18.16) in (5.18.15), we get

$$\frac{d}{dt} M\mathbf{r}_c = \mathbf{P}$$

$$\text{i.e.,}\quad \dot{\mathbf{r}}_c = \frac{\mathbf{P}}{M} = \mathbf{C} = \text{a constant vector.} \tag{5.18.17}$$

If $\mathbf{C} = \mathbf{0}$, (5.18.17) shows that

$$\dot{\mathbf{r}}_c = \mathbf{0},$$

i.e., the center of mass is at rest.

If $\mathbf{C} \neq \mathbf{0}$, Eq. (5.18.17) gives

$$\mathbf{r}_c = \mathbf{C}t + C_1. \tag{5.18.18}$$

Therefore, locus of $\mathbf{r}_c$ is

$$\mathbf{r} = \mathbf{C}t + C_1.$$

which represents a straight line.

Therefore, the center of mass moves in a straight line with constant velocity. i.e., the center of mass is either at rest or moves in a straight line with uniform velocity.

Example 5.4 Let m_1, m_2, m_3 be the masses of three particles and $\mathbf{V}_{12}, \mathbf{V}_{23}, \mathbf{V}_{31}$ be their relative velocities. Prove that the total kinetic energy of the system about the center of mass is

$$\frac{1}{2}\frac{m_1 m_2 V_{12}^2 + m_2 m_3 V_{23}^2 + m_3 m_1 V_{31}^2}{m_1 + m_2 + m_3}.$$

Solution In Fig. 5.7, let A, B, and C be the positions of the particles with masses m_1, m_2, and m_3, respectively. Let G be the center of mass of the system. Let $\mathbf{r}_1$, $\mathbf{r}_2$, $\mathbf{r}_3$, $\mathbf{r}_c$ be the position vectors of A, B, C, and G, respectively, relative to some origin O.

$$\begin{aligned} \mathbf{r}_c &= \frac{m_1\mathbf{r}_1 + m_2\mathbf{r}_2 + m_3\mathbf{r}_3}{m_1 + m_2 + m_3} \\ &= \frac{m_1\mathbf{r}_1 + m_2\mathbf{r}_2 + m_3\mathbf{r}_3}{M} \end{aligned} \tag{5.18.19}$$

where $M = m_1 + m_2 + m_3$.

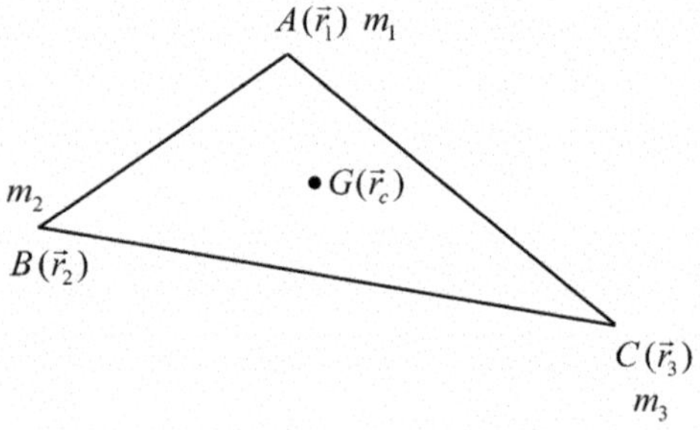

Fig. 5.7 Illustrated diagram for Example 5.4

Equation (5.18.19) gives

$$\dot{\mathbf{r}}_c = \frac{m_1\dot{\mathbf{r}}_1 + m_2\dot{\mathbf{r}}_2 + m_3\dot{\mathbf{r}}_3}{M},$$

$$\text{i.e.,}\quad \mathbf{V}_0 = \frac{1}{M}(m_1\mathbf{V}_1 + m_2\mathbf{V}_2 + m_3\mathbf{V}_3).$$

The total kinetic energy of the system about O is given by

$$T = T_0 + T', \tag{5.18.20}$$

where

$$T = \frac{1}{2}(m_1\mathbf{V}_1^2 + m_2\mathbf{V}_2^2 + m_3\mathbf{V}_3^2),$$

$$T_0 = \frac{1}{2}M\mathbf{V}_0^2 = \text{KE of the motion of center of mass,}$$

and

$$T' = \text{KE of the system about G.}$$

Equation (5.18.20) gives

$$\begin{aligned}
T' &= T - T_0 \\
&= \frac{1}{2}(m_1V_1^2 + m_2V_2^2 + m_3V_3^2) - \frac{1}{2}MV_0^2 \\
&= \frac{1}{2}\left[m_1V_1^2 + m_2V_2^2 + m_3V_3^2 - MV_0^2\right] \\
&= \frac{1}{2}\left[m_1V_1^2 + m_2V_2^2 + m_3V_3^2 - M\frac{1}{M^2}(m_1\mathbf{V}_1 + m_2\mathbf{V}_2 + m_3\mathbf{V}_3)^2\right] \\
&= \frac{1}{2M}\left[Mm_1V_1^2 + Mm_2V_2^2 + Mm_3V_3^2 - (m_1\mathbf{V}_1 + m_2\mathbf{V}_2 + m_3\mathbf{V}_3)^2\right] \\
&= \frac{1}{2M}\Big[V_1^2(Mm_1 - m_1^2) + V_2^2(Mm_2 - m_2^2) + V_3^2(Mm_3 - m_3^2) \\
&\qquad - 2m_1m_2\mathbf{V}_1\cdot\mathbf{V}_2 - 2m_2m_3\mathbf{V}_2\cdot\mathbf{V}_3 - 2m_3m_1\mathbf{V}_3\cdot\mathbf{V}_1\Big] \\
&= \frac{1}{2M}\Big[m_1(M - m_1)V_1^2 + m_2(M - m_2)V_2^2 + m_3(M - m_3)V_3^2 \\
&\qquad - 2m_1m_2\mathbf{V}_1\cdot\mathbf{V}_2 - 2m_2m_3\mathbf{V}_2\cdot\mathbf{V}_3 - 2m_3m_1\mathbf{V}_3\cdot\mathbf{V}_1\Big] \\
&= \frac{1}{2M}\Big[m_1(m_2 + m_3)V_1^2 + m_2(m_3 + m_1)V_2^2 + m_3(m_1 + m_2)V_3^2 \\
&\qquad - 2m_1m_2\mathbf{V}_1\cdot\mathbf{V}_2 - 2m_2m_3\mathbf{V}_2\cdot\mathbf{V}_3 - 2m_3m_1\mathbf{V}_3\cdot\mathbf{V}_1\Big].
\end{aligned} \tag{5.18.21}$$

Now,

$$
\begin{aligned}
&\frac{1}{2}\frac{m_1m_2V_{12}^2+m_2m_3V_{23}^2+m_3m_1V_{31}^2}{m_1+m_2+m_3}\\
&=\frac{1}{2M}\left[m_1m_2\mathbf{V}_{12}^2+m_2m_3\mathbf{V}_{23}^2+m_3m_1\mathbf{V}_{31}^2\right]\\
&=\frac{1}{2M}\left[m_1m_2(\mathbf{V}_1-\mathbf{V}_2)^2+m_2m_3(\mathbf{V}_2-\mathbf{V}_3)^2+m_3m_1(\mathbf{V}_3-\mathbf{V}_1)^2\right]\\
&=\frac{1}{2M}\Big[V_1^2(m_1m_2+m_3m_1)+V_2^2(m_2m_3+m_1m_2)+V_3^2(m_3m_1+m_2m_3)\\
&\qquad-2m_1m_2\mathbf{V}_1\cdot\mathbf{V}_2-2m_2m_3\mathbf{V}_2\cdot\mathbf{V}_3-2m_3m_1\mathbf{V}_3\cdot\mathbf{V}_1\Big]\\
&=\frac{1}{2M}\Big[m_1(m_2+m_3)V_1^2+m_2(m_3+m_1)V_2^2+m_3(m_1+m_2)V_3^2\\
&\qquad-2m_1m_2\mathbf{V}_1\cdot\mathbf{V}_2-2m_2m_3\mathbf{V}_2\cdot\mathbf{V}_3-2m_3m_1\mathbf{V}_3\cdot\mathbf{V}_1\Big].
\end{aligned}
\tag{5.18.22}
$$

From (5.18.21) and (5.18.22), we have

$$
T'=\frac{1}{2}\frac{m_1m_2V_{12}^2+m_2m_3V_{23}^2+m_3m_1V_{31}^2}{m_1+m_2+m_3}.
$$

Hence proved.

Example 5.5 Use D′Alembert's principle to determine the equation of motion of a simple pendulum.

Solution Let OA be the strength of length L(say) with O as the point of suspension as shown in Fig. 5.8. Let P be the position of the particle at time t with position vector $\mathbf{r}$ relative to O. Introduce the reversed effective force $-m\ddot{\mathbf{r}}$ to the particle.

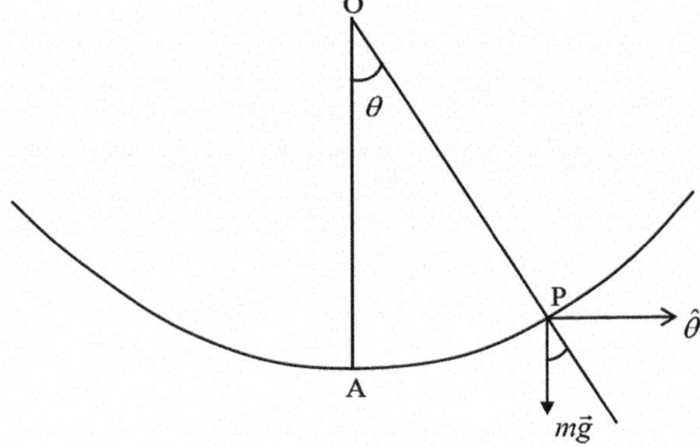

Fig. 5.8 Motion of a simple pendulum

By D′Alembert's principle

$$(m\mathbf{g} - m\ddot{\mathbf{r}}) \cdot \delta\mathbf{r} = 0,$$

$$\text{i.e.,}\quad \mathbf{g} \cdot \delta\mathbf{r} = \ddot{\mathbf{r}} \cdot \delta\mathbf{r},$$

$$\text{i.e.,}\quad g\,\delta r \cos\left(\frac{\pi}{2} + \theta\right) = \ddot{\mathbf{r}} \cdot \delta\mathbf{r},$$

$$\text{i.e.,}\quad -g\,\delta r \sin\theta = \ddot{\mathbf{r}} \cdot \delta\mathbf{r}. \tag{5.18.23}$$

Now,

$$\begin{aligned}\ddot{\mathbf{r}} &= \frac{1}{r}\frac{d}{dt}(r^2\dot{\theta})\hat{\theta}\\ &= \frac{1}{L}\frac{d}{dt}(L^2\dot{\theta})\hat{\theta}\\ &= L\,\ddot{\theta}\,\hat{\theta}.\end{aligned}$$

Therefore, Eq. (5.18.23) gives

$$\begin{aligned}-g\delta r \sin\theta &= L\,\ddot{\theta}\,\hat{\theta} \cdot \delta\mathbf{r}\\ &= L\,\ddot{\theta}\,\delta r,\\ \text{i.e.,}\quad L\ddot{\theta} &= -g\sin\theta,\\ \therefore\quad \ddot{\theta} &= -\frac{g\sin\theta}{L},\end{aligned}$$

which is the required equation of motion of a simple pendulum.

Example 5.6 An inextensible string of negligible mass hanging over a smooth peg at B connects one mass m_1 on a frictionless inclined plane of angle α to another mass m_2. Use D′Alembert's principle to obtain the equation of motion of masses m_1 and m_2.

Solution Let P and Q be the positions of the masses m_1 and m_2 respectively. From Fig. 5.9, let BP $= r_1$, BQ $= r_2$ and $\mathbf{BP} = \mathbf{r}_1$, $\mathbf{BQ} = \mathbf{r}_2$.

Introduce the reversed effective forces $-m_1\ddot{\mathbf{r}}_1$ and $-m_2\ddot{\mathbf{r}}_2$ on m_1 and m_2 respectively.

By D′Alembert's principle,

$$\begin{aligned}&(m_1\mathbf{g} - m_1\ddot{\mathbf{r}}_1) \cdot \delta\mathbf{r}_1 + (m_2\mathbf{g} - m_2\ddot{\mathbf{r}}_2) \cdot \delta\mathbf{r}_2 = 0,\\ \text{i.e.,}\quad &m_1\mathbf{g} \cdot \delta\mathbf{r}_1 - m_1\ddot{\mathbf{r}}_1 \cdot \delta\mathbf{r}_1 + m_2\mathbf{g} \cdot \delta\mathbf{r}_2 - m_2\ddot{\mathbf{r}}_2 \cdot \delta\mathbf{r}_2 = 0,\\ \text{i.e.,}\quad &m_1\,g\,\delta r_1 \sin\alpha - m_1\,\ddot{r}_1\,\delta r_1 + m_2\,g\,\delta r_2 - m_2\,\ddot{r}_2\,\delta r_2 = 0.\end{aligned} \tag{5.18.24}$$

Since the string is inextensible,

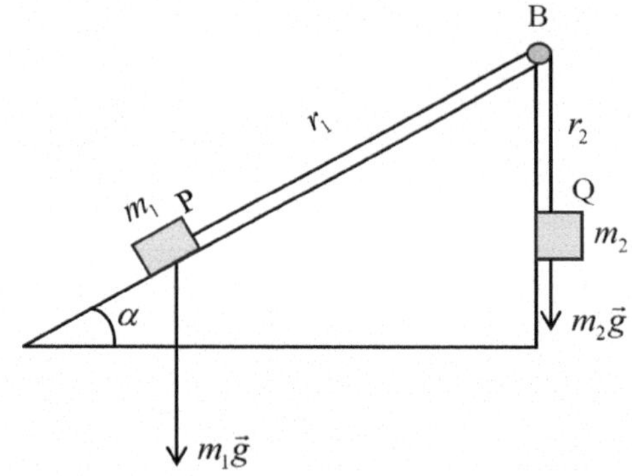

Fig. 5.9 Illustration for Example 5.6

$$\begin{aligned} \therefore\ & r_1 + r_2 = L, \\ \therefore\ & \delta r_1 + \delta r_2 = 0, \\ \text{i.e.,}\ & \delta r_2 = -\delta r_1. \end{aligned} \tag{5.18.25}$$

Further

$$\begin{aligned} & r_1 + r_2 = L, \\ \therefore\ & \ddot{r}_1 + \ddot{r}_2 = 0, \\ \therefore\ & \ddot{r}_2 = -\ddot{r}_1. \end{aligned} \tag{5.18.26}$$

Therefore (5.18.24) gives

$$\begin{aligned} & m_1\, g\, \delta r_1 \sin\alpha - m_1\, \ddot{r}_1\, \delta r_1 - m_2\, g\, \delta r_1 - m_2\, \ddot{r}_1\, \delta r_1 = 0, \\ \therefore\ & m_1\, g \sin\alpha - m_1\, \ddot{r}_1 - m_2\, g - m_2\, \ddot{r}_1 = 0, \\ \therefore\ & (m_1 + m_2)\ddot{r}_1 = (m_1 \sin\alpha - m_2), \\ \therefore\ & \ddot{r}_1 = \frac{m_1 \sin\alpha - m_2}{m_1 + m_2}\, g. \end{aligned}$$

Therefore (5.18.26) gives

$$\ddot{r}_2 = \frac{m_2 - m_1 \sin\alpha}{m_1 + m_2}\, g,$$

which are the required equations of motion.

Example 5.7 Three particles of masses 2, 1, and 3 respectively have position vectors $\mathbf{r}_1 = 5t\hat{i} - 2t^2\hat{j} + (3t - 2)\hat{k}$, $\mathbf{r}_2 = (2t - 3)\hat{i} + (12 - 5t^2)\hat{j} + (4 + 6t - 3t^3)\hat{k}$, $\mathbf{r}_3 = (2t - 1)\hat{i} + (t^2 + 2)\hat{j} - t^3\hat{k}$, where t is the time. Find (a) the velocity of the center of mass at $t = 1$ and (b) the total linear momentum of the system at $t = 1$.

Solution Given

$$M = m_1 + m_2 + m_3 = 2 + 1 + 3 = 6,$$

$$\mathbf{r}_1 = 5t\hat{i} - 2t^2\hat{j} + (3t - 2)\hat{k},$$

$$\mathbf{r}_2 = (2t - 3)\hat{i} + (12 - 5t^2)\hat{j} + (4 + 6t - 3t^3)\hat{k},$$

$$\mathbf{r}_3 = (2t - 1)\hat{i} + (t^2 + 2)\hat{j} - t^3\hat{k}.$$

Therefore

$$\mathbf{v}_1 = \dot{\mathbf{r}}_1 = 5\hat{i} - 4t\hat{j} + 3\hat{k},$$
$$\mathbf{v}_2 = \dot{\mathbf{r}}_2 = 2\hat{i} - 10t\hat{j} + (6 - 9t^2)\hat{k},$$
$$\mathbf{v}_3 = \dot{\mathbf{r}}_3 = 2\hat{i} + 2t\hat{j} - 3t^2\hat{k}$$

and

$$\mathbf{v}_1\big]_{t=1} = 5\hat{i} - 4\hat{j} + 3\hat{k},$$
$$\mathbf{v}_2\big]_{t=1} = 2\hat{i} - 10\hat{j} - 3\hat{k},$$
$$\mathbf{v}_3\big]_{t=1} = 2\hat{i} + 2\hat{j} - 3\hat{k}.$$

$$\begin{aligned}
\therefore\quad \dot{\mathbf{r}}_c\big]_{t=1} &= \frac{m_1\mathbf{v}_1^1 + m_2\mathbf{v}_2^1 + m_3\mathbf{v}_3^1}{M} \\
&= \frac{2(5\hat{i} - 4\hat{j} + 3\hat{k}) + (2\hat{i} - 10\hat{j} - 3\hat{k}) + 3(2\hat{i} + 2\hat{j} - 3\hat{k})}{6} \\
&= \frac{18\hat{i} - 12\hat{j} - 6\hat{k}}{6} \\
&= 3\hat{i} - 2\hat{j} - \hat{k}.
\end{aligned}$$

$$\begin{aligned}
\therefore\quad \mathbf{P} &= M\dot{\mathbf{r}}_c^1 \\
&= 6(3\hat{i} - 2\hat{j} - \hat{k}) \\
&= 18\hat{i} - 12\hat{j} - 6\hat{k}.
\end{aligned}$$

Example 5.8 Three particles of masses 2, 3, and 5 move under the influence of force field so that their position vectors relative to a fixed coordinate system are given respectively by $\mathbf{r}_1 = 2t\hat{i} - 3\hat{j} + t^2\hat{k}$, $\mathbf{r}_2 = (t + 1)\hat{i} + 3t\hat{j} - 4\hat{k}$, $\mathbf{r}_3 = t^2\hat{i} - t\hat{j} + (2t - 1)\hat{k}$, where t is the time. Find (a) the total angular momentum of the system and (b) the total external torque applied to the system with respect to the origin.

Solution Given

$$\left.\begin{aligned}\mathbf{r}_1 &= 2t\hat{i} - 3\hat{j} + t^2\hat{k}\\ \mathbf{r}_2 &= (t+1)\hat{i} + 3t\hat{j} - 4\hat{k}\\ \mathbf{r}_3 &= t^2\hat{i} - t\hat{j} + (2t-1)\hat{k}\end{aligned}\right\}.$$

Therefore

$$\left.\begin{aligned}\mathbf{v}_1 &= \dot{\mathbf{r}}_1 = 2\hat{i} + 2t\hat{k}\\ \mathbf{v}_2 &= \dot{\mathbf{r}}_2 = \hat{i} + 3\hat{j}\\ \mathbf{v}_3 &= \dot{\mathbf{r}}_3 = 2t\hat{i} - \hat{j} + 2\hat{k}\end{aligned}\right\}.$$

Now

$$\begin{aligned}\boldsymbol{\Omega}_1 &= \mathbf{r}_1 \times m\mathbf{v}_1\\ &= 2\,\mathbf{r}_1 \times \mathbf{v}_1\\ &= 2\begin{vmatrix}\hat{i} & \hat{j} & \hat{k}\\ 2t & -3 & t^2\\ 2 & 0 & 2t\end{vmatrix}\\ &= 2[-6t\hat{i} - (4t^2 - 2t^2)\hat{j} + 6\hat{k}]\\ &= -12t\hat{i} - 4t^2\hat{j} + 12\hat{k}.\end{aligned}$$

$$\begin{aligned}\boldsymbol{\Omega}_2 &= \mathbf{r}_2 \times m_2\mathbf{v}_2\\ &= 3\,\mathbf{r}_2 \times \mathbf{v}_2\\ &= 3\begin{vmatrix}\hat{i} & \hat{j} & \hat{k}\\ t+1 & 3t & -4\\ 1 & 3 & 0\end{vmatrix}\\ &= 3[12\hat{i} - 4\hat{j} + (3t - 3 - 3t)\hat{k}]\\ &= 3[12\hat{i} - 4\hat{j} + 3\hat{k}]\\ &= 36\hat{i} - 12\hat{j} + 9\hat{k}.\end{aligned}$$

$$\begin{aligned}\boldsymbol{\Omega}_3 &= \mathbf{r}_3 \times m_3\mathbf{v}_3\\ &= 5\,\mathbf{r}_3 \times \mathbf{v}_3\\ &= 5\begin{vmatrix}\hat{i} & \hat{j} & \hat{k}\\ t^2 & -t & 2t-1\\ 2t & -1 & 2\end{vmatrix}\\ &= 5[\hat{i}(-2t + 2t - 1) - \hat{j}(2t^2 - 4t^2 + 2t) + \hat{k}(-t^2 + 2t^2)]\end{aligned}$$

$$= 5[-\hat{i} - \hat{j}(-2t^2 + 2t) + t^2\hat{k}]$$
$$= -5\hat{i} + 10(t^2 - t)\hat{j} + 5t^2\hat{k}.$$

$$\begin{aligned}\therefore \quad \boldsymbol{\Omega} &= \boldsymbol{\Omega}_1 + \boldsymbol{\Omega}_2 + \boldsymbol{\Omega}_3 \\ &= -12t\hat{i} - 4t^2\hat{j} + 12\hat{k} + 36\hat{i} - 12\hat{j} + 9\hat{k} - 5\hat{i} + 10(t^2 - t)\hat{j} + 5t^2\hat{k} \\ &= (31 - 12t)\hat{i} + (6t^2 - 10t - 12)\hat{j} + (5t^2 + 21)\hat{k}.\end{aligned}$$

$$\therefore \quad \boldsymbol{\Lambda} = \frac{d\boldsymbol{\Omega}}{dt} = -12\hat{i} + (12t - 10)\hat{j} + 10t\hat{k}.$$

Example 5.9 Three particles of masses 2, 3, and 5 move under the influence of a force field so that their position vectors relative to a fixed coordinate system are given respectively by $\mathbf{r}_1 = 2t\hat{i} - 3\hat{j} + t^2\hat{k}$, $\mathbf{r}_2 = (t + 1)\hat{j} + 3t\hat{j} - 4\hat{k}$, $\mathbf{r}_3 = t^2\hat{i} - t\hat{j} + (2t - 1)\hat{k}$, where t is the time. Find the total work done by moving the particles from their positions at time $t = 1$ to their positions at time t_2.

Solution Given

$$\left.\begin{aligned}\mathbf{r}_1 &= 2t\hat{i} - 3\hat{j} + t^2\hat{k} \\ \mathbf{r}_2 &= (t + 1)\hat{i} + 3t\hat{j} - 4\hat{k} \\ \mathbf{r}_3 &= t^2\hat{i} - t\hat{j} + (2t - 1)\hat{k}\end{aligned}\right\}.$$

Therefore

$$\left.\begin{aligned}\mathbf{v}_1 &= \dot{\mathbf{r}}_1 = 2\hat{i} + 2t\hat{k} \\ \mathbf{v}_2 &= \dot{\mathbf{r}}_2 = \hat{i} + 3\hat{j} \\ \mathbf{v}_3 &= \dot{\mathbf{r}}_3 = 2t\hat{i} - \hat{j} + 2\hat{k}\end{aligned}\right\}.$$

$$\begin{aligned}T &= \frac{1}{2}2\,v_1^2 + \frac{1}{2}3\,v_2^2 + \frac{1}{2}5\,v_3^2 \\ &= \frac{1}{2}\left[2(4 + 4t^2) + 3(1 + 9) + 5(4t^2 + 1 + 4)\right] \\ &= \frac{1}{2}\left[8 + 8t^2 + 30 + 20t^2 + 25\right] \\ &= \frac{1}{2}[28t^2 + 63].\end{aligned}$$

$$\therefore \quad W = T_2 - T_1$$
$$= \frac{1}{2}[28 \times 4 + 63] - \frac{1}{2}[28 + 63]$$
$$= \frac{1}{2}[28 \times 3]$$
$$= 14 \times 3$$
$$= 42 \text{ units.}$$

5.19 Exercise-V

1. A quadrilateral ABCD has masses 1, 2, 3, and 4 units located at the points A(−1, −2, 2), B(3, 2, −1), C(1, −2, 4), and D(3, 1, 2). Find the coordinate of the center of mass.
   Hints : $\bar{x} = \frac{\sum mx}{\sum m}, \quad \bar{y} = \frac{\sum my}{\sum m}, \quad \bar{z} = \frac{\sum mz}{\sum m}$
2. A system consists of two particles of masses m_1 and m_2. Prove that the center of mass of the system divides the line joining m_1 and m_2 into two segments whose lengths are in the ratio $m_2 : m_1$.
   Hints: $m_1 x = m_2(a - x)$, i.e., $x = \frac{am_2}{m_1+m_2}, \quad a - x = \frac{am_1}{m_1+m_2}$.
3. A bomb dropped from an airplane explodes in midair. Prove that, if air resistance is neglected, then the center of mass describes a parabola.
   Hints: Equation of motion of the bomb is $M\ddot{\mathbf{r}} = Mg\hat{j}$, i.e., $(\ddot{\bar{x}}, \ddot{\bar{y}}) = (0, g) \implies \bar{x} = Ut, \ \& \ \bar{y} = \frac{1}{2}gt^2$.
4. Three particles of masses 1, 2, and 3 respectively, have position vectors $\mathbf{r}_1 = (4t, -t^2, t-2)$, $\mathbf{r}_2 = (t-3, 12-5t^2, 2+3t-2t^3)$, and $\mathbf{r}_3 = (2t-1, t^2+3, -2t^3)$, where t is the time. Find (a) the velocity of the center of mass at time $t = 1$, (b) the Total linear momentum of the system at $t = 1$.
   Hints: $\mathbf{p} = M\dot{\mathbf{r}}_c \cdot \dot{\mathbf{r}}_c = \dfrac{m_1\dot{\mathbf{r}}_1 + m_2\dot{\mathbf{r}}_2 + m_3\dot{\mathbf{r}}_3}{6}$.
5. Find the center of mass of a uniform road bounded by $y = \sin x$, the X-axis, and the ordinates $x = 0$ and $x = \pi$.
   Answer: $\left(\frac{\pi}{2}, \frac{\pi}{8}\right)$,
6. A system of particles consists of a mass of 2 gm at (0, −1, −2) 4 gm mass at (−3, 0, 2) and 1 gm at (2, −2, 0). Find the coordinates of the center of mass.
   Answer: $\left(-\frac{10}{7}, -\frac{4}{7}, \frac{4}{7}\right)$
7. Find the centroid of a semi-circular region of radius a.
   Hints: $\bar{x} = 0$ due to symmetry, $\bar{y} = \dfrac{1}{2}\dfrac{\int_{-a}^{a} y^2\,dx}{\int_{-a}^{a} y\,dx} = \frac{4a}{3\pi}$; $y^2 = a^2 - x^2$.
8. Find the centroid of a semi-circular wire of radius a.
   Answer: $\left(0, \frac{2a}{\pi}\right)$.

9. Three particles of masses 2, 3, and 5 move under the influence of a force field so that their position vectors $\mathbf{r}_1 = (2t, -3, t^2)$, $\mathbf{r}_2 = (t+1, 3t, -4)$, and $\mathbf{r}_3 = (t^2, -t, 2t-1)$. Find (a) the total angular momentum, (b) the total external torque taken about a point whose position vector is given by $\mathbf{r}_0 = (t, -2t, 3)$.
Answer: $\boldsymbol{\Omega} = (4t^2 - 12t + 137, -40t + 14, 21 - 5t^2)$, $\boldsymbol{\Lambda} = (8t - 12, 16t - 40, -10t)$.

Chapter 6
Introduction to Dynamics of Rigid Bodies

6.1 Basic Concepts

6.1.1 Rigid Body

A rigid body is a system of particles in which the distance between any two particles remains unchanged regardless of the forces acting on it.

6.1.2 Displacement, Translation, Rotation

A displacement of a rigid body is a change of its position from one to another. If during the motion of a rigid body, all points of the body on the same line remain fixed, the displacement is termed as a rotation of the body about that line. On the other hand, if during a displacement, all points of the body move in lines parallel to each other, the displacement is called a translation.

6.1.3 General Motion of a Rigid Body

The general motion of a rigid body can be considered as a translation plus rotation about the center of mass (Chasle's theorem).

N. Ahmed et al., *Classical Dynamics*, University Texts in the Mathematical Sciences,
https://doi.org/10.1007/978-981-95-6394-4_6

6.1.4 Plane Motion of a Rigid Body

If all the points of a rigid body move parallel to a fixed plane, the motion of the body is regarded as plane motion or two-dimensional motion. The fixed plane is termed as the plane of the motion of the body.

6.1.5 Rotation About a Fixed Axis

In this case, the rigid body rotates about a fixed axis $\perp^r$ to a fixed plane. Here, the motion of the body is two dimensional with the fixed plane as the plane of motion.

6.1.6 General Plane Motion

In this case, the motion of the body is considered as a translation parallel to a fixed plane together with a rotation about an axis $\perp^r$ to the plane passing through the center of mass of the body.

6.1.7 Distance of a Point P (r) from a Direction $\hat{a}$

Let $\hat{a}$ be the unit vector along **OA**.

Let **OP** $= \mathbf{r}$. Let θ be the angle between the vectors $\mathbf{r}$ and $\hat{a}$. Draw a perpendicular PN from P on OA. Suppose that PN $= p$.

Now from Fig. 6.1, we have

$$p = r \sin\theta. \tag{6.1.1}$$

By definition of cross product $\mathbf{r} \times \hat{a} = r \ \sin\theta \ \hat{n}$,

$$\therefore \quad \left|\mathbf{r} \times \hat{a}\right| = r \sin\theta.$$

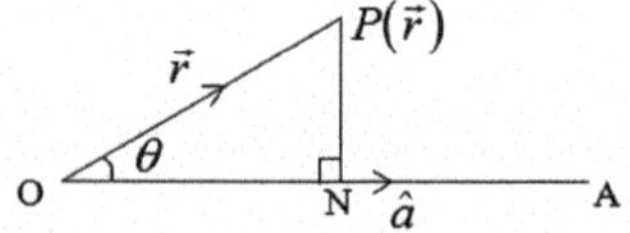

Fig. 6.1 Perpendicular distance of a point from a unit vector

Equation (6.1.1) gives

$$p = \left|\mathbf{r} \times \hat{a}\right|.$$

Thus, the distance of the point $p = \mathrm{P}(\mathbf{r})$ from $\hat{a}$ is given by

$$p = \left|\mathbf{r} \times \hat{a}\right|.$$

6.1.8 Moments and Products of Inertia

If p is the $\perp^r$ distance of an element m of a rigid body form a given line AB, then mp^2 is called the moment of inertia of the element about AB and $\sum mp^2$ is termed as the moment inertia of the body about AB.

Let $\hat{a}$ be the unit vector along **AB** and suppose that $\mathbf{OP} = \mathbf{r}$, where O is a given point on AB and P is a particle of mass m as shown in Fig. 6.2.

Now $\mathbf{r} \times \hat{a} = r \sin\theta\, \hat{n}$, $\hat{n}$ being a unit vector $\perp^r$ to both $\mathbf{r}$ and $\mathbf{a}$ and θ is the angle between the vectors $\mathbf{r}$ and $\mathbf{a}$.

We have,

$$\left|\mathbf{r} \times \hat{a}\right| = \left|r \sin\theta\, \hat{n}\right| = r \sin\theta = p, \quad p = PN = \perp^r \text{ distance of P from AB.}$$

$$\therefore \quad I = \text{MI of the body about AB} = \sum mp^2 = \sum m\left|\mathbf{r} \times \hat{a}\right|^2 = \sum m\left(\mathbf{r} \times \hat{a}\right)^2.$$

Introduce a tri-rectangular coordinate system, with O as the origin as depicted in Fig. 6.3. Let P (x, y, z) be a single particle of mass m of a rigid body.

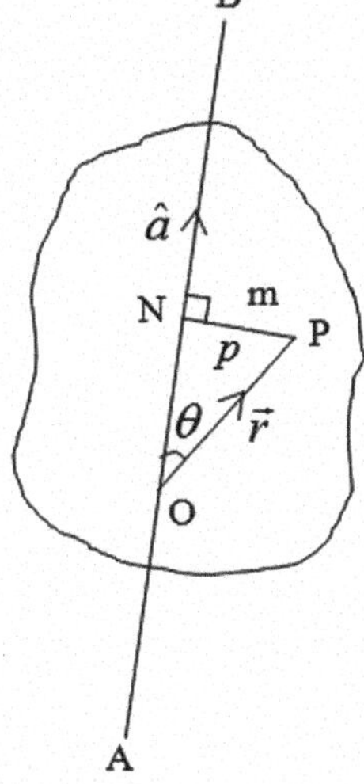

Fig. 6.2 Illustration of moment of inertia about a line

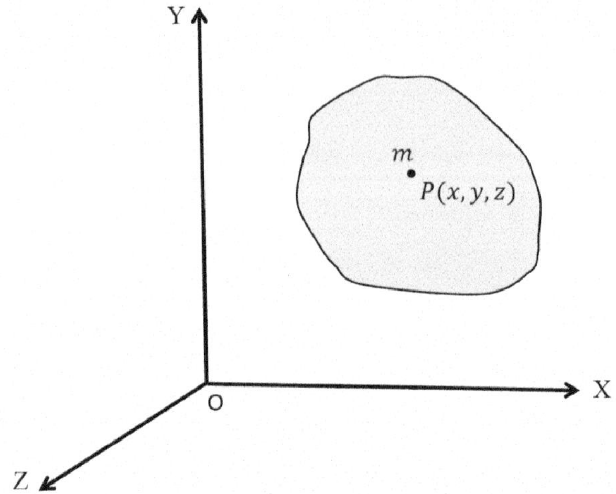

Fig. 6.3 Moment of inertia about the coordinate axes

M.I. of the body about the line AB is given by $I = \sum mp^2 = \sum m(\mathbf{r} \times \hat{a})^2$. Let the symbols A, B, C, D, E, F be defined as follows:

$$A = \sum m\left(y^2 + z^2\right), \quad B = \sum m\left(z^2 + x^2\right), \quad C = \sum m\left(x^2 + y^2\right),$$

$$D = \sum myz, \quad E = \sum mzx, \quad F = \sum mxy.$$

The quantities A, B, and C are the moments of inertia about the axes OX, OY, and OZ, respectively. The quantities D, E, and F are termed as the products of inertia with respect to the pairs (OY, OZ); (OZ, OX); (OX, OY), respectively.

6.2 Parallel Axis Theorem on Moment of Inertia

If I is the moment of inertia of a rigid body about an axis OQ, I_0 the moment of inertia of the body about an axis parallel to OQ and passing through the center of mass and d is the distance between the axes, then $I = I_0 + Md^2$, where M is the mass of the body.

Proof Let G be the C.I. and GL be a line parallel to OQ. Let $\hat{a}$ be the unit vector along **OQ**. Let P be a particle of mass of the body. In Fig. 6.4, let **OP** $= \mathbf{r}$, **OG** $= \mathbf{r}_c$ and **GP** $= \mathbf{r}'$. Now,

$$\text{I} = \text{M.I. of the body about OQ} = \sum m\left(\mathbf{r} \times \hat{a}\right)^2. \tag{6.2.1}$$

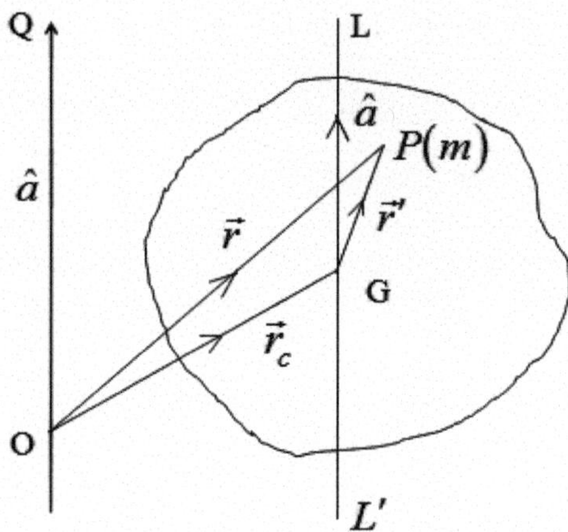

Fig. 6.4 Geometry for parallel axis theorem

We have,

$$\mathbf{r} = \mathbf{r}_c + \mathbf{r}',$$

$$\therefore \quad \mathbf{r} \times \hat{a} = \mathbf{r}_c \times \hat{a} + \mathbf{r}' \times \hat{a}.$$

Equation (6.2.1) gives

$$\mathrm{I} = \sum m\,(\mathbf{r}_c \times \hat{a} + \mathbf{r}' \times \hat{a})^2 = \sum m\,\left\{(\mathbf{r}_c \times \hat{a})^2 + 2\,(\mathbf{r}_c \times \hat{a}) \cdot (\mathbf{r}' \times \hat{a}) + (\mathbf{r}' \times \hat{a})^2\right\},$$

$$\text{i.e.,}\quad \mathrm{I} = \sum m(\mathbf{r}_c \times \hat{a})^2 + 2\sum m\,(\mathbf{r}_c \times \hat{a}) \cdot (\mathbf{r}' \times \hat{a}) + \sum m(\mathbf{r}' \times \hat{a})^2,$$

$$\text{i.e.,}\quad \mathrm{I} = (\mathbf{r}_c \times \hat{a})^2 \sum m + 2\,(\mathbf{r}_c \times \hat{a}) \cdot \sum m\,(\mathbf{r}' \times \hat{a}) + \sum m(\mathbf{r}' \times \hat{a})^2. \tag{6.2.2}$$

Now

$|\mathbf{r}_c \times \hat{a}| = \perp^r$ distance of the point G from the line OQ = d,

$$\therefore \quad d^2 = (\mathbf{r}_c \times \hat{a})^2. \tag{6.2.3}$$

Again

$$\sum m\left(\mathbf{r}' \times \hat{a}\right) = \left(\sum m\mathbf{r}'\right) \times \hat{a}. \tag{6.2.4}$$

Now

$$\frac{\sum m\mathbf{r}'}{M} = \text{p.v. of G relative to G} = \mathbf{0}$$

$$\therefore \quad \sum m\mathbf{r}' = \mathbf{0}.$$

Hence

$$\sum m\left(\mathbf{r}' \times \hat{a}\right) = \mathbf{0}. \tag{6.2.5}$$

Thus Eq. (6.2.2) reduces to

$$\text{I} = Md^2 + \sum m\left(\mathbf{r}' \times \hat{a}\right)^2. \tag{6.2.6}$$

Now

$$\sum m\left(\mathbf{r}' \times \hat{a}\right)^2 = \text{M.I. of the body about GL} = I_0.$$

Equation (6.2.6) reduces to

$$\text{I} = \text{I}_o + Md^2.$$

6.3 Six Constant Theorem

If A, B, C are the moment of inertia of a rigid body about three mutually perpendicular axes OX, OY, OZ; and D, E, F are the products of inertia of the body about OY-OZ, OZ-OX, OX-OY, then the moment of inertia of the body about an axis OQ having direction cosines λ, μ, ν is given by

$$I = A\lambda^2 + B\mu^2 + C\nu^2 - 2D\mu\nu - 2E\nu\lambda - 2F\lambda\mu.$$

Proof Let OQ be the axis about which the moment of inertia of the body is to be obtained. Let $\hat{a}$ be the unit vector along OQ (see Fig. 6.5).

$$\therefore \quad \hat{a} = \lambda\hat{i} + \mu\hat{j} + \nu\hat{k}.$$

Let P (x, y, z) be a single particle of mass m of the body with position vector $\mathbf{r}$ relative to O.

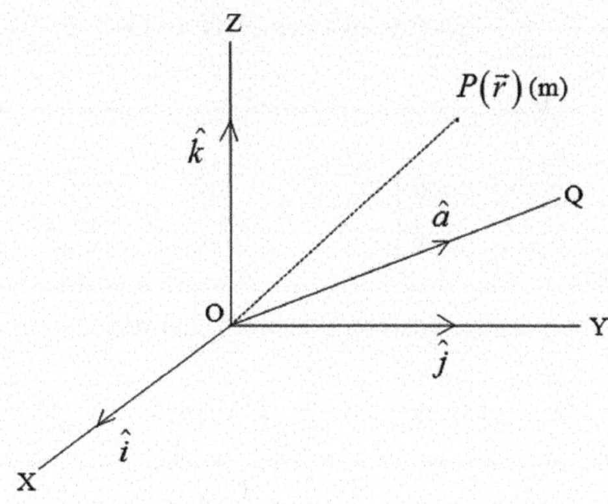

Fig. 6.5 Geometry for six constant theorem

$$\therefore \quad \mathbf{r} = x\hat{i} + y\hat{j} + z\hat{k}.$$

The moment of inertia of the body about OQ is given by

$$I = \sum m \left(\mathbf{r} \times \hat{a}\right)^2. \tag{6.3.1}$$

Now

$$\mathbf{r} \times \hat{a} = \begin{vmatrix} \hat{i} & \hat{j} & \hat{k} \\ x & y & z \\ \lambda & \mu & \nu \end{vmatrix} = \hat{i}\,(y\nu - z\mu) + \hat{j}\,(z\lambda - x\nu) + \hat{k}\,(x\mu - y\lambda),$$

$$\therefore \quad \left(\mathbf{r} \times \hat{a}\right)^2 = (y\nu - z\mu)^2 + (z\lambda - x\nu)^2 + (x\mu - y\lambda)^2$$

i.e., $\left(\mathbf{r} \times \hat{a}\right)^2 = \lambda^2\left(y^2 + z^2\right) + \mu^2\left(z^2 + x^2\right) + \nu^2\left(x^2 + y^2\right) - 2yz\mu\nu - 2zx\nu\lambda - 2xy\lambda\mu.$

Equation (6.3.1) gives

$$I = \sum m\left[\lambda^2\left(y^2 + z^2\right) + \mu^2\left(z^2 + x^2\right) + \nu^2\left(x^2 + y^2\right) - 2\mu\nu yz - 2\nu\lambda zx - 2\lambda\mu xy\right]$$

i.e., $I = A\lambda^2 + B\mu^2 + C\nu^2 - 2\mu\nu D - 2\nu\lambda E - 2\lambda\mu F.$

6.4 Perpendicular Axes Theorem

Let xy be the plane of the lamina in an xyz coordinate system. Let I_x, I_y, I_z denote moments of inertia about x, y, and z axes respectively. Then

$$I_z = I_x + I_y.$$

Let OX and OY be two $\perp^r$ axes on the plane of the lamina. Let OZ be an axis $\perp^r$ to the plane of the lamina.

Consider an element of mass m of the body located at the point $P(x, y, 0)$ in the plane of the lamina. It may be noted that

$$\begin{aligned} I_x &= \text{moment of inertia of the lamina about OX} \\ &= \sum my^2, \\ I_y &= \text{moment of inertia of the lamina about OY} \\ &= \sum mx^2. \end{aligned}$$

In Fig. 6.6, join OP and suppose that OP $= r$.

$$\therefore \quad r^2 = x^2 + y^2.$$

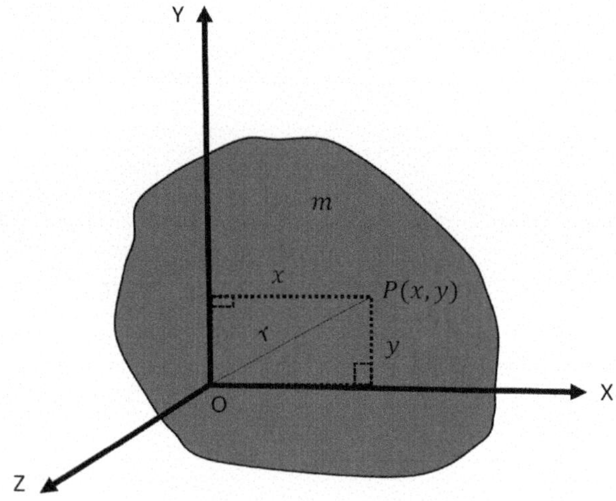

Fig. 6.6 Illustration of perpendicular axes theorem

Now,

$$\begin{aligned} I_z &= \text{moment of inertia of the lamina about OZ} \\ &= \sum m\,(\text{OP})^2 \\ &= \sum m\,r^2 \\ &= \sum m\,(x^2 + y^2) \\ &= \sum mx^2 + \sum my^2 \\ &= I_y + I_x = I_x + I_y \quad . \end{aligned}$$

Hence the theorem.

Example 6.1 Two particles of masses m_1 and m_2 respectively are connected by a rigid massless rod of length a and move freely in a plane. Show that the moment of inertia of the system about an axis $\perp^r$ to the plane and passing through the center of mass is μa^2, where the reduced mass

$$\mu = \frac{m_1 m_2}{m_1 + m_2}.$$

Solution In Fig. 6.7, let A and B be the positions of two masses m_1 and m_2 respectively. Let $\mathbf{r}_1 = -r_1\hat{i}$ and $\mathbf{r}_2 = r_2\hat{i}$ be the position vectors of A and B respectively relative to the center of mass C, $\hat{i}$ being the unit vector along CB or OX. Let CY be a line $\perp^r$ to plane through C.

Now, position vector of C relative to C $= \mathbf{0}$,

$$\text{i.e.,} \quad \frac{m_1\mathbf{r}_1 + m_2\mathbf{r}_2}{m_1 + m_2} = \mathbf{0},$$

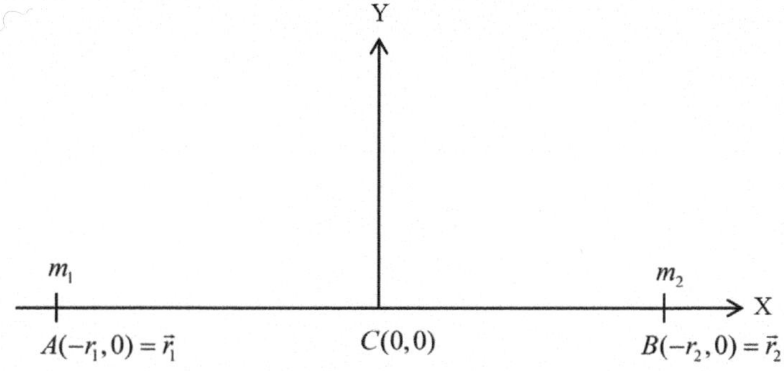

Fig. 6.7 Illustrated diagram for Example 6.1

$$\text{i.e.,}\quad m_1\mathbf{r}_1 + m_2\mathbf{r}_2 = \mathbf{0},$$
$$\text{i.e.,}\quad -m_1 r_1 + m_2 r_2 = 0,$$
$$\therefore\quad m_1 r_1 = m_2 r_2,$$
$$\text{i.e.,}\quad \frac{r_1}{r_2} = \frac{m_2}{m_1} = K\ \text{(say)},$$
$$\therefore\quad r_1 = K m_2,\quad r_2 = K m_1.$$

But

$$\text{AB} = a,$$
$$\text{i.e.,}\quad r_1 + r_2 = a,$$
$$\text{i.e.,}\quad K m_1 + K m_2 = a,$$
$$\text{i.e.,}\quad K(m_1 + m_2) = a,$$
$$\text{i.e.,}\quad K = \frac{a}{m_1 + m_2}.$$

Now,

$$\begin{aligned} I &= \text{moment of inertia of the system of masses about CY} \\ &= m_1 r_1^2 + m_2 r_2^2 \\ &= m_1 K^2 m_2^2 + m_2 K^2 m_1^2 \\ &= m_1\, m_2\, K^2 (m_2 + m_1) \\ &= m_1\, m_2\, (m_1 + m_2)\, \frac{a^2}{(m_1 + m_2)^2} \\ &= \frac{m_1\, m_2\, a^2}{m_1 + m_2} \\ &= \frac{m_1\, m_2}{m_1 + m_2}\, a^2 \\ &= \mu a^2. \end{aligned}$$

Proved.

6.5 Some Simple Cases of Moments of Inertia

I. A thin uniform rod of mass M and length $2a$.

In Fig. 6.8, let AB be the rod of mass M and length $2a$. Let O be the midpoint (center of mass) of the rod. Take O as the origin, X-axis along **OB**, and Y-axis through

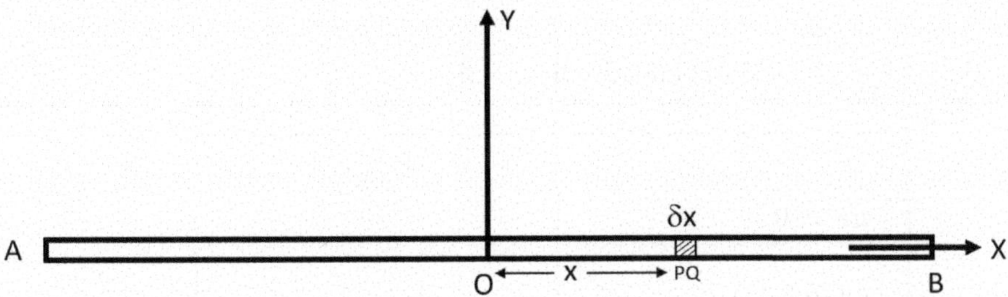

Fig. 6.8 Moment of inertia of a uniform rod

O and $\perp^r$ to AB. Consider an element PQ of length δx at a distance x from O as shown in the figure. Let m be the mass per unit length of the rod.

$\therefore$ Mass of the element PQ $= m\,\delta x$.

$$\text{Moment of inertia of the element PQ about OY} = m\,\delta x\,x^2 = m\,x^2\,\delta x.$$

$$\therefore\quad I_y = \text{moment of inertia of the rod about OY}$$

$$= \int_{x=-a}^{x=a} m\,x^2\,dx$$

$$= 2m \int_{x=0}^{x=a} x^2 dx$$

$$= 2m\,\frac{a^3}{3}$$

$$= \frac{2}{3}\frac{M}{2a}\,a^3 = \frac{1}{3}Ma^2.$$

II. Rectangular lamina

Let ABCD be the lamina, such that AB $= 2a$ and AD $= 2b$ whose center is O.

Let m be the mass per unit area of the lamina. Take O as the origin, X-axis $\parallel$ to AB, Y-axis $\parallel$ to AD, and Z-axis $\perp^r$ to the lamina.

In Fig. 6.9, consider an elementary strip PQ of breadth δx $\perp^r$ to OX at a distance x from OY.

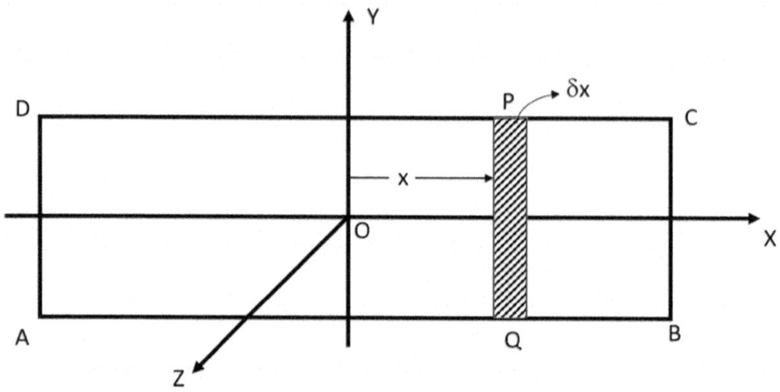

Fig. 6.9 Moment of inertia of a rectangular lamina

$$\begin{aligned}\text{Mass of the strip PQ} &= \text{PQ}\,\delta x\, m\\ &= 2b\,\delta x\, m\\ &= 2b\, m\,\delta x.\end{aligned}$$

$$\begin{aligned}\text{Moment of inertia of the strip PQ about OX} &= \frac{1}{3}2b\, m\,\delta x\, b^2\\ &= \frac{2}{3}mb^3\delta x.\end{aligned}$$

$$\begin{aligned}\therefore\quad I_x &= \text{MI of the lamina about OX}\\ &= \int_{-a}^{a}\frac{2}{3}mb^3dx\\ &= \frac{4}{3}mb^3\int_0^a dx\\ &= \frac{4}{3}mb^3a\\ &= \frac{4}{3}\frac{M}{2a\,2b}ab^3\\ &= \frac{1}{3}Mb^2.\end{aligned}$$

$$\begin{aligned}\text{Similarly,}\quad I_y &= \text{MI about OY}\\ &= \frac{1}{3}Ma^2.\end{aligned}$$

By $\perp^r$ axes theorem,

$$\begin{aligned} I_z &= \text{MI of the lamina about OZ} \\ &= I_x + I_y \\ &= \frac{1}{3}Mb^2 + \frac{1}{3}Ma^2 \\ &= \frac{1}{3}M(a^2 + b^2). \end{aligned}$$

III. Rectangular parallelepiped:

Let the lengths of the sides be $2a$, $2b$, and $2c$.

In Fig. 6.10, take the center O of the parallelepiped as the origin, X-axis $\parallel$ to the side $2a$, Y-axis $\parallel$ to the side $2b$, and Z-axis $\parallel$ to the side $2c$. Consider a slice PQRS $\perp^r$ to OX of the thickness δx at a distance x from O. Let m be the mass per unit volume of the parallelepiped.

$$\text{Mass of the slice PQRS} = 2b\,2c\,\delta x\,m = 4\,b\,c\,m\,\delta x.$$

Now,

$$\text{MI of the slice about OX} = \frac{1}{3}4\,b\,c\,m\,\delta x(b^2 + c^2) = \frac{4}{3}bcm(b^2 + c^2)\delta x.$$

$$\begin{aligned} \therefore \quad I_x &= \text{MI of the rectangular parallelepiped about OX} \\ &= \int_{-a}^{a} \frac{4}{3}bcm(b^2 + c^2)dx \end{aligned}$$

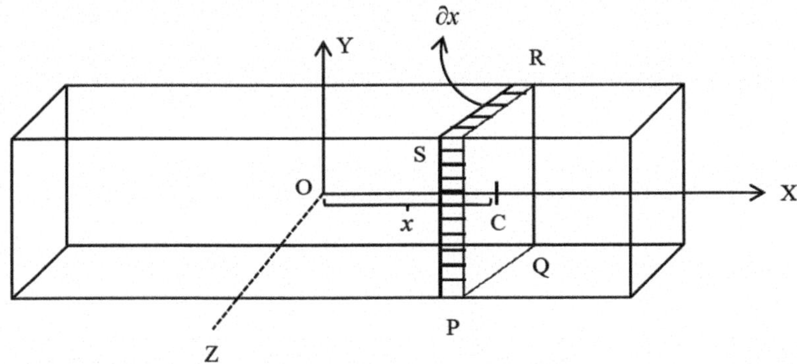

Fig. 6.10 Moment of inertia of a rectangular parallelepiped

$$= \frac{8}{3} bcm(b^2 + c^2)\, a$$

$$= \frac{8}{3} abc(b^2 + c^2)\, \frac{M}{2a\, 2b\, 2c}$$

$$= \frac{1}{3}(b^2 + c^2)M$$

$$= \frac{1}{3} M(b^2 + c^2).$$

Similarly, $I_y =$ MI of the rectangular parallelepiped about OY

$$= \frac{1}{3} M(c^2 + a^2),$$

and $I_z = \frac{1}{3} M(a^2 + b^2).$

IV. A circular ring of mass M and radius a.

In Fig. 6.11, let O be the center of the ring and A′A and B′B be two $\perp^r$ diameters of the ring. Take O as the origin, X-axis along **OA** and Y-axis along **OB**, Z-axis $\perp^r$ to the XY plane. Let m be the mass per unit length of the circle.

Consider an element PQ $= a\delta\theta$ of the ring where $\angle AOP = \theta$ and $\angle AOQ = \theta + \delta\theta$.

Draw $\perp^r$ PM from P on OX.

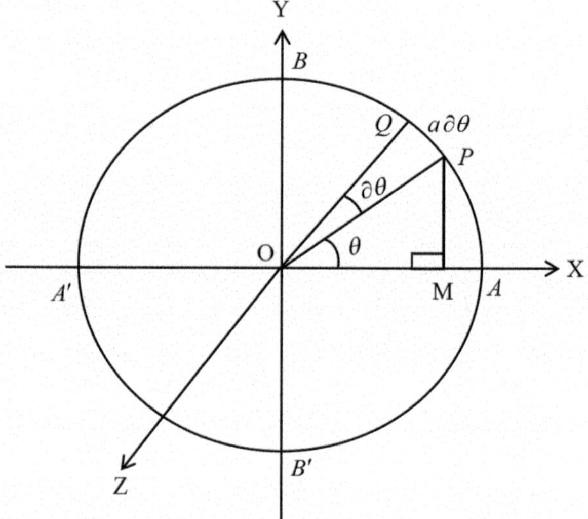

Fig. 6.11 Illustration of moment of inertia of a circular ring

$$\text{Mass of the element PQ} = a\delta\theta \cdot m$$
$$= ma\delta\theta.$$

$$\text{MI of the element PQ about OX} = ma\delta\theta\, PM^2$$
$$= ma\,\delta\theta\, a^2 \sin^2\theta$$
$$= ma^3 \sin^2\theta\delta\theta.$$

$$\therefore \quad I_x = \text{MI of the ring about OX}$$
$$= \int_{\theta=0}^{2\pi} ma^3 \sin^2\theta\, d\theta$$
$$= 2ma^3 \int_{\theta=0}^{\pi} \sin^\theta\, d\theta$$
$$= 4ma^3 \int_{\theta=0}^{\frac{\pi}{2}} \sin^2\theta\, d\theta$$
$$= 4ma^3 \cdot \frac{1}{2} \cdot \frac{\pi}{2}$$
$$= \pi ma^3 = \pi a^3 \frac{M}{2\pi a} = \frac{1}{2}Ma^2.$$

Similarly, $I_y = \frac{1}{2}Ma^2.$

By $\perp^r$ axes theorem,

$$I_z = I_x + I_y$$
$$= \frac{1}{2}Ma^2 + \frac{1}{2}Ma^2 = Ma^2.$$

V. A circular disc of mass M and radius a.

Method I: Let O be the center of the disc and A′A and B′B be two $\perp^r$ diameters of the disc as shown in Fig. 6.12. Take O as the origin, X-axis along **OA**, Y-axis along **OB**, and Z-axis $\perp^r$ to the plane of the disc. Let m be the mass per unit area of the disc.

Consider a ring of radius r and thickness δr $(0 < r < a)$ of the disc.

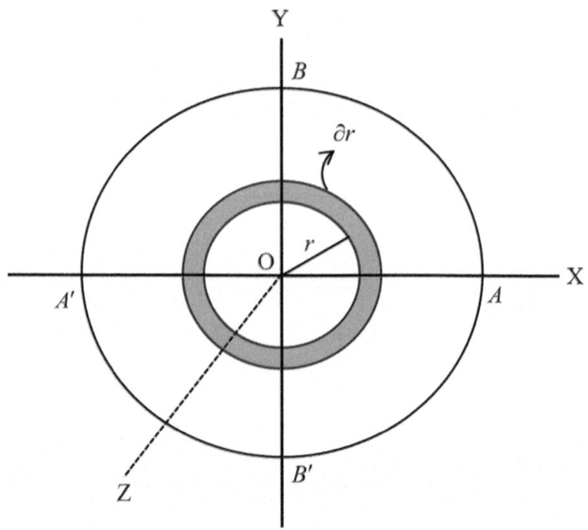

Fig. 6.12 Moment of inertia of a circular disc of radius a

$$\begin{aligned}\text{Mass of the ring} &= 2\pi\, r\, \delta r\, m\\ &= 2\pi m r \delta r.\\ \text{MI of the ring about OX} &= \frac{1}{2} 2\pi\, m\, r\, \delta r\, r^2\\ &= \pi\, m\, r^3 \delta r.\end{aligned}$$

$$\begin{aligned}\therefore\quad I_x &= \text{MI of the disc about OX}\\ &= \int_{r=0}^{a} \pi\, m\, r^3 dr\\ &= \pi\, m\, \frac{a^4}{4} = \pi \frac{a^4}{4}\frac{M}{\pi a^2} = \frac{1}{4} M a^2.\end{aligned}$$

Similarly, $I_y = \frac{1}{4} M a^2.$

By $\perp^r$ axes theorem,

$$I_z = I_x + I_y = \frac{1}{2} M a^2.$$

Method II: In Fig. 6.13, consider an elementary area $\delta x \delta y$ at a point P(x, y) in the disc.

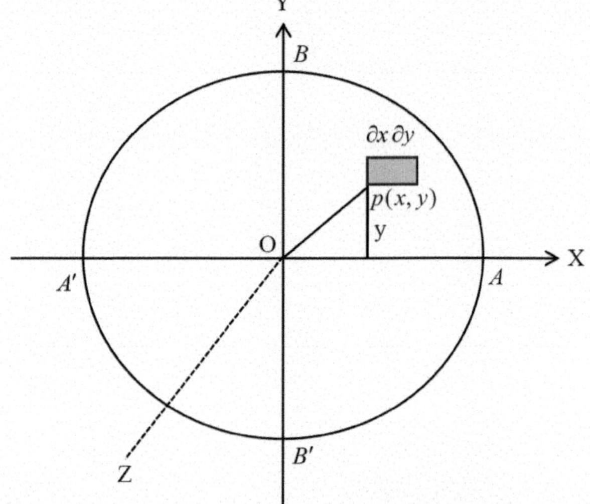

Fig. 6.13 Illustration for the moment of inertia of a circular disc by considering elementary area

Mass of the element $= m\delta x\delta y$.

$$\text{MI of the element about OX} = m\delta x\delta y y^2$$
$$= my^2\delta x\delta y$$

$$\therefore \quad I_x = \text{MI of the disc about OX}$$
$$= \iint\limits_R my^2 dxdy,$$

where $R = \{(x, y) : x^2 + y^2 \le a^2\}$.

Consider the transformation

$$x = r\cos\theta, \quad y = r\sin\theta.$$

$$\therefore \quad J = r.$$

$$I_x = \int_{r=0}^{a}\int_{\theta=0}^{2\pi} mr^2 \sin^2\theta \cdot r dr d\theta$$

$$= m\int_{r=0}^{a} r^3 dr \int_{\theta=0}^{2\pi} \sin^2\theta d\theta$$

$$= m\frac{a^4}{4} 4\int_{\theta=0}^{\frac{\pi}{2}} \sin^2\theta d\theta$$

$$= ma^4\frac{1}{2}\frac{\pi}{2} = \frac{\pi m a^4}{4} = \frac{\pi a^4}{4}\frac{M}{\pi a^2} = \frac{1}{4}Ma^2.$$

Similarly, $I_y = \frac{1}{4}Ma^2.$

By $\perp^r$ axes theorem,

$$I_z = I_x + I_y = \frac{1}{2}Ma^2.$$

VI. Elliptic disc of mass M and axes $2a$ and $2b$.

From Fig. 6.14, take the center O as the origin, X-axis along **OA**, Y-axis along **OB**, and Z-axis is $\perp^r$ to the plane of the disc. Let m be the mass per unit area. Consider an elementary area $\delta x \delta y$ at the point P(x, y) in the disc.

Mass of the element $= m\,\delta x\,\delta y$.

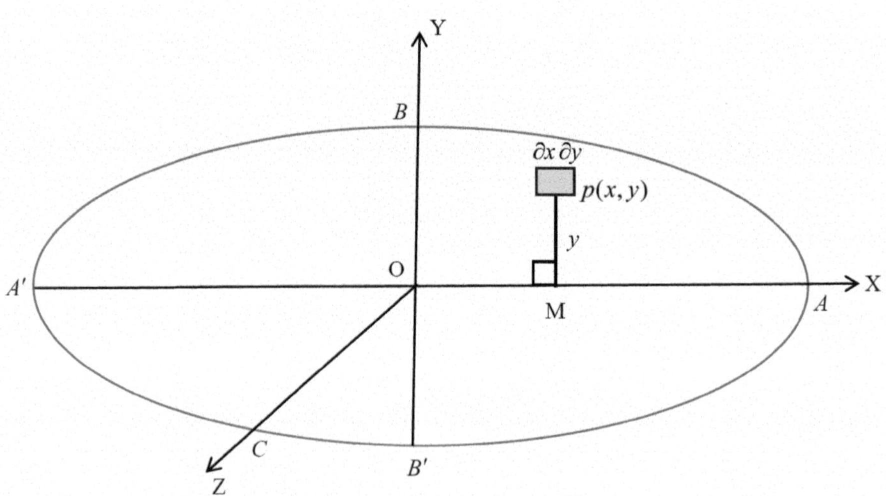

Fig. 6.14 Moment of inertia of a elliptic disc

$$\text{MI of the element about OX} = m\,\delta x\,\delta y y^2$$
$$= my^2\delta x\delta y.$$

$$\therefore \quad I_x = \text{MI of the disc about OX}$$
$$= \iint_R my^2 dxdy,$$

where $R = \left\{(x, y) : \frac{x^2}{a^2} + \frac{y^2}{b^2} \le 1\right\}$.

Consider the transformations

$$x = ar\cos\theta, \quad y = br\sin\theta.$$

$$\therefore \quad J = abr.$$

$$\therefore \quad I_x = \int_{r=0}^{1}\int_{\theta=0}^{2\pi} m\,b^2\,r^2\,\sin^2\theta.abr\,drd\theta$$
$$= mab^3 \int_{r=0}^{1} r^3 dr \int_{\theta=0}^{2\pi} \sin^2\theta d\theta$$
$$= mab^3\,\frac{1}{4}\,4\,\frac{1}{2}\,\frac{\pi}{2}$$
$$= \frac{1}{4}\pi mab^3 = \frac{1}{4}\pi ab^3\,\frac{M}{\pi ab} = \frac{1}{4}Mb^2.$$

Similarly $I_y = \frac{1}{4}Ma^2$.

By $\perp^r$ axes theorem,

$$I_z = I_x + I_y = \frac{1}{4}M(a^2 + b^2).$$

VII. Hollow Sphere: It is formed by the revolution of a semi-circular arc about its bounding diameter

In Fig. 6.15, consider a ring PQ of the hollow sphere with thickness $a\delta\theta$, where $\angle AOP = \theta$. Let m be the mass per unit surface area of the shell.

$$\text{Mass of the ring PQ} = 2\pi a\ \sin\theta\ a\ \delta\theta\ m$$
$$= 2\pi m\,a^2\sin\theta\,\delta\theta.$$

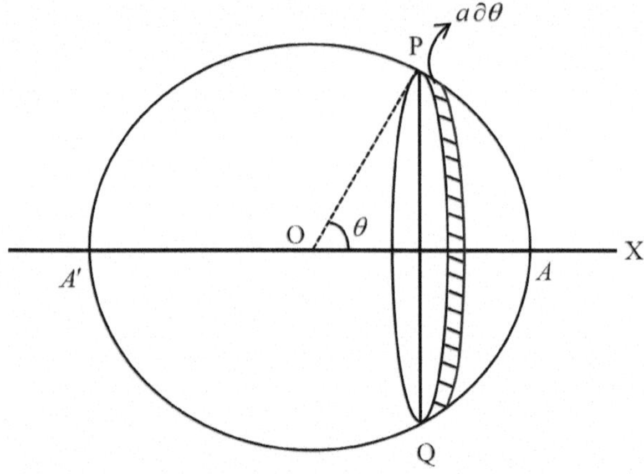

Fig. 6.15 Moment of inertia of a hollow sphere

$$
\begin{aligned}
\text{MI of the ring PQ about OX} &= 2\pi m\, a^2 \sin\theta\, \delta\theta\, a^2 \sin^2\theta \\
&= 2\pi m\, a^4 \sin^3\theta\, \delta\theta.
\end{aligned}
$$

$$
\begin{aligned}
I &= \text{MI of the hollow sphere about OX} \\
&= \int_{\theta=0}^{\pi} 2\pi m\, a^4 \sin^3\theta\, d\theta \\
&= 4\pi m\, a^4 \int_{\theta=0}^{\frac{\pi}{2}} \sin^3\theta\, d\theta \\
&= 4\pi m\, a^4 \frac{2}{3} = \frac{8}{3}\pi\, a^4 \frac{M}{4\pi a^2} = \frac{2}{3} M a^2.
\end{aligned}
$$

$\therefore$ MI of the hollow sphere of mass M and radius a about a diameter is given by

$$I = \frac{2}{3} M a^2.$$

VIII. Solid sphere:

Method I: Let M be the mass and a be the radius of a solid sphere with the point O as the center.

From Fig. 6.16, consider a thin shell of radius r and thickness δr such that $0 < r < a$. Let m be the mass per unit volume.

$$
\begin{aligned}
\text{Mass of the elementary shell} &= 4\pi\, r^2\, \delta r\, m \\
&= 4\pi\, m\, r^2\, \delta r.
\end{aligned}
$$

Fig. 6.16 Moment of inertia of the solid sphere

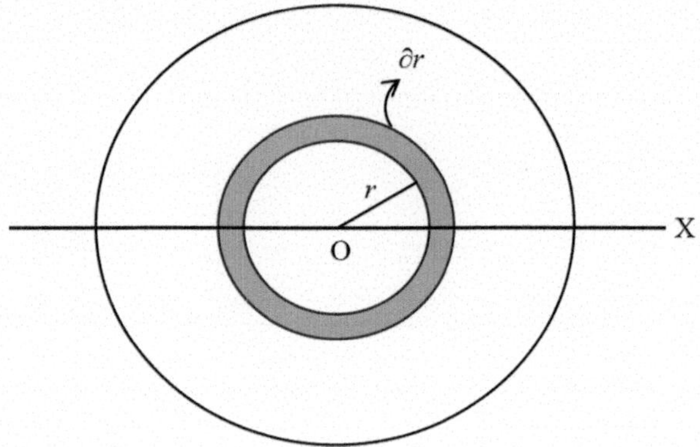

$$\text{MI of the shell about OX} = \frac{2}{3}4\pi\, m\, r^2\, \delta r\, r^2$$
$$= \frac{8}{3}\pi\, m\, r^4\, \delta r.$$

$$\therefore \quad I = \text{MI of the solid sphere about OX}$$
$$= \int_{r=0}^{a} \frac{8}{3}\pi\, m\, r^4\, dr$$
$$= \frac{8}{3}\pi m\, \frac{a^5}{5}$$
$$= \frac{8}{15}\pi a^5 \frac{3M}{4\pi a^3} = \frac{2}{5}Ma^2.$$

$\therefore$ MI of a solid sphere about a diameter of given by

$$I = \frac{2}{5}Ma^2.$$

Method II: Let O be the center of the solid sphere with mass M and radius a. Let OA, OB, and OC be the three mutually perpendicular semi-diameters of the sphere as shown in Fig. 6.17. Take O as the origin, X-axis along **OA**, Y-axis along **OB**, and Z-axis along **OC**. Consider an elementary volume $\delta v = \delta x \delta y \delta z$ at a point P(x, y, z) in the sphere. Let m be the mass per unit volume of the sphere.

$$\therefore \quad \text{Mass of the element} = m\ \delta x \delta y \delta z.$$

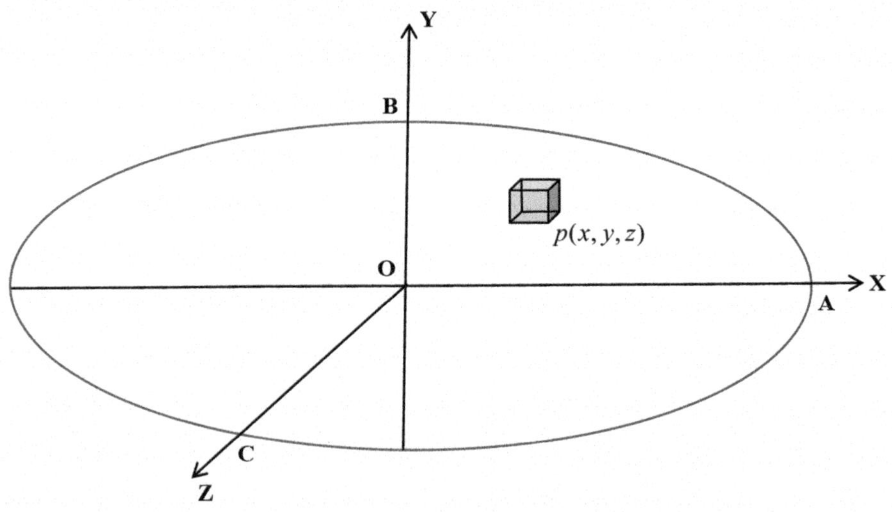

Fig. 6.17 Illustrative diagram of solid sphere with elementary volume element at P

$$\text{MI of the element about OX} = m\,\delta x \delta y \delta z (y^2 + z^2) = m(y^2 + z^2)\delta x \delta y \delta z.$$

$$\therefore \quad I_x = \text{MI of the solid sphere about OX} = \iiint\limits_R m(y^2 + z^2)dxdydz,$$

where $R = \{(x, y, z) : x^2 + y^2 + z^2 \le a^2\}$.

Consider the transformations

$$x = r\sin\theta\cos\phi, \quad y = r\sin\theta\sin\phi, \quad z = r\cos\theta.$$

$$\therefore \quad J = r^2\sin\theta.$$

$$\therefore \quad I_x = m\int\limits_{r=0}^{a}\int\limits_{\theta=0}^{\pi}\int\limits_{\phi=0}^{2\pi}\left[r^2\sin^2\theta\sin^2\phi + r^2\cos^2\theta\right]r^2\sin\theta.drd\theta d\phi$$

$$= m\int\limits_{r=0}^{a} r^4 dr \int\limits_{\theta=0}^{\pi}\int\limits_{\phi=0}^{2\pi}(\sin^2\theta\sin^2\phi + \cos^2\theta)\sin\theta\, d\theta d\phi$$

$$= m\frac{a^5}{5}(I_1 + I_2).$$

$$\begin{aligned}
I_1 &= \int_{\theta=0}^{\pi}\int_{\phi=0}^{2\pi} \sin^3\theta \sin^2\phi \, d\theta d\phi \\
&= \int_{\theta=0}^{\pi} \sin^3\theta \, d\theta \int_{\phi=0}^{2\pi} \sin^2\phi \, d\phi \\
&= 2\int_{0}^{\frac{\pi}{2}} \sin^3\theta \, d\theta \; 4\int_{0}^{\frac{\pi}{2}} \sin^2\phi \, d\phi \\
&= 8\,\frac{2}{3}\,\frac{1}{2}\,\frac{\pi}{2} = \frac{8\pi}{6} = \frac{4\pi}{3}.
\end{aligned}$$

$$\begin{aligned}
I_2 &= \int_{\theta=0}^{\pi}\int_{\phi=0}^{2\pi} \cos^2\theta \sin\theta \, d\theta d\phi \\
&= \int_{\theta=0}^{\pi} \cos^2\theta \sin\theta \, d\theta \int_{\phi=0}^{2\pi} d\phi \\
&= 2\int_{\theta=0}^{\frac{\pi}{2}} \cos^2\theta \sin\theta \, d\theta \; 2\pi \\
&= 2\;2\pi\;\frac{1}{3} = \frac{4\pi}{3}.
\end{aligned}$$

$$\begin{aligned}
\therefore \quad I_x &= m\,\frac{a^5}{5}\left(\frac{4\pi}{3} + \frac{4\pi}{3}\right) \\
&= m\frac{a^5}{5}\,\frac{8\pi}{3} \\
&= \frac{8}{15}\pi a^5 m \\
&= \frac{8}{15}\pi a^5\,\frac{M \times 3}{4\pi a^3} = \frac{2}{5}Ma^2.
\end{aligned}$$

Therefore, MI of a solid sphere about a diameter $= \frac{2}{5}Ma^2$.

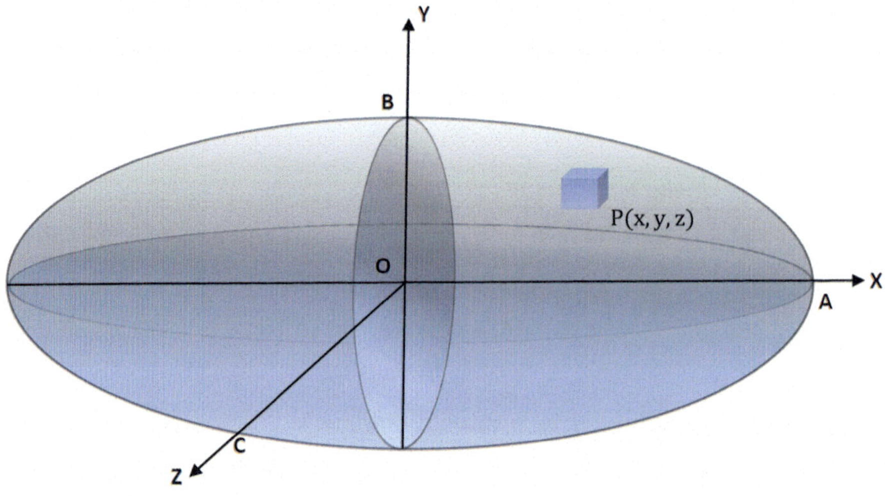

Fig. 6.18 Schematic diagram of solid ellipsoid

IX. Solid ellipsoid about any principal axis

Let the equations of the ellipsoid be

$$\frac{x^2}{a^2} + \frac{y^2}{b^2} + \frac{z^2}{c^2} = 1.$$

From Fig. 6.18, consider a volume element $\delta v = \delta x \delta y \delta z$ at a point P(x, y, z) in the solid. Let m be the mass per unit volume of the solid.

$$\text{Mass of the volume element} = m\ \delta x \delta y \delta z$$

$$\begin{aligned}\text{MI of the element about OX} &= m\ \delta x \delta y \delta z (y^2 + z^2)\\ &= m(y^2 + z^2)\delta x \delta y \delta z\end{aligned}$$

$$\begin{aligned}I_x &= \text{MI of the ellipsoid about OX}\\ &= \iiint_R m(y^2 + z^2) dx dy dz,\end{aligned}$$

where $R = \{(x, y, z) : \frac{x^2}{a^2} + \frac{y^2}{b^2} + \frac{z^2}{c^2} \leq 1.$

Consider the transformations

$$x = ar \sin\theta \cos\phi, \quad y = br \sin\theta \sin\phi, \quad z = cr \cos\theta.$$

$$\therefore \quad J = abc\, r^2 \sin\theta.$$

$$\therefore \quad I_x = m \int_{r=0}^{1} \int_{\theta=0}^{\pi} \int_{\phi=0}^{2\pi} (b^2 r^2 \sin^2\theta \sin^2\phi + c^2 r^2 \cos^2\theta) abc\, r^2 \sin\theta\, dr d\theta d\phi$$

$$= mabc \int_{r=0}^{1} r^4 dr \int_{\theta=0}^{\pi} \int_{\phi=0}^{2\pi} (b^2 \sin^2\theta \sin^2\phi + c^2 \cos^2\theta) \sin\theta\, d\theta d\phi$$

$$= mabc\, \frac{1}{5}(I_1 + I_2)$$

$$= \frac{1}{5} mabc (I_1 + I_2).$$

$$I_1 = \int_{\theta=0}^{\pi} \int_{\phi=0}^{2\pi} b^2 \sin^2\theta \sin^2\phi .\, \sin\theta\, d\theta d\phi$$

$$= b^2 \int_{\theta=0}^{\pi} \sin^3\theta\, d\theta \int_{\phi=0}^{2\pi} \sin^2\phi\, d\phi$$

$$= 2b^2 \int_{\theta=0}^{\frac{\pi}{2}} \sin^3\theta\, d\theta \cdot 4 \int_{\phi=0}^{\frac{\pi}{2}} \sin^2\phi\, d\phi$$

$$= 8b^2\, \frac{2}{3}\, \frac{1}{2}\, \frac{\pi}{2} = \frac{4}{3}\pi b^2.$$

$$I_2 = \int_{\theta=0}^{\pi} \int_{\phi=0}^{2\pi} c^2 \cos^2\theta \sin\theta\, d\theta d\phi$$

$$= c^2 \int_{\theta=0}^{\pi} \cos^2\theta \sin\theta\, d\theta \int_{\phi=0}^{2\pi} d\phi$$

$$= 2c^2\, \frac{1}{3}\, 2\pi = \frac{4}{3}\pi c^2.$$

$$\therefore \quad I_x = \frac{1}{5} mabc \left(\frac{4}{3}\pi b^2 + \frac{4}{3}\pi c^2\right)$$

$$= \frac{4}{15} mabc\, \pi (b^2 + c^2)$$

$$= \frac{4}{15}\pi a\, b\, c\, (b^2 + c^2)\, \frac{3M}{4\pi abc} = \frac{1}{5} M\, (b^2 + c^2).$$

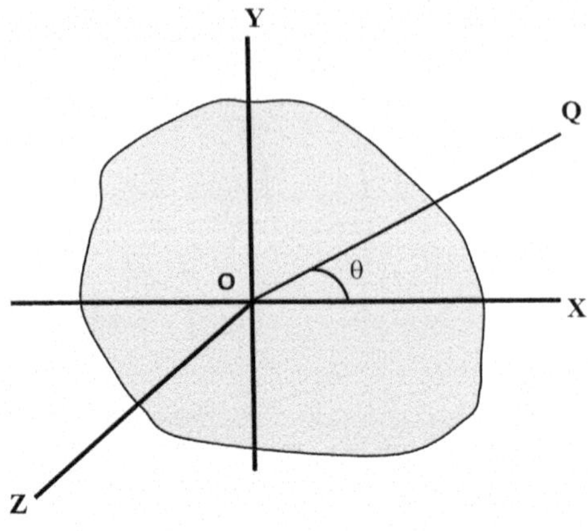

Fig. 6.19 An application of six constant theorem: MI of a plane lamina about a line

$$\text{Similarly,}\quad I_y = \frac{1}{5}M\,(c^2 + a^2),$$
$$I_z = \frac{1}{5}M\,(a^2 + b^2).$$

§. A corollary to six constant theorem.

Let XY be the plane of a lamina. Let OQ be a line on the lamina such that $\angle XOQ = \theta$ as displayed in Fig. 6.19. Let A and B denote the moments of inertia about OX and OY and let F be the product of the moment of inertia about OX-OY, then find the moment of inertia of the lamina about OQ.

For the line OQ,

$$\alpha = \theta,\quad \beta = \frac{\pi}{2} - \theta,\quad \gamma = \frac{\pi}{2}$$

$$\therefore\quad l = \cos\theta,\quad m = \sin\theta,\quad n = 0.$$

By the six constant theorem,

$$\begin{aligned} I &= \text{MI of the lamina about OQ} \\ &= Al^2 + Bm^2 + Cn^2 - 2Dmn - 2Enl - 2Flm \\ &= A\cos^2\theta + B\sin^2\theta - 2F\cos\theta\sin\theta. \end{aligned}$$

6.5.1 *KE and Angular Momentum of a Rigid Body About a Fixed Axis*

Derivation : Let a rigid body rotate with uniform angular velocity ω about a fixed axis OZ as shown in Fig. 6.20. In this case, the motion of the body is two dimensional,

Fig. 6.20 Rigid body rotation about a fixed axis

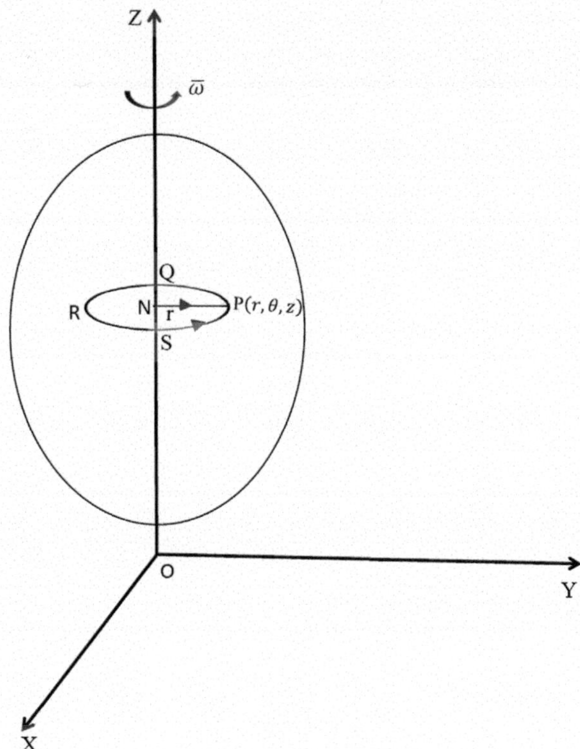

with the Z-plane as the plane of motion. Let m be the mass of a single particle P of the body.

Let (r, θ, z) be the cylindrical polar coordinates of P. As the body rotates about OZ, therefore the particle P describes a circle PQRS about OZ with linear velocity $\mathbf{v} = \boldsymbol{\omega} \times \mathbf{r}$, where $\mathbf{r} = \mathbf{NP}$,N being the center of the circle PQRS.

If T is the KE of the body and $\boldsymbol{\Omega}$ is the angular momentum of the body about OZ, then

$$T = \frac{1}{2}\sum m\mathbf{v}^2, \tag{6.5.1}$$

&

$$\boldsymbol{\Omega} = \sum \mathbf{r} \times m\mathbf{v}. \tag{6.5.2}$$

It is noted that $\mathbf{r} = \mathbf{NP} = r\hat{r}$ and $\boldsymbol{\omega} = \omega\hat{k}$, where $r = N$; $\hat{r}$, $\hat{k}$ are the unit vectors along r and z directions.

Now

$$\mathbf{v} = \boldsymbol{\omega} \times \mathbf{r} = \omega\hat{k} \times r\hat{r} = \omega r\hat{\theta}, \tag{6.5.3}$$

$\hat{\theta}$ being the unit vector in θ direction (transverse direction)

Equation (6.5.1) gives,

$$T = \frac{1}{2}\sum m\omega^2 r^2 = \frac{1}{2}\omega^2\sum mr^2 = \frac{1}{2}\omega^2 I, \tag{6.5.4}$$

where $I = \sum m\, r^2 =$ M.I. of the body about OZ.

Thus,

$$T = \frac{1}{2}I\omega^2. \tag{6.5.5}$$

Equation (6.5.2) gives,

$$\boldsymbol{\Omega} = \sum r\hat{r} \times m\left(\omega r\hat{\theta}\right) = \sum m\omega r^2\left(\hat{r} \times \hat{\theta}\right) = \sum m\omega r^2\hat{k} = \omega I\hat{k} = I\boldsymbol{\omega}. \tag{6.5.6}$$

6.5.2 Linear Momentum for a Rigid Body Moving in Space

Let m be the mass of a typical particle of a rigid body moving in space. In Fig. 6.21, let $\mathbf{OP} = \mathbf{r}$ and G be the center of mass of the body.

If $\mathbf{r}_c = \mathbf{OG}$, then $\mathbf{r}_c =$ P.V. of G relative to O $= \frac{\sum m\mathbf{r}}{\sum m} = \frac{\sum m\mathbf{r}}{M}$,
where $M = \sum m =$ total mass of the body.

$\therefore \quad \sum m\mathbf{r} = M\mathbf{r}_c$, and hence

The total linear momentum of the body is given by

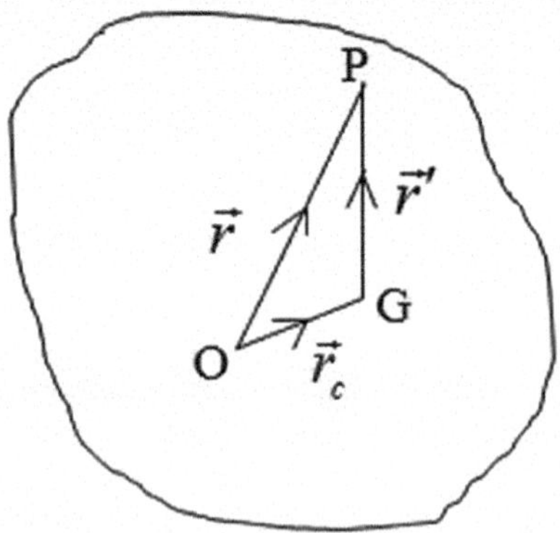

Fig. 6.21 Illustrative diagram for the linear momentum of a rigid body motion

$$\bar{P} = \sum m\dot{\mathbf{r}} = M\dot{\mathbf{r}}_c = M\mathbf{V}_0, \tag{6.5.7}$$

where $\mathbf{V}_0 = \dot{\mathbf{r}}_c =$ velocity of G relative to O. Equation (6.5.7) shows that at any instant the total linear momentum of a body moving in space is the same as that of a single particle of mass M concentrated at the centroid of the body and having a velocity equal to that of the centroid.

6.5.3 Angular Momentum and External Torque for the General Motion of a Rigid Body

Let O be a fixed point in space and m_α denote the mass of a typical particle P_α of the body. Let the α -th particle P_α be acted on by an external force $\mathbf{F}_\alpha$ and an internal force $\mathbf{I}_{\alpha\beta}$ due to the β -th particle P_β of mass m_β.

In Fig. 6.22, let $\mathbf{r}_\alpha$ and $\mathbf{r}_\beta$ be the position vectors of P_α and P_β respectively relative to O.

The angular momentum of the body about the fixed point O is given by

$$\boldsymbol{\Omega} = \sum_\alpha \mathbf{r}_\alpha \times m_\alpha \mathbf{v}_\alpha = \sum_\alpha \mathbf{r}_\alpha \times m_\alpha \dot{\mathbf{r}}_\alpha. \tag{6.5.8}$$

The torque $\boldsymbol{\Lambda}$ of the system of external forces acting on the body about O is given by

$$\boldsymbol{\Lambda} = \sum_\alpha \mathbf{r}_\alpha \times \mathbf{F}_\alpha. \tag{6.5.9}$$

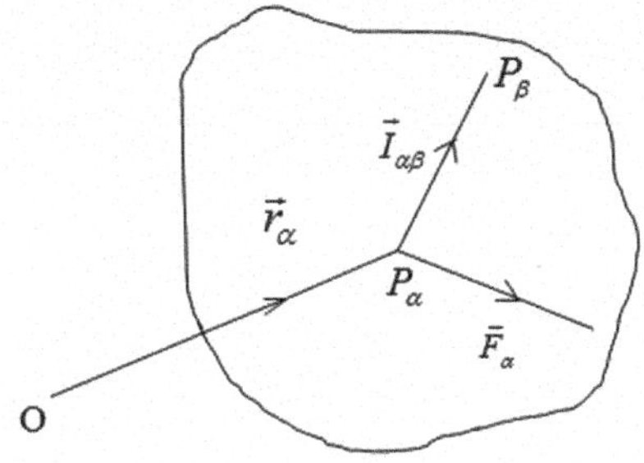

Fig. 6.22 Illustrative diagram for the angular momentum and external torque of a rigid body motion

6.5.4 Principle of Angular Momentum

For the general motion of a rigid body

$$\frac{d\mathbf{\Omega}}{dt} = \mathbf{\Lambda}.$$

Proof By definition

$$\mathbf{\Omega} = \sum_{\alpha} \mathbf{r}_{\alpha} \times m_{\alpha}\mathbf{v}_{\alpha}, \tag{6.5.10}$$

$$\mathbf{\Lambda} = \sum_{\alpha} \mathbf{r}_{\alpha} \times \mathbf{F}_{\alpha}. \tag{6.5.11}$$

Equation (6.5.10) gives,

$$\frac{d\mathbf{\Omega}}{dt} = \sum_{\alpha} [\mathbf{r}_{\alpha} \times m_{\alpha}\mathbf{a}_{\alpha} + \mathbf{v}_{\alpha} \times m_{\alpha}\mathbf{v}_{\alpha}] = \sum_{\alpha} \mathbf{r}_{\alpha} \times m_{\alpha}\mathbf{a}_{\alpha}, \tag{6.5.12}$$

where $\mathbf{a}_{\alpha} = \ddot{\mathbf{r}}_{\alpha}$ = acceleration of the α -th in particle.
Newton's second law of motion leads

$$m_{\alpha}\mathbf{a}_{\alpha} = \mathbf{F}_{\alpha} + \sum_{\beta} \mathbf{I}_{\alpha\beta}. \tag{6.5.13}$$

On use of (6.5.13), (6.5.12) reduces to

$$\frac{d\mathbf{\Omega}}{dt} = \sum_{\alpha} \mathbf{r}_{\alpha} \times \left[\mathbf{F}_{\alpha} + \sum_{\beta} \mathbf{I}_{\alpha\beta}\right] = \mathbf{\Lambda} + \sum_{\alpha}\sum_{\beta} \mathbf{r}_{\alpha} \times \mathbf{I}_{\alpha\beta}. \tag{6.5.14}$$

Now $\mathbf{I}_{\alpha\beta}$ = Force of attraction of the β -th on particle on the α -th particle,

$$\text{i.e.,}\quad \mathbf{I}_{\alpha\beta} = I_{\alpha\beta}\hat{I}_{\alpha\beta} = I_{\alpha\beta}\frac{\mathbf{PP}_{\text{fi}}}{|\mathbf{PP}_{\text{fi}}|} = \lambda_{\alpha\beta}(\mathbf{r}_{\beta} - \mathbf{r}_{\alpha});\quad \lambda_{\alpha\beta} = \frac{I_{\alpha\beta}}{|\mathbf{PP}_{\text{fi}}|}. \tag{6.5.15}$$

From (6.5.15) and Newton's third law it follows that

$$\lambda_{\alpha\beta} = \lambda_{\beta\alpha}. \tag{6.5.16}$$

Now

$$\sum_{\alpha}\sum_{\beta}\mathbf{r}_{\alpha}\times\mathbf{I}_{\alpha\beta}=\sum_{\alpha}\sum_{\beta}\mathbf{r}_{\alpha}\times\lambda_{\alpha\beta}\left(\mathbf{r}_{\beta}-\mathbf{r}_{\alpha}\right)=\sum_{\alpha}\sum_{\beta}\lambda_{\alpha\beta}\left(\mathbf{r}_{\alpha}\times\mathbf{r}_{\beta}\right). \tag{6.5.17}$$

We have

$$\begin{aligned}\sum_{\alpha}\sum_{\beta}\lambda_{\alpha\beta}\left(\mathbf{r}_{\alpha}\times\mathbf{r}_{\beta}\right)&=\sum_{\beta}\sum_{\alpha}\lambda_{\beta\alpha}\left(\mathbf{r}_{\beta}\times\mathbf{r}_{\alpha}\right)\\&=\sum_{\beta}\sum_{\alpha}\lambda_{\alpha\beta}\left(\mathbf{r}_{\beta}\times\mathbf{r}_{\alpha}\right)\\&=-\sum_{\alpha}\sum_{\beta}\lambda_{\alpha\beta}\left(\mathbf{r}_{\alpha}\times\mathbf{r}_{\beta}\right)\quad\text{(by 6.5.16)}\end{aligned}$$

$$\therefore\quad\sum_{\alpha}\sum_{\beta}\lambda_{\alpha\beta}(\mathbf{r}_{\alpha}\times\mathbf{r}_{\beta})=\mathbf{0}. \tag{6.5.18}$$

Equation (6.5.17) gives,

$$\sum_{\alpha}\sum_{\beta}\mathbf{r}_{\alpha}\times\mathbf{I}_{\alpha\beta}=\mathbf{0}. \tag{6.5.19}$$

Equation (6.5.14) reduces to

$$\frac{d\mathbf{\Omega}}{dt}=\mathbf{\Lambda}.$$

Hence the theorem is proved. □

6.5.5 Conservation of Angular Momentum

If the resultant external torque acting on a body is zero, then the angular momentum of the body is conserved.

Proof The principle of angular momentum of a rigid body gives

$$\frac{d\mathbf{\Omega}}{dt}=\mathbf{\Lambda}. \tag{6.5.20}$$

If $\mathbf{\Lambda}=\mathbf{0}$, then Equation (6.5.20) reduces

$$\frac{d\boldsymbol{\Omega}}{dt} = \mathbf{0},$$

which yields

$$\boldsymbol{\Omega} = \mathbf{C} = \text{a constant vector.}$$

Thus, the angular momentum of the body is conserved. □

6.5.6 *Velocity of a Particle of a Rigid Body*

Let G be the C.I. of a rigid body with $\mathbf{r}_c$ as the position vector relative to some origin O. Let m be an element of the body located at a point P. Join GP and OP. From Fig. 6.23, let $\mathbf{OG} = \mathbf{r}_c$ & $\mathbf{GP} = \mathbf{r}'$. By addition of vectors,

$$\mathbf{OP} = \mathbf{OG} + \mathbf{GP},$$

$$\text{i.e.,}\quad \mathbf{r} = \mathbf{r}_c + \mathbf{r}',$$

$$\text{i.e.,}\quad \dot{\mathbf{r}} = \dot{\mathbf{r}}_c + \dot{\mathbf{r}}',$$

$$\text{i.e.,}\quad \mathbf{v} = \mathbf{v}_0 + \mathbf{v}'. \tag{6.5.21}$$

Now,

$$\mathbf{v}_0 = \dot{\mathbf{r}}_c = \text{Velocity of the C.I. G relative to O},$$

&

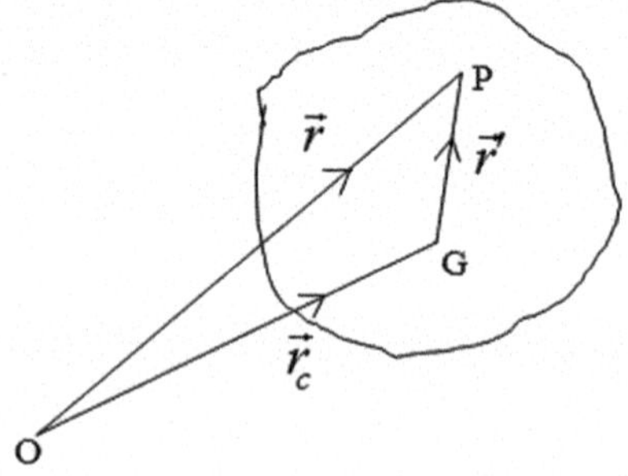

Fig. 6.23 Illustrative diagram for the particle velocity in a rigid body

$\mathbf{v}' = \dot{\mathbf{r}}' =$ Velocity of P relative to G$=\boldsymbol{\omega} \times \mathbf{r}' = \boldsymbol{\omega} \times \mathbf{GP}$.

Equation (6.5.21) gives,

$$\mathbf{v}_P = \mathbf{v}_G + \boldsymbol{\omega} \times \mathbf{GP},$$

i.e., $\mathbf{v} = \mathbf{v}_0 + \boldsymbol{\omega} \times \mathbf{GP}$.

Theorem 6.1 *The KE of a rigid body moving in space is equal to the sum of the KE of a single particle of mass equal to the mass of the body concentrated at its centroid and moving with the centroid's velocity, together with the KE of the body in its motion relative to the centroid.*

Proof Consider a particle of mass m with position vector $\mathbf{r}$ relative to a point O fixed in space.

Let G be the centroid of the body (Fig. 6.24). Suppose that $\mathbf{OG} = \mathbf{r}_c$ and $\mathbf{GP} = \mathbf{r}'$.

$$\therefore \quad \mathbf{r} = \mathbf{r}_c + \mathbf{r}'.$$

The KE of the body is given by

$$T = \frac{1}{2}\sum m\mathbf{v}^2 = \frac{1}{2}\sum m\dot{\mathbf{r}}^2 = \frac{1}{2}\sum m(\dot{\mathbf{r}}_c + \dot{\mathbf{r}}')^2 = \frac{1}{2}\sum m\left(\dot{\mathbf{r}}_c^2 + 2\dot{\mathbf{r}}_c \cdot \dot{\mathbf{r}}' + (\dot{\mathbf{r}}')^2\right),$$

$$\text{i.e.,} \quad T = \frac{1}{2}\sum m\dot{\mathbf{r}}_c^2 + \sum m\dot{\mathbf{r}}_c \cdot \dot{\mathbf{r}}' + \frac{1}{2}\sum m(\dot{\mathbf{r}}')^2. \tag{6.5.22}$$

Now $\frac{\sum m\mathbf{r}'}{M} =$ P.V. of G relative to G $= \mathbf{0}$,

$$\therefore \quad \sum m\mathbf{r}' = \mathbf{0} \ \& \ \sum m\dot{\mathbf{r}}' = \mathbf{0}. \tag{6.5.23}$$

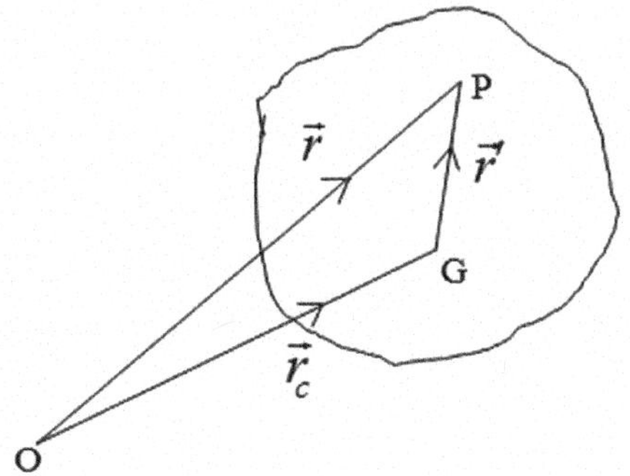

Fig. 6.24 Schematic diagram for Theorem 6.1

By use of (6.5.23), (6.5.22) can be written as

$$T = \frac{1}{2}\left(\sum m\right)\dot{\mathbf{r}}_c^2 + \dot{\mathbf{r}}_c \cdot \sum m\dot{\mathbf{r}}' + \frac{1}{2}\sum m\left(\dot{\mathbf{r}}'\right)^2 = \frac{1}{2}M\mathbf{V}_0^2 + \frac{1}{2}\sum mv'^2, \tag{6.5.24}$$

where $\mathbf{V}_0 = \dot{\mathbf{r}}_c =$ velocity of the centroid relative to O, $\mathbf{v}' = \dot{\mathbf{r}}' =$ velocity of the particle P relative to G.

Equation (6.5.24) gives KE of the body=KE of a particle of mass M placed at G moving with velocity $\mathbf{V}_0 = \dot{\mathbf{r}}_c$ +KE of the body in its motion relative to the centroid. □

Theorem 6.2 *The KE for the general motion of a rigid body is given by*

$$T = \frac{1}{2}MV_0^2 + \frac{1}{2}I_0\omega^2,$$

where M is the mass of the body, I_0 is the moment of inertia of the body about the instantaneous axis of rotation through the centroid and $\boldsymbol{\omega}$ is the angular velocity of the body about the instantaneous axis of rotation through G.

Proof Proceeding as in the previous article, we get

$$T = \frac{1}{2}M\mathbf{V}_0^2 + \frac{1}{2}\sum m\mathbf{v}'^2. \tag{6.5.25}$$

Dropping the dash in (6.5.25), we can write

$$T = \frac{1}{2}MV_0^2 + \frac{1}{2}\sum m\mathbf{v}^2. \tag{6.5.26}$$

Now

$$\mathbf{v} = \boldsymbol{\omega} \times \mathbf{r},$$

$$\therefore \sum m\mathbf{v}^2 = \sum m(\boldsymbol{\omega} \times \mathbf{r})^2 = \sum m\left(\omega\hat{e} \times \mathbf{r}\right)^2 = \sum m\omega^2\left(\mathbf{r} \times \hat{e}\right)^2, \tag{6.5.27}$$

where $\hat{e}$ is the unit vector along the instantaneous axis of rotation.

Equation (6.5.27) gives

$$\sum m\mathbf{v}^2 = \omega^2\sum m\left(\mathbf{r} \times \hat{e}\right)^2 = \omega^2 I_0,$$

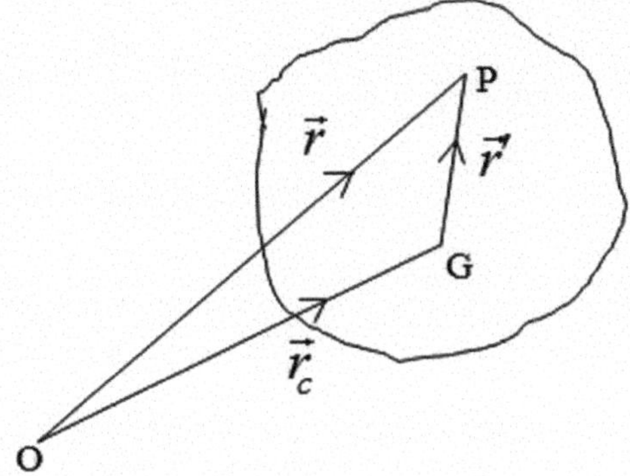

Fig. 6.25 Schematic diagram for Theorem 6.3

where $I_0 = \sum m(\mathbf{r} \times \hat{e})^2$ = Moment of inertia of the body about the instantaneous axis of rotation through the centroid.

Thus, Equation (6.5.26) gives

$$T = \frac{1}{2}MV_0^2 + \frac{1}{2}I_0\omega^2.$$

□

Theorem 6.3 *The angular momentum of a rigid body about a fixed point O is equal to that about O of a single particle of mass equal to the body, concentrated at its centroid and moving with the centroid's velocity, together with the angular momentum about the centroid of the body in its motion relative to the centroid (Fig. 6.25).*

Proof Let m be the mass of a typical particle P of the body, with position vector $\mathbf{r}$ relative to a fixed point O. Let G be the centroid of the body and suppose that $\mathbf{OG} = \mathbf{r}_c$ and $\mathbf{GP} = \mathbf{r}'$.

Addition law of vectors gives $\mathbf{r} = \mathbf{r}_c + \mathbf{r}'$ and so, $\dot{\mathbf{r}} = \dot{\mathbf{r}}_c + \dot{\mathbf{r}}'$. The angular momentum of the body about O is given by

$$\boldsymbol{\Omega} = \sum \mathbf{r} \times m\mathbf{v} = \sum \mathbf{r} \times m\dot{\mathbf{r}} = \sum (\mathbf{r}_c + \mathbf{r}') \times m\left(\dot{\mathbf{r}}_c + \dot{\mathbf{r}}'\right),$$

$$\text{i.e., } \boldsymbol{\Omega} = \sum \left[\mathbf{r}_c \times m\dot{\mathbf{r}}_c + \mathbf{r}_c \times m\dot{\mathbf{r}}' + \mathbf{r}' \times m\dot{\mathbf{r}}_c + \mathbf{r}' \times m\dot{\mathbf{r}}'\right],$$

$$\text{i.e., } \boldsymbol{\Omega} = \sum \mathbf{r}_c \times m\dot{\mathbf{r}}_c + \sum \mathbf{r}_c \times m\dot{\mathbf{r}}' + \sum \mathbf{r}' \times m\dot{\mathbf{r}}_c + \sum \mathbf{r}' \times m\dot{\mathbf{r}}',$$

$$\text{i.e., } \boldsymbol{\Omega} = \mathbf{r}_c \times \dot{\mathbf{r}}_c \sum m + \mathbf{r}_c \times \sum m\dot{\mathbf{r}}' + \left(\sum m\mathbf{r}'\right) \times \dot{\mathbf{r}}_c + \sum \mathbf{r}' \times m\dot{\mathbf{r}}'. \tag{6.5.28}$$

It is noted that $\sum m = M =$ mass of the body,

$$\frac{\sum m\mathbf{r}'}{M} = \text{P.V.of G relative to G} = \mathbf{0},$$

$$\left.\begin{array}{l}\sum m\mathbf{r}' = \mathbf{0}\\ \sum m\dot{\mathbf{r}}' = \mathbf{0}\end{array}\right\}. \tag{6.5.29}$$

By virtue of (6.5.29), (6.5.28) reduces to

$$\boldsymbol{\Omega} = \mathbf{r}_c \times M\dot{\mathbf{r}}_c + \sum \mathbf{r}' \times m\dot{\mathbf{r}}'$$

$$\text{i.e.,}\quad \boldsymbol{\Omega} = \mathbf{r}_c \times M\dot{\mathbf{r}}_c + \sum \mathbf{r}' \times m\dot{\mathbf{r}}' = \boldsymbol{\Omega}_c + \boldsymbol{\Omega}', \tag{6.5.30}$$

where $\boldsymbol{\Omega}_c = \mathbf{r}_c \times M\dot{\mathbf{r}}_c$ =angular momentum about O, of a single particle of mass M(mass of the body) placed at G and moving with the velocity of G, and $\boldsymbol{\Omega}' = \sum \mathbf{r}' \times m\mathbf{r}' =$ angular momentum about G, of the body in its motion relative to G. □

6.5.7 D′Alembert's Principle

The reversed effective forces acting on each particle of a rigid body and the external forces of the system are in equilibrium.

6.5.8 General Equations of Motion of a Rigid Body

Let m_α be the mass of the αth particle P_α of the body as depicted in Fig. 6.26. Let this particle be acted on by an external force $\mathbf{F}_\alpha$. If $\mathbf{r}_\alpha$ is the position vector of P_α relative to a fixed origin O, then the resultant of the reversed effective force and the external force acting on P_α is $-m_\alpha\ddot{\mathbf{r}}_\alpha + \mathbf{F}_\alpha$. By D′Alembert's principle, it follows that the system of forces $\mathbf{R}_\alpha = -m_\alpha\ddot{\mathbf{r}}_\alpha + \mathbf{F}_\alpha,\ \alpha = 1, 2, \ldots$ are in equilibrium. Hence, from the ordinary conditions of equilibrium, we obtain

$$\sum_\alpha \mathbf{R}_\alpha = \mathbf{0}, \tag{6.5.31}$$

&

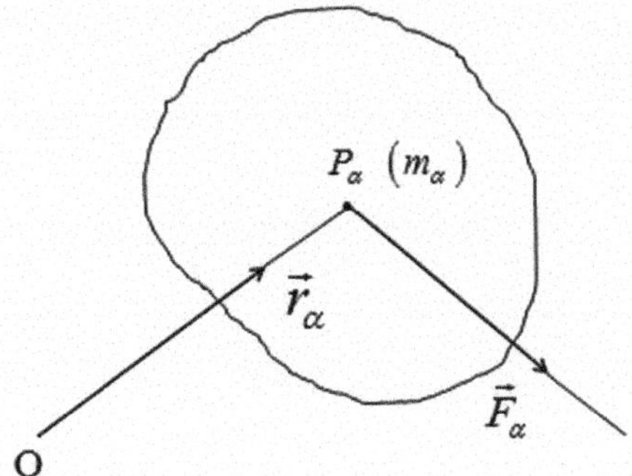

Fig. 6.26 Diagram showing the position vectors and external forces acted on a system of particles of a rigid body

$$\sum_\alpha \mathbf{r}_\alpha \times \mathbf{R}_\alpha = \mathbf{0}. \tag{6.5.32}$$

Equation (6.5.31) gives,

$$\sum_\alpha [-m_\alpha \ddot{\mathbf{r}}_\alpha + \mathbf{F}_\alpha] = \mathbf{0}$$

or

$$\sum_\alpha m_\alpha \ddot{\mathbf{r}}_\alpha = \sum_\alpha \mathbf{F}_\alpha. \tag{6.5.33}$$

From equation (6.5.32), we get

$$\sum_\alpha \mathbf{r}_\alpha \times (-m_\alpha \ddot{\mathbf{r}}_\alpha + \mathbf{F}_\alpha) = \mathbf{0}$$

or

$$\sum_\alpha \mathbf{r}_\alpha \times m_\alpha \ddot{\mathbf{r}}_\alpha = \sum_\alpha \mathbf{r}_\alpha \times \mathbf{F}_\alpha. \tag{6.5.34}$$

Dropping the suffixes in (6.5.33) and (6.5.34), we derive

$$\sum m\,\ddot{\mathbf{r}} = \sum \mathbf{F}, \tag{6.5.35}$$

and

$$\sum \mathbf{r} \times m\,\ddot{\mathbf{r}} = \sum \mathbf{r} \times \mathbf{F}. \tag{6.5.36}$$

Equation (6.5.35) determines the translatory motion of the body, and equation (6.5.36) represents the rotational motion of the body about the fixed point O.

Equations (6.5.35) and (6.5.36) are the general equations of motion of a rigid body.

6.5.9 *Motion of the Center of Inertia*

Let $\mathbf{r}_c$ be the position vector of the center of inertia G at time t relative to a given point O. Let $\mathbf{r}$ stipulate the position vector of a typical particle of mass m at time t, of a rigid body of mass M.

$$\therefore \quad \frac{\sum m\mathbf{r}}{\sum m} = \frac{\sum m\mathbf{r}}{M} = \text{p.v. of the C.I. G relative to O} = \mathbf{r}_c.$$

$$\therefore \quad \sum m\mathbf{r} = M\,\mathbf{r}_c. \tag{6.5.37}$$

The translatory motion of the body is governed by the equation

$$\sum m\ddot{\mathbf{r}} = \sum \mathbf{F},$$

$$\text{i.e.,} \quad M\ddot{\mathbf{r}}_c = \sum \mathbf{F}. \tag{6.5.38}$$

Equation (6.5.38) is the equation of motion of a single particle of mass placed at the center of inertia of the body and acted on by the external forces $\sum \mathbf{F}$.

Thus, the translatory motion of a rigid body is equivalent to the motion of a single particle of mass M (total mass of the body) under the action of the external force $\sum \mathbf{F}$ which is the resultant of the external forces acting on the body.

Hence the equation for the translatory motion of a body is

$$M\,\ddot{\mathbf{r}}_c = \sum \mathbf{F}.$$

The above equation is termed as the equation of motion of the C.I.

6.5.10 *Motion of a Body Relative to the Center of Inertia*

Let G be the center of inertia of a rigid body and O be given pt. Let P be a particle of mass m of the body. Let P be acted on by an external force $\mathbf{F}$. From Fig. 6.27, we write, $\mathbf{OG} = \mathbf{r}_c$, $\mathbf{GP} = \mathbf{r}'$ and $\mathbf{OP} = \mathbf{r}$.

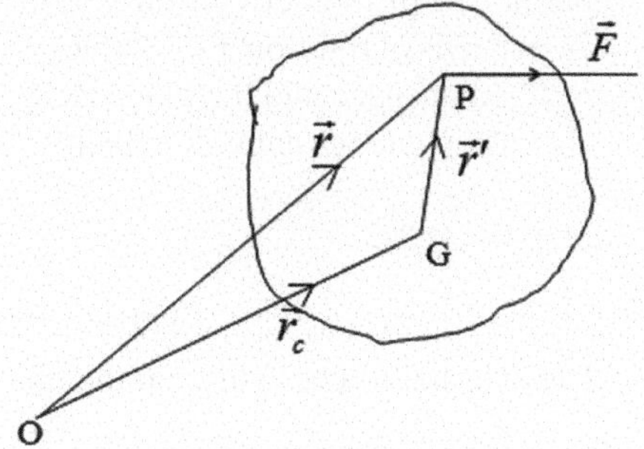

Fig. 6.27 Schematic diagram for the rigid body motion relative to the center of inertia

By the addition law of vectors

$$\mathbf{r} = \mathbf{r}_c + \mathbf{r}'. \tag{6.5.39}$$

The equation motion of the C.I. of the body

$$M\,\ddot{\mathbf{r}}_c = \sum \mathbf{F}. \tag{6.5.40}$$

The rotational motion of the body about O is specified by the equation

$$\sum \mathbf{r} \times m\ddot{\mathbf{r}} = \sum \mathbf{r} \times \mathbf{F}. \tag{6.5.41}$$

Equation (6.5.41) gives,

$$\sum \left(\mathbf{r}_c + \mathbf{r}'\right) \times m\left(\ddot{\mathbf{r}}_c + \ddot{\mathbf{r}}'\right) = \sum \left(\mathbf{r}_c + \mathbf{r}'\right) \times \mathbf{F},$$

i.e.,
$$\begin{aligned} &\sum \mathbf{r}_c \times m\ddot{\mathbf{r}}_c + \sum \mathbf{r}_c \times m\ddot{\mathbf{r}}' + \sum \mathbf{r}' \times m\ddot{\mathbf{r}}_c + \sum \mathbf{r}' \times m\ddot{\mathbf{r}}' \\ &= \sum \mathbf{r}_c \times \mathbf{F}_c + \sum \mathbf{r}' \times \mathbf{F}. \end{aligned} \tag{6.5.42}$$

Now

$$\begin{aligned} \mathbf{T}_1 &= \text{1st term of the LHS of (6.5.42)} \\ &= \sum \mathbf{r}_c \times m\ddot{\mathbf{r}}_c \\ &= \mathbf{r}_c \times \ddot{\mathbf{r}}_c \sum m = \mathbf{r}_c \times M\ddot{\mathbf{r}}_c, \\ \mathbf{T}_2 &= \text{2nd term of the LHS of (6.5.42)} \\ &= \sum \mathbf{r}_c \times m\ddot{\mathbf{r}}' \end{aligned}$$

$$= \mathbf{r}_c \times \sum m\ddot{\mathbf{r}}' = \mathbf{r}_c \times \frac{d^2}{dt^2} \sum m\mathbf{r}'.$$

But $\dfrac{\sum m\mathbf{r}'}{M}$ = P.V. of G relative to G $= \mathbf{0}$

$$\therefore \quad \sum m\mathbf{r}' = \mathbf{0}. \tag{6.5.43}$$

By virtue of (6.5.43),

$$\mathbf{T}_2 = \mathbf{0}. \tag{6.5.44}$$

$\mathbf{T}_3 = 3^{\text{rd}}$ term of LHS of (6.5.42)

$$= \sum \mathbf{r}' \times m\ddot{\mathbf{r}}_c = \left(\sum m\mathbf{r}'\right) \times \ddot{\mathbf{r}}_c = \mathbf{0}. \tag{6.5.45}$$

$\mathbf{U}_1 = 1^{\text{st}}$ term of the RHS of (6.5.42)

$$= \sum \mathbf{r}_c \times \mathbf{F} = \mathbf{r}_c \times \sum \mathbf{F} = \mathbf{r}_c \times M\ddot{\mathbf{r}}_c = \mathbf{T}_1, \tag{6.5.46}$$

$$\mathbf{U}_2 = \sum \mathbf{r}' \times \mathbf{F}. \tag{6.5.47}$$

Equation (6.5.42) gives

$$\mathbf{T}_1 + \mathbf{T}_2 + \mathbf{T}_3 + \mathbf{T}_4 = \mathbf{U}_1 + \mathbf{U}_2$$

i.e., $\mathbf{T}_1 + \mathbf{T}_4 = \mathbf{T}_1 + \mathbf{U}_2$

i.e., $\mathbf{T}_4 = \mathbf{U}_2$

$$\text{i.e.,} \quad \sum \mathbf{r}' \times m\ddot{\mathbf{r}}' = \sum \mathbf{r}' \times \mathbf{F}. \tag{6.5.48}$$

This is the equation of motion of a rigid body with the center of inertia as a fixed point. Dropping the dashes in (6.5.48), we get

$$\sum \mathbf{r} \times m\ddot{\mathbf{r}} = \sum \mathbf{r} \times \mathbf{F}. \tag{6.5.49}$$

Let $\boldsymbol{\Omega}_c$ be the angular momentum of the rigid body about G, the C.I. of the body. By definition of angular momentum $\boldsymbol{\Omega}_c = \sum \mathbf{r} \times m\dot{\mathbf{r}}$,

$$\therefore \quad \frac{d\boldsymbol{\Omega}_c}{dt} = \sum \dot{\mathbf{r}} \times m\dot{\mathbf{r}} + \sum \mathbf{r} \times m\ddot{\mathbf{r}} = \sum \mathbf{r} \times m\ddot{\mathbf{r}}.$$

Equation (6.5.49) reduces to

$$\therefore \quad \frac{d\boldsymbol{\Omega}_c}{dt} = \sum \mathbf{r} \times \mathbf{F} = \boldsymbol{\Lambda}_c, \tag{6.5.50}$$

where $\boldsymbol{\Lambda}_c = \sum \mathbf{r} \times \mathbf{F}$ = resultant external torque about G of the external forces acting on the body.

Therefore, motion of a rigid body about the C.I. is given by

$$\frac{d\boldsymbol{\Omega}_c}{dt} = \boldsymbol{\Lambda}_c.$$

Remarks: The general equations of motion of a rigid body are

$$M\ddot{\mathbf{r}}_c = \sum \mathbf{F} \tag{A}$$

$$\frac{d\boldsymbol{\Omega}_c}{dt} = \boldsymbol{\Lambda}_c. \tag{B}$$

Equation (A) is known as the equation of motion of the C.I. or equation for the translatory motion, and Eq. (B) is the equation of motion of a rigid body about the C.I. or the equation for the rotational motion of a body. In case of plane motion of a rigid body, $\boldsymbol{\Omega}_c = I_0\omega$, where I_0 is M.I. of the body about the instantaneous axis of rotation through the C.I., $\perp^r$ to the plane of motion and ω is the angular velocity of the body about the instantaneous axis of rotation.

Equation (B) takes the form

$$I_0\frac{d\omega}{dt} = \boldsymbol{\Lambda}_c.$$

6.5.11 Angular Momentum for the General Plane Motion of a Rigid Body

Following Theorem (6.3), the angular momentum about a fixed point O, for the general motion of a rigid body is given by

$$\boldsymbol{\Omega} = \mathbf{r}_c \times M\dot{\mathbf{r}}_c + \sum \mathbf{r}' \times m\dot{\mathbf{r}}'. \tag{6.5.51}$$

Here the rotations have their usual meanings.

Dropping the dashes in (6.5.51), we obtain

$$\boldsymbol{\Omega} = \mathbf{r}_c \times M\dot{\mathbf{r}}_c + \sum \mathbf{r} \times m\dot{\mathbf{r}} = \mathbf{r}_c \times M\dot{\mathbf{r}}_c + \boldsymbol{\Omega}_c. \tag{6.5.52}$$

Let the Z-plane be the plane of motion. The fixed point O is the origin of the coordinate system.

It is recalled that the motion of a rigid body consists of

- the motion of the C.I.(Translatory motion),
- the motion about the C.I. (rotational motion).

These two are independent of each other. Further, in the case of the two-dimensional motion of a rigid body, the motion of the body about the CI is equivalent to the motion of the rigid body about a fixed axis through G $\perp^r$ to the plane of motion.

As the Z-plane is the plane of motion, therefore the instantaneous axis of rotation through G is parallel to OZ. That is, the instantaneous axis of rotation through G is fixed.

Let $\boldsymbol{\omega} = \omega\hat{k}$ be the angular velocity of the body about G, where $\hat{k}$ is the unit vector along **OZ**. Now

$$\boldsymbol{\Omega}_c = \sum \mathbf{r} \times m\dot{\mathbf{r}} = \sum r\hat{r} \times m\mathbf{v} = \sum r\hat{r} \times m\,(\boldsymbol{\omega} \times \mathbf{r}),$$

$$\text{i.e., } \boldsymbol{\Omega}_c = \sum r\hat{r} \times m\left(\omega\hat{k} \times r\hat{r}\right) = \sum r\hat{r} \times m\left(\omega r \times \hat{\theta}\right) = \sum mr^2\omega\left(\hat{r} \times \hat{\theta}\right),$$

$$\text{i.e., } \boldsymbol{\Omega}_c = \sum mr^2\omega\hat{z} = \sum mr^2\omega\hat{k} = \left(\sum mr^2\right)\omega\hat{k} = I_0\omega\hat{k} = I_0\boldsymbol{\omega},$$

where $I_0 = \sum mr^2 = MI$ of the body about the instantaneous axis of rotation through G. Thus,

$$\boldsymbol{\Omega} = \mathbf{r}_c \times M\dot{\mathbf{r}}_c + I_0\boldsymbol{\omega}.$$

6.5.12 Motion of a Rigid Body About a Fixed Axis

In Fig. 6.28, let OZ be the fixed axis of rotation. Let $\boldsymbol{\omega}$ stipulate the constant angular velocity of the body about OZ.

If $\boldsymbol{\Omega}$ is the angular momentum of the body about OZ, then

$$\boldsymbol{\Omega} = I\boldsymbol{\omega}, \qquad \left[\text{From §6.5.1}\right] \tag{6.5.53}$$

where I stands for the moment of inertia of the body about OZ.

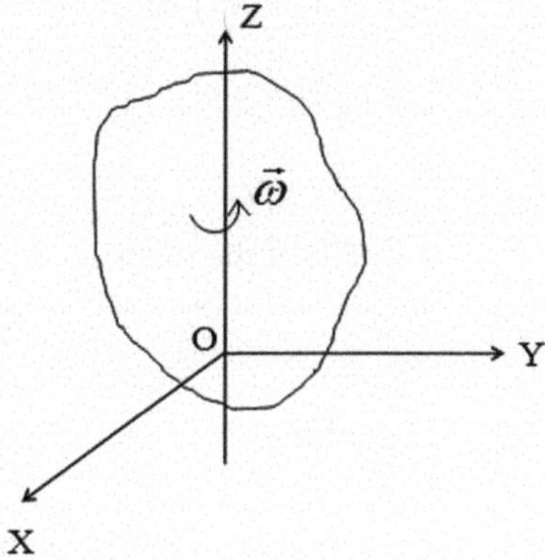

Fig. 6.28 Rigid body rotation about a fixed axis with uniform angular velocity

By the principle of the angular moment, we have

$$\frac{d\boldsymbol{\Omega}}{dt} = \boldsymbol{\Lambda}, \tag{6.5.54}$$

where $\boldsymbol{\Lambda}$ is the torque of all external forces about OZ. Using (6.5.53) in (6.5.54), we get

$$\frac{d}{dt} I\omega = \boldsymbol{\Lambda}$$

$$\text{i.e.,} \quad I\dot{\omega} = \boldsymbol{\Lambda},$$

which is the equation of motion of a rigid body about a fixed axis.

Theorem 6.4 *The total work done in moving a rigid body from one position to another with kinetic energies T_1 and T_2 respectively is given by $W = T_2 - T_1$.*

Proof Let m_α be the mass of the α-th particle of a rigid body in motion. Let the particle be acted on by external force $\mathbf{F}_\alpha$, and the internal force $\mathbf{I}_{\alpha\beta}$ due to the β -th particle of mass m_β of the body. Let $\mathbf{r}_\alpha$ be the position vector of the α -th particle relative to some origin O.

The equation of motion of the α -th particle is

$$m_\alpha \ddot{\mathbf{r}}_\alpha = \mathbf{F}_\alpha + \sum_\beta \mathbf{I}_{\alpha\beta},$$

$$\text{i.e.,} \quad \frac{d}{dt}(m_\alpha \mathbf{v}_\alpha) = \mathbf{F}_\alpha + \sum_\beta \mathbf{I}_{\alpha\beta} = \mathbf{R}_\alpha,$$

$$\text{i.e.,} \quad \dot{\mathbf{r}}_\alpha \cdot \frac{d}{dt}(m_\alpha \mathbf{v}_\alpha) = \dot{\mathbf{r}}_\alpha \cdot \mathbf{F}_\alpha + \dot{\mathbf{r}}_\alpha \cdot \sum_\beta \mathbf{I}_{\alpha\beta} = \dot{\mathbf{r}}_\alpha \cdot \mathbf{R}_\alpha, \tag{6.5.55}$$

where $\mathbf{R}_\alpha = \mathbf{F}_\alpha + \sum_\beta \mathbf{I}_{\alpha\beta}$=Impressed force on the α-th in particle.

(6.5.55) can be written as

$$m_\alpha \mathbf{v}_\alpha \cdot \frac{d\mathbf{v}_\alpha}{dt} = \mathbf{r}_\alpha \cdot \mathbf{R}_\alpha. \tag{6.5.56}$$

We note that

$$\frac{d}{dt} v_\alpha^2 = \frac{d}{dt} \mathbf{v}_\alpha^2 = 2\mathbf{v}_\alpha \cdot \frac{d\mathbf{v}_\alpha}{dt}$$

$$\therefore \quad \mathbf{v}_\alpha \cdot \frac{d\mathbf{v}_\alpha}{dt} = \frac{1}{2} \frac{d}{dt} v_\alpha^2. \tag{6.5.57}$$

Using (6.5.57) in (6.5.56), we obtain

$$\frac{m_\alpha}{2} \frac{d}{dt} v_\alpha^2 = \dot{\mathbf{r}}_\alpha \cdot \mathbf{R}_\alpha,$$

$$\text{i.e.,} \quad \frac{1}{2} \frac{d}{dt} \left(m_\alpha v_\alpha^2\right) = \dot{\mathbf{r}}_\alpha \cdot \mathbf{R}_\alpha,$$

$$\text{i.e.,} \quad \frac{d}{dt} \left(\frac{1}{2} m_\alpha v_\alpha^2\right) = \dot{\mathbf{r}}_\alpha \cdot \mathbf{R}_\alpha,$$

$$\text{i.e.,} \quad \frac{d}{dt} \left(\sum_\alpha \frac{1}{2} m_\alpha v_\alpha^2\right) = \sum_\alpha \dot{\mathbf{r}}_\alpha \cdot \mathbf{R}_\alpha,$$

$$\text{i.e.,} \quad \frac{d}{dt} T = \sum_\alpha \dot{\mathbf{r}}_\alpha \cdot \mathbf{R}_\alpha, \tag{6.5.58}$$

where $T = \sum_\alpha \frac{1}{2} m_\alpha v_\alpha^2$=total KE of the body at the t.

Equation (6.5.58) reduces to

$$\frac{dT}{dt} = \sum_\alpha \mathbf{R}_\alpha \cdot \dot{\mathbf{r}}_\alpha = \sum_\alpha \mathbf{R}_\alpha \cdot \frac{d\mathbf{r}_\alpha}{dt},$$

$$\text{i.e.,}\quad dT = \sum_{\alpha} \mathbf{R}_{\alpha} \cdot d\mathbf{r}_{\alpha} = \sum_{\alpha} dW_{\alpha} = d\left(\sum_{\alpha} W_{\alpha}\right) = dW, \tag{6.5.59}$$

where W is the work function.

Equation(6.5.59) can be rewritten as

$$dW = \frac{dT}{dt}dt. \tag{6.5.60}$$

Integrating (6.5.60) w.r.t t from $t = t_1$ to $t = t_2$, we find

$$W_{12} = \int_{t=t_1}^{t_2} \frac{dT}{dt}dt = \int_{t=t_1}^{t_2} dT = T_2 - T_1,$$

where $T_1 = T]_{t=t_1}$ = KE of the body at time $t = t_1$, of the body at time $t = t_1$ and $T_2 = T]_{t=t_2}$ = KE of the body at time $t = t_2$.

Thus, W_{12} = Work done by the external and internal forces in moving the rigid body from 1st position to 2nd position $= T_2 - T_1 =$ gain in KE, which is the required result. □

Theorem 6.5 *If both the external and internal forces acting on a rigid body are conservative, then the total energy of the body is conserved.*

Proof If the α-th particle with mass m_α of the body be acted upon by an external force $\mathbf{F}_\alpha$ and the internal force $\mathbf{I}_{\alpha\beta}$ due to the β-th particle, then the resultant force acting on the α-th particle is given by

$$\mathbf{R}_{\alpha} = \mathbf{F}_{\alpha} + \sum_{\beta} \mathbf{I}_{\alpha\beta}. \tag{6.5.61}$$

Let δW_α be the work done by the force $\mathbf{R}_\alpha$ on the α -th particle for the displacement $\delta\mathbf{r}_\alpha$, where $\mathbf{r}_\alpha$ is the position of the α-th particle relative to some origin O.

$$\therefore\quad \delta W_{\alpha} = \mathbf{R}_{\alpha} \cdot \delta\mathbf{r}_{\alpha} = \mathbf{F}_{\alpha} \cdot \delta\mathbf{r}_{\alpha} + \sum_{\beta} \mathbf{I}_{\alpha\beta} \cdot \delta\mathbf{r}_{\alpha}.$$

Thus, the total work done by the external and internal forces for an elementary displacement of the body is given by

$$\delta W = \sum_{\alpha} \mathbf{F}_{\alpha} \cdot \delta\mathbf{r}_{\alpha} + \sum_{\alpha} \sum_{\beta} \mathbf{I}_{\alpha\beta} \cdot \delta\mathbf{r}_{\alpha}. \tag{6.5.62}$$

Letting $\delta\mathbf{r}_\alpha \to \mathbf{0}$, (6.5.62) can be written as

$$dW = \sum_{\alpha} \mathbf{F}_\alpha \cdot d\mathbf{r}_\alpha + \sum_{\alpha}\sum_{\beta} \mathbf{I}_{\alpha\beta} \cdot d\mathbf{r}_\alpha. \tag{6.5.63}$$

Let V_α be the force potential corresponding to the external force $\mathbf{F}_\alpha$ and $V_{\alpha\beta}$ be that corresponding to the internal force $\mathbf{I}_{\alpha\beta}$.

$$\therefore \quad \mathbf{F}_\alpha = -\nabla V_\alpha, \tag{6.5.64}$$

&

$$\mathbf{I}_{\alpha\beta} = -\nabla_\alpha V_{\alpha\beta}. \tag{6.5.65}$$

Now

$$\begin{aligned}
&\mathbf{I}_{\alpha\beta} \cdot d\mathbf{r}_\alpha + \mathbf{I}_{\beta\alpha} \cdot d\mathbf{r}_\beta = -\nabla_\alpha V_{\alpha\beta} \cdot d\mathbf{r}_\alpha + \left(-\nabla_\beta V_{\beta\alpha}\right) \cdot d\mathbf{r}_\beta,\\
&\text{i.e., } \mathbf{I}_{\alpha\beta} \cdot d\mathbf{r}_\alpha + \mathbf{I}_{\beta\alpha} \cdot d\mathbf{r}_\beta = -\nabla_\alpha V_{\alpha\beta} \cdot d\mathbf{r}_\alpha - \nabla_\beta V_{\alpha\beta} \cdot d\mathbf{r}_\beta,\\
&\text{i.e., } \mathbf{I}_{\alpha\beta} \cdot d\mathbf{r}_\alpha + \mathbf{I}_{\beta\alpha} \cdot d\mathbf{r}_\beta = -\left[\frac{\partial V_{\alpha\beta}}{\partial x_\alpha} dx_\alpha + \frac{\partial V_{\alpha\beta}}{\partial y_\alpha} dy_\alpha + \frac{\partial V_{\alpha\beta}}{\partial z_\alpha} dz_\alpha\right.\\
&\qquad \left. + \frac{\partial V_{\alpha\beta}}{\partial x_\beta} dx_\beta + \frac{\partial V_{\alpha\beta}}{\partial y_\beta} dy_\beta + \frac{\partial V_{\alpha\beta}}{\partial z_\beta} dz_\beta\right],\\
&\text{i.e., } \mathbf{I}_{\alpha\beta} \cdot d\mathbf{r}_\alpha + \mathbf{I}_{\beta\alpha} \cdot d\mathbf{r}_\beta = -dV_{\alpha\beta} \quad \left[\because \ V_{\alpha\beta} = V_{\alpha\beta}\left(x_\alpha, y_\alpha, z_\alpha, x_\beta, y_\beta, z_\beta\right)\right].
\end{aligned}$$

Thus

$$\sum_{\alpha}\sum_{\beta} \mathbf{I}_{\alpha\beta} \cdot d\mathbf{r}_\alpha = -\sum_{\alpha}\sum_{\beta} dV_{\alpha\beta} = -d\left(\sum_{\alpha}\sum_{\beta} V_{\alpha\beta}\right) = -dV_I,$$

where $V_I = \sum_{\alpha}\sum_{\beta} V_{\alpha\beta}$ =total internal potential.

Again

$$\sum_{\alpha} \mathbf{F}_\alpha \cdot d\mathbf{r}_\alpha = -\sum_{\alpha} \nabla V_\alpha \cdot d\mathbf{r}_\alpha = -\sum_{\alpha} dV_\alpha = -d\left(\sum_{\alpha} V_\alpha\right) = -dV_E,$$

where $V_E = \sum_{\alpha} V_\alpha$ = total external potential.

Equation (6.5.63) reduces to

$$dW = -dV_E - dV_I = -d\left(V_E + V_I\right) = -dV, V = V_E + V_I. \tag{6.5.66}$$

$$\therefore \quad W_{12} = -\int_{\text{1st pos}}^{\text{2nd pos}} dV = V_1 - V_2. \tag{6.5.67}$$

Again by Theorem 6.4

$$W_{12} = T_2 - T_1. \tag{6.5.68}$$

From (6.5.67) and (6.5.68), we have

$$V_1 - V_2 = T_2 - T_1,$$

$$\text{i.e.,} \quad T_1 + V_1 = T_2 + V_2,$$

i.e., $T + V =$ constant, which is the principle of conservation of energy. □

6.6 Momental Ellipsoid

Let O be a fixed point in a rigid body. Let OX, OY, and OZ be three mutually perpendicular axes at O. Let OQ be a variable line with direction cosines l, m, n. Then with usual notations, the moment of inertia of the body about OQ is given by

$$I = Al^2 + Bm^2 + Cn^2 - 2Dmn - 2Enl - 2Flm.$$

Let the point Q be such that

$$I \propto \frac{1}{OQ^2},$$

$$\text{i.e.,} \quad I = \frac{Mk^4}{OQ^2},$$

$$\text{i.e.,} \quad Al^2 + Bm^2 + Cn^2 - 2Dmn - 2Enl - 2Flm = \frac{Mk^4}{OQ^2}. \tag{6.6.1}$$

Let $(\bar{x}, \bar{y}, \bar{z})$ be the coordinates of Q.
Equations of line OQ are

$$\frac{x}{l} = \frac{y}{m} = \frac{z}{n}. \tag{6.6.2}$$

Since $Q(\bar{x}, \bar{y}, \bar{z})$ is on (6.6.2),

$$\frac{\bar{x}-0}{l} = \frac{\bar{y}-0}{m} = \frac{\bar{z}-0}{n} = r = OQ,$$

$$\therefore \quad \frac{\bar{x}}{l} = \frac{\bar{y}}{m} = \frac{\bar{z}}{n} = r = OQ,$$

$$\text{i.e.,} \quad \bar{x} = lr, \quad \bar{y} = mr, \quad \bar{z} = nr. \tag{6.6.3}$$

Equation (6.6.1) gives

$$Al^2 + Bm^2 + Cn^2 - 2Dmn - 2Enl - 2Flm = \frac{Mk^4}{r^2}$$

$$\text{i.e.,} \quad A(lr)^2 + B(mr)^2 + C(nr)^2 - 2D(mr)(nr) - 2E(nr)(lr) - 2F(lr)(mr) = Mk^4,$$

$$\text{i.e.,} \quad A\bar{x}^2 + B\bar{y}^2 + C\bar{z}^2 - 2D\bar{y}\bar{z} - 2E\bar{z}\bar{x} - 2F\bar{x}\bar{y} = Mk^4. \tag{6.6.4}$$

$\therefore$ Locus of $Q(\bar{x}, \bar{y}, \bar{z})$ is

$$Ax^2 + By^2 + Cz^2 - 2Dyz - 2Ezx - 2Fxy = Mk^4, \tag{6.6.5}$$

which represents an ellipsoid.

It is seen that for every ellipsoid, there exist three mutually perpendicular diameters such that, if they are taken as the axes of coordinates, the resulting equation of the ellipsoid has no term involving yz, zx, xy. These axes of coordinates are called the principal axes of the ellipsoid.

Let the equation of the momental ellipsoid referred to as the principal axes by

$$A'x^2 + B'y^2 + C'z^2 = Mk^4. \tag{6.6.6}$$

The products of inertia with respect to these new axes are zero.

Thus, for anybody whatever there exists at each point O a set of three mutually perpendicular axes such that the products of inertia of the body about them, taken two at a time, all vanish.

These three axes are called the principal axes of the body at point O; also a plane through any two of these axes is called a principal plane of the body.

Theorem 6.6 *The total work done in rotating a rigid body from an angle θ_1 where the angular speed is ω_1 to angle θ_2 where the angular speed is ω_2 is the difference in the KE of the rotation at ω_1 and ω_2.*

$$\text{i.e.,}\quad \int_{\theta_1}^{\theta_2} \Lambda\, d\theta = \frac{1}{2} I\, \omega_2^2 - \frac{1}{2} I\, \omega_1^2. \tag{6.6.7}$$

Proof With usual notations,

$$
\begin{aligned}
dw &= \Lambda\, d\theta \\
\therefore \quad w &= \int_{\theta_1}^{\theta_2} \Lambda\, d\theta \\
&= \int_{\theta_1}^{\theta_2} I\dot{\omega}\, d\theta \\
&= I \int_{t_1}^{t_2} \dot{\omega}\, \frac{d\theta}{dt}\, dt \\
&= I \int_{t_1}^{t_2} \dot{\omega}\, \omega\, dt \\
&= I \int_{t_1}^{t_2} \omega\, \frac{d\omega}{dt}\, dt \\
&= I \int_{t_1}^{t_2} \omega\, d\omega \\
&= I\, \frac{\omega^2}{2}\Big]_{\omega_1}^{\omega_2} \\
&= \frac{1}{2} I\, (\omega_2^2 - \omega_1^2) \\
&= \frac{1}{2} I\, \omega_2^2 - \frac{1}{2} I\, \omega_1^2.
\end{aligned}
$$

□

6.7 Illustrations

Example 6.2 Show that the moment of inertia of a rectangular plate of mass M and sides $2a$, $2b$ about a diagonal is $\frac{2M}{3}\frac{a^2b^2}{a^2+b^2}$.

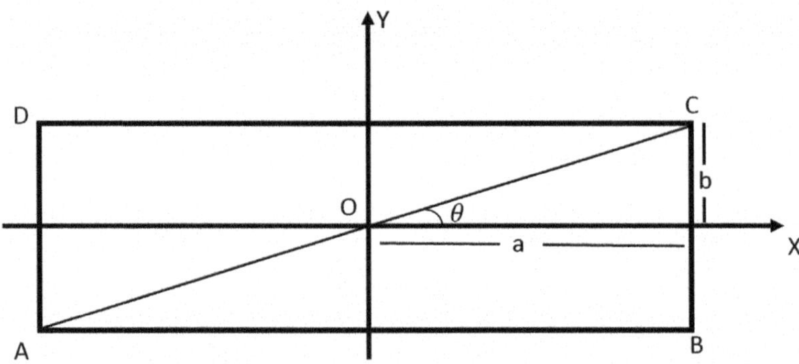

Fig. 6.29 Geometry for Example 6.2

Solution Let ABCD be the rectangular plate and suppose that $AB = DC = 2a$, $AD = BC = 2b$. Take the center O of the plate as the origin, the X-axis parallel to AB, and the Y-axis parallel to AD.

Now,

$$\begin{aligned} A &= \text{MI of the lamina about OX} \\ &= \frac{1}{3}Mb^2, \end{aligned}$$

$$\begin{aligned} B &= \text{MI of the lamina about OY} \\ &= \frac{1}{3}Ma^2, \end{aligned}$$

$$\begin{aligned} F &= \text{Product of inertia about OX-OY} \\ &= 0 \quad \text{(due to symmetry)}. \end{aligned}$$

Let $\angle XOC = \theta$.

Therefore, from the Fig. 6.29, we have

$$\therefore \quad \cos\theta = \frac{a}{\sqrt{a^2+b^2}}, \quad \text{and} \quad \sin\theta = \frac{b}{\sqrt{a^2+b^2}}.$$

Now,

$$\begin{aligned} I &= \text{MI of the plate about OC} \\ &= A\cos^2\theta + B\sin^2\theta \end{aligned}$$

$$= \frac{1}{3}Mb^2\frac{a^2}{a^2+b^2}+\frac{1}{3}Ma^2\frac{b^2}{a^2+b^2}$$
$$= \frac{2M}{3}\frac{a^2b^2}{a^2+b^2}.$$

Example 6.3 Show that the moment of inertia of an elliptic area of mass M and semi-axes a and b about a diameter of length $2r$ is $\frac{M}{4}\frac{a^2b^2}{r^2}$.

Solution Take the center O of the area as the origin, the X-axis along the major axis, and the Y-axis along the minor axis. Let P′P be a diameter of length $2r$ inclined at an angle θ to OX as depicted in Fig. 6.30.

The equation of the ellipse is

$$\frac{x^2}{a^2}+\frac{y^2}{b^2}=1. \tag{6.7.1}$$

Let $\mathrm{P} = (r\cos\theta, r\sin\theta)$.

Since P is on equation (6.7.1)

$$\therefore \quad \frac{r^2\cos^2\theta}{a^2}+\frac{r^2\sin^2\theta}{b^2}=1.$$
i.e., $b^2r^2\cos^2\theta + a^2r^2\sin^2\theta = a^2b^2$,

i.e., $r^2(b^2\cos^2\theta + a^2\sin^2\theta) = a^2b^2$,

i.e., $b^2\cos^2\theta + a^2\sin^2\theta = \dfrac{a^2b^2}{r^2}$.

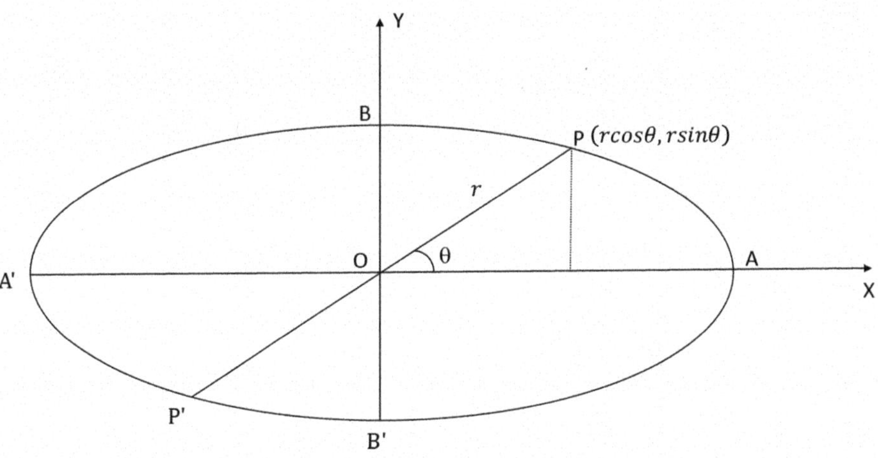

Fig. 6.30 Geometry for Example 6.3

Now,

$$\begin{aligned} A &= \text{MI of the plate about OX} \\ &= \frac{1}{4}Mb^2. \end{aligned}$$

$$\begin{aligned} B &= \text{MI of the plate about OY} \\ &= \frac{1}{4}Ma^2. \end{aligned}$$

$$\begin{aligned} F &= \text{Product of inertia about OX-OY} \\ &= 0 \quad \text{(due to symmetry)}. \end{aligned}$$

Now,

$$\begin{aligned} I &= \text{MI of the area about P}'\text{P} \\ &= \frac{1}{4}Mb^2\cos^2\theta + \frac{1}{4}Ma^2\sin^2\theta \\ &= \frac{M}{4}(b^2\cos^2\theta + a^2\sin^2\theta) \\ &= \frac{M}{4}\frac{a^2b^2}{r^2}. \end{aligned}$$

Example 6.4 Show that the moment of inertia of an ellipse of mass M and semi-axes a and b about a tangent is $\frac{5M}{4}p^2$ where p is the $\perp^r$ from the center on the tangent.

Solution Let the equation of the ellipse be

$$\frac{x^2}{a^2} + \frac{y^2}{b^2} = 1. \tag{6.7.2}$$

In Fig. 6.31, let T′QT be a tangent to the ellipse at Q. Draw a line P′P through the center O of the ellipse parallel to T′QT.

Let θ be the angle of inclination of P′P to OX.

$$\begin{aligned} \therefore \quad m &= \text{slope of the tangent} \\ &= \tan\theta. \end{aligned}$$

The equation of the tangent to the ellipse at Q is

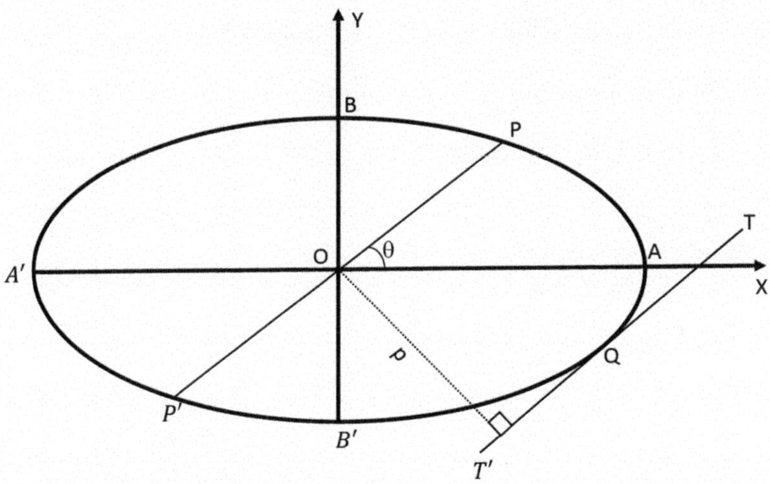

Fig. 6.31 Geometry for Example 6.4

$$y = mx + \sqrt{a^2m^2 + b^2}\,,$$

i.e., $y = x\tan\theta + \sqrt{a^2\tan^2\theta + b^2}\,,$

i.e., $x\tan\theta - y + \sqrt{a^2\tan^2\theta + b^2} = 0.$

$$\therefore \quad p = \frac{\sqrt{a^2\tan^2\theta + b^2}}{\sqrt{\tan^2\theta + 1}}$$

$$= \frac{\sqrt{a^2\tan^2\theta + b^2}}{\sec\theta}.$$

$$\therefore \quad p^2 = \frac{a^2\tan^2\theta + b^2}{\sec^2\theta}$$

$$= \frac{a^2\dfrac{\sin^2\theta}{\cos^2\theta} + b^2}{\dfrac{1}{\cos^2\theta}}$$

$$= a^2\sin^2\theta + b^2\cos^2\theta. \tag{6.7.3}$$

Now, $A =$ MI of the ellipse about OX $= \frac{1}{4}Mb^2,$

$B =$ MI of the ellipse about OY $= \frac{1}{4}Ma^2.$

$$\begin{aligned}
\therefore\quad I_0 &= \text{MI of the ellipse about P}'\text{P} \\
&= A\cos^2\theta + B\sin^2\theta \\
&= \frac{1}{4}Mb^2\cos^2\theta + \frac{1}{4}Ma^2\sin^2\theta \\
&= \frac{M}{4}(b^2\cos^2\theta + a^2\sin^2\theta) \\
&= \frac{M}{4}p^2. \qquad \text{[by (6.7.3)]}
\end{aligned}$$

By $\|$ axes theorem,

$$\begin{aligned}
I &= \text{Mi of the ellipse about T}'\text{QT} \\
&= I_0 + Md^2 \\
&= \frac{M}{4}p^2 + Mp^2 \\
&= \frac{5}{4}Mp^2.
\end{aligned}$$

Hence Proved.

Example 6.5 If k_1 and k_2 are the radii of gyration of an elliptic lamina about two conjugate diameters, then prove that

$$\frac{1}{k_1^2} + \frac{1}{k_2^2} = 4\left(\frac{1}{a^2} + \frac{1}{b^2}\right).$$

Let the equation of the ellipse be

$$\frac{x^2}{a^2} + \frac{y^2}{b^2} = 1. \tag{6.7.4}$$

In Fig. 6.32, let $P'P = 2r_1$ and $Q'Q = 2r_2$ be two conjugate diameters of the ellipse.

Therefore, P$(r_1\cos\theta, r_1\sin\theta)$ lies on Eq. (6.7.4)

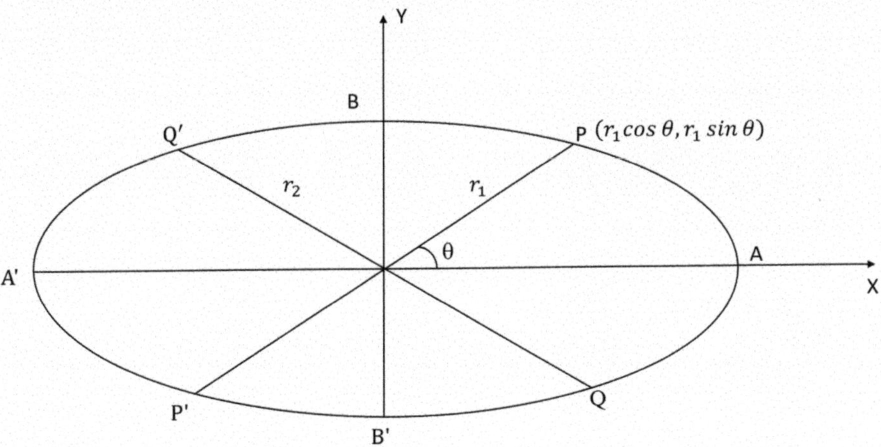

Fig. 6.32 Illustration for Example 6.5

$$\therefore\quad \frac{r_1^2\cos^2\theta}{a^2}+\frac{r_1^2\sin^\theta}{b^2}=1,$$

$$\text{i.e.,}\quad b^2r_1^2\cos^2\theta+a^2r_1^2\sin^2\theta=a^2b^2,$$

$$\text{i.e.,}\quad r_1^2(b^2\cos^2\theta+a^2\sin^2\theta)=a^2b^2,$$

$$\therefore\quad b^2\cos^2\theta+a^2\sin^2\theta=\frac{a^2b^2}{r_1^2}. \tag{6.7.5}$$

$$\begin{aligned}\text{Now,}\quad Mk_1^2&=\text{MI of the ellipse about P}'\text{P}\\&=A\cos^2\theta+B\sin^2\theta\\&=\frac{1}{4}Mb^2\cos^2\theta+\frac{1}{4}Ma^2\sin^2\theta\\&=\frac{M}{4}(b^2\cos^2\theta+a^2\sin^2\theta)\\&=\frac{M}{4}\frac{a^2b^2}{r_1^2}.\end{aligned}$$

$$\therefore\quad k_1^2=\frac{a^2b^2}{4r_1^2},$$

$$\therefore \quad \frac{1}{k_1^2} = \frac{4r_1^2}{a^2b^2}.$$

Similarly $\dfrac{1}{k_2^2} = \dfrac{4r_2^2}{a^2b^2}.$

$$\frac{1}{k_1^2} + \frac{1}{k_2^2} = \frac{4}{a^2b^2}(r_1^2 + r_2^2). \tag{6.7.6}$$

Since P′P and Q′Q are two conjugate diameters,

$$\therefore \quad r_1^2 + r_2^2 = a^2 + b^2. \tag{6.7.7}$$

Therefore, Eqs. (6.7.6) and (6.7.7) give

$$\frac{1}{k_1^2} + \frac{1}{k_2^2} = \frac{4}{a^2b^2}(a^2 + b^2),$$

i.e., $\dfrac{1}{k_1^2} + \dfrac{1}{k_2^2} = 4\left(\dfrac{1}{b^2} + \dfrac{1}{a^2}\right),$

i.e., $\dfrac{1}{k_1^2} + \dfrac{1}{k_2^2} = 4\left(\dfrac{1}{a^2} + \dfrac{1}{b^2}\right).$

Proved.

Example 6.6 Show that the moment of inertia of a uniform cube about any axis through its center is the same.

Solution Consider a cube of edge $2a$ and with center at the point O. Let OX, OY, and OZ be three mutually perpendicular axes parallel to the edges.

$$\begin{aligned}\text{Now,} \quad A &= \text{MI of the cube about OX}\\ &= \frac{M}{3}(a^2 + a^2) = \frac{2}{3}Ma^2.\end{aligned}$$

$$\begin{aligned}\text{Similarly} \quad B = C &= \frac{2}{3}Ma^2\\ D &= \text{product of inertia about OY-OZ}\\ &= 0 \quad \text{(due to symmetry)}.\end{aligned}$$

Similarly $E = F = 0$.

Let OQ be any line through O with direction cosines l, m, and n.
By the six constant theorem,

$$\begin{aligned} I &= Al^2 + Bm^2 + Cn^2 - 2Dmn - 2Enl - 2Flm \\ &= \frac{2}{3}Ma^2l^2 + \frac{2}{3}Ma^2m^2 + \frac{2}{3}Ma^2n^2 \\ &= \frac{2}{3}Ma^2(l^2 + m^2 + n^2) \\ &= \frac{2}{3}Ma^2 = \text{a constant.} \end{aligned}$$

Example 6.7 Show that the momental ellipsoid at the center of an elliptic plate is

$$\frac{x^2}{a^2} + \frac{y^2}{b^2} + z^2\left(\frac{1}{a^2} + \frac{1}{b^2}\right) = \text{constant.}$$

Solution Let the equation of the ellipse be

$$\frac{x^2}{a^2} + \frac{y^2}{b^2} = 1.$$

$$\begin{aligned} A &= \text{MI of the ellipse about OX} \\ &= \frac{1}{4}Mb^2. \\ B &= \text{MI of the ellipse about OY} \\ &= \frac{1}{4}Ma^2. \end{aligned}$$

$$\begin{aligned} C &= \text{MI of the ellipse about OZ} \\ &= A + B. \\ &= \frac{M}{4}(a^2 + b^2). \end{aligned}$$

Since the ellipse in Fig. 6.33 is symmetric about any axis.

$$\therefore \quad D = E = F = 0.$$

Therefore, equation of the momental ellipsoid at O is

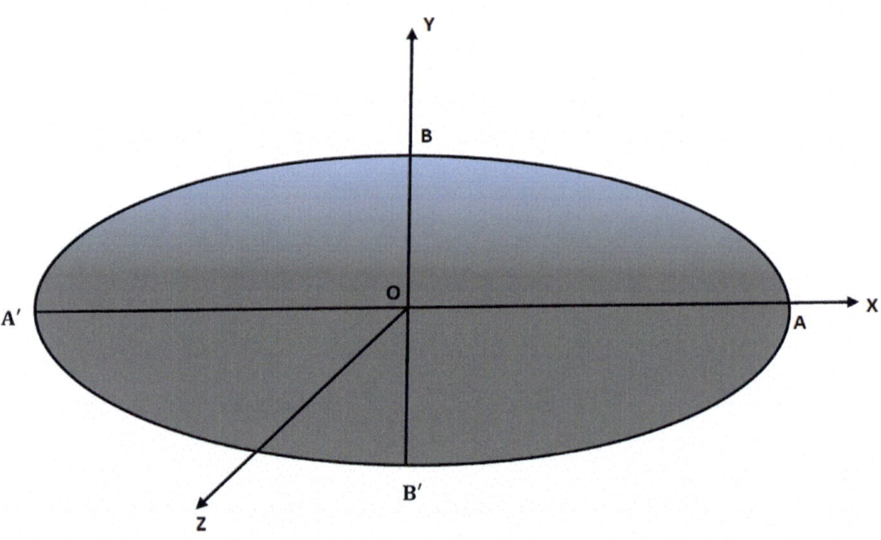

Fig. 6.33 Illustration for Example 6.7

$$\begin{aligned}
&Ax^2 + By^2 + Cz^2 = \text{constant},\\
\text{i.e.,}\quad &\frac{1}{4}Mb^2x^2 + \frac{1}{4}Ma^2y^2 + \frac{M}{4}(a^2+b^2)z^2 = \text{constant},\\
\text{i.e.,}\quad &b^2x^2 + a^2y^2 + (a^2+b^2)z^2 = \text{constant},\\
\text{i.e.,}\quad &\frac{x^2}{a^2} + \frac{y^2}{b^2} + (\frac{1}{a^2} + \frac{1}{b^2})z^2 = \text{constant}.
\end{aligned}$$

Example 6.8 Show that the equation of the momental ellipsoid at the corner of a cube of side $2a$, referred to as its principal axes, is $2x^2 + 11(y^2 + z^2) = \text{constant}$.

Solution Let OABCA′B′C′O′ be the cube with G as the center.

We have, $$OG^2 = a^2 + a^2 + a^2 = 3a^2.$$

Take O as the origin, X-axis along **OG**, Y-axis, and Z-axis as shown in the Fig. 6.34. It is clear that OX, OY, and OZ are the principal axes for the cube at O.

$$\begin{aligned}
A &= \text{MI of the cube about OX}\\
&= \frac{2}{3}Ma^2.
\end{aligned}$$

$$\begin{aligned}
B &= \text{MI of the cube about OY}\\
&= I_0 + Md^2\\
&= \frac{2}{3}Ma^2 + M\,OG^2
\end{aligned}$$

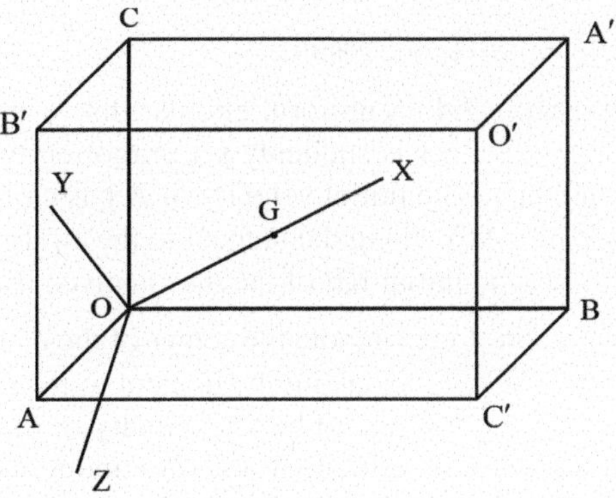

Fig. 6.34 Illustration for Example 6.8

$$= \frac{2}{3}Ma^2 + 3Ma^2$$
$$= (\frac{2}{3} + 3)\, Ma^2 = \frac{11}{3}Ma^2.$$

Similarly, $C =$ MI of the cube about OZ

$$= \frac{11}{3}Ma^2.$$

Therefore, the equation of the momental ellipsoid at O is

$$Ax^2 + By^2 + Cz^2 = \text{constant},$$

i.e., $\frac{2}{3}Ma^2x^2 + \frac{11}{3}Ma^2y^2 + \frac{11}{3}Ma^2z^2 = \text{constant},$

i.e., $2x^2 + 11y^2 + 11z^2 = \text{constant},$

i.e., $2x^2 + 11(y^2 + z^2) = \text{constant}.$

Proved.

Example 6.9 Show that the equation of the momental ellipsoid at the center of a solid ellipsoid is $(b^2 + c^2)x^2 + (c^2 + a^2)y^2 + (a^2 + b^2)z^2 = \text{constant}$.

Solution In Fig. 6.35, let O be the center of the ellipsoid with semi-axes $OA = a$, $OB = b$, $OC = c$, and M be the mass of the ellipsoid. Take O as the origin, X-axis along **OA**, Y-axis along **OB**, and Z-axis along **OC**.

Now, $A =$ MI of the body about OX

$$= \frac{M}{5}(b^2 + c^2),$$

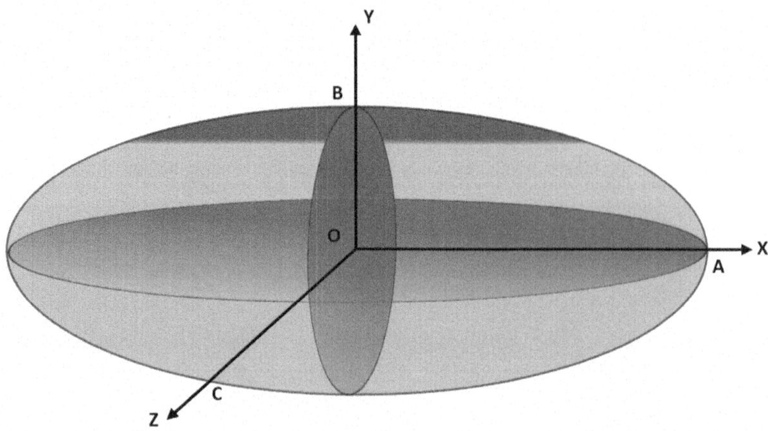

Fig. 6.35 A solid ellipsoid with center at O

$$\begin{aligned} B &= \text{MI of the body about OY} \\ &= \frac{M}{5}(c^2 + a^2), \end{aligned}$$

$$\begin{aligned} C &= \text{MI of the ellipsoid about OZ} \\ &= \frac{M}{5}(a^2 + b^2). \end{aligned}$$

Therefore, the equation of the momental ellipsoid at O is

$$Ax^2 + By^2 + Cz^2 = \text{constant},$$

$$\text{i.e.,}\quad \frac{M}{5}(b^2 + c^2)x^2 + \frac{M}{5}(c^2 + a^2)y^2 + \frac{M}{5}(a^2 + b^2)z^2 = \text{constant},$$

$$\text{i.e.,}\quad (b^2 + c^2)x^2 + (c^2 + a^2)y^2 + (a^2 + b^2)z^2 = \text{constant}.$$

Example 6.10 Show that the moment of inertia of a right solid cone, whose height is h and the radius of whose base is a, is $\frac{3Ma^2}{20}\frac{6h^2+a^2}{h^2+a^2}$ about a slanted side, and $\frac{3M}{80}(h^2 + 4a^2)$ about a line through the center of gravity of the cone $\perp^r$ to its axis.

Solution Let OAB be the solid circular cone with O as the vertex and h as the height. Let C be the center of the base and α be the semi-vertical angle of the cone.

Consider an elementary disc PQR of thickness δx and R as the center. Let OR $= x$. Let G be the center of gravity of the cone.

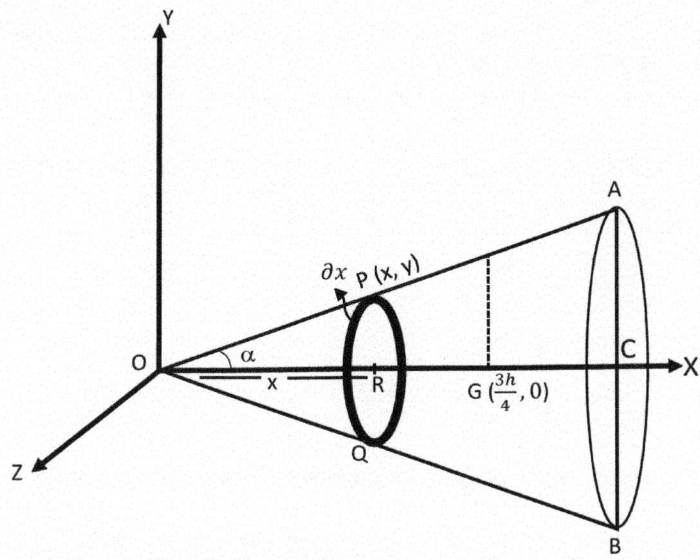

Fig. 6.36 A solid cone with vertex at O and height h

$$\therefore \quad \mathrm{OG} = \frac{3}{4}h.$$

In Fig. 6.36, take the X-axis along **OC**, the Y-axis parallel to the diameter BA of the base.

$$\begin{aligned}\text{Mass of the disc} &= \pi y^2 \delta x\, m \\ &= \pi\, x^2 \tan^2 \alpha\, \delta x\, m \\ &= \pi m \tan^2 \alpha\, x^2\, \delta x.\end{aligned}$$

$$\begin{aligned}\text{MI of the disc about OX} &= \frac{1}{2}(\pi m \tan^2 \alpha\, x^2\, \delta x)\, y^2 \\ &= \frac{1}{2}\,\pi m \tan^2 \alpha\, x^2\, \delta x\, x^2 \tan^2 \alpha \\ &= \frac{1}{2}\pi m \tan^4 \alpha\, x^4\, \delta x.\end{aligned}$$

$$
\begin{aligned}
A = I_x &= \int\limits_{x=0}^{h} \frac{1}{2}\pi m \tan^4\alpha\, x^4\, dx \\
&= \frac{1}{2}\pi m \tan^4\alpha\, \frac{h^5}{5} \\
&= \frac{1}{10}\pi m \tan^4\alpha\, h^5 \\
&= \frac{1}{10}\pi\, h^5\, \frac{a^4}{h^4}\, \frac{3.M}{\pi a^2 h} \\
&= \frac{3}{10} M a^2.
\end{aligned}
$$

$$
\begin{aligned}
\text{MI of the disc about OY} &= \frac{1}{4}\pi m \tan^4\alpha\, x^4\, \delta x + \pi m \tan^2\alpha\, x^2\, \delta x\, x^2 \\
&= \frac{1}{4}\pi m \tan^4\alpha\, x^4\, \delta x + \pi m \tan^2\alpha\, x^4\, \delta x.
\end{aligned}
$$

$$
\begin{aligned}
\therefore\quad B &= \frac{1}{4}\pi m \tan^4\alpha \frac{h^5}{5} + \pi m \tan^2\alpha \frac{h^5}{5} \\
&= \frac{3}{20} M a^2 + \pi m \tan^2\alpha\, \frac{h^5}{5} \\
&= \frac{3}{20} M a^2 + \pi \frac{h^5}{5}\, \frac{a^2}{h^2}\, \frac{3M}{\pi a^2 h} \\
&= \frac{3}{20} M a^2 + \frac{3M}{5} h^2.
\end{aligned}
$$

Further,

$$C = B.$$

$$l = \cos\alpha, \quad m = \sin\alpha, \quad n = 0, \quad D = E = F = 0.$$

$$
\begin{aligned}
I &= Al^2 + Bm^2 \\
&= \frac{3}{10} M a^2 \cos^2\alpha + \left(\frac{3}{20} M a^2 + \frac{3M}{5} h^2\right) \sin^2\alpha \\
&= \frac{3}{10} M a^2 \frac{h^2}{a^2 + h^2} + \left(\frac{3}{20} M a^2 + \frac{3M}{5} h^2\right) \frac{a^2}{a^2 + h^2} \\
&= \frac{M a^2}{(a^2 + h^2)} \left[\frac{3h^2}{10} + \left(\frac{3}{20} a^2 + \frac{3h^2}{5}\right)\right]
\end{aligned}
$$

$$= \frac{3Ma^2}{(a^2+h^2)}\left[\frac{h^2}{10}+\frac{a^2}{20}+\frac{h^2}{5}\right]$$

$$= \frac{3Ma^2}{(a^2+h^2)}\frac{2h^2+a^2+4h^2}{20}$$

$$= \frac{3Ma^2}{20}\frac{6h^2+a^2}{a^2+h^2}.$$

2nd part: By parallel axes theorem,

$$B = I_0 + Md^2 = I_0 + M\frac{9h^2}{16}.$$

$$\therefore \quad I_0 = B - \frac{9Mh^2}{16}$$

$$= \frac{3}{20}Ma^2 + \frac{3M}{5}h^2 - \frac{9Mh^2}{16}$$

$$= \frac{3}{20}Ma^2 + \frac{Mh^2}{80}(48-45)$$

$$= \frac{3}{20}Ma^2 + \frac{3Mh^2}{80}$$

$$= \frac{3M}{80}(h^2+4a^2).$$

Example 6.11 Consider a rigid body R capable of rotating in a plane about an axis through O perpendicular to the plane. If Λ is the magnitude of the torque applied to the body under the influence of a force **F** at a point A, show the work done in a rotating the body through an angle $d\theta$ is given by $dW = \Lambda d\theta$ and the instantaneous power developed is $P = \frac{dW}{dt} = \Lambda\,\omega$, where ω is the angular speed.

Solution Consider a rigid body as seen in Fig. 6.37, which rotates with an angular velocity ω about an axis that passes through O and is perpendicular to the plane of the motion. Let **F** be the force acting on any arbitrary point of a rigid body.

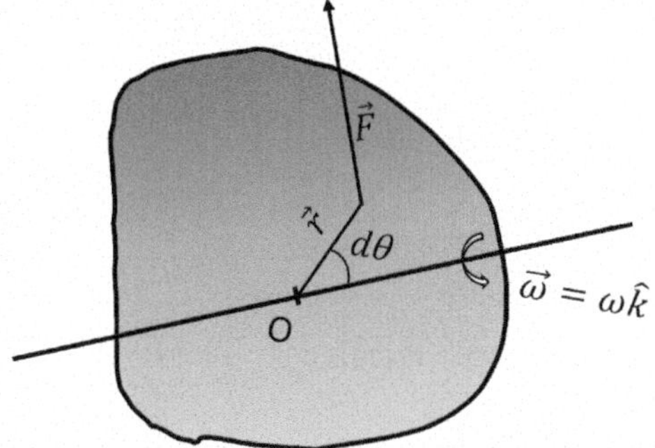

Fig. 6.37 Illustration for Example 6.11

Therefore, the work done by the force $\mathbf{F}$ is given by

$$\begin{aligned}
dW &= \mathbf{F} \cdot d\mathbf{r} \\
&= \mathbf{F} \cdot \frac{d\mathbf{r}}{dt}\, dt \\
&= \mathbf{F} \cdot \mathbf{v}\, dt \\
&= \mathbf{F} \cdot \boldsymbol{\omega} \times \mathbf{r}\, dt \\
&= \boldsymbol{\omega} \times \mathbf{r} \cdot \mathbf{F}\, dt \\
&= \boldsymbol{\omega} \cdot \mathbf{r} \times \mathbf{F}\, dt \\
&= \boldsymbol{\omega} \cdot \boldsymbol{\Lambda}\, dt \\
&= \omega\hat{k} \cdot \Lambda\hat{k}\, dt \\
&= \omega\, \Lambda\, dt \\
&= \Lambda\, \frac{d\theta}{dt}\, dt = \Lambda\, d\theta.
\end{aligned}$$

Hence proved.
Instantaneous power applied is given as

$$P = \frac{dW}{dt} = \Lambda\, \frac{d\theta}{dt} = \Lambda\, \omega.$$

Example 6.12 If the moment of inertia of a rigid body about the axis through O having direction cosines $[\lambda, \mu, \nu]$ in terms of these direction cosines and A, B, C, D, E, F.

Solution Let OQ be the axis about which the moment of inertia of the body is to be obtained. Let $\hat{a}$ be the unit vector along OQ (see Fig. 6.38).

$$\therefore \quad \hat{a} = \lambda\hat{i} + \mu\hat{j} + \nu\hat{k}.$$

Let P (x, y, z) be a single particle of mass m of the body with position vector $\mathbf{r}$ relative to O.

$$\therefore \quad \mathbf{r} = x\hat{i} + y\hat{j} + z\hat{k}.$$

The moment of inertia of the body about OQ is given by

$$I = \sum m \left(\mathbf{r} \times \hat{a}\right)^2. \tag{6.7.8}$$

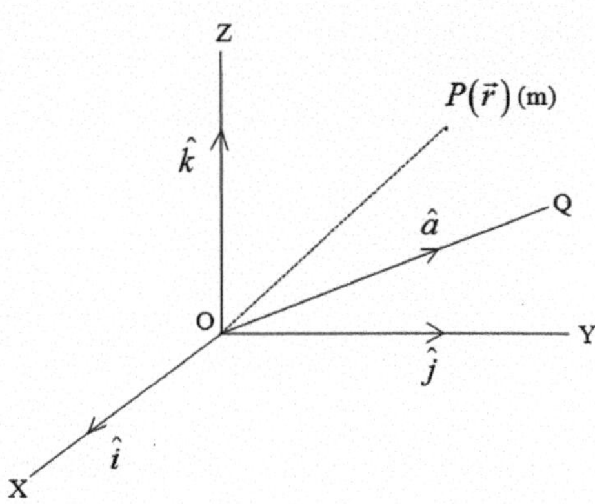

Fig. 6.38 Geometry for Example 6.12

Now

$$\mathbf{r} \times \hat{a} = \begin{vmatrix} \hat{i} & \hat{j} & \hat{k} \\ x & y & z \\ \lambda & \mu & \nu \end{vmatrix} = \hat{i}\,(y\nu - z\mu) + \hat{j}\,(z\lambda - x\nu) + \hat{k}\,(x\mu - y\lambda)\,,$$

$$\therefore \quad (\mathbf{r} \times \hat{a})^2 = (y\nu - z\mu)^2 + (z\lambda - x\nu)^2 + (x\mu - y\lambda)^2,$$

$$\text{i.e., } (\mathbf{r} \times \hat{a})^2 = \lambda^2\left(y^2 + z^2\right) + \mu^2\left(z^2 + x^2\right) + \nu^2\left(x^2 + y^2\right) - 2yz\mu\nu - 2zx\nu\lambda - 2xy\lambda\mu.$$

Equation (6.7.8) gives

$$I = \sum m\left[\lambda^2\left(y^2 + z^2\right) + \mu^2\left(z^2 + x^2\right) + \nu^2\left(x^2 + y^2\right) - 2\mu\nu yz - 2\nu\lambda zx - 2\lambda\mu xy\right]$$

$$\text{i.e., } I = A\lambda^2 + B\mu^2 + C\nu^2 - 2\mu\nu D - 2\nu\lambda E - 2\lambda\mu F.$$

Example 6.13 A uniform rigid rod AB moves so that A and B have velocities $\mathbf{u}_A$, $\mathbf{u}_B$ at any instant. Show that the KE is then $T = \frac{M}{6}\left[\mathbf{u}_A^2 + \mathbf{u}_A \cdot \mathbf{u}_B + \mathbf{u}_B^2\right]$, M being the mass of the rod.

Solution Let G be the center of mass of the rod. Let GQ be the instantaneous axis of rotation of the rod through G.

Referring to Fig. 6.39, take the X-axis along GB, Y-axis through G in such a way that GQ lies in the XY plane. Let XY plane be the plane of the paper. Take the Z-axis through G, which is $\perp^r$ to the plane of the paper. Let θ be the angle of inclination of GQ to GX.

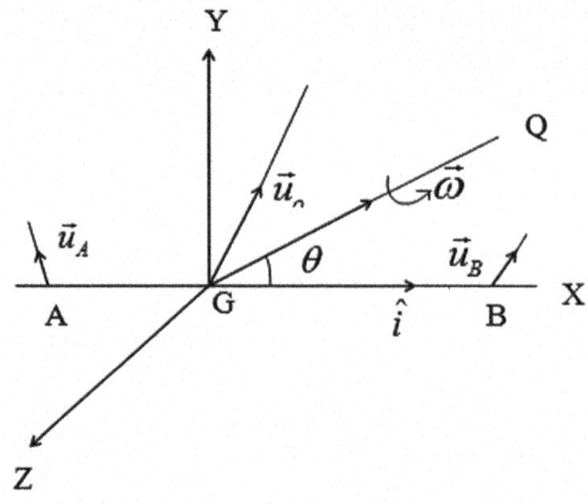

Fig. 6.39 Illustration for Example 6.13

If $(\lambda,\ \mu,\ \nu)$ are the d.c.s of OQ, then

$$\lambda = \cos\theta,\ \mu = \cos\left(\frac{\pi}{2} - \theta\right) = \sin\theta,\ \nu = \cos 90^\circ = 0.$$

A = M.I. of the rod about GX = 0,
B = M.I. of rod about GY = $\frac{1}{3}Ma^2$; where $2a$ is the length of the rod AB,
C = M.I. of the rod about GZ = $\frac{1}{3}Ma^2$.

$$\left.\begin{array}{l} D = \sum myz = 0 \\ E = \sum mzx = 0 \\ F = \sum mxy = 0 \end{array}\right\}. \tag{6.7.9}$$

Now the MI of the rod about GQ is given by

$$I_0 = A\lambda^2 + B\mu^2 + C\nu^2 - 2D\mu\nu - 2E\nu\lambda - 2F\lambda\mu = \frac{1}{3}Ma^2\sin^2\theta. \tag{6.7.10}$$

Let ω be the angular velocity of the rod about GQ. Let $\mathbf{u}_0$ be the velocity of the C.I. We have

$$\mathbf{u}_B = \mathbf{u}_0 + \omega \times \mathbf{GB} = \mathbf{u}_0 + \omega \times a\hat{i} = \mathbf{u}_0 + a\omega \times \hat{i}, \tag{6.7.11}$$

$$\mathbf{u}_A = \mathbf{u}_0 + \omega \times \mathbf{GA} = \mathbf{u}_0 + \omega \times a\left(-\hat{i}\right) = \mathbf{u}_0 - a\omega \times \hat{i}. \tag{6.7.12}$$

(6.7.11)+(6.7.12) gives,

$$\mathbf{u}_A + \mathbf{u}_B = 2\mathbf{u}_0,$$

$$\text{i.e., } \mathbf{u}_0 = \frac{\mathbf{u}_A + \mathbf{u}_B}{2}. \tag{6.7.13}$$

(6.7.11) – (6.7.12) yields,

$$\mathbf{u}_B - \mathbf{u}_A = 2a\boldsymbol{\omega} \times \hat{i} = -2a\omega \sin\theta\, \hat{k},$$

$$\therefore \quad (\mathbf{u}_B - \mathbf{u}_A)^2 = 4a^2\omega^2 \sin^2\theta,$$

$$\text{i.e., } a^2\omega^2 \sin^2\theta = \frac{(\mathbf{u}_B - \mathbf{u}_A)^2}{4}. \tag{6.7.14}$$

Now, $T = KE$ of the rod at time $t = \frac{1}{2}M\mathbf{V}_0^2 + \frac{1}{2}I_0\boldsymbol{\omega}^2$,

$$\text{i.e., } T = \frac{1}{2}M\mathbf{u}_0^2 + \frac{1}{6}Ma^2 \sin^2\theta\, \omega^2,$$

$$\text{i.e., } T = \frac{1}{2}M\left(\frac{\mathbf{u}_A + \mathbf{u}_B}{2}\right)^2 + \frac{1}{6}M\omega^2 a^2 \sin^2\theta,$$

$$\text{i.e., } T = \frac{1}{2}M\frac{(\mathbf{u}_A + \mathbf{u}_B)^2}{4} + \frac{1}{6}M\frac{(\mathbf{u}_B - \mathbf{u}_A)^2}{4},$$

$$\text{i.e., } T = \frac{M}{8}(\mathbf{u}_A + \mathbf{u}_B)^2 + \frac{M}{24}(\mathbf{u}_B - \mathbf{u}_A)^2,$$

$$\text{i.e., } T = \frac{M}{24}\left[3(\mathbf{u}_A + \mathbf{u}_B)^2 + (\mathbf{u}_B - \mathbf{u}_A)^2\right],$$

$$\text{i.e., } T = \frac{M}{24}\left[4\mathbf{u}^2{}_A + 4\mathbf{u}^2{}_B + 4\mathbf{u}_A \cdot \mathbf{u}_B\right],$$

$$\text{i.e., } T = \frac{M}{6}\left[\mathbf{u}^2{}_A + \mathbf{u}_A \cdot \mathbf{u}_B + \mathbf{u}^2{}_B\right].$$

Example 6.14 Obtain the equation of motion of a compound pendulum.

Solution The motion of a compound pendulum is a particular case of the motion of a rigid body about a fixed axis.

Let O be the point of suspension, and G be the center of mass of the body with mass M. Let OA be the downward vertical line through O. Let the vertical plane of the motion of the body through G be chosen as the XY plane, where OZ is the horizontal axis of rotation. Let $\angle AOG = \theta$, therefore θ is the angle between the plane ZOA, fixed in space and the plane ZOG, fixed in the body at time t.

The instantaneous angular velocity of the body about OZ is

$$\boldsymbol{\omega} = \omega\hat{k} = \dot{\theta}\hat{k}, \tag{6.7.15}$$

$\hat{k}$ being the unit vector along **OZ**.

The external forces acting on the body are

- the weight $M\mathbf{g}$ through G,
- the reaction of the axis of rotation OZ.

Let $G = (\bar{x}, \bar{y}, 0)$.

Therefore, from Fig. 6.40, we have

$$\bar{x} = h\sin\theta, \ \ \bar{y} = -h\cos\theta.$$

Thus,

$$\boldsymbol{\Lambda} = \mathbf{r} \times \mathbf{F} = \left(\bar{x}\hat{i} + \bar{y}\hat{j}\right) \times Mg\left(-\hat{j}\right) = Mg\left(-\bar{x}\hat{k}\right) = -Mg\bar{x}\hat{k} = -Mgh\sin\theta\hat{k}$$

The equation of motion of the compound pendulum is

$$I\dot{\boldsymbol{\omega}} = \boldsymbol{\Lambda},$$

$$\text{i.e., } I\ddot{\theta}\hat{k} = -Mgh\sin\theta\hat{k},$$

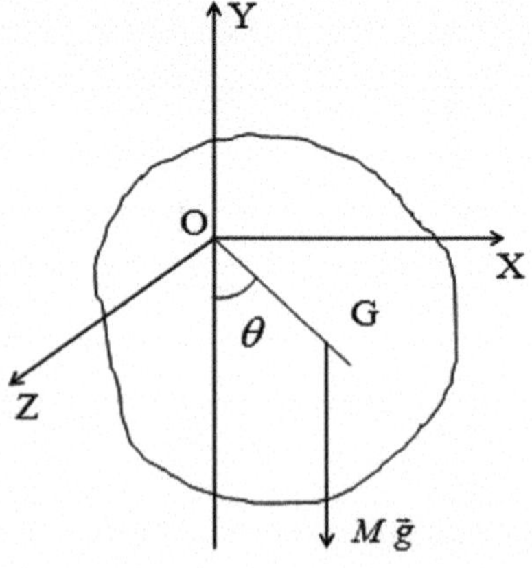

Fig. 6.40 Schematic diagram for the motion of compound pendulum

$$\text{i.e.,}\quad I\ddot{\theta} = -Mgh\sin\theta, \tag{6.7.16}$$

where I is the M.I. of the body about OZ. If K is the radius of gyration of the body about OZ, then

$$I = MK^2.$$

Then Eq. (6.7.16) reduces to

$$MK^2\ddot{\theta} = -Mgh\sin\theta,$$

$$\text{i.e.,}\quad K^2\ddot{\theta} = -gh\sin\theta,$$

$$\text{i.e.,}\qquad \ddot{\theta} = -\frac{gh}{K^2}\sin\theta,$$

which is the required equation of motion of a compound pendulum.

Example 6.15 The angular momentum of a rigid body about a point O at time t is given by $\Omega = 2e^t\hat{i} - (3t+1)\,\hat{j} + (7\sin t - t)\,\hat{k}$. Find the torque $\mathbf{\Lambda}_0$ at time $t = 0$. Also obtain $|\mathbf{\Lambda}_0|$.

Solution By the principle of angular momentum,

$$\frac{d\mathbf{\Omega}}{dt} = \mathbf{\Lambda}. \tag{6.7.17}$$

Given that

$$\mathbf{\Omega} = 2e^t\hat{i} - (3t+1)\,\hat{j} + (7\sin t - t)\,\hat{k}.$$

$$\therefore\quad \frac{d\mathbf{\Omega}}{dt} = 2e^t\hat{i} - 3\hat{j} + (7\cos t - 1)\,\hat{k}.$$

Hence (6.7.17) gives,

$$\mathbf{\Lambda}_0 = \mathbf{\Lambda}]_{t=0} = \left.\frac{d\mathbf{\Omega}}{dt}\right]_{t=0} = 2\hat{i} - 3\hat{j} + 6\hat{k}.$$

$$\therefore\quad |\mathbf{\Lambda}_0| = \sqrt{4+9+36} = 7 \text{ units}.$$

Example 6.16 A uniform rod of length 4 m and mass 6 kg rotates with angular speed 10 radians per second about an axis $\perp^r$ to it and passes through the center. Find (a) the KE of rotation, (b) Angular momentum.

Solution Here $2a$ = length of the rod =4m, $\omega = 10/\text{s}$.
$\therefore \quad a = 2m.$

Now

$$I = \text{MI of the rod about the axis of rotation}=\frac{1}{3}Ma^2 = 8\text{kgm}^2.$$

Thus (a) the KE of rotation is given by

$$T = \frac{1}{2}I\omega^2 = \frac{1}{2}8\,\text{kgm}^2 100/\text{s}^2 = 400\,\text{Nm} = 400\,\text{J}.$$

(b) Magnitude of the angular momentum $\mathbf{\Omega}$ is given by

$$\Omega = |\mathbf{\Omega}| = I\omega = 8\,\text{kgm}^2 100/\text{s} = 800\,\text{kg}.$$

Example 6.17 An elliptic disc of mass M and axes $2a$, $2b\,(a > b)$ rotates with angular velocity ω about a diagonal of the disc inclined at an angle $\frac{\pi}{4}$ to the major axis. Find the KE of rotation.

Solution Consider an elliptic disc as described in Fig. 6.41 with center at O, and A′A the major axis, B′B the minor axis, and PQ be the axis of rotation inclined at an angle $\frac{\pi}{4}$ to OA.

Now,

$$\text{A=MI of the disc about X}'\text{X} = \frac{1}{4}Mb^2,$$

$$\text{B=MI of the disc about Y}'\text{Y} = \frac{1}{4}Ma^2,$$

$$\text{F=PI of the disc w.r.t OX-OY=0} \quad \text{(due to symmetry)}.$$

By the six constant theorem,

$$I = \text{MI of the disc about PQ} = \text{A}\cos^2\theta + \text{B}\sin^2\theta - 2\text{F}\sin\theta\cos\theta,$$

$$\text{i.e.,} \quad I = \frac{1}{4}Mb^2\cos^2\frac{\pi}{4} + \frac{1}{4}Ma^2\sin^2\frac{\pi}{4} = \frac{1}{8}M(a^2 + b^2).$$

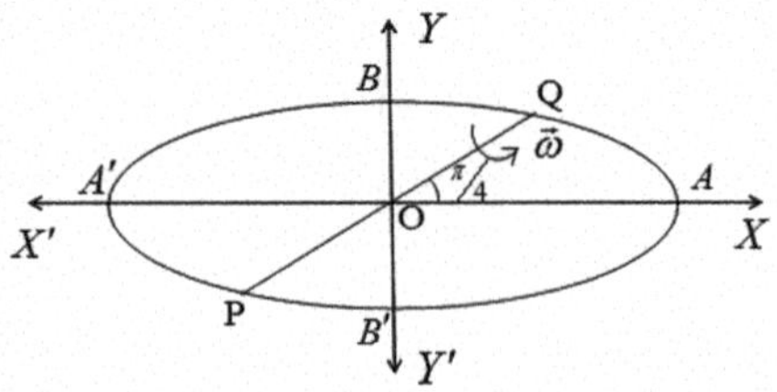

Fig. 6.41 Illustrative diagram for Example 6.17

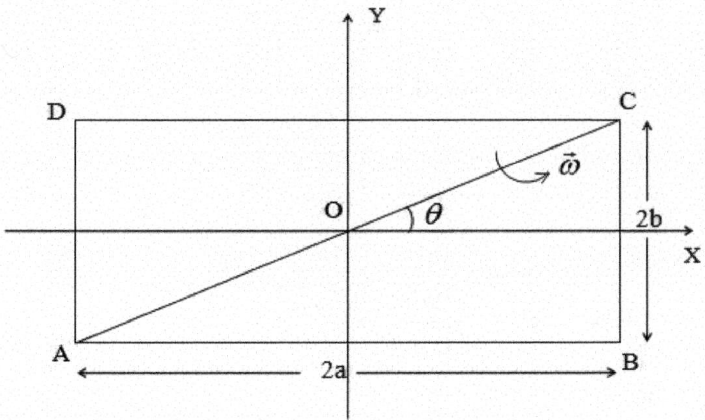

Fig. 6.42 Schematic diagram for the rotation of rectangular plate with an angular velocity ω

Hence the KE of rotation of the disc is given by

$$T = \frac{1}{2} I \omega^2 = \frac{1}{16} M \left(a^2 + b^2\right) \omega^2.$$

Example 6.18 A rectangular plate of mass M and sides $2a$, $2b$ ($a > b$) rotates about a diagonal with angular velocity ω. Find the KE of rotation of the plate.

Solution Let ABCD be the rectangular plate of mass M and sides AB = DC = $2a$ and BC = AD = $2b$.

Let the diagonal AC be the axis of rotation, which is inclined at an angle θ (say) to AB. Let O be the center of the plate. Take the X-axis and Y-axis through O $\parallel$ to AB & AD, respectively. Now from Fig. 6.42, we have

$$\cos\theta = \frac{a}{\sqrt{a^2 + b^2}}, \quad \sin\theta = \frac{b}{\sqrt{a^2 + b^2}}.$$

$$\text{A} = \text{MI of the plate about OX} = \frac{1}{3} M b^2,$$

$$\text{B} = \text{MI of the plate about OY} = \frac{1}{3} M a^2,$$

$$\text{F} = \text{PI of the plate w.r.t OX - OY} = 0 \quad \text{(due to symmetry)}.$$

Now

$$I = \text{MI of the plate about AC} = \text{A}\cos^2\theta + \text{B}\sin^2\theta - 2\text{F}\sin\theta\cos\theta,$$
$$= \frac{1}{3}Mb^2\cos^2\theta + \frac{1}{3}Ma^2\sin^2\theta = \frac{2M}{3}\frac{a^2b^2}{a^2+b^2},$$

$$\therefore \quad T = \text{KE of the rotation}\frac{1}{2}I\omega^2 = \frac{1}{3}\frac{Ma^2b^2\omega^2}{a^2+b^2}.$$

6.8 Exercise—VI

1. Show that in case of a general motion of a rigid body, the angular momentum $\boldsymbol{\Omega}$ and the external torque $\boldsymbol{\Lambda}$ are connected by the relation $\frac{d\boldsymbol{\Omega}}{dt} = \boldsymbol{\Lambda}$.
2. With usual rotation, establish the relation $\frac{d\boldsymbol{\Omega}}{dt} = \boldsymbol{\Lambda}$ in case of the general motion of a rigid body.
3. A rigid body moves in space. Show that its KE at any instant is given by

$$T = \frac{1}{2}M\mathbf{V}_0^2 + \frac{1}{2}I_0\omega^2,$$

where the symbols have their usual meanings.
4. The angular momentum of a rigid body about the origin O at time t is given by

$$\boldsymbol{\Omega} = 2t^2\hat{i} + (3t+2)\,\hat{j} + \left(t^3 - 4t^2\right)\hat{k}.$$

Find the torque at time $t = 0$.
Answer: $\boldsymbol{\Lambda}]_{t=1} = 4\hat{i} + 3\hat{j} - 5\hat{k},\ |\boldsymbol{\Lambda}|_{t=1} = 5\sqrt{2}$
5. A uniform rod of length 0.4 m and mass 6 kg rotates with an angular speed of 10 radians/s about an axis $\perp^r$ to it and passes through the center. Find (a) the KE and (b) the angular momentum of the rod.
Answer: $T = 4J$, $\Omega = 0.8\,\text{kgm}^2/\text{s}$
6. An elliptic disc of mass M and axes $2a$, $2b$ $(a > b)$ rotates with angular velocity $\boldsymbol{\omega}$ about a diagonal of the disc inclined at an angle $\frac{\pi}{4}$ to the major axis. Find the angular momentum of the disc about O.
Answer: $\Omega = \frac{1}{8}M\left(a^2+b^2\right)\omega$
7. An elliptic disc of mass M and axes $2a$, $2b$ $(a > b)$ rotates with angular velocity $\boldsymbol{\omega}$ about a diagonal of the disc inclined at an angle $\frac{\pi}{4}$ to the major axis. Find the angular momentum of the plate about the center.
Answer: $\Omega = \frac{2Ma^2b^2\omega}{3(a^2+b^2)}$

Chapter 7
Motion of a Rigid Body Under Impulsive Forces

7.1 General Equations of Motion of a Rigid Body Under Impulsive Forces

With usual rotations, equations of motion of a rigid body under finite forces are

$$M\,\ddot{\mathbf{r}}_c = \sum \mathbf{F}, \tag{7.1.1}$$

$$\frac{d\mathbf{\Omega}_c}{dt} = \sum \mathbf{r}' \times \mathbf{F}. \tag{7.1.2}$$

Let τ be the time during which the impulsive forces act on the body. Integrating (7.1.1) with respect to t from $t = 0$ to $t = \tau$, we obtain

$$\int_0^\tau M\ddot{\mathbf{r}}_c dt = \int_0^\tau \left(\sum \mathbf{F}\right) dt,$$

$$\text{i.e.,}\quad M\dot{\mathbf{r}}_c\Big]_0^\tau = \sum \int_0^\tau \mathbf{F} dt,$$

$$\text{i.e.,}\quad M\left(\mathbf{q}' - \mathbf{q}\right) = \sum \mathbf{I}, \tag{7.1.3}$$

where $\mathbf{q} = \dot{\mathbf{r}}_c]_{t=0}$ = velocity of the C.I. just before impulse,
$\mathbf{q}' = \dot{\mathbf{r}}_c]_{t=\tau}$ =velocity of the C.I. just after impulse,

N. Ahmed et al., *Classical Dynamics*, University Texts in the Mathematical Sciences,
https://doi.org/10.1007/978-981-95-6394-4_7

$\mathbf{I} = \int_0^{\tau} \mathbf{F} dt$ = impulse applied to a typical particle of the body.

Again by integrating Eq. (7.1.2) with respect to t from $t = 0$ to $t = \tau$, we obtain

$$\int_0^{\tau} \frac{d\boldsymbol{\Omega}_c}{dt} dt = \int_0^{\tau} \left(\sum \mathbf{r}' \times \mathbf{F} \right) dt,$$

$$\text{i.e.,} \quad \boldsymbol{\Omega}_c \big]_0^{\tau} = \sum \int_0^{T} \mathbf{r}' \times \mathbf{F} dt,$$

$$\text{i.e.,} \quad (\boldsymbol{\Omega}_c)_{\tau} - (\boldsymbol{\Omega}_c)_0 = \sum \mathbf{r}' \times \int_0^{T} \mathbf{F} dt = \sum \mathbf{r}' \times \mathbf{I}. \tag{7.1.4}$$

(7.1.3) and (7.1.4) are the equations of motion of a rigid body under impulsive forces. (7.1.3) represents the motion of the C.I., and (7.1.4) gives the motion about the C.I. under impulsive forces.

7.2 Equations of Motion of a Rigid Body Under Impulsive Forces in Two Dimensions

Equations of motion of a rigid body under impulsive forces are

$$M \left(\mathbf{q}' - \mathbf{q} \right) = \sum \mathbf{I}, \tag{7.2.1}$$

$$(\boldsymbol{\Omega}_c)_{\tau} - (\boldsymbol{\Omega}_c)_0 = \sum \mathbf{r}' \times \mathbf{I}, \tag{7.2.2}$$

where

$$\boldsymbol{\Omega}_c = \sum \mathbf{r}' \times m \dot{\mathbf{r}}'. \tag{7.2.3}$$

Let XY be the plane of motion, and suppose that

$$\left. \begin{aligned} \mathbf{q} &= \hat{i} u + \hat{j} v \\ \mathbf{q}' &= \hat{i} u' + \hat{j} v' \\ \mathbf{I} &= \hat{i} X + \hat{j} Y \end{aligned} \right\}, \tag{7.2.4}$$

where $\hat{i},\ \hat{j},\ \hat{k}$ stipulate the unit vectors along **OX, OY, OZ**, respectively.

(7.2.1) and (7.2.4) yield

$$\left.\begin{array}{l} M\left(u'-u\right)=\sum X \\ M\left(v'-v\right)=\sum Y \end{array}\right\}. \tag{7.2.5}$$

Let $\hat{r}$ be the unit vector along $\mathbf{r}'$ and so

$$\mathbf{r}'=r\hat{r}, \tag{7.2.6}$$

where $r=\left|\mathbf{r}'\right|$.

Now

$$\dot{\mathbf{r}}'=\boldsymbol{\omega}\times\mathbf{r}', \tag{7.2.7}$$

$\boldsymbol{\omega}$ being the angular velocity about the C.I. G (say).

Equations (7.2.6) and (7.2.7) give

$$\dot{\mathbf{r}}'=\boldsymbol{\omega}\times r\hat{r}=\omega\hat{z}\times r\hat{r}=\omega r\left(\hat{z}\times\hat{r}\right)=\omega r\hat{\theta}, \tag{7.2.8}$$

$\hat{\theta}$ is the unit vector along the azimuthal direction.

Equation (7.2.3) gives

$$\boldsymbol{\Omega}_c=\sum r\hat{r}\times m\omega r\hat{\theta}=\sum m\omega r^2\hat{z}=\omega\left(\sum mr^2\right)\hat{z}=MK^2\omega\hat{z}, \tag{7.2.9}$$

K being the radius of gyration of the body about the instantaneous axis of rotation through the C.I. ∥ to Z-axis.

By virtue of Eq. (7.2.9), (7.2.2) reduces to

$$\left(MK^2\omega\hat{z}\right)_\tau-\left(MK^2\omega\hat{z}\right)_0=\sum\mathbf{r}'\times\mathbf{I}=L\hat{z}$$

$$MK^2\left(\omega'-\omega\right)=L, \tag{7.2.10}$$

where $\omega'=\omega]_{t=\tau}=$ angular velocity just after impulse,

$\omega=\omega]_{t=0}=$ angular velocity just before impulse,

$L=$ total external torque of the net impulse about C.I.

(7.2.5) and (7.2.10) are the equations of motion of a rigid body under impulsive forces in two dimensions.

7.3 Applications of the Principle of Virtual Works

Let m_α denote the mass of the α-th particle of a system, with position vector $\mathbf{r}_\alpha$ relative to some origin O. Let a number of impulses be applied to the system simultaneously. Let $\mathbf{I}_\alpha$ denote the impulse applied to the particle m_α and $\mathbf{q}_\alpha$, $\mathbf{q}'_\alpha$ be the velocities of this particle just before and just after impulse. Let $\mathbf{k}_{\alpha\beta}$ be the impulse of the internal action of the β-th particle on the α-th particle.

Now, change in momentum = impulse gives

$$m_\alpha\left(\mathbf{q}'_\alpha - \mathbf{q}_\alpha\right) = \mathbf{I}_\alpha + \sum_\beta \mathbf{k}_{\alpha\beta},$$

$$\text{i.e.,}\quad \sum m_\alpha\left(\mathbf{q}_{\alpha'} - \mathbf{q}_\alpha\right) = \sum \mathbf{I}_\alpha + \sum_\alpha\sum_\beta \mathbf{k}_{\alpha\beta} = \sum \mathbf{I}_\alpha,$$

$$\text{i.e.,}\quad \sum\left[-m_\alpha\left(\mathbf{q}_{\alpha'} - \mathbf{q}_\alpha\right) + \mathbf{I}_\alpha\right] = \mathbf{0}. \tag{7.3.1}$$

Equation (7.3.1) shows that the impulses $-m_\alpha\left(\mathbf{q}_{\alpha'} - \mathbf{q}_\alpha\right) + \mathbf{I}_\alpha \;\; (\alpha = 1, 2, \ldots\ldots)$ are in equilibrium. Let the system be given a virtual displacement consistent with the same geometrical conditions.

Then the principle of virtual works gives

$$\sum\left\{-m_\alpha\left(\mathbf{q}_{\alpha'} - \mathbf{q}_\alpha\right) + \mathbf{I}_\alpha\right\} \cdot \delta\,\mathbf{r}_\alpha = 0,$$

$$\text{i.e.,}\quad \sum m_\alpha\left(\mathbf{q}_{\alpha'} - \mathbf{q}_\alpha\right) \cdot \delta\mathbf{r}_\alpha = \sum \mathbf{I}_\alpha \cdot \delta\mathbf{r}_\alpha. \tag{7.3.2}$$

Equation (7.3.2) is termed the equation of virtual work under impulsive forces. Omitting the dashes, (7.3.2) can be rewritten as

$$\sum m\left(\mathbf{q}' - \mathbf{q}\right) \cdot \delta\mathbf{r} = \sum \mathbf{I} \cdot \delta\mathbf{r}. \tag{7.3.3}$$

7.4 Carnot's Theorem on Collision

Statement The KE of a system of inelastic bodies after collision is less than the KE of the system before collision by the KE of the relative motion before and after.

Proof Let us consider two colliding bodies of a system of inelastic bodies. For two impinging bodies,

$$\text{the velocity of separation} = e \ \text{(velocity of approach)}\,. \tag{7.4.1}$$

As the bodies are inelastic, therefore $e = 0$ and hence (7.4.1) gives velocity of separation $=0$. That is, there is no relative displacement of the point of contact of two impinging bodies after collision. Thus, immediately after impacts, the points of contact of the bodies have a common velocity along the common normal, and hence the work done by mutual action and reaction is zero. Let the virtual displacement be the actual displacement immediately after a collision. Let $\mathbf{r}$ be the position vector of the point of contact. Let $\mathbf{q}$ and $\mathbf{q}'$ denote the velocities of the common point of contact just before and after collision, respectively. For a small time interval δt, the displacement of the point of contact is given by

$$\delta \mathbf{r} = \mathbf{q}'\,\delta t. \tag{7.4.2}$$

We assume that there are no external impulses applied to the system.
The equation of virtual work under impulsive forces is

$$\sum m\,(\mathbf{q}' - \mathbf{q}) \cdot \delta\mathbf{r} = \sum \mathbf{I} \cdot \delta\mathbf{r} = 0 \quad \left[\because \quad \mathbf{I} = \mathbf{0}\right]$$

$$\text{i.e.,} \quad \sum m\,(\mathbf{q}' - \mathbf{q}) \cdot \mathbf{q}'\delta\mathbf{r} = 0,$$

$$\text{i.e.,} \quad \sum m\,(\mathbf{q}' - \mathbf{q}) \cdot \mathbf{q}' = 0,$$

$$\text{i.e.,} \quad \sum mq^{1^2} = \sum m\mathbf{q} \cdot \mathbf{q}',$$

$$\text{i.e.,} \quad 2\sum mq'^2 = \sum m\,(2\mathbf{q} \cdot \mathbf{q}'). \tag{7.4.3}$$

We note that

$$|\mathbf{q}' - \mathbf{q}|^2 = (\mathbf{q}' - \mathbf{q})^2 = q'^2 + q^2 - 2\mathbf{q} \cdot \mathbf{q}',$$

$$\text{i.e.,} \quad 2\mathbf{q} \cdot \mathbf{q}' = q'^2 + q^2 - |\mathbf{q}' - \mathbf{q}|^2. \tag{7.4.4}$$

Equation (7.4.4) leads Eq. (7.4.3) to take the form

$$2\sum mq'^2 = \sum m\left\{q'^2 + q^2 - |\mathbf{q}' - \mathbf{q}|^2\right\}. \tag{7.4.5}$$

On simplification of Eq. (7.4.5), we obtain

$$\sum mq'^2 = \sum mq^2 - \sum m|\mathbf{q}' - \mathbf{q}|^2,$$

$$\text{i.e.,}\quad \sum \frac{1}{2}mq'^2 = \sum \frac{1}{2}mq^2 - \sum \frac{1}{2}m\,|\mathbf{q}' - \mathbf{q}|^2,$$

$$\text{i.e.,}\quad T' = T - T_0, \tag{7.4.6}$$

where

$T = \sum \frac{1}{2} m\, q^2 =$ KE of the system just before collision,

$T' = \sum \frac{1}{2} m\, q'^2 =$ K.E. of the system just after collision,

$T_0 = \sum \frac{1}{2} m\, |\mathbf{q}' - \mathbf{q}|^2 =$ K.E. of the system due to relative motion before and after impact.

Equation (7.4.6) shows that $T' < T$ by T_0 which confirms the proof of the theorem. □

7.5 Carnot's Theorem on Explosion

Statement The KE of a system after an explosion is increased by the KE of the relative motion before and after.

Proof Consider two particles that are about to be separated by an explosion. Thus, just before the explosion, there is no relative displacement between the particle considered, and so there is no virtual work done by their mutual actions and reactions.

Let the virtual displacement be the actual displacement in a short interval of time just before the explosion. Let $\mathbf{r}$ be the position vector of either particle, and $\mathbf{q}$ and $\mathbf{q}'$ be the velocities of the same particle just before and just after the explosion. Therefore,

$$\delta \mathbf{r} = \mathbf{q}\, \delta t. \tag{7.5.1}$$

The equation of virtual work gives

$$\sum m\,(\mathbf{q}' - \mathbf{q}) \cdot \delta \mathbf{r} = \sum \mathbf{I} \cdot \delta \mathbf{r}. \tag{7.5.2}$$

As there is no external impulse applied to the system,

$$\therefore\quad \mathbf{I} = \mathbf{0}, \tag{7.5.3}$$

for all particles.

Relations (7.5.1) and (7.5.3) lead Eq. (7.5.2) to reduce to the form:

$$\sum m\left(\mathbf{q}' - \mathbf{q}\right) \cdot \mathbf{q}\delta t = 0,$$

i.e., $$\sum m\left(\mathbf{q}' - \mathbf{q}\right) \cdot \mathbf{q} = 0,$$

i.e., $$\sum m\mathbf{q} \cdot \mathbf{q}' = \sum mq^2,$$

i.e., $$\sum m\left(2\mathbf{q} \cdot \mathbf{q}'\right) = 2\sum mq^2. \tag{7.5.4}$$

We have

$$\left|\mathbf{q}' - \mathbf{q}\right|^2 = \left(\mathbf{q}' - \mathbf{q}\right)^2 = q'^2 + q^2 - 2\mathbf{q} \cdot \mathbf{q}',$$

i.e., $$2\mathbf{q} \cdot \mathbf{q}' = q'^2 + q^2 - \left|\mathbf{q}' - \mathbf{q}\right|^2. \tag{7.5.5}$$

On use of (7.5.5), (7.5.4) reduces to

$$\sum m\left\{q'^2 + q^2 - \left|\mathbf{q}' - \mathbf{q}\right|^2\right\} = 2\sum mq^2,$$

i.e., $$\sum mq'^2 = \sum mq^2 + \sum m\left|\mathbf{q}' - \mathbf{q}\right|^2,$$

i.e., $$\sum \frac{1}{2}mq'^2 = \sum \frac{1}{2}mq^2 + \sum \frac{1}{2}m\left|\mathbf{q} - \mathbf{q}\right|^2,$$

i.e., $$T' = T + T_o, \tag{7.5.6}$$

where

$T = \sum \frac{1}{2}m\,q^2 =$ KE of the system just before explosion,

$T' = \sum \frac{1}{2}m\,q'^2 =$ K.E. of the system just after explosion,

$T_0 = \sum \frac{1}{2}m\left|\mathbf{q}' - \mathbf{q}\right|^2 =$ K.E. of the system due to relative motion before and after impact.

Equation (7.5.6) clearly shows that $T' > T$ by T_0, and it establishes the proof of the theorem. □

7.6 Bertrand's Theorem

Statement If a system in motion is acted upon by any impulses, the kinetic energy of the subsequent motion is greater than, if this system were subject to any additional constraints and acted upon by the same impulses.

Proof Let $\mathbf{q}_\alpha$ denote the velocity of the α-th particle of mass m_α just before impulse is applied. Let $\mathbf{q}'_\alpha$ be the velocity of that particle just after the impulse. Let $\mathbf{q}''_\alpha$ denote the velocity of the same particle immediately after impact if an additional constraint is applied to the system. It may be noted that no work is done by constraints for small displacements.

With usual notations, the principle of virtual works gives

$$\sum m_\alpha \left(\mathbf{q}'_\alpha - \mathbf{q}_\alpha\right) \cdot \delta \mathbf{r}_\alpha = \sum_\alpha \mathbf{I}_\alpha \cdot \delta \mathbf{r}_\alpha, \tag{7.6.1}$$

&

$$\sum m_\alpha \left(\mathbf{q}''_\alpha - \mathbf{q}_\alpha\right) \cdot \delta \mathbf{r}_\alpha = \sum_\alpha (\mathbf{I}_\alpha + \mathbf{C}_\alpha) \cdot \delta \mathbf{r}_\alpha = \sum \mathbf{I}_\alpha \cdot \delta \mathbf{r}_\alpha. \tag{7.6.2}$$

Equations (7.6.1) and (7.6.2) together give

$$\sum m_\alpha \left(\mathbf{q}'_\alpha - \mathbf{q}_\alpha\right) \cdot \delta \mathbf{r}_\alpha = \sum m_\alpha \left(\mathbf{q}''_\alpha - \mathbf{q}_\alpha\right) \cdot \delta \mathbf{r}_\alpha. \tag{7.6.3}$$

Omitting the suffixes in (7.6.3), we obtain

$$\sum m \left(\mathbf{q}' - \mathbf{q}\right) \cdot \delta\, \mathbf{r} = \sum m \left(\mathbf{q}'' - \mathbf{q}\right) \cdot \delta \mathbf{r}. \tag{7.6.4}$$

Let the virtual displacement be assumed to be the actual displacement just after impact, when the additional constraint is imposed. To the first order of approximation,

$$\mathbf{q}'' = \frac{\delta \mathbf{r}}{\delta t},$$

$$\text{i.e.,} \quad \delta \mathbf{r} = \mathbf{q}'' \delta t. \tag{7.6.5}$$

By use of (7.6.5), Eq. (7.6.4) reduces to

$$\sum m\mathbf{q}' \cdot \mathbf{q}'' = \sum m\mathbf{q}'' \cdot \mathbf{q}'',$$

$$\text{i.e.,}\quad \sum m\,(2\mathbf{q}'\cdot\mathbf{q}'') = 2\sum m\mathbf{q}''^2. \tag{7.6.6}$$

We have

$$|\mathbf{q}' - \mathbf{q}''|^2 = q'^2 + q''^2 - 2\mathbf{q}'\cdot\mathbf{q}'',$$

$$\text{i.e.,}\quad 2\mathbf{q}''\cdot\mathbf{q}' = q''^2 + q'^2 - |\mathbf{q}'' - \mathbf{q}'|^2. \tag{7.6.7}$$

By virtue of (7.6.7), Eq. (7.6.6) becomes

$$\sum m\left\{q''^2 + q'^2 - |\mathbf{q}'' - \mathbf{q}|^2\right\} = 2\sum mq''^2,$$

$$\text{i.e.,}\quad \sum mq'^2 = \sum mq''^2 + \sum m|\mathbf{q}'' - \mathbf{q}'|^2,$$

$$\text{i.e.,}\quad \sum \frac{1}{2}mq'^2 = \sum \frac{1}{2}mq''^2 + \sum \frac{1}{2}m|\mathbf{q}'' - \mathbf{q}'|^2,$$

$$\text{i.e.,}\quad T' = T'' + T_0, \tag{7.6.8}$$

where
$T' = \sum \frac{1}{2}mq'^2$ =KE of the system after impulse without additional constraint,
$T'' = \sum \frac{1}{2}mq''^2$ = KE of the system after impulse with additional constraint,
$T_0 = \sum \frac{1}{2}m|\mathbf{q}'' - \mathbf{q}'|^2$ = KE of the relative motion with or without additional constraint.

Equation (7.6.8) shows that $T' > T''$ by T_0. □

7.7 Kelvin's Theorem

Statement If any points of a connected system are suddenly set in motion with prescribed velocities, the KE of the resulting motion is less than that of any other kinematically possible motion in which the given points have the prescribed velocities.

Proof Let the symbol $\sum$ be used to denote a summation extending to all, the particles of the system and $\sum'$ to be used for denoting the summation extending to those particles only whose motion is prescribed. Let **I** be the external impulse applied to any one of those particles having prescribed velocities. Let **q** be the velocity of a particle in the actual motion and $\mathbf{q}'$ denote the velocity of the same particle in any other kinematically possible motion. Let $\mathbf{q}''$ be the prescribed velocity of an assigned particle.

The equation of virtual works gives

$$\sum m\,(\mathbf{q}-\mathbf{0})\cdot\delta\mathbf{r}=\sum{}'\mathbf{I}\cdot\delta\mathbf{r}''. \tag{7.7.1}$$

In actual motion, let the virtual displacement be considered as real displacement. To the first order of approximation

$$\frac{\delta\mathbf{r}}{dt}=\mathbf{q},\quad \frac{\delta\mathbf{r}''}{\delta t}=\mathbf{q}''. \tag{7.7.2}$$

Equation (7.7.2) leads Eq. (7.7.1) to take the form:

$$\sum m\mathbf{q}\cdot\mathbf{q}=\sum{}'\mathbf{I}\cdot\mathbf{q}'',$$

$$\text{i.e.,}\quad \sum mq^2=\sum{}'\mathbf{I}\cdot\mathbf{q}''. \tag{7.7.3}$$

Now by taking a virtual displacement in the second kinematically possible motion as actual displacement, it is obtained that

$$\delta\mathbf{r}=\mathbf{q}'\,\delta t, \tag{7.7.4}$$

and so (7.7.1) reduces to

$$\sum m\mathbf{q}\cdot\mathbf{q}'=\sum{}'\mathbf{I}\cdot\mathbf{q}''. \tag{7.7.5}$$

Equations (7.7.3) and (7.7.5) yield

$$\sum mq^2=\sum m\,(\mathbf{q}\cdot\mathbf{q}'),$$

$$\text{i.e.,}\quad \sum 2mq^2=\sum m\,(2\mathbf{q}\cdot\mathbf{q}'). \tag{7.7.6}$$

We note that

$$|\mathbf{q}'-\mathbf{q}|^2=q'^2+q^2-2\mathbf{q}\cdot\mathbf{q}',$$

or

$$2\mathbf{q}\cdot\mathbf{q}'=q'^2+q^2-|\mathbf{q}'-\mathbf{q}|^2. \tag{7.7.7}$$

By the use of (7.7.7), Eq. (7.7.6) becomes

$$2\sum mq^2 = \sum m\left\{q'^2 + q^2 - |\mathbf{q}' - \mathbf{q}|^2\right\},$$

i.e., $\sum mq^2 = \sum mq'^2 - \sum m|\mathbf{q}' - \mathbf{q}|^2,$

i.e., $\frac{1}{2}\sum mq^2 = \frac{1}{2}\sum mq'^2 - \frac{1}{2}\sum m|\mathbf{q}' - \mathbf{q}|^2,$

i.e., $\sum \frac{1}{2}mq^2 = \sum \frac{1}{2}mq'^2 - \sum \frac{1}{2}m\,|\mathbf{q}' - \mathbf{q}|^2,$

i.e., $T = T' - T_0,$ (7.7.8)

where

$T' = \sum \frac{1}{2}mq'^2$ =KE of any kinematically possible motion,

$T = \sum \frac{1}{2}mq^2$ =KE of any kinematically possible motion,

$T_0 = \sum \frac{1}{2}m\,|\mathbf{q}' - \mathbf{q}|^2$ =KE of any kinematically possible motion.

Equation (7.7.8) establishes the fact that $T < T'$ and thus theorem is established. □

7.8 Illustrations

Example 7.1 AB, BC are two equal similar rods freely hinged at B and lie in a straight line on a smooth table. The end A is struck by a blow $\perp^r$ to AB; show that the resulting velocity at A is $\frac{7}{2}$ times that of B.

Solution In Fig. 7.1, let $2a$ be the length of each of the rods with mass m, P the impulse of the blow applied at A, X the impulsive action and reaction at B, u_1 the velocity of the center of inertia G_1 of the rod AB $\perp^r$ to AB after the blow, ω_1 the angular velocity of the rod AB about G_1 after impulse, u_2 the velocity of the center of inertia G_2 of the rod BC $\perp^r$ to BC after the blow, and ω_2 be the angular velocities of the rod BC about G_2.

Equations of motion of the rods AB and BC under impulsive forces are

$m\,(u_1 - 0) = P + X,$ (7.8.1)

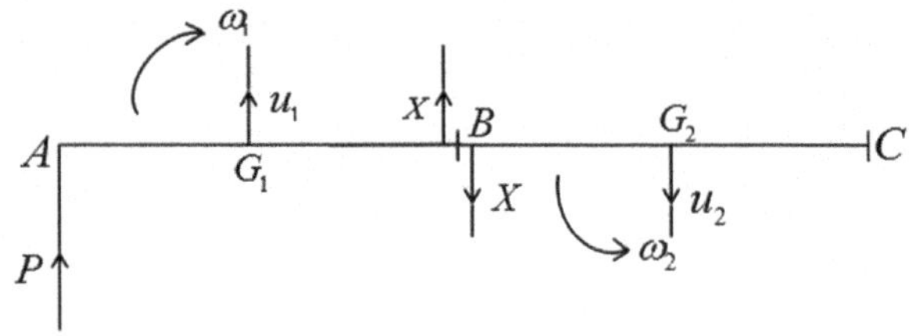

Fig. 7.1 Illustration for Example 7.1

$$m\frac{1}{3}a^2\left(\omega_1-0\right)=\left(P-X\right)a, \tag{7.8.2}$$

$$m\left(u_2-0\right)=X, \tag{7.8.3}$$

$$m\frac{1}{3}a^2\left(\omega_2-0\right)=Xa. \tag{7.8.4}$$

By equating the velocity of the joint B from both rods, we get

$$u_1-a\omega_1=-\left(u_2+a\omega_2\right). \tag{7.8.5}$$

Substituting the expressions for $u_1,\ \omega_1,\ u_2,\ \omega_2$ from Eqs. (7.8.1), (7.8.2), (7.8.3), (7.8.4) in Eq. (7.8.5), we derive

$$\frac{P+X}{m}-\frac{3\left(P-X\right)}{m}=-\frac{X}{m}-\frac{3X}{m},$$

which gives $P=4X$,

$$\text{i.e.,}\quad \frac{P}{4}=\frac{X}{1}=K\ (\text{say}),$$

$$\text{i.e.,}\quad P=4K,\quad X=K. \tag{7.8.6}$$

Now

$$\frac{v_A}{v_B}=\frac{\text{velocity of the end A}}{\text{velocity of the joint B}}=\frac{u_1+a\omega_1}{u_2+a\omega_2}=\frac{\frac{P+X}{m}+\frac{3(P-X)}{m}}{\frac{X}{m}+\frac{3X}{m}}$$

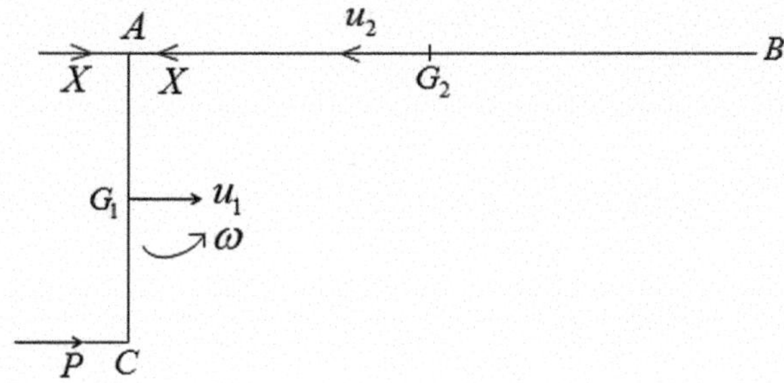

Fig. 7.2 Illustration for Example 7.2

$$\text{i.e.,}\quad \frac{v_A}{v_B} = \frac{4P - 2X}{4X} = \frac{16K - 2K}{4K} = \frac{7}{2}. \tag{7.8.7}$$

Thus, we have

$$v_A = \frac{7}{2} v_B.$$

Example 7.2 Two equal uniform rods, AB and AC, are freely jointed at A and are placed on a smooth table so as to be at right angles. The rod AC is struck by a blow at C in a direction $\perp^r$ to itself. Show that the resulting velocities of the middle points of AB and AC are in the ratio of 2:7.

Solution In Fig. 7.2, let $2a$ be the length of each of the rods AB and AC with mass m, G_1 the c.g of AC, G_2 that of AB, u_1 be the linear velocity of AC $\perp^r$ to AC just after impulse, u_2 be that of AB along **BA** just after impulse, ω the angular velocity of AC about G_1 immediately after impulse, P the impulsive action at C, and X the impulsive reaction at A.

Equations of motion of the rods AC and AB are

$$m\,(u_1 - 0) = P + X, \tag{7.8.8}$$

$$m\frac{1}{3}a^2\,(\omega - 0) = (P - X)\,a, \tag{7.8.9}$$

$$m\,(u_2 - 0) = X. \tag{7.8.10}$$

Equation (7.8.8) gives

$$u_1 = \frac{P + X}{m}.$$

Equation (7.8.9) gives

$$a\omega = \frac{3(P - X)}{m}.$$

Equation (7.8.10) yields

$$u_2 = \frac{X}{m}.$$

By equating the velocity at A for both rods, we get

$$u_1 - a\omega = -u_2,$$

i.e., $\dfrac{P + X}{m} - \dfrac{3\,(P - X)}{m} = \dfrac{-X}{m},$

i.e., $5X = 2P,$

i.e., $\dfrac{X}{2} = \dfrac{P}{5} = K\,(\text{say})$

i.e., $X = 2K, \;\; P = 5K.$

Now

$$\frac{\text{velocityofG}_1}{\text{velocityofG}_2} = \frac{u_1}{u_2} = \frac{P + X}{X} = \frac{5K + 2K}{2K} = \frac{7}{2},$$

i.e., $\dfrac{u_2}{u_1} = \dfrac{2}{7},$

i.e., $u_2 : u_1 = 2 : 7.$

Example 7.3 A uniform rod AB of mass $2m$ is freely jointed at B to a second rod BC of mass m. The rods lie on a smooth horizontal plane at right angles to each other, and an impulse I is applied to AB at A in a direction $\|$ to BC. Prove that the KE of the system is $\frac{5}{6}\frac{I^2}{m}$.

Solution Let AB and BC be the rods of the same length $2a$, G_1 the c.g. of AB, G_2 the c.g. of BC, u_1 the linear velocity of AB just after impulse, u_2 that of BC just after impulse, ω the angular velocity of AB about G_1, and X be the impulsive action and reaction at B as shown in Fig. 7.3.

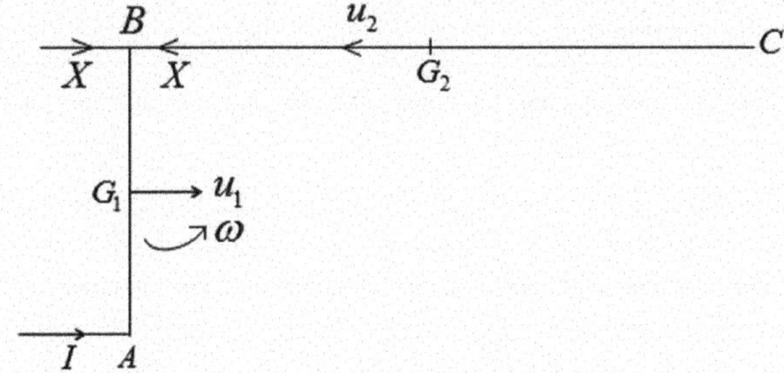

Fig. 7.3 Illustration for Example 7.3

Equations of the motions of the rods are

$$2mu_1 = I + X, \tag{7.8.11}$$

$$\frac{2}{3}ma^2\omega = (I - X)\,a, \tag{7.8.12}$$

$$mu_2 = X. \tag{7.8.13}$$

Equations (7.8.11), (7.8.12), and (7.8.13) give

$$u_1 = \frac{I + X}{2m},\; a\omega = \frac{3\,(I - X)}{2m},\; u_2 = \frac{X}{m}.$$

The velocity of the joint B due to both rods gives

$$u_1 - a\omega = -u_2,$$

i.e., $\dfrac{I + X}{2m} - \dfrac{3\,(I - X)}{2m} = -\dfrac{X}{m},$

i.e., $6X = 2I,$

i.e., $I = 3X.$

The KE generated for the system is given by

$$T = \text{KE of AB+KE of BC} = \frac{1}{2}2mu_1^2 + \frac{1}{2}2m\frac{a^2}{3}\omega^2 + \frac{1}{2}mu_2^2,$$

i.e., $T == m\left(\dfrac{I + X}{2m}\right)^2 + \dfrac{m}{3}\dfrac{9(I - X)^2}{4m^2} + \dfrac{1}{2}m\dfrac{X^2}{m^2},$

i.e., $T = m\dfrac{(3X + X)^2}{4m} + \dfrac{3}{4m}(3X - X)^2 + \dfrac{1}{2}\dfrac{X^2}{m},$

$$\text{i.e.,}\quad T=\frac{30X^2}{4m}=\frac{15}{2}\frac{X^2}{m}=\frac{15}{2m}\left(\frac{I}{3}\right)^2=\frac{15I^2}{2m\times 9}=\frac{5I^2}{6m},$$

which proves the result.

Example 7.4 A uniform circular disc of mass M initially at rest but free to move in any manner is suddenly set in motion by an impulse applied to a point A on its boundary. The initial velocity of A has components U along the radius of the circle through A, V along the tangent to the circle at A, and W along a third $\perp^r$ at A; prove that T, the initial KE of the disc is given by $2T = M\left(U^2+\frac{1}{3}V^2+\frac{1}{5}W^2\right)$.

Solution Let O denote the center of the disc. In Fig. 7.4, take A as the origin, X-axis along **AO**, Y-axis along the tangent to the circle at A, and Z-axis along a third $\perp^r$.

Let $\mathbf{q}=(u,\ v,\ w)$ be the velocity of the center of inertia and $\boldsymbol{\omega}=(\omega_1,\ \omega_2,\ \omega_3)$ be the angular velocity of the disc about the C.I. just after impulse.

We have

$$\begin{aligned}\mathbf{q}' &= \mathbf{q}+\boldsymbol{\omega}\times\mathbf{r},\\ \text{i.e.,}\quad U\hat{i}+V\hat{j}+W\hat{k} &= u\hat{i}+v\hat{j}+w\hat{k}+\boldsymbol{\omega}\times\mathbf{OA}\\ &= u\hat{i}+v\hat{j}+w\hat{k}+\left(\omega_1\hat{i}+\omega_2\hat{j}+\omega_3\hat{k}\right)\times\left(-a\hat{i}\right),\\ &= u\hat{i}+v\hat{j}+w\hat{k}+a\omega_2\hat{k}-a\omega_3\hat{j},\end{aligned}$$

which gives

$$\left.\begin{aligned}U&=u\\ V&=v-a\omega_3\\ W&=w+a\omega_2\end{aligned}\right\},\tag{7.8.14}$$

$$\text{i.e.,}\quad \left.\begin{aligned}u&=U\\ v&=V+a\omega_3\\ w&=W-a\omega_2\end{aligned}\right\}.\tag{7.8.15}$$

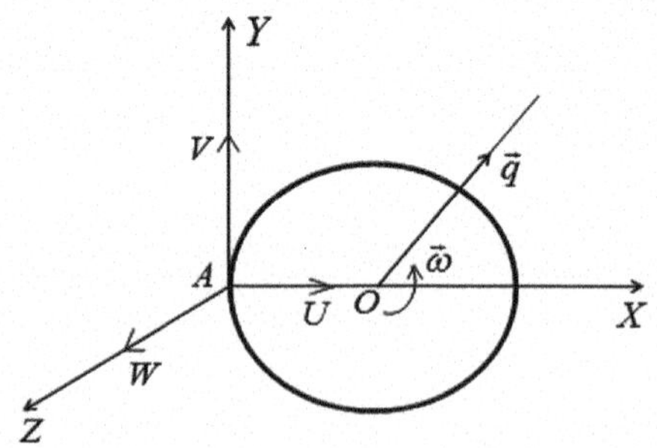

Fig. 7.4 Geometry for Example 7.4

Now,

T = KE of the disc just after impulse $= \frac{1}{2}MV_0^2 + \frac{1}{2}I\omega^2$,

$$\text{i.e.,} \quad T = \frac{1}{2}Mq^2 + \frac{1}{2}\left(Al^2 + Bm^2 + Cn^2 - 2Dmn - 2Enl - 2Flm\right)\omega^2, \tag{7.8.16}$$

with usual meanings of the symbols.

In the present case,

$$D = E = 0 \quad (\because \quad z = 0 \text{ is on the disc}),$$

and $F = 0$ (due to symmetry).

Therefore, Eq. (7.8.16) gives

$$\begin{aligned} 2T &= Mq^2 + \left(Al^2 + Bm^2 + Cn^2\right)\omega^2, \\ &= M\left(u^2 + v^2 + w^2\right) + A\omega_1^2 + B\omega_2^2 + C\omega_3^2, \\ &= M\left\{U^2 + (V + a\omega_3)^2 + (W - a\omega_2)^2\right\} + A\omega_1^2 + B\omega_2^2 + C\omega_3^2. \end{aligned} \tag{7.8.17}$$

As the motion is started suddenly, therefore, by Kelvin's theorem, it follows that T is minimum. Hence

$$\frac{\partial T}{\partial \omega_1} = 0, \tag{7.8.18}$$

$$\frac{\partial T}{\partial \omega_2} = 0, \tag{7.8.19}$$

$$\frac{\partial T}{\partial \omega_3} = 0. \tag{7.8.20}$$

Equation (7.8.18) gives

$$\omega_1 = 0. \tag{7.8.21}$$

Equation (7.8.19) gives

$$\omega_2 = \frac{4W}{5a}. \tag{7.8.22}$$

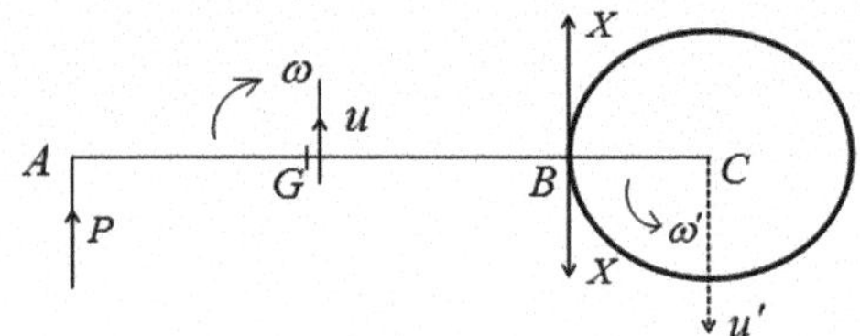

Fig. 7.5 Illustration for Example 7.5

Equation (7.8.20) yields

$$\omega_3 = -\frac{2V}{3a}. \tag{7.8.23}$$

Substituting (7.8.21), (7.8.22), (7.8.23) in (7.8.17), and replacing A by $\frac{1}{4}Ma^2$,

$$B \text{ by } \frac{1}{4}Ma^2 \; \& \; C \text{ by } \frac{1}{2}Ma^2,$$

we obtain

$$2T = M\left[U^2 + \frac{1}{3}V^2 + \frac{1}{5}W^2\right],$$

which proves the proposition.

Example 7.5 A circular disc of mass M and radius r is smoothly linked to a rod AB at a point on the circumference so that when AB is produced, it passes through the center of the disc. The rod and the disc lie on a smooth horizontal table. A blow P is given to the rod at the extreme A at right angles to it, proving that the KE produced is $\frac{9P^2}{10M}$, $2M$ being the mass of the rod.

Solution Let G be the center of gravity of the rod AB, C be the center of the disc (Fig. 7.5). Let $2a$ be the length of the rod AB, and X be the impulsive action and reaction at B. Let $\langle u, \omega \rangle$ & $\langle u', \omega' \rangle$ be the velocity system of the rod and the disc just after the impulse.

Equations of motion of the systems under impulsive forces are

$$2Mu = P + X, \tag{7.8.24}$$

$$2M\frac{a^2}{3}\omega = (P - X)\,a, \tag{7.8.25}$$

$$M\,u' = X, \tag{7.8.26}$$

$$\frac{1}{2}Mr^2\omega' = Xr. \tag{7.8.27}$$

Equations (7.8.24), (7.8.25); (7.8.26), (7.8.27) give

$$u = \frac{P + X}{2M}, \tag{7.8.28}$$

$$a\omega = \frac{3(P - X)}{2M}, \tag{7.8.29}$$

$$u' = \frac{X}{M}, \tag{7.8.30}$$

$$r\omega' = \frac{2X}{M}. \tag{7.8.31}$$

By equating the velocity of B from two sides, we obtain

$$u - a\omega = -\left(r\omega' + u'\right),$$

i.e., $\dfrac{P + X}{2M} - \dfrac{3(P - X)}{2M} = -\dfrac{2X}{M} - \dfrac{X}{M},$

i.e., $P = 5X.$ (7.8.32)

Now

$$T_1 = \text{KE of the rod AB} = \frac{1}{2}2Mu^2 + \frac{1}{2}2M\frac{1}{3}a^2\omega^2 = Mu^2 + \frac{1}{3}Ma^2\omega^2. \tag{7.8.33}$$

In view of (7.8.28) and (7.8.29), Eq. (7.8.33) reduces to

$$T_1 = M\left(\frac{P + X}{2M}\right)^2 + \frac{1}{3}M\left\{\frac{3(P - X)}{2M}\right\}^2,$$

and by the use of (7.8.32), we get

$$T_1 = \frac{21P^2}{25M}. \tag{7.8.34}$$

Again

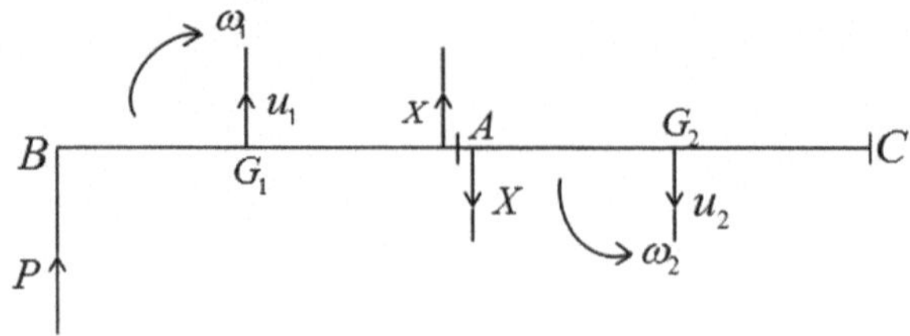

Fig. 7.6 Rods are attached freely at A

$$
\begin{aligned}
T_2 &= \text{KE of the disc} \\
&= \frac{1}{2}Mu'^2 + \frac{1}{2}\frac{1}{2}Mr^2\omega'^2 \\
&= \frac{3X^2}{2M} \quad \left[\text{From (7.8.30) and (7.8.31)}\right] \\
&= \frac{3P^2}{50M}. \quad \left[\text{From (7.8.32)}\right]
\end{aligned}
$$

Therefore, T = KE of the system after impulse $= T_1 + T_2 = \frac{9P^2}{10M}$, which is the required result.

Example 7.6 Two equal uniform rods AB and AC are freely hinged at A and rest on a straight table. A blow is struck at B $\perp^r$ to the rods; show that the KE generated is $\frac{7}{4}$ times what it would be if the rods were rigidly fastened at A.

Solution Two cases arise.

Case I: When the rods are freely hinged at A.

Let G_1 be the center of inertia of the rod AB and G_2 that of the rod BC (Fig. 7.6). Let P denote the impulse applied at A, and X be the impulsive action and reaction at the joint A.

Let u_1 and u_2 be the velocities of the rods AB and AC, respectively, just after the impulse. Let ω_1 and ω_2 denote the angular velocities of AB and AB, respectively, just after the impulse is applied. Let M be the mass of each of the rods.

Equations of motion of the rods AB and AC under impulsive forces are

$$M\,(u_1 - 0) = P + X, \tag{7.8.35}$$

$$\frac{1}{3}Ma^2\,(\omega_1 - 0) = (P - X)\,a, \tag{7.8.36}$$

$$M\,(u_2 - 0) = X, \tag{7.8.37}$$

$$\frac{1}{3}Ma^2\,(\omega_2 - 0) = Xa. \tag{7.8.38}$$

Equations (7.8.35), (7.8.36), (7.8.37), (7.8.38) give

$$u_1 = \frac{P + X}{M}, \tag{7.8.39}$$

$$a\omega_1 = \frac{3\,(P - X)}{M}, \tag{7.8.40}$$

$$u_2 = \frac{X}{M}, \tag{7.8.41}$$

$$a\omega_2 = \frac{3X}{M}. \tag{7.8.42}$$

Now, by equating the velocity of the common point A for the rods, we obtain

$$u_1 - a\omega_1 = -\,(u_2 + a\omega_2)\,,$$

$$\text{i.e.,}\quad \frac{P + X}{M} - \frac{3\,(P - X)}{M} = -\left[\frac{X}{M} + \frac{3X}{M}\right],$$

$$\text{i.e.,}\quad P + X - 3P + 3X = -4X,$$

$$\text{i.e.,}\quad 2P = 8X,$$

$$\text{i.e.,}\quad P = 4X,$$

$$\text{i.e.,}\quad \frac{P}{4} = \frac{X}{1} = K\ (\text{say})\,.$$

$$\therefore\quad P = 4K,\quad X = K. \tag{7.8.43}$$

On substitution of (7.8.43), Eqs. (7.8.39), (7.8.40), (7.8.41), (7.8.42) yield

$$u_1 = \frac{4K + K}{M} = \frac{5K}{M}, \tag{7.8.44}$$

$$a\omega_1 = \frac{3\,(4K - K)}{M} = \frac{9K}{M}, \tag{7.8.45}$$

$$u_2 = \frac{K}{M}, \tag{7.8.46}$$

$$a\omega_2 = \frac{3K}{M}. \tag{7.8.47}$$

Therefore, the KE of the system of rods is given by

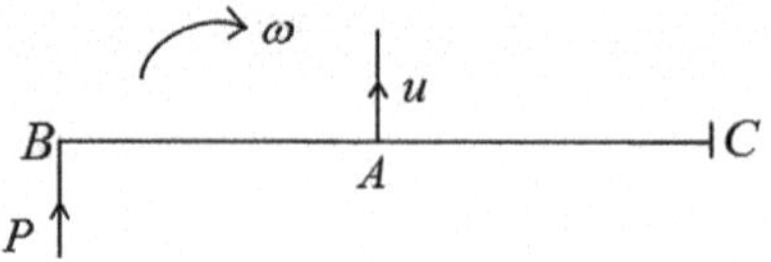

Fig. 7.7 Rods are joined rigidly at A

$$T = \text{KE of the rod AB} + \text{KE of the rod AC}$$
$$= \frac{1}{2}Mu_1^2 + \frac{1}{2}MK_1^2\omega_1^2 + \frac{1}{2}Mu_2^2 + \frac{1}{2}MK_2^2\omega_2^2,$$

$$\text{i.e.,}\quad T = \frac{1}{2}M\left(\frac{5K}{M}\right)^2 + \frac{1}{2}M\frac{1}{3}a^2\omega_1^2 + \frac{1}{2}M\left(\frac{K}{M}\right)^2 + \frac{1}{2}M\frac{1}{3}a^2\omega_2^2,$$

$$\text{i.e.,}\quad T = \frac{25K^2}{2M} + \frac{1}{6}M\left(\frac{9K}{M}\right)^2 + \frac{1}{2}\frac{K^2}{M} + \frac{1}{6}M\left(\frac{3K}{M}\right)^2,$$

$$\text{i.e.,}\quad T = \frac{25K^2}{2M} + \frac{81K^2}{6M} + \frac{K^2}{2M} + \frac{3K^2}{2M},$$

$$\text{i.e.,}\quad T = \frac{28K^2}{M},$$

$$\text{i.e.,}\quad T = \frac{28}{M}\left(\frac{P}{4}\right)^2,$$

$$\text{i.e.,}\quad T = \frac{7P^2}{4M}. \tag{7.8.48}$$

Case II: When the rods are rigidly fastened at A (Fig. 7.7). Let u be the linear velocity and ω the angular velocity of BC just after the impulse.

Equations of motion of the rods are

$$2Mu = P, \tag{7.8.49}$$

$$2M\frac{1}{3}(2a)^2\omega = 2aP. \tag{7.8.50}$$

Equation (7.8.49) gives

$$u = \frac{P}{2M}. \tag{7.8.51}$$

Equation (7.8.50) yields

$$\frac{M}{3}4a^2\omega = Pa,$$

$$\therefore \quad a\omega = \frac{3P}{4M}. \tag{7.8.52}$$

The KE of the rod is given by

$$T' = \frac{1}{2}2Mu^2 + \frac{1}{2}\frac{1}{3}2M(2a)^2\omega^2,$$

$$\text{i.e.,} \quad T' = Mu^2 + \frac{4}{3}Ma^2\omega^2,$$

$$\text{i.e.,} \quad T' = M\left[u^2 + \frac{4}{3}(a\omega)^2\right],$$

$$\text{i.e.,} \quad T' = M\left[\frac{P^2}{4M^2} + \frac{3P^2}{4M^2}\right] = \frac{P^2}{M}. \tag{7.8.53}$$

Now

$$\frac{T}{T'} = \frac{7P^2}{4M} \times \frac{M}{P^2} = \frac{7}{4},$$

$$\text{i.e.,} \quad T = \frac{7}{4}T',$$

which confirms the proof.

Example 7.7 A rectangular lamina, whose sides are of length $2a$ and $2b$, is at rest when one corner is caught and suddenly made to move with prescribed speed V in the plane of the lamina. Show by using Kelvin's theorem that the greater angular velocity which can thus be imparted to the lamina is $\frac{3V}{4\sqrt{a^2+b^2}}$.

Solution Consider a rectangle ABCD as seen in Fig. 7.8, with center at O and suppose that AB=DC $= 2a$, AD=BC $= 2b$. Take O as the origin, X-axis $\parallel$ **AB**, and Y-axis $\parallel$ **AD**.

Let the corner C (a, b) be made suddenly to move with velocity $\mathbf{V} = V\cos\theta\hat{i} + V\sin\theta\hat{j}$. Let $\mathbf{q} = \hat{i}u + \hat{j}v$ be the linear velocity of the C.I. and $\boldsymbol{\omega} = \omega\hat{k}$ be the angular velocity of the lamina about the C.I. just after impulse.

$$\therefore \quad \mathbf{V} = \mathbf{q} + \omega \times \mathbf{OC} = u\hat{i} + v\hat{j} + \omega\hat{k} \times \left(a\hat{i} + b\hat{j}\right) = (u - b\omega)\,\hat{i} + (v + a\omega)\,\hat{j}$$

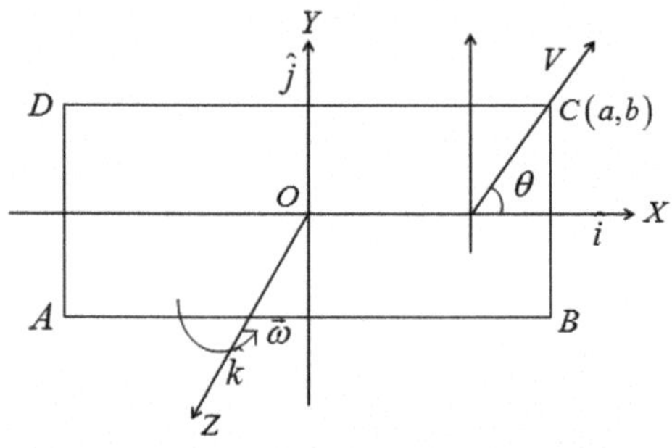

Fig. 7.8 Illustration for Example 7.7

i.e., $V\cos\theta\hat{i} + V\sin\theta\hat{j} = (u - b\omega)\,\hat{i} + (v + a\omega)\,\hat{j}$

$$\text{i.e.,}\quad \left.\begin{array}{l} u = V\cos\theta + b\omega \\ v = V\sin\theta - a\omega \end{array}\right\}. \tag{7.8.54}$$

Now

$$T = \text{KE of the lamina after impulse} = \frac{1}{2}MV_0^2 + \frac{1}{2}I\omega^2,$$

$$\text{i.e.,}\quad T = \frac{1}{2}Mq^2 + \frac{1}{2}\frac{1}{3}M\left(a^2 + b^2\right)\omega^2,$$

$$\text{i.e.,}\quad T = \frac{1}{2}M\left(u^2 + v^2\right) + \frac{1}{6}M\left(a^2 + b^2\right)\omega^2,$$

$$\text{i.e.,}\quad T = \frac{1}{2}M\left\{(V\cos\theta + b\omega)^2 + (V\sin\theta - a\omega)^2\right\} + \frac{1}{6}M\left(a^2 + b^2\right)\omega^2.$$

By Kelvin's theorem, it follows that T is minimum and hence

$$\frac{\partial T}{\partial\omega} = 0, \tag{7.8.55}$$

which gives

$$3\,(V\cos\theta + b\omega)\,b - 3\,(V\sin\theta - a\omega)\,a + \left(a^2 + b^2\right)\omega = 0,$$

$$\omega = \frac{3V\,(a\sin\theta - b\cos\theta)}{4\left(a^2 + b^2\right)}. \tag{7.8.56}$$

Fig. 7.9 Illustration for Example 7.8

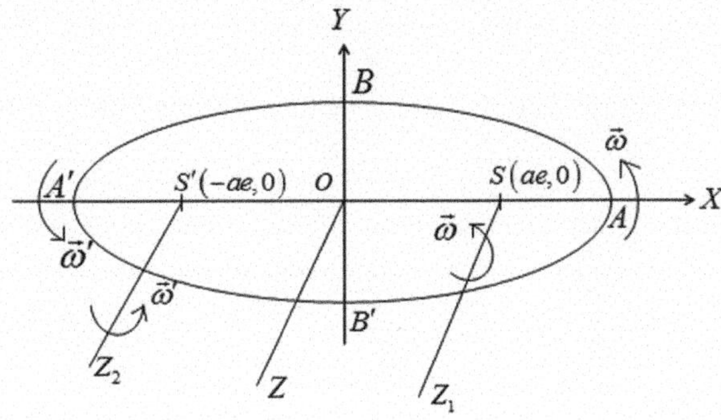

Equation (7.8.56) gives

$$\omega_{\max} = \frac{3\,V}{4(a^2 + b^2)}\sqrt{a^2 + b^2} = \frac{3\,V}{\sqrt{a^2 + b^2}}.$$

Example 7.8 A lamina in the form of an ellipse is rotating in its own plane about one of its foci with angular velocity ω. The focus is set free, and the other focus at the same instant is fixed; show that the ellipse now rotates about it with angular velocity

$$\omega = \frac{2 - 5e^2}{2 + 3e^2}.$$

Solution Let AA′ $= 2a$ and BB′ $= 2b$ be the major and minor axes of the ellipse with center. Let S and S′ be the foci of the ellipse (Fig. 7.9). Take O as the origin, X-axis along **OA**, Y-axis along **OB**, and Z-axis through O $\perp^r$ the plane of the lamina . Let SZ_1 and $S'Z_2$ be two lines ∥ OZ. Suppose that initially, the focus S was fixed and the lamina rotated about S (i.e., SZ_1) with angular velocity ω. Now, let this focus S be made free and immediately the other focus S′ be made fixed. Let ω' be new angular velocity of the lamina about S′ $\left(\text{i.e., } S'Z_2\right)$.

The impulse is applied at the point S', and so the moment of the impulse about S′ $= \mathbf{0}$. Hence the angular momentum of the lamina about S′ is conserved, i.e., the angular momentum of the lamina about S′ just before impulse = angular momentum of the lamina about S′ just after the impulse.

$$\text{i.e.,}\quad \boldsymbol{\Omega} = \boldsymbol{\Omega}'. \tag{7.8.57}$$

Now

$$\boldsymbol{\Omega} = \text{angular momentum of the lamina about just before impulse} = \mathbf{r} \times M\mathbf{v} + I_O\omega,$$

$$\text{i.e.,}\quad \boldsymbol{\Omega} = \mathbf{S'O} \times M\mathbf{v} + I_O\omega\hat{k} = ae\hat{i} \times M\left(\omega \times \mathbf{r}'\right) + I_O\omega\hat{k},$$

$$\text{i.e.,}\quad \boldsymbol{\Omega} = ae\hat{i} \times M\left\{\omega \times \mathbf{SO}\right\} + I_O\omega\hat{k} = ae\hat{i} \times M\left\{\omega\hat{k} \times \left(-ae\hat{i}\right)\right\} + I_O\omega\hat{k},$$

i.e., $\boldsymbol{\Omega} = -ae\hat{i} \times M\left\{\omega ae\hat{j}\right\} + I_O\omega\hat{k} = -a^2e^2M\omega\hat{k} + I_O\omega\hat{k},$

i.e., $\boldsymbol{\Omega} = -Ma^2e^2\omega\hat{k} + \frac{1}{4}M\left(a^2+b^2\right)\omega\hat{k}.$

$\boldsymbol{\Omega}'$ = angular momentum of the lamina about just after impulse,

i.e., $\boldsymbol{\Omega}' = \mathbf{r} \times M\mathbf{v} + I_O\boldsymbol{\omega}' = \mathbf{S'O} \times M\mathbf{v} + I_O\omega'\hat{k},$

i.e., $\boldsymbol{\Omega}' = ae\hat{i} \times M\mathbf{v} + I_O\omega'\hat{k} = ae\hat{i} \times M\left(\boldsymbol{\omega}' \times \mathbf{r}\right) + I_O\omega'\hat{k},$

i.e., $\boldsymbol{\Omega}' = ae\hat{i} \times M\left(\boldsymbol{\omega}' \times \mathbf{S'O}\right) + I_O\omega'\hat{k} = ae\hat{i} \times M\left(\omega'\hat{k} \times ae\hat{i}\right) + I_O\omega'\hat{k},$

i.e., $\boldsymbol{\Omega}' = Ma^2e^2\omega'\hat{i} \times \hat{j} + I_O\omega'\hat{k} = Ma^2e^2\omega'\hat{k} + I_O\omega'\hat{k}.$

Equation (7.8.57) gives

$$-Ma^2e^2\omega\hat{k} + \frac{1}{4}M\left(a^2+b^2\right)\omega\hat{k} = Ma^2e^2\omega'\hat{k} + \frac{1}{4}M\left(a^2+b^2\right)\omega'\hat{k},$$

i.e., $-a^2e^2\omega + \frac{1}{4}\left(a^2+b^2\right)\omega = a^2e^2\omega' + \frac{1}{4}\left(a^2+b^2\right)\omega',$

i.e., $-4a^2e^2\omega + \left(a^2+b^2\right)\omega = 4a^2e^2\omega' + \left(a^2+b^2\right)\omega',$

i.e., $\omega\left[a^2 + a^2\left(1-e^2\right) - 4a^2e^2\right] = \omega'\left[a^2 + a^2\left(1-e^2\right) + 4a^2e^2\right],$

i.e., $\omega\left(2a^2 - 5a^2e^2\right) = \omega'\left(2a^2 + 3a^2e^2\right),$

i.e., $\omega\left(2-5e^2\right) = \omega'\left(2+3e^2\right),$

$$\therefore\quad \omega' = \frac{2-5e^2}{2+3e^2}\omega.$$

Example 7.9 A corner of a uniform rectangular plate at rest is set in motion with velocity $V \perp^r$ to the plane of the plate. Show that the initial velocity of the center of the plate is $\frac{V}{7}$ in the same direction.

Solution Let ABCD be the rectangular plate with sides AB=DC $= 2a$ & BC=AD $= 2b$ and mass M. Let the velocity V be imposed to the plate at C $\perp^r$ to the plane of the plate. Let J be the impulse at C. Let U be the velocity of the center of the plate and ω be the angular velocity of the plate about the axis of rotation BD. Let p denote the $\perp^r$ distance of C from BD.

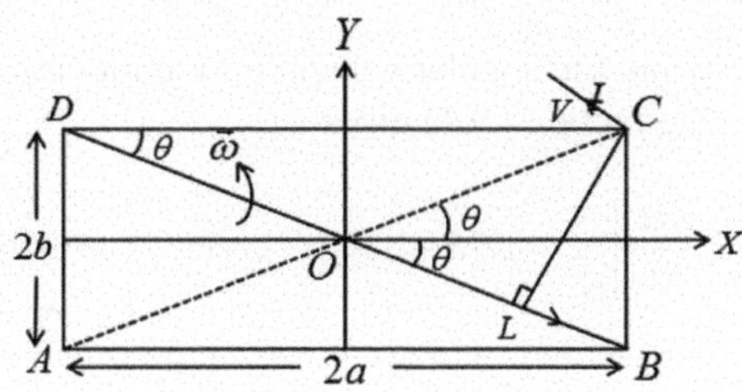

Fig. 7.10 Illustration for Example 7.9

Now from Fig. 7.10, we get

$$p = \text{CL}{=}\text{CD} \sin \theta {=} 2a \frac{2b}{\sqrt{4a^2 + 4b^2}} = \frac{2ab}{\sqrt{a^2 + b^2}}.$$

Equations of motion of the plate under impulsive forces are

$$MU = J, \tag{7.8.58}$$

$$I\omega = Jp = \frac{2abJ}{\sqrt{a^2 + b^2}}. \tag{7.8.59}$$

We have

$$I = \text{MI of the plate about BD} = \text{A} \cos^2 \theta + \text{B} \sin^2 \theta - 2\text{F} \sin \theta \cos \theta$$

$$\text{i.e.,} \quad I = \frac{1}{3} Mb^2 \cos^2 \theta + \frac{1}{3} Ma^2 \sin^2 \theta = \frac{1}{3} Mb^2 \frac{a^2}{a^2 + b^2} + \frac{1}{3} Ma^2 \frac{b^2}{a^2 + b^2} = \frac{2Ma^2b^2}{3(a^2 + b^2)}.$$

Equation (7.8.59) gives

$$\frac{2}{3} \frac{Ma^2b^2\omega}{a^2 + b^2} = J \frac{2ab}{\sqrt{a^2 + b^2}},$$

$$\text{i.e.,} \quad \frac{Mab\omega}{3\left(a^2 + b^2\right)} = \frac{J}{\sqrt{a^2 + b^2}} = \frac{MU}{\sqrt{a^2 + b^2}},$$

$$\text{i.e.,} \quad \frac{ab\omega}{3\left(a^2 + b^2\right)} = \frac{U}{\sqrt{a^2 + b^2}},$$

$$\text{i.e.,} \quad \omega = \frac{3U\left(a^2 + b^2\right)}{ab\sqrt{a^2 + b^2}} = \frac{3U\sqrt{a^2 + b^2}}{ab}.$$

As V is the velocity of the corner C,

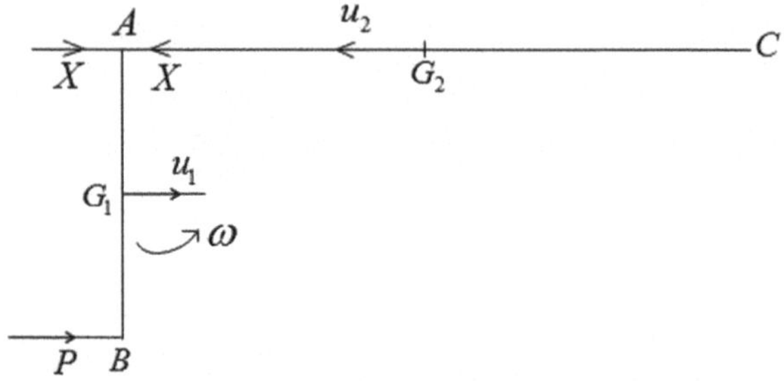

Fig. 7.11 Illustration for Example 7.10

$$\therefore \quad V = U + p\omega = U + \frac{2ab}{\sqrt{a^2+b^2}} \frac{3U\sqrt{a^2+b^2}}{ab} = 7U,$$

$$\text{i.e.,} \quad U = \frac{V}{7}.$$

Example 7.10 Two uniform rods, AB, AC, are freely jointed at A and laid on a smooth horizontal table so that the angle BAC is a right angle. The rod is struck by a blow P and B in a direction $\perp^r$ to AB; show that the initial velocity of A is $\frac{2P}{4m'+m}$, where m and m' are the masses of AB,AC, respectively.

Solution Let X be the impulsive action and reaction at the joint A. Let G_1 be the c.g. of AB and G_2 that of AC. Let $2a$ be the length of the rod AB. Let $\langle u_1, \omega\rangle$, $\langle u_2, 0\rangle$ be the velocity systems of the rods AB and AC immediately after impulse as shown in Fig. 7.11.

Equations of motion of the rods are

$$mu_1 = P + X, \tag{7.8.60}$$

$$\frac{1}{3}ma^2\omega = Pa - Xa, \tag{7.8.61}$$

$$m'u_2 = X. \tag{7.8.62}$$

Equation (7.8.60) gives

$$u_1 = \frac{P+X}{m}. \tag{7.8.63}$$

Equation (7.8.61) gives

$$a\omega = \frac{3(P-X)}{m}. \tag{7.8.64}$$

Equation (7.8.62) gives

$$u_2 = \frac{X}{m'}. \tag{7.8.65}$$

Equating the velocity at A for both rods, we have

$$u_1 - a\omega = -u_2,$$

$$\text{i.e.,}\quad \frac{P+X}{m} - \frac{3(P-X)}{m} = -\frac{X}{m'},$$

$$\text{i.e.,}\quad \frac{P+X-3P+3X}{m} = -\frac{X}{m'},$$

$$\text{i.e.,}\quad \frac{4X-2P}{m} = -\frac{X}{m'},$$

$$\text{i.e.,}\quad 4Xm' - 2Pm' = -mX,$$

$$\text{i.e.,}\quad \left(4m'+m\right)X = 2Pm',$$

$$\text{i.e.,}\quad X = \frac{2Pm'}{4m'+m}.$$

∴ The initial velocity of A is given by

$$u_2 = \frac{X}{m'} = \frac{2P}{4m'+m},$$

which conforms the proof.

Example 7.11 A rectangular plate ABCD of mass m has corner A fixed. An impulse J is applied at the opposite corner C at right angles to the plate. Show that the KE generated is $\frac{12J^2}{7m}$.

Solution Let AMN be the axis of rotation. Let AB=DC $= 2a$ & AD=BC $= 2b$. Take A as the origin, X-axis along **AB** and Y-axis along **AD**.

From Fig. 7.12, we have

$$\text{Slope of AMN=Slope of BD=}\frac{2b}{-2a} = -\frac{b}{a}.$$

Hence equation of the axis of rotation is

$$y = -\frac{b}{a}x,$$

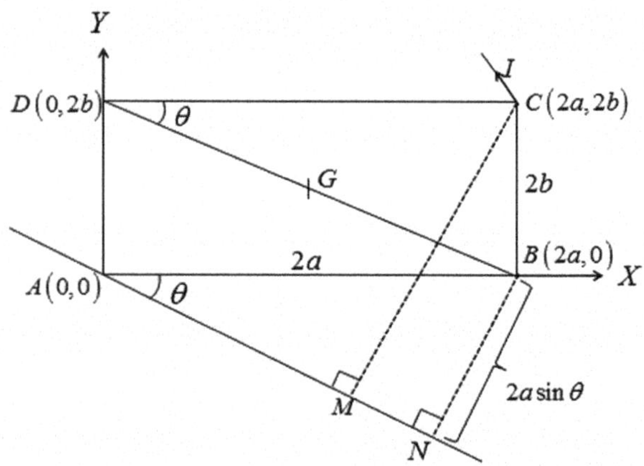

Fig. 7.12 Illustration for Example 7.11

$$\text{i.e.,}\quad bx + ay = 0. \tag{7.8.66}$$

The $\perp^r$ distance p of the point C $(2a, 2b)$ from the line (7.8.66) is given by

$$p = \frac{2ab + 2ab}{\sqrt{a^2 + b^2}} = \frac{4ab}{\sqrt{a^2 + b^2}}.$$

Let I_0 be the moment of inertia of the plate about BD.
i.e., I_0 is the moment of inertia of the plate about AC.
Hence

$$I_0 = \frac{1}{3}mb^2\cos^2\theta + \frac{1}{3}ma^2\sin^2\theta = \frac{1}{3}mb^2\frac{a^2}{a^2+b^2} + \frac{1}{3}ma^2\frac{b^2}{a^2+b^2} = \frac{2ma^2b^2}{a^2+b^2}.$$

Now

$$I = \text{MI of the plate about AMN} = \text{MI about BD} + m(2a\sin\theta)^2,$$

$$\text{i.e.,}\quad I = \text{MI about AC} + 4ma^2\sin^2\theta,$$

$$\text{i.e.,}\quad I = I_0 + 4\,ma^2\sin^2\theta,$$

$$\text{i.e.,}\quad I = \frac{2\,ma^2b^2}{3(a^2+b^2)} + 4\,ma^2\frac{b^2}{a^2+b^2} = \frac{14\,ma^2b^2}{3(a^2+b^2)}.$$

Let ω be the angular velocity of the plate about just after impulse.
The equation of the rotational motion of the plate under impulsive forces is

$$I\omega = Jp = \frac{4abJ}{\sqrt{a^2+b^2}},$$

$$\text{i.e.,}\quad \frac{14ma^2b^2}{3\left(a^2+b^2\right)}\omega = \frac{4abJ}{\sqrt{a^2+b^2}},$$

$$\text{i.e.,}\quad \frac{7mab\omega}{3\sqrt{a^2+b^2}} = 2J,$$

$$\text{i.e.,}\quad \omega = \frac{6J\sqrt{a^2+b^2}}{7mab}.$$

$$\therefore\quad \text{KE of the plate} = \frac{1}{2}I\omega^2 = \frac{1}{2}\frac{14ma^2b^2\omega^2}{3\left(a^2+b^2\right)} = \frac{7ma^2b^2}{3\left(a^2+b^2\right)}\frac{36J^2\left(a^2+b^2\right)}{49m^2a^2b^2},$$

$$\text{i.e.,}\quad \text{KE of the plate}=\frac{12J^2}{7m}.$$

Example 7.12 Three equal rods, AB, BC, and CD, are freely joined and placed in a straight line on a smooth table. The rod AB is struck at its end A by a blow which is $\perp^r$ to its length; find the resulting motion, and show that the velocity of the center of AB is 19 times that of CD, and its angular velocity 11 times that of CD.

Solution Equations of motion of the system of rods as described in Fig. 7.13 are

$$mu_1 = P + X, \tag{7.8.67}$$

$$\frac{1}{3}ma^2\omega_1 = Pa - Xa, \tag{7.8.68}$$

$$mu_2 = X + Y, \tag{7.8.69}$$

$$\frac{1}{3}ma^2\omega_2 = Xa - Ya, \tag{7.8.70}$$

$$mu_3 = Y, \tag{7.8.71}$$

$$\frac{1}{3}ma^2\omega_3 = Ya. \tag{7.8.72}$$

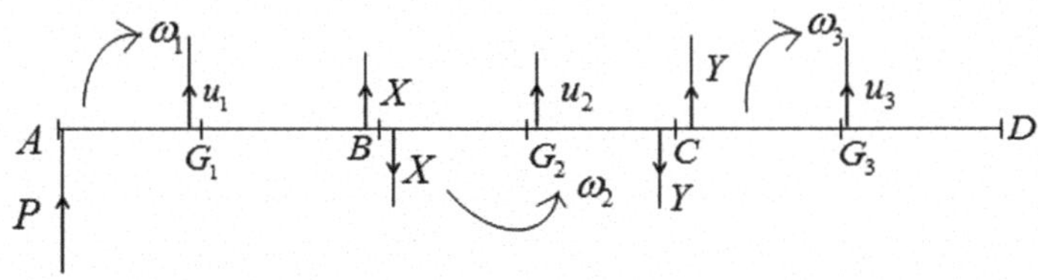

Fig. 7.13 Illustration for Example 7.12

Equations (7.8.67), (7.8.68); (7.8.69), (7.8.70); (7.8.71), (7.8.72) give

$$u_1 = \frac{P+X}{m}, \quad a\omega_1 = \frac{3(P-X)}{m};$$
$$u_2 = \frac{X+Y}{m}, \quad a\omega_2 = \frac{3(X+Y)}{m};$$
$$u_3 = \frac{Y}{m}, \qquad a\omega_3 = \frac{3Y}{m}.$$

Equating the velocity of B and C for both connecting rods, we get

$$u_1 - a\omega_1 = -(u_2 + a\omega_2), \tag{7.8.73}$$

$$-u_2 + a\omega_2 = u_3 + a\omega_3. \tag{7.8.74}$$

Equation (7.8.73) gives

$$\frac{P+X}{m} - \frac{3(P-X)}{m} = -\left[\frac{X+Y}{m} + \frac{3(X-Y)}{m}\right],$$

i.e., $P + X - 3P + 3X = -[X + Y + 3X - 3Y]$,

i.e., $-2P + 4X = -[4X - 2Y] = -4X + 2Y$,

i.e., $8X = 2P + 2Y$,

i.e., $4X = P + Y$. (7.8.75)

Equation (7.8.74) yields

$$-\frac{X+Y}{m} + \frac{3(X-Y)}{m} = \frac{Y}{m} + \frac{3Y}{m},$$

i.e., $-X - Y + 3X - 3Y = Y + 3Y$,

i.e., $X = 4Y$,

i.e., $\frac{X}{4} = \frac{Y}{1} = K$ (say)

$\therefore \quad X = 4K, \quad Y = K.$

Equation (7.8.75) gives

$$16K = P + K$$
$$\therefore \quad P = 15K.$$

Thus

$$P = 15K, \quad X = 4K, \quad Y = K.$$

Now

$$\frac{u_1}{u_3} = \frac{\frac{P+X}{m}}{\frac{Y}{m}} = \frac{P+X}{Y} = \frac{15K+4K}{K} = \frac{19K}{K} = 19.$$

Again

$$\frac{\omega_1}{\omega_3} = \frac{\frac{3(P+X)}{ma}}{\frac{3Y}{ma}} = \frac{P+X}{Y} = \frac{15K-4K}{K} = \frac{11K}{K} = 11,$$

$$\therefore \quad \omega_1 = 11\omega_3.$$

Example 7.13 A uniform square lamina, of mass M and side $2a$, is moving freely about a diagonal with uniform angular velocity ω. When one of the corners not in that diagonal becomes fixed, show that the new angular velocity is $\frac{\omega}{7}$, and that the impulse of the force on the fixed point is $\frac{\sqrt{2}}{7}Ma\omega$.

Solution Let ABCD be the square lamina of mass M and sides $2a$. Let G be the center of the lamina as shown in Fig. 7.14. Take G as the origin, X-axis parallel to **AB**, and Y-axis || **AD**. Let the diagonal be the initial axis of rotation. Let Q′BQ || AC be the axis of rotation after the impulse is applied.

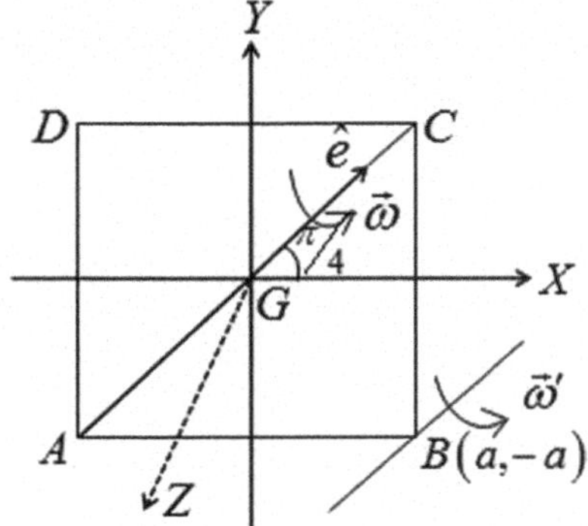

Fig. 7.14 Illustration for Example 7.13

Now,

$$\mathbf{AC} = 2a\hat{\imath} + 2a\hat{\jmath}.$$

Therefore, unit vector along **AC** is given by

$$\hat{e} = \frac{\hat{\imath} + \hat{\jmath}}{\sqrt{2}}$$
$$\therefore \ \hat{\imath} + \hat{\jmath} = \sqrt{2}\hat{e}.$$

$$\begin{aligned} I_0 &= \text{MI of the lamina about AC} \\ &= \frac{1}{3}Ma^2\cos^2\frac{\pi}{4} + \frac{1}{3}Ma^2\sin^2\frac{\pi}{4} \\ &= \frac{1}{3}Ma^2. \end{aligned}$$

Now,

$$\begin{aligned} \mathbf{\Omega} &= \text{angular momentum of the lamina about B just before impulse} \\ &= \text{angular momentum of the lamina about A} \\ &= I_0\boldsymbol{\omega} = I_0\omega\hat{e}. \end{aligned}$$

$$\begin{aligned} I &= \text{MI of the lamina about B} \\ &= \text{MI of the lamina about AC} + Md^2 \\ &= I_0 + M\,2a^2 \\ &= \frac{1}{3}Ma^2 + 2Ma^2 \\ &= Ma^2(\frac{1}{3} + 2) \\ &= \frac{7}{3}Ma^2. \end{aligned}$$

$$\begin{aligned} \mathbf{\Omega}' &= \text{angular momentum of the lamina about B just after impulse} \\ &= I\boldsymbol{\omega}'; \qquad \text{where } \boldsymbol{\omega}' \text{ being the angular velocity about Q}'\text{BQ} \\ &= I\omega'\hat{e} \\ &= \frac{7}{3}Ma^2\omega'\hat{e}. \end{aligned}$$

Since B is made fixed. As the corner B is made fixed. Thus, angular momentum about B is conserved with the sentence: As the corner B is made fixed, the angular momentum about B is conserved.

$$\therefore \quad \boldsymbol{\omega} = \boldsymbol{\omega}',$$

$$\text{i.e.,} \quad I_0\omega\hat{e} = \frac{7}{3}Ma^2\omega'\hat{e},$$

$$\text{i.e.,} \quad \frac{1}{3}Ma^2\omega = \frac{7}{3}Ma^2\omega',$$

$$\text{i.e.,} \quad \omega = 7\omega',$$

$$\text{i.e.,} \quad \omega' = \frac{\omega}{7}.$$

2nd part: Let **J** be the impulse applied at B. Equation of motion of the C.I. under the impulsive force is

$$M(\mathbf{V}' - \mathbf{V}) = \mathbf{J}. \tag{7.8.76}$$

Now,

$$\begin{aligned}\mathbf{V} &= \text{Velocity of the C.I. just before impulse} \\ &= \mathbf{0}.\end{aligned}$$

$$\begin{aligned}\mathbf{V}' &= \text{Velocity of the C.I. just after impulse} \\ &= \boldsymbol{\omega}' \times \mathbf{r} \\ &= \boldsymbol{\omega}' \times \mathbf{BG}.\end{aligned}$$

Now,

$$\begin{aligned}\boldsymbol{\omega}' &= \omega'\hat{e} \\ &= \omega'\frac{\hat{i} + \hat{j}}{\sqrt{2}}.\end{aligned}$$

And,

$$\mathbf{BG} = -a\hat{i} + a\hat{j}.$$

$$\begin{aligned}\therefore \quad \boldsymbol{\omega}' \times \mathbf{BG} &= \frac{a\omega'}{\sqrt{2}}\begin{vmatrix}\hat{i} & \hat{j} & \hat{k} \\ 1 & 1 & 0 \\ -1 & 1 & 0\end{vmatrix} \\ &= \frac{a\omega'\hat{k}}{\sqrt{2}}2 = \sqrt{2}a\omega'\hat{k}\end{aligned}$$

$$\therefore \quad \mathbf{V}' = \sqrt{2}a\omega'\hat{k}.$$

Equation (7.8.76) gives

$$M\sqrt{2}a\omega'\hat{k} = \mathbf{J} = J\hat{k}.$$

$$\begin{aligned}\therefore \quad J &= M\sqrt{2}a\omega' \\ &= \sqrt{2}aM\frac{\omega}{7} \\ &= \frac{\sqrt{2}}{7}a\,\omega M = \frac{\sqrt{2}}{7}Ma\,\omega.\end{aligned}$$

Example 7.14 A cube is turning with angular velocity ω about the diagonal when suddenly the diagonal is let go and one of the edges which does not meet this diagonal is fixed, then show that the resulting angular velocity about this edge is $\frac{\omega\sqrt{3}}{12}$.

Solution Let OABCA′B′C′O′ be the cube of mass M and edges $2a$. In Fig. 7.15, take O as the origin, the X-axis along **OA**, the Y-axis along **OB**, and Z-axis along **OC**.

Let the diagonal **OO′** be the initial axis of rotation and **AC′** be the new axis of rotation. Let ω' be the angular velocity of the cube at time t about **AC′**

$$\mathbf{OO'} = 2a\hat{i} + 2a\hat{j} + 2a\hat{k}.$$

$$\therefore \quad \hat{e} = \text{unit vector along } \mathbf{OO'} = \frac{2a\hat{i} + 2a\hat{j} + 2a\hat{k}}{\sqrt{4a^2 + 4\dot{a}^2 + 4a^2}} = \frac{\hat{i} + \hat{j} + \hat{k}}{\sqrt{3}}$$

$$\mathbf{AC'} = 2a\hat{j}.$$

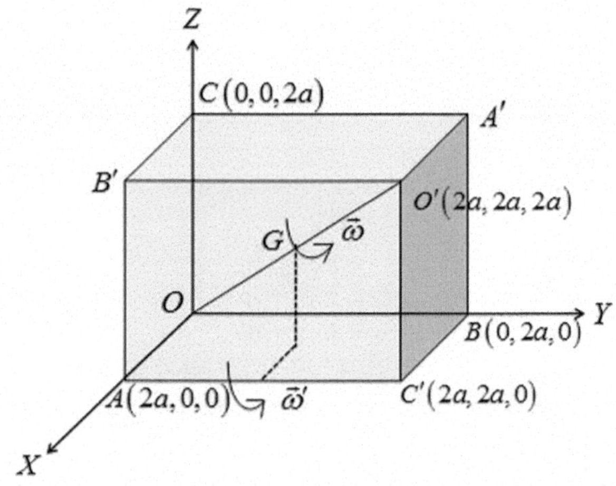

Fig. 7.15 Schematic diagram of a cube for Example 7.14

Therefore, unit vector along $\mathbf{AC}' = \hat{j}$.

Now

$$\boldsymbol{\omega} = \text{angular velocity about } \mathbf{OO}' = \omega\hat{e} = \omega\frac{\hat{i}+\hat{j}+\hat{k}}{\sqrt{3}} = \frac{\omega}{\sqrt{3}}\left(\hat{i}+\hat{j}+\hat{k}\right).$$

$$\boldsymbol{\omega}' = \text{angular velocity about } \mathbf{AC}' = \omega'\hat{j}.$$

As AC′ is made fixed, therefore impulse is applied along **AC′** and hence

$$\boldsymbol{\Lambda} = \text{external torque about } \mathbf{AC}' = \mathbf{0}.$$

Thus, the angular momentum about **AC′** is conserved.

$$\therefore \quad \boldsymbol{\Omega} = \boldsymbol{\Omega}'. \tag{7.8.77}$$

Now

$$\boldsymbol{\Omega} = \text{angular momentum of the cube about AC}' \text{ just before impulse,}$$

$$\text{i.e., } \boldsymbol{\Omega} = \left(I_0\boldsymbol{\omega}\cdot\hat{j}\right)\hat{j} = \frac{2}{3}Ma^2\left(\boldsymbol{\omega}\cdot\hat{j}\right)\hat{j} = \frac{2}{3}Ma^2\frac{\omega}{\sqrt{3}}\hat{j} = \frac{2Ma^2\omega}{3\sqrt{3}}\hat{j}.$$

Again

$$\boldsymbol{\Omega}' = \text{angular momentum of the cube about } \mathbf{AC}'$$

just after impulse

$$\text{i.e., } \boldsymbol{\Omega}' = I\boldsymbol{\omega}' = I\omega'\hat{j}.$$

Now

$$I = \text{M.I. of the cube about AC}' = \frac{2}{3}Ma^2 + M\left(a^2+a^2\right) = \frac{8}{3}Ma^2.$$

$$\Omega' = \frac{8}{3}Ma^2\omega'\hat{j}.$$

Therefore, Eq. (7.8.77) gives

$$\frac{2Ma^2\omega}{3\sqrt{3}} = \frac{8}{3}Ma^2\omega',$$

$$\therefore \quad \omega' = \frac{\omega}{4\sqrt{3}} = \frac{\omega\sqrt{3}}{12}.$$

Example 7.15 A uniform cube is spinning freely with angular velocity ω_0 about a diagonal through a corner O. When suddenly the diagonal OP of one of the faces through O becomes fixed, show that the new angular velocity is $\frac{2\sqrt{2}\omega_0}{5\sqrt{3}}$.

Solution Let OABCA′B′PO′ be the cube of mass M and edge $2a$. OO′ is the initial axis of rotation of the cube and OP is the new axis of rotation of the cube as depicted in Fig. 7.16.

Now ω_0 = initial angular velocity of the cube about **OO′** and let ω be the new angular velocity of the cube about **OP**.

$$\mathbf{OO'} = 2a\hat{i} + 2a\hat{j} + 2a\hat{k},$$

$$\therefore \quad \hat{e} = \text{Unit vector along } \mathbf{OO'} = \frac{2a\hat{i} + 2a\hat{j} + 2a\hat{k}}{\sqrt{4a^2 + 4a^2 + 4a^2}} = \frac{\hat{i} + \hat{j} + \hat{k}}{\sqrt{3}}.$$

$$\mathbf{OP} = 2a\hat{i} + 2a\hat{j},$$

$$\therefore \quad \hat{u} = \text{Unit vector along } \mathbf{OP} = \frac{2a\hat{i} + 2a\hat{j}}{\sqrt{4a^2 + 4a^2}} = \frac{\hat{i} + \hat{j}}{\sqrt{2}}.$$

As the diagonal **OP** is made fixed. Therefore, impulse is applied along **OP** and hence the external torque about **OP** = **0** = **Λ**.

Thus

$$\mathbf{\Lambda} = \mathbf{0}.$$

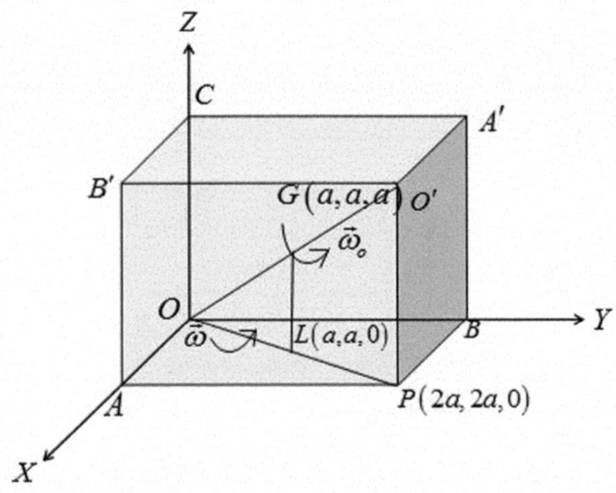

Fig. 7.16 Schematic diagram of a cube for Example 7.15

Therefore, by the principle of angular momentum, the angular momentum of the cube about **OP** is conserved.

$$\therefore \quad \boldsymbol{\Omega} = \boldsymbol{\Omega}'. \tag{7.8.78}$$

Now

$\boldsymbol{\Omega}$ = angular momentum of the cube about OP just before impulse,

$$\text{i.e.,}\quad \boldsymbol{\Omega} = (I_0\omega_0 \cdot \hat{u})\,\hat{u} = I_0\,(\omega_0 \cdot \hat{u})\,\hat{u} = I_0\,(\omega_0\hat{e} \cdot \hat{u})\,\hat{u} = \frac{2}{3}Ma^2\omega_0\frac{2}{\sqrt{6}}\hat{u} = \frac{4}{3\sqrt{6}}Ma^2\omega_0\hat{u}.$$

Again

Ω' = angular momentum of the cube about OPjust after impulse $= I\omega'\hat{u}$.

Now

$$I = \text{MI of the cube about OP} = \frac{2}{3}Ma^2 + Ma^2 = \frac{5}{3}Ma^2,$$

$$\therefore \quad \Omega' = \frac{5}{3}Ma^2\omega'\hat{u}.$$

Therefore, Eq. (7.8.78) gives

$$\frac{4}{3\sqrt{6}}Ma^2\omega_0 = \frac{5}{3}Ma^2\omega'$$

$$\text{or}\quad \omega' = \frac{4\omega_0}{5\sqrt{6}} = \frac{4\omega_0}{5\sqrt{2}\sqrt{3}} = \frac{2\sqrt{2}\omega_0}{5\sqrt{3}}.$$

Thus the result is proved.

7.9 Exercise-VII

1. An elliptic area, of eccentricity e, is rotating with angular velocity ω about one latus rectum; suddenly this latus rectum is loosed and the other fixed. Show that the new angular velocity is

$$\frac{1-4e^2}{1+4e^2}\omega.$$

2. A uniform rod of mass m and length $2a$ lies at rest on a smooth horizontal table. It is struck by a horizontal blow of impulse P at right angles to its length at a distance b from the center. Find the point about which it will begin to turn.
3. A uniform solid falls from rest, and strikes a smooth horizontal inelastic plane. Find the change of KE produced by the impact. If V is the velocity with which the body strikes the plane and if (f, g, h) are the coordinates of the point of contact and l, m, n are the direction cosines of the upward vertical with reference to the principal axes of inertia at the center of gravity, show the loss of KE is

$$\frac{1}{2}\frac{V^2}{\frac{1}{M}+\frac{\lambda^2}{A^2}+\frac{\mu^2}{B^2}+\frac{\nu^2}{C^2}},$$

where$\lambda = gn - hm, \mu = hl - fn, \nu = fm - gl,\ A, B, C$ are the principal moments of inertia about the center of gravity. (M being the mass of the body.)
4. A uniform rod AB, of length $2a$, is lying on a smooth horizontal plane and is struck by a horizontal blow, of impulse P, in a direction $\perp^r$ to the rod at a point distant b from its center; to find the motion.
5. Two equal rods AB, BC smoothly hinged at B rotate round B with angular velocity n, the rods being at right angles. If A is suddenly fixed, show that the rods are to close up at a rate $\frac{9n}{8}$.
6. Two uniform rods AB, BC are freely jointed at B and laid on a horizontal table; AB is truck by a horizontal blow of impulse P in a direction $\perp^r$ to AB at a distance C from its center; the length of AB, BC being $2a$ and $2b$ and their masses M and M', find the motion immediately after the blow.
7. A uniform circular disc of mass m initially at rest but free to move in any manner is suddenly set in motion by an impulse applied to a point A of its circumference. The initial velocity of A has components U along the radius of the circle through A,3U along the tangent to the circle at A, and 5U along a third $\perp^r$. Prove that the initial KE of the disc is given by $2T = 9mU^2$.
8. Two equal and uniform rods, AB and BC, are smoothly jointed at B and placed in a horizontal line; the rod BC is struck at G by a blow at right angles to it; find the position of G so that the angular velocities of AB and BC may be equal in magnitude.
9. Two rods, AB and BC, of length $2a$ and $2b$ of masses proportional to their lengths are freely jointed at B and lie in a straight line. A blow is communicated to the end A; show that the resulting kinetic energy when the system is free is to the energy when C is fixed as $(4a + 3b)\ (3a + 4b) : 12(a + b)^2$.
10. Compare the kinetic energies imparted to an elliptic disc by a blow J in its plane through a focus at right angles to the major axis (I) when the disc is free and (II) when the center of the disc is fixed.
11. A uniform rod AB of mass M and length $2a$ lies at rest on a smooth horizontal table. An impulse J is applied at A in the plane of the table and $\perp^r$ to the rod. Determine the velocity of the centroid and the angular velocity of the rod.

12. An elliptic lamina is rotating about its center on a smooth horizontal table. If ω_1, ω_2, ω_3 be its angular velocities when the extremity of its major axis, its focus and the extremity of its minor axis, respectively, become fixed, prove that $\frac{7}{\omega_1} = \frac{6}{\omega_2} + \frac{5}{\omega_3}$.
13. A uniform circular disc is spinning with angular velocity ω about a diameter when a point P on its rim is suddenly fixed. If the radius vector to P makes an angle α with this diameter, show that the angular velocities after fixing about the tangent and normal at P are $\frac{1}{5}\omega \sin\alpha$ & $\omega \cos\alpha$.
14. A uniform circular plate is turning in its own plane about a point A on its circumference with uniform angular velocity ω; suddenly A is released and another point B of the circumference is fixed; show that the angular velocity about B is $\frac{\omega}{3}(1 + 2\cos\alpha)$, where α is the angle that AB subtends at the center.
15. A uniform circular disc is rotating about a diameter when a point P of its edge is suddenly fixed. Prove that the velocity of the center O of the disc immediately afterward is $\frac{1}{5}$ of the velocity of P before it was fixed.
16. A circular plane rotates about an axis through its center $\perp^r$ to its plane with angular velocity ω. The axis is set free and a point in the circumference of the plane is fixed, showing that the resulting angular velocity is $\frac{1}{3}\omega$.

Chapter 8
Rigid Body Motion in Space with One Point Fixed

8.1 Moving Axes and Fixed Axes

Theorem 8.1 *Suppose that a rigid body turns about a point O fixed in the body and space. Let XYZ denote an inertial coordinate system with O as the origin. Consider another coordinate system xyz having the same origin O, rotating with respect to the XYZ system. If* $\frac{d\mathbf{A}}{dt}\Big]_S$ *and* $\frac{d\mathbf{A}}{dt}\Big]_M$ *denote, respectively, the time derivatives of* $\mathbf{A}$ *relative to the systems XYZ and xyz, respectively, then there exists a vector* $\boldsymbol{\omega}$ *such that*

$$\frac{d\mathbf{A}}{dt}\Big]_S = \frac{d\mathbf{A}}{dt}\Big]_M + \boldsymbol{\omega} \times \mathbf{A}.$$

Proof Let

$$\mathbf{A} = A_1\hat{i} + A_2\hat{j} + A_3\hat{k}, \tag{8.1.1}$$

where $\hat{i}$, $\hat{j}$, $\hat{k}$ are the unit vectors along **Ox**, **Oy**, **Oz**, respectively (see Fig. 8.1).

Equation (8.1.1) gives

$$\frac{d\mathbf{A}}{dt}\Big]_M = \dot{A}_1\hat{i} + \dot{A}_2\hat{j} + \dot{A}_3\hat{k}, \tag{8.1.2}$$

&

$$\frac{d\mathbf{A}}{dt}\Big]_S = \left(\dot{A}_1\hat{i} + \dot{A}_2\hat{j} + \dot{A}_3\hat{k}\right) + \left(A_1\dot{\hat{i}} + A_2\dot{\hat{j}} + A_3\dot{\hat{k}}\right) = \frac{d\mathbf{A}}{dt}\Big]_M + \left(A_1\dot{\hat{i}} + A_2\dot{\hat{j}} + A_3\dot{\hat{k}}\right). \tag{8.1.3}$$

N. Ahmed et al., *Classical Dynamics*, University Texts in the Mathematical Sciences,
https://doi.org/10.1007/978-981-95-6394-4_8

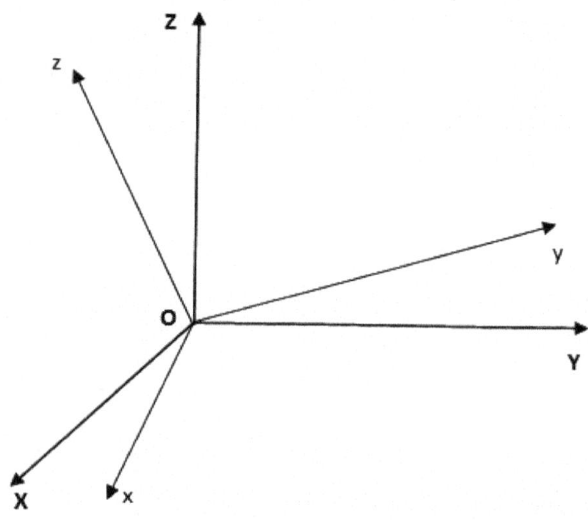

Fig. 8.1 Rotation of a coordinate system relative to another coordinate system with same origin

We note that

$\hat{i}^2 = 1,$

i.e., $\hat{i} \cdot \hat{i} = 1,$

$$\text{i.e.,}\quad \hat{i} \cdot \frac{d\hat{i}}{dt} = 0. \tag{8.1.4}$$

Equation (8.1.4) shows that $\frac{d\hat{i}}{dt}$ is perpendicular to $\hat{i}$.
That is $\frac{d\hat{i}}{dt}$ is a vector lying on the plane of $\hat{j}$ & $\hat{k}$ and so $\frac{d\hat{i}}{dt}$ can be expressed as

$$\frac{d\hat{i}}{dt} = \alpha_1 \hat{j} + \alpha_2 \hat{k}. \tag{8.1.5}$$

Similarly, we can write

$$\frac{d\hat{j}}{dt} = \alpha_3 \hat{k} + \alpha_4 \hat{i}, \tag{8.1.6}$$

&

$$\frac{d\hat{k}}{dt} = \alpha_5 \hat{i} + \alpha_6 \hat{j}. \tag{8.1.7}$$

Further, it is recalled that $\hat{i} \cdot \hat{j} = 0$ which gives

$$\hat{i} \cdot \frac{d\hat{j}}{dt} + \frac{d\hat{i}}{dt} \cdot \hat{j} = 0$$

$$\text{i.e.,} \quad \hat{i} \cdot \left(\alpha_3\hat{k} + \alpha_4\hat{i}\right) + \hat{j} \cdot \left(\alpha_1\hat{j} + \alpha_2\hat{k}\right) = 0,$$

$$\text{i.e.,} \quad \alpha_1 + \alpha_4 = 0,$$

$$\left.\begin{array}{ll} \text{i.e.,} & \alpha_4 = -\alpha_1 \\ \text{Similarly} & \alpha_5 = -\alpha_2 \\ \text{and} & \alpha_6 = -\alpha_3 \end{array}\right\}. \tag{8.1.8}$$

By the use of (8.1.8), Eqs. (8.1.5), (8.1.6), and (8.1.7) reduce to

$$\left.\begin{array}{l} \frac{d\hat{i}}{dt} = \alpha_1\hat{j} + \alpha_2\hat{k} \\ \frac{d\hat{j}}{dt} = \alpha_3\hat{k} - \alpha_1\hat{i} \\ \frac{d\hat{k}}{dt} = -\alpha_2\hat{i} - \alpha_3\hat{j} \end{array}\right\}. \tag{8.1.9}$$

Replacing α_1 by ω_3, α_2 by ω_2, and α_3 by ω_1, we obtain

$$\left.\begin{array}{l} \dot{\hat{i}} = \omega_3\hat{j} - \omega_2\hat{k} \\ \dot{\hat{j}} = \omega_1\hat{k} - \omega_3\hat{i} \\ \dot{\hat{k}} = \omega_2\hat{i} - \omega_1\hat{j} \end{array}\right\}. \tag{8.1.10}$$

Now Eq. (8.1.3) takes the form

$$\left.\frac{d\mathbf{A}}{dt}\right]_S = \left.\frac{d\mathbf{A}}{dt}\right]_M + A_1\left(\omega_3\hat{j} - \omega_2\hat{k}\right) + A_2\left(\omega_1\hat{k} - \omega_3\hat{i}\right) + A_3\left(\omega_2\hat{i} - \omega_1\hat{j}\right),$$

$$\text{i.e.,} \quad \left.\frac{d\mathbf{A}}{dt}\right]_S = \left.\frac{d\mathbf{A}}{dt}\right]_M + \hat{i}\left(\omega_2A_3 - \omega_3A_1\right) + \hat{j}\left(\omega_3A_1 - \omega_1A_2\right) + \hat{k}\left(\omega_1A_2 - \omega_2A_3\right),$$

$$\text{i.e.,} \quad \left.\frac{d\mathbf{A}}{dt}\right]_S = \left.\frac{d\mathbf{A}}{dt}\right]_M + \begin{vmatrix} \hat{i} & \hat{j} & \hat{k} \\ \omega_1 & \omega_2 & \omega_3 \\ A_1 & A_2 & A_3 \end{vmatrix}$$

$$\text{i.e.,} \quad \left.\frac{d\mathbf{A}}{dt}\right]_S = \left.\frac{d\mathbf{A}}{dt}\right]_M + \boldsymbol{\omega} \times \mathbf{A}, \tag{8.1.11}$$

where $\boldsymbol{\omega} = \omega_1 \hat{i} + \omega_2 \hat{j} + \omega_3 \hat{k}$.

In (8.1.11), $\boldsymbol{\omega}$ is the angular velocity of the moving system xyz relative to system XYZ fixed in space. □

8.2 Velocity and Angular Velocity of a Rigid Body with One Point Fixed

Let a rigid body rotate with angular velocity $\boldsymbol{\omega}$ about a point O fixed in the body. In Fig. 8.2, let P be a particle of the body with position vector $\mathbf{r}$ relative to O. If $\mathbf{v}$ denote the velocity of the particle, then

$$\mathbf{v} = \left.\frac{d\mathbf{r}}{dt}\right]_S = \left.\frac{d\mathbf{r}}{dt}\right]_M + \boldsymbol{\omega} \times \mathbf{r}. \tag{8.2.1}$$

We note that $\mathbf{r}$ is unchanged relative to the body.

$$\text{i.e.,} \quad \left.\frac{d\mathbf{r}}{dt}\right]_M = \mathbf{0}. \tag{8.2.2}$$

Using (8.2.2) in (8.2.1), it is obtained

$$\mathbf{v} = \boldsymbol{\omega} \times \mathbf{r}. \tag{8.2.3}$$

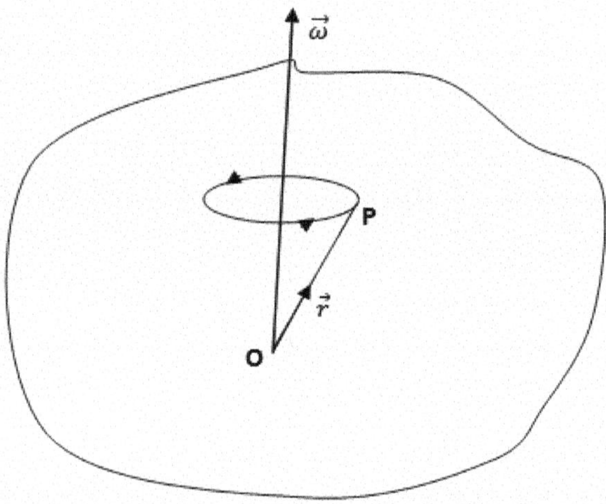

Fig. 8.2 Rigid body rotation about a fixed point in the body

8.3 Angular Momentum

Consider a rigid body rotating with angular velocity $\boldsymbol{\omega}$ about point O fixed in the body and space. Let $\mathbf{r}$ and $\mathbf{v}$ denote, respectively, the position vector and velocity of a typical particle P of the body. The angular momentum of the body about O is given by

$$\boldsymbol{\Omega} = \sum \mathbf{r} \times m\mathbf{v}, \tag{8.3.1}$$

where m is the mass of the particle P.

In Fig. 8.3, let $\mathbf{OX}, \ \mathbf{OY}, \ \mathbf{OZ}$ be three mutually perpendicular axes at O fixed in the body. Let $\hat{i}, \ \hat{j}, \ \hat{k}$ stipulate the unit vectors along $\mathbf{OX}, \ \mathbf{OY}, \ \mathbf{OZ}$, respectively. If $(x, \ y, \ z)$ are the coordinates of P, then

$$\mathbf{r} = \hat{i}x + \hat{j}y + \hat{k}z. \tag{8.3.2}$$

Let

$$\boldsymbol{\omega} = \omega_1\hat{i} + \omega_2\hat{j} + \omega_3\hat{k}. \tag{8.3.3}$$

Now

$$\mathbf{v} = \boldsymbol{\omega} \times \mathbf{r} = \begin{vmatrix} \hat{i} & \hat{j} & \hat{k} \\ \omega_1 & \omega_2 & \omega_3 \\ x & y & z \end{vmatrix} = v_1\hat{i} + v_2\hat{j} + v_3\hat{k}, \tag{8.3.4}$$

where

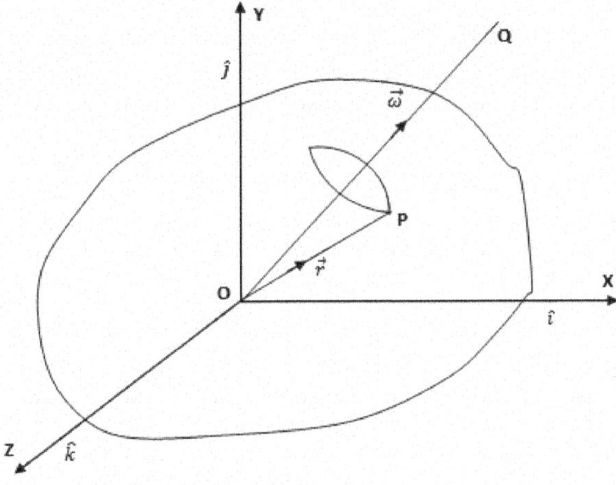

Fig. 8.3 Illustrative diagram for the angular momentum of a rigid body which rotates about a fixed point O

$$\left.\begin{aligned} v_1 &= z\omega_2 - y\omega_3 \\ v_2 &= x\omega_3 - z\omega_1 \\ v_3 &= y\omega_1 - x\omega_2 \end{aligned}\right\}. \tag{8.3.5}$$

We have

$$\mathbf{r} \times \mathbf{v} = \begin{vmatrix} \hat{i} & \hat{j} & \hat{k} \\ x & y & z \\ v_1 & v_2 & v_3 \end{vmatrix} = \hat{i}\,(yv_3 - zv_2) + \hat{j}\,(zv_1 - xv_3) + \hat{k}\,(xv_2 - yv_1)\,. \tag{8.3.6}$$

Equation (8.3.1) gives

$$\mathbf{\Omega} = \sum m \left\{\hat{i}\,(yv_3 - zv_2) + \hat{j}\,(zv_1 - xv_3) + \hat{k}\,(xv_2 - yv_1)\right\} = \Omega_1\hat{i} + \Omega_2\hat{j} + \Omega_3\hat{k}. \tag{8.3.7}$$

Now

$$\Omega_1 = \sum m\,(yv_3 - zv_2) = \sum m\,\{y\,(y\omega_1 - x\omega_2) - z\,(x\omega_3 - z\omega_1)\},$$

$$\text{i.e.,}\quad \Omega_1 = \sum m\left\{\omega_1\left(y^2 + z^2\right) - xy\omega_2 - zx\omega_3\right\} = A\omega_1 - F\omega_2 - E\omega_3. \tag{8.3.8}$$

Similarly

$$\Omega_2 = B\omega_2 - D\omega_3 - F\omega_1, \tag{8.3.9}$$

$$\Omega_3 = C\omega_3 - E\omega_1 - D\omega_2, \tag{8.3.10}$$

where

$$A = \sum m\left(y^2 + z^2\right) = \text{M.I. of the body about OX},$$

$$B = \sum m\left(z^2 + x^2\right) = \text{M.I. of the body about OY},$$

$$C = \sum m\left(x^2 + y^2\right) = \text{M.I. of the body about OZ},$$

$$D = \sum myz = \text{P.I. of the body about OY} - \text{OZ},$$

$$E = \sum mzx = \text{P.I. of the body about OZ} - \text{OX},$$

$$F = \sum mxy = \text{P.I. of the body about OX} - \text{OY}.$$

Equation (8.3.7) yields

$$\mathbf{\Omega} = (A\omega_1 - F\omega_2 - E\omega_3)\,\hat{i} + (B\omega_2 - D\omega_3 - F\omega_1)\,\hat{j} + (C\omega_3 - E\omega_1 - D\omega_2)\,\hat{k}.$$

8.4 Inertia Matrix and Inertia Tensor

Equations (8.3.8), (8.3.9), and (8.3.10) can be rewritten as follows:

$$\left.\begin{aligned} \Omega_1 &= \quad A\omega_1 - F\omega_2 - E\omega_3 \\ \Omega_2 &= -F\omega_1 + B\omega_2 - D\omega_3 \\ \Omega_3 &= -E\omega_1 - D\omega_2 + C\omega_3 \end{aligned}\right\}. \tag{8.4.1}$$

The set of equations (8.4.1) can be expressed by a single matrix equation as given below:

$$\begin{bmatrix} \Omega_1 \\ \Omega_2 \\ \Omega_3 \end{bmatrix} = \begin{bmatrix} A & -F & -E \\ -F & B & -D \\ -E & -D & C \end{bmatrix} \begin{bmatrix} \omega_1 \\ \omega_2 \\ \omega_3 \end{bmatrix},$$

$$\text{i.e., } \bar{\Omega} = J\bar{\omega}, \tag{8.4.2}$$

where

$$\mathbf{\Omega} = \begin{bmatrix} \Omega_1 \\ \Omega_2 \\ \Omega_3 \end{bmatrix}, \quad J = \begin{bmatrix} A & -F & -E \\ -F & B & -D \\ -E & -D & C \end{bmatrix}, \quad \boldsymbol{\omega} = \begin{bmatrix} \omega_1 \\ \omega_2 \\ \omega_3 \end{bmatrix}.$$

In Eq. (8.4.2), J is termed as the **Inertia Matrix**.
Let the symbols $J_{\alpha\beta}$ be introduced as follows:

$$\left.\begin{aligned} J_{11} = A, \quad J_{12} = -F, \; J_{13} = -E \\ J_{21} = -F, \quad J_{22} = B, \quad J_{23} = -D \\ J_{31} = -E, \; J_{32} = -D, \quad J_{33} = C \end{aligned}\right\}. \tag{8.4.3}$$

Then Eq. (8.4.1) can be expressed as

$$\left.\begin{aligned}\Omega_1 &= J_{11}\omega_1 + J_{12}\omega_2 + J_{13}\omega_3 = J_{1\beta}\omega_\beta \\ \Omega_2 &= J_{21}\omega_1 + J_{22}\omega_2 + J_{23}\omega_3 = J_{2\beta}\omega_\beta \\ \Omega_3 &= J_{31}\omega_1 + J_{32}\omega_2 + J_{33}\omega_3 = J_{3\beta}\omega_\beta\end{aligned}\right\}. \tag{8.4.4}$$

In tensor form, Eq. (8.4.4) may be written as

$$\Omega_\alpha = J_{\alpha\beta}\omega_\beta. \tag{8.4.5}$$

In (8.4.5), $J_{\alpha\beta}$ is called the **Inertia Tensor**. It is a symmetric tensor of rank 2.

8.5 Principal Axes of Inertia

A set of three rectangular axes having origin O which are fixed in a rigid body and rotating with it, and which are such that the products of inertia about them taken two at a time are zero, are called the principal axes of the body at O. If OX, OY, OZ are the principal axes for the body at O, then D=E=F=O and so in such case,

$$\boldsymbol{\Omega} = A\omega_1\hat{i} + B\omega_2\hat{j} + C\omega_3\hat{k}.$$

In the case of principal axes,

$$\left.\begin{aligned}D &= -J_{23} = -J_{32} = 0 \\ E &= -J_{31} = -J_{13} = 0 \\ F &= -J_{12} = -J_{21} = 0\end{aligned}\right\}. \tag{8.5.1}$$

Equation (8.5.1) depicts that, for principal axes, $J_{\alpha\beta} = 0 \; \forall \; \alpha \neq \beta$. Let a rigid body with one point fixed, rotate with angular velocity $\boldsymbol{\omega}$ about an axis OQ. Let $\boldsymbol{\Omega}$ be the angular momentum of the body about OQ.

If $\boldsymbol{\Omega} \parallel$ (like) $\boldsymbol{\omega}$, then OQ is said to be principal axis of the body at O. As $\boldsymbol{\Omega} \parallel$ (like) $\boldsymbol{\omega}$, therefore, $\exists$ a positive scalar I such that

$$\boldsymbol{\Omega} = I\boldsymbol{\omega}. \tag{8.5.2}$$

In (8.5.2), I is termed as a **principal moment of inertia** of the body at O.

8.6 Number of Principal Axes for a Rigid Body at the Point of Rotation

Let a rigid body with one point O fixed, rotate with angular velocity $\boldsymbol{\omega}$ about the instantaneous axis of rotation OQ. Let OX, OY, OZ be three mutually $\perp^r$ axes at O. Let $\hat{i}$, $\hat{j}$, $\hat{k}$ stipulate unit vectors along **OX**, **OY**, **OZ**, respectively.

Suppose that

$$\boldsymbol{\omega} = \omega_1 \hat{i} + \omega_2 \hat{j} + \omega_3 \hat{k}. \tag{8.6.1}$$

The angular momentum of the body about O is given by

$$\boldsymbol{\Omega} = \Omega_1 \hat{i} + \Omega_2 \hat{j} + \Omega_3 \hat{k}, \tag{8.6.2}$$

where

$$\left.\begin{aligned} \Omega_1 &= \quad A\omega_1 - F\omega_2 - E\omega_3 \\ \Omega_2 &= -F\omega_1 + B\omega_2 - D\omega_3 \\ \Omega_3 &= -E\omega_1 - D\omega_2 + C\omega_3 \end{aligned}\right\}.$$

Let the axis of rotation OQ be a principal axis.

$$\therefore \quad \boldsymbol{\Omega} = I\boldsymbol{\omega},$$

i.e., $(\Omega_1, \Omega_2, \Omega_3) = I\,(\omega_1, \omega_2, \omega_3)$,

$$\therefore \left.\begin{aligned} \Omega_1 &= I\omega_1 \\ \Omega_2 &= I\omega_2 \\ \Omega_3 &= I\omega_3 \end{aligned}\right\}. \tag{8.6.3}$$

First equation of (8.6.3) gives

$$A\omega_1 - F\omega_2 - E\omega_3 = I\omega_1,$$

i.e., $(A - I)\,\omega_1 - F\omega_2 - E\omega_3 = 0.$ (8.6.4)

Similarly second and third equations of (8.6.3) yield

$$-F\omega_1 + (B - I)\,\omega_2 - D\omega_3 = 0, \tag{8.6.5}$$

$$-E\omega_1 - D\omega_2 + (C - I)\,\omega_3 = 0. \tag{8.6.6}$$

Eliminating $\omega_1, \omega_2, \omega_3$ from (8.6.4), (8.6.5), (8.6.6), we get

$$\begin{vmatrix} A-I & -F & -E \\ -F & B-I & -D \\ -E & -D & C-I \end{vmatrix} = 0, \tag{8.6.7}$$

which is a cubic equation in I, and hence we get three values of I in general. That is, there exist three principal axes of the body at O. The roots of equation (8.6.7) are called the principal moments of inertia of the body at O.

8.7 Kinetic Energy of Rotation

Let a rigid body with one point O fixed to rotate with angular velocity $\boldsymbol{\omega}$. Let OQ stipulate the instantaneous axis of rotation. As displayed in Fig. 8.4, let OX, OY, OZ denote three mutually $\perp^r$ axes through O. Consider a particle P of mass m in the body and let $\mathbf{r}$ denote the position vector of P relative to O.

The velocity $\mathbf{v}$ of the particle P is given by

$$\mathbf{v} = \boldsymbol{\omega} \times \mathbf{r}. \tag{8.7.1}$$

The K.E of the body is given by

$$T = \frac{1}{2}\sum m\mathbf{v}^2 = \frac{1}{2}\sum m\mathbf{v}\cdot\mathbf{v} = \frac{1}{2}\sum m\boldsymbol{\omega}\times\mathbf{r}\cdot\mathbf{v} = \frac{1}{2}\sum m\boldsymbol{\omega}\cdot(\mathbf{r}\times\mathbf{v}),$$

$$\text{i.e.,}\quad T = \frac{1}{2}\boldsymbol{\omega}\cdot\sum \mathbf{r}\times m\mathbf{v} = \frac{1}{2}\boldsymbol{\omega}\cdot\boldsymbol{\Omega}. \tag{8.7.2}$$

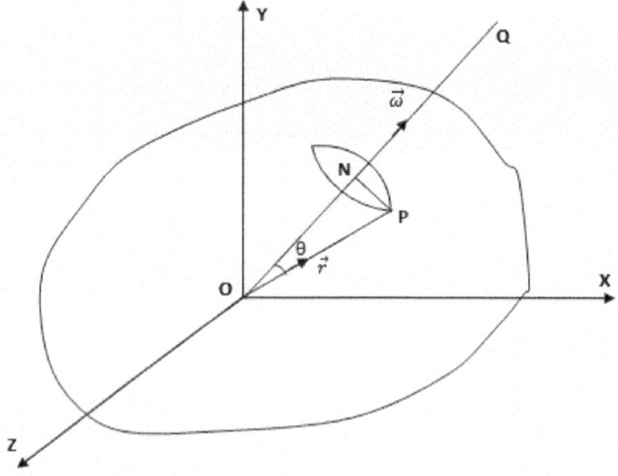

Fig. 8.4 Illustration for the KE of rotation of a rigid body about a fixed point

Let

$$\boldsymbol{\omega} = \omega_1 \hat{i} + \omega_2 \hat{j} + \omega_3 \hat{k}. \tag{8.7.3}$$

Further

$$\boldsymbol{\Omega} = \Omega_1 \hat{i} + \Omega_2 \hat{j} + \Omega_3 \hat{k}, \tag{8.7.4}$$

where

$$\left.\begin{aligned} \Omega_1 &= A\omega_1 - F\omega_2 - E\omega_3 \\ \Omega_2 &= B\omega_2 - D\omega_3 - F\omega_1 \\ \Omega_3 &= C\omega_3 - E\omega_1 - D\omega_2 \end{aligned}\right\}. \tag{8.7.5}$$

From Eqs. (8.7.3) and (8.7.4), we have

$$\boldsymbol{\omega} \cdot \boldsymbol{\Omega} = \omega_1 \Omega_1 + \omega_2 \Omega_2 + \omega_3 \Omega_3. \tag{8.7.6}$$

On simplification of (8.7.6), it is derived that

$$\boldsymbol{\omega} \cdot \boldsymbol{\Omega} = \omega_1 \Omega_1 + \omega_2 \Omega_2 + \omega_3 \Omega_3 = A\omega_1^2 + B\omega_2^2 + C\omega_3^2 - 2D\omega_2\omega_3 - 2E\omega_3\omega_1 - 2F\omega_1\omega_2$$

and hence the KE of rotation is given by

$$T = \frac{1}{2}\left(A\omega_1^2 + B\omega_2^2 + C\omega_3^2 - 2D\omega_2\omega_3 - 2E\omega_3\omega_1 - 2F\omega_1\omega_2\right).$$

KE of rotation in terms of inertia tensor:

$$T = \frac{1}{2}\boldsymbol{\omega} \cdot \boldsymbol{\Omega} = \frac{1}{2}\omega_\alpha \Omega_\alpha = \frac{1}{2} I_{\alpha\beta}\omega_\alpha\omega_\beta.$$

8.8 Euler's Dynamical Equations for the Motion of a Rigid Body with One Point Fixed

Statement Let a rigid body turn about a point O fixed in the body and space. Suppose that

- OX, OY, OZ are the principal axes through O fixed in the body.
- L, M, N are the components of the external torque about OX, OY, and OZ, respectively.
- $\omega_1, \omega_2, \omega_3$ are the resolved parts of the angular velocity $\boldsymbol{\omega}$ along **OX**, **OY**, **OZ**, respectively,
- $\hat{i}, \hat{j}, \hat{k}$ are the unit vectors along **OX**, **OY**, **OZ**, respectively.
- A, B, C are the principal moments of inertia about these axes,

then

$$\left.\begin{aligned} A\dot{\omega}_1 - (B - C)\,\omega_2\omega_3 = L \\ B\dot{\omega}_2 - (C - A)\,\omega_3\omega_1 = M \\ C\dot{\omega}_3 - (A - B)\ \omega_1\omega_2 = N \end{aligned}\right\}.$$

Proof The angular momentum $\boldsymbol{\Omega}$ of the body about O is given by

$$\boldsymbol{\Omega} = \Omega_1\hat{i} + \Omega_2\hat{j} + \Omega_3\hat{k}, \tag{8.8.1}$$

where

$$\left.\begin{aligned} \Omega_1 = A\omega_1 - F\omega_2 - E\omega_3 \\ \Omega_2 = B\omega_2 - D\omega_3 - F\omega_1 \\ \Omega_3 = C\omega_3 - E\omega_1 - F\omega_2 \end{aligned}\right\}. \tag{8.8.2}$$

As OX, OY, OZ are the principal axes for the body at O,

$$\therefore \quad D = E = F = 0. \tag{8.8.3}$$

Equation (8.8.3) leads Eq. (8.8.2) to take the form:

$$\Omega_1 = A\omega_1,\ \Omega_2 = B\omega_2,\ \Omega_3 = C\omega_3. \tag{8.8.4}$$

By use of (8.8.4), Eq. (8.8.1) reduces to

$$\boldsymbol{\Omega} = A\omega_1\hat{i} + B\omega_2\hat{j} + C\omega_3\hat{k}. \tag{8.8.5}$$

By the principle of angular momentum

$$\left.\frac{d\boldsymbol{\Omega}}{dt}\right]_F = \boldsymbol{\Lambda}, \tag{8.8.6}$$

where $\boldsymbol{\Lambda}$ is the external torque about O.

Equation (8.8.6) gives

$$\left.\frac{d\boldsymbol{\Omega}}{dt}\right]_M + \boldsymbol{\omega} \times \boldsymbol{\Omega} = \boldsymbol{\Lambda}. \tag{8.8.7}$$

Equation (8.8.5) yields

$$\left.\frac{d\boldsymbol{\Omega}}{dt}\right]_M = A\dot{\omega}_1\hat{i} + B\dot{\omega}_2\hat{j} + C\dot{\omega}_3\hat{k}. \tag{8.8.8}$$

Further

$$\boldsymbol{\omega} \times \boldsymbol{\Omega} = \begin{vmatrix} i & j & k \\ \omega_1 & \omega_2 & \omega_3 \\ A\omega_1 & B\omega_2 & C\omega_3 \end{vmatrix} = (C - B)\,\omega_2\omega_3\hat{i} + (A - C)\,\omega_3\omega_1\hat{j} + (B - A)\,\omega_1\omega_2\hat{k}. \tag{8.8.9}$$

We note that

$$\boldsymbol{\Lambda} = L\hat{i} + M\hat{j} + N\hat{k}. \tag{8.8.10}$$

On substitutions of Eqs. (8.8.8), (8.8.9), and (8.8.10) in Eq. (8.8.7) and equating the coefficients of $\hat{i},\ \hat{j},\ \hat{k}$ we obtain

$$\left.\begin{aligned} A\dot{\omega}_1 - (B - C)\,\omega_2\omega_3 &= L \\ B\dot{\omega}_2 - (C - A)\,\omega_3\omega_1 &= M \\ C\dot{\omega}_3 - (A - B)\,\omega_1\omega_2 &= N \end{aligned}\right\}. \tag{8.8.11}$$

Equation (8.8.11) is termed Euler's dynamical equations of motion of a rigid body with one point fixed. □

8.9 Euler's Equations in Case of the Axes Not Necessarily the Principal Axes

Suppose that the body turns about a point O fixed in the body. Let us introduce a coordinate system (x, y, z) fixed in the body. The angular momentum of the body about O is denoted by $\boldsymbol{\Omega}$ is given by

$$\boldsymbol{\Omega} = (\Omega_1,\ \Omega_2,\ \Omega_3)\,, \tag{8.9.1}$$

where

$$\left.\begin{aligned} \Omega_1 &= A\omega_1 - F\omega_2 - E\omega_3 \\ \Omega_2 &= B\omega_2 - D\omega_3 - F\omega_1 \\ \Omega_3 &= C\omega_3 - E\omega_1 - D\omega_2 \end{aligned}\right\}. \tag{8.9.2}$$

The angular momentum $\boldsymbol{\Omega}$ and external torque $\boldsymbol{\Lambda}$ about O are connected by

$$\left.\frac{d\boldsymbol{\Omega}}{dt}\right]_F = \boldsymbol{\Lambda},$$

$$\text{i.e.,} \quad \left.\frac{d\mathbf{\Omega}}{dt}\right]_M + \boldsymbol{\omega} \times \mathbf{\Omega} = \mathbf{\Lambda}, \tag{8.9.3}$$

$\boldsymbol{\omega}$ being the angular velocity of the moving system relative to the fixed system.

Equation (8.9.1) gives

$$\left.\frac{d\mathbf{\Omega}}{dt}\right]_M = \dot{\Omega}_1\hat{i} + \dot{\Omega}_2 j + \dot{\Omega}_3\hat{k}. \tag{8.9.4}$$

$$\boldsymbol{\omega} \times \mathbf{\Omega} = \begin{vmatrix} \hat{i} & \hat{j} & \hat{k} \\ \omega_1 & \omega_2 & \omega_3 \\ \Omega_1 & \Omega_2 & \Omega_3 \end{vmatrix} = \hat{i}\left(\omega_2\Omega_3 - \omega_3\Omega_2\right) + \hat{j}\left(\omega_3\Omega_1 - \omega_1\Omega_3\right) + \hat{k}\left(\omega_1\Omega_2 - \omega_2\Omega_1\right). \tag{8.9.5}$$

Let

$$\mathbf{\Lambda} = (L,\ M,\ N)\,. \tag{8.9.6}$$

Substituting Eqs. (8.9.4), (8.9.5), (8.9.6) in (8.9.3) and by equating the coefficients of $\hat{i},\ \hat{j},\ \hat{k}$, we derive the following equations:

$$\left.\begin{aligned} \dot{\Omega}_1 + (\omega_2\Omega_3 - \omega_3\Omega_2) &= L \\ \dot{\Omega}_2 + (\omega_3\Omega_1 - \omega_1\Omega_3) &= M \\ \dot{\Omega}_3 + (\omega_1\Omega_2 - \omega_2\Omega_1) &= N \end{aligned}\right\}. \tag{8.9.7}$$

If T denotes the KE of the body, then

$$T = \frac{1}{2}\left(A\omega_1^2 + B\omega_2^2 + C\omega_3^2 - 2D\omega_2\omega_3 - 2E\omega_3\omega_1 - 2F\omega_1\omega_2\right). \tag{8.9.8}$$

Equation (8.9.8) gives

$$\left.\begin{aligned} \frac{\partial T}{\partial\omega_1} &= A\omega_1 - F\omega_2 - E\omega_3 = \Omega_1 \\ \frac{\partial T}{\partial\omega_2} &= B\omega_2 - D\omega_3 - F\omega_1 = \Omega_2 \\ \frac{\partial T}{\partial\omega_3} &= C\omega_3 - E\omega_1 - D\omega_2 = \Omega_3 \end{aligned}\right\}. \tag{8.9.9}$$

Using (8.9.9) in (8.9.7), we obtain

$$\left.\begin{aligned} \frac{d}{dt}\left(\frac{\partial T}{\partial\omega_1}\right) - \omega_3\frac{\partial T}{\partial\omega_2} + \omega_2\frac{\partial T}{\partial\omega_3} &= L \\ \frac{d}{dt}\left(\frac{\partial T}{\partial\omega_2}\right) - \omega_1\frac{\partial T}{\partial\omega_3} + \omega_3\frac{\partial T}{\partial\omega_1} &= M \\ \frac{d}{dt}\left(\frac{\partial T}{\partial\omega_3}\right) - \omega_2\frac{\partial T}{\partial\omega_1} + \omega_1\frac{\partial T}{\partial\omega_2} &= N \end{aligned}\right\}. \tag{8.9.10}$$

The set of Equations (8.9.10) is Euler's equations of motion of a rigid body with one point fixed, in which the axes OX, OY, and OZ are not necessarily the principal axes.

8.10 Integrals of Energy and Angular Momentum

Suppose that a rigid body rotates under no external forces, with one point O of it fixed in the body and space. Let OX, OY, and OZ be the principal axes for the body at O and $\boldsymbol{\omega} = (\omega_1, \omega_2, \omega_3)$ denote the angular velocity of the body about O. Let A, B, and C denote the moments of inertia of the body about OX, OY, and OZ, respectively.

Euler's dynamical equations for the motion of a body under no external forces are

$$A\dot{\omega}_1 - (B - C)\ \omega_2\omega_3 = 0, \tag{8.10.1}$$

$$B\dot{\omega}_2 - (C - A)\ \omega_3\ \omega_1 = 0, \tag{8.10.2}$$

$$C\dot{\omega}_3 - (A - B)\ \omega_1\ \omega_2 = 0. \tag{8.10.3}$$

Now (8.10.1) $\times\ \omega_1$ + (8.10.2) $\times\ \omega_2$ + (8.10.3) $\times\ \omega_3$ gives

$$A\omega_1\dot{\omega}_1 + B\omega_2\dot{\omega}_2 + C\omega_3\dot{\omega}_3 = 0. \tag{8.10.4}$$

Integrating (8.10.4), we get

$$A\omega_1^2 + B\omega_2^2 + C\omega_3^2 = \text{a constant} = 2\lambda\ (\text{say}). \tag{8.10.5}$$

The kinetic energy of the body is given by

$$T = \frac{1}{2}\left(A\omega_1^2 + B\omega_2^2 + C\omega_3^2\right) = \frac{1}{2}2\lambda = \lambda = \text{a Constant}$$

and hence Eq. (8.10.5) shows that the KE of the body is conserved. Equation (8.10.5) is termed as the **Integral of Energy**.

Again (8.10.1) $\times\ A\omega_1$ + (8.10.2) $\times\ B\omega_2$ + (8.10.3) $\times\ C\omega_3$ gives

$$A^2\omega_1\dot{\omega}_1 + B^2\omega_2\dot{\omega}_2 + C^2\omega_3\dot{\omega}_3 = 0. \tag{8.10.6}$$

Integrating (8.10.6), we derive

$$A^2\omega_1^2 + B^2\omega_2^2 + C^2\omega_3^2 = \text{a constant} = \mu^2\ (\text{say}). \tag{8.10.7}$$

The angular momentum $\boldsymbol{\Omega}$ about O is given by

$$\boldsymbol{\Omega} = (\Omega_1, \Omega_2, \Omega_3) = (A\omega_1, B\omega_2, C\omega_3) .$$

Thus,

$$\Omega^2 = A^2\omega_1^2 + B^2\omega_2^2 + C^2\omega_3^2 = \mu^2,$$

or

$$|\boldsymbol{\Omega}| = \mu = \text{a constant.}$$

Equation (8.10.7) shows that the magnitude of the angular momentum of the body about O is conserved and so (8.10.7) is called the **Integral of Angular Momentum**.

8.11 Invariable Line

Let a rigid body with one point O fixed rotate with angular velocity $\boldsymbol{\omega}$ under no forces. Let the three mutually perpendicular axes OX, OY, and OZ fixed in the body be the principal axes for the body. Let $\boldsymbol{\omega} = (\omega_1, \omega_2, \omega_3)$. Under usual notations, the kinetic energy of the body is given by

$$T = \frac{1}{2}\left(A\omega_1^2 + B\omega_2^2 + C\omega_3^2\right) . \tag{8.11.1}$$

The angular momentum $\boldsymbol{\Omega}$ of the body about O is given by

$$\boldsymbol{\Omega} = A\omega_1\hat{i} + B\omega_2\hat{j} + C\omega_3\hat{k}, \tag{8.11.2}$$

where $\hat{i},\ \hat{j},\ \hat{k}$ stipulate the unit vectors along **OX**, **OY**, **OZ**, respectively.

As there is no force acting on the body, it follows from the integrals of energy and angular momentum that T and $|\boldsymbol{\Omega}| = \Omega$ are constants. In the absence of an external force

$$\boldsymbol{\Lambda} = \text{external torque about the point O} = \mathbf{0}. \tag{8.11.3}$$

Further, the angular momentum $\boldsymbol{\Omega}$ and external torque $\boldsymbol{\Lambda}$ are connected by the relation

$$\left.\frac{d\boldsymbol{\Omega}}{dt}\right]_F = \boldsymbol{\Lambda}, \tag{8.11.4}$$

where F refers to the system fixed in space.

Equation (8.11.3) leads Eq. (8.11.4) to take the form:

$$\left.\frac{d\boldsymbol{\Omega}}{dt}\right]_F = \mathbf{0}. \tag{8.11.5}$$

Equation (8.11.5) shows that $\boldsymbol{\Omega}$ is a constant vector in space, and so its direction is fixed in space. The line through point O indicating the direction of $\boldsymbol{\Omega}$ is called the invariable line. It is also observed that

$$T = \frac{1}{2}\left(\boldsymbol{\omega} \cdot \boldsymbol{\Omega}\right),$$

i.e., $\boldsymbol{\omega} \cdot \boldsymbol{\Omega} = 2T$,

i.e., $\boldsymbol{\omega} \cdot \Omega\hat{\Omega} = 2T$,

i.e., $\boldsymbol{\omega} \cdot \hat{\Omega} = \dfrac{2T}{\Omega} = \text{a constant}$,

i.e., $\boldsymbol{\omega} \cdot \hat{\Omega} = p$, (8.11.6)

where $p = \dfrac{2T}{\Omega}$.

With usual notations, the equation of a plane in normal form is

$$\mathbf{r} \cdot \hat{n} = p. \tag{8.11.7}$$

Comparing Eq. (8.11.6) with (8.11.7), it can be concluded that the locus of $\boldsymbol{\omega}$ is a plane with $\hat{\Omega}$ as the unit normal vector. The distance of the plane (8.11.6) from the point O, the point of rotation, is $p = \frac{2T}{\Omega}$ = a constant.

The plane represented by Eq. (8.11.6) which is perpendicular to a constant vector $\hat{\Omega}$ and which is at a constant distance from the point of rotation O is called the **invariable plane** as it is fixed in space. As the invariable line is parallel to the vector $\boldsymbol{\Omega} = (A\omega_1,\ B\omega_2,\ C\omega_3)$.

Therefore, the direction ratios of the invariable line are $A\omega_1,\ B\omega_2,\ C\omega_3$, and so the invariable line through the point O is given by

$$\frac{x}{A\omega_1} = \frac{y}{B\omega_2} = \frac{z}{C\omega_3}. \tag{8.11.8}$$

Again since $\boldsymbol{\omega} = (\omega_1,\ \omega_2,\ \omega_3)$ therefore the instantaneous axis of rotation is given by

$$\frac{x}{\omega_1} = \frac{y}{\omega_2} = \frac{z}{\omega_3}. \tag{8.11.9}$$

Definition If a rigid body with one point O fixed rotates with angular velocity $\boldsymbol{\omega}$ under no external forces, then the line through the point O parallel to the vector $\boldsymbol{\Omega}$, where $\boldsymbol{\Omega}$ is the angular momentum of the body about O, is called **invariable line** for the body. The plane which is perpendicular to the invariable line and which is at a constant distance $\frac{2T}{\Omega}$ from O, T being the KE of the body, is called the **invariable plane** for the body.

8.12 Euler's Geometrical Equations

8.12.1 Eulerian Angles

Let a rigid body rotate about a point O fixed in the body and space. In Fig. 8.5, let Ox, Oy, Oz be a set of tri-rectangular axes fixed in space and Ox', Oy', Oz' be the principal axes of the body at O. The axes Ox', Oy', Oz' are fixed in the body and so they move when the body moves. Initially Ox', Oy', Oz' coincide with Ox, Oy, Oz. In order to know the position of the body at any instant referred to the initial position three angles θ, ϕ, ψ are chosen to define the position of the principal axes and therefore of the body itself. The angles θ, ϕ, ψ are termed as the Eulerian angles.

Let OA be the line of intersection of the planes xOy & x'Oy'. θ is the angle between Oz & Oz'. ϕ is the angle between Ox and OA and ψ is the angle between Ox' and OA.

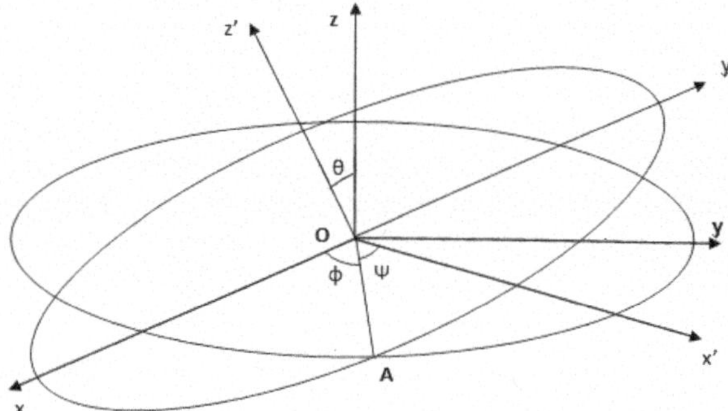

Fig. 8.5 Geometrical interpretation of θ, ϕ, ψ

8.12.2 Euler's Geometrical Equations

Let a rigid body turn about a point O fixed in the body. If ω_1, ω_2, ω_3 be the components of the angular velocity about the principal axes Ox', Oy', Oz' to determine ω_1, ω_2, ω_3 in terms of Eulerian angles θ, ϕ, ψ.

We consider three separate figures how xyz coordinate system is rotated into $x'y'z'$ system by successive rotations through the Eulerian angles θ, ϕ, ψ.

Let

$$\hat{\imath},\ \hat{\jmath},\ \hat{k};\ \hat{I},\ \hat{J},\ \hat{K};\ \hat{I}',\ \hat{J}',\ \hat{K}';\ \hat{\imath}',\ \hat{\jmath}',\ \hat{k}'$$

stipulate unit vectors along **Ox**, **Oy**, **Oz**; **OX**, **OY**, **OZ**; **OX**$'$, **OY**$'$, **OZ**$'$; **Ox**$'$, **Oy**$'$, **Oz**$'$, respectively.

From Fig. 8.6:

$$\left.\begin{aligned} \hat{\imath} &= \hat{I}\cos\phi + \hat{J}\cos\left(\tfrac{\pi}{2}+\phi\right) + \hat{K}\cos 90^0 = \hat{I}\cos\phi - \hat{J}\sin\phi \\ \hat{\jmath} &= \hat{I}\cos\left(\tfrac{\pi}{2}-\phi\right) + \hat{J}\cos\phi + \hat{K}\cos 90^0 = \hat{I}\sin\phi + \hat{J}\cos\phi \\ \hat{k} &= \hat{K} \end{aligned}\right\}. \tag{8.12.1}$$

From Fig. 8.7:

$$\left.\begin{aligned} \hat{I} &= \hat{I}' \\ \hat{J} &= \hat{I}'\cos 90^0 + \hat{J}'\cos\theta + \hat{K}'\cos\left(90^0+\theta\right) = \hat{J}'\cos\theta - \hat{K}'\sin\theta \\ \hat{K} &= \hat{I}'\cos 90^0 + \hat{J}'\cos\left(90^0-\theta\right) + \hat{K}'\cos\theta = \hat{J}'\sin\theta + \hat{K}'\cos\theta \end{aligned}\right\}. \tag{8.12.2}$$

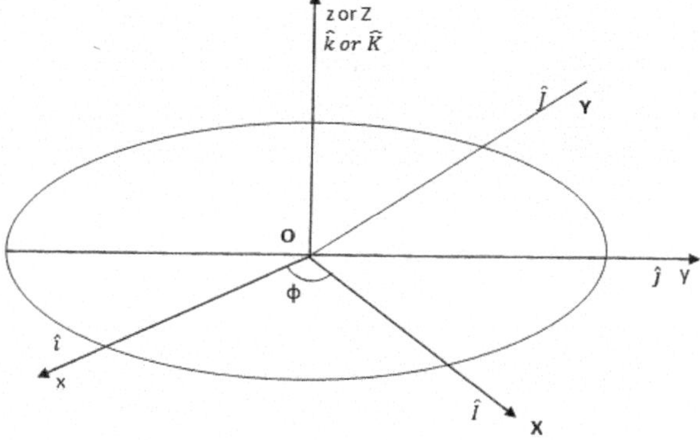

Fig. 8.6 Rotation of coordinate system through angle ϕ about the z-axis

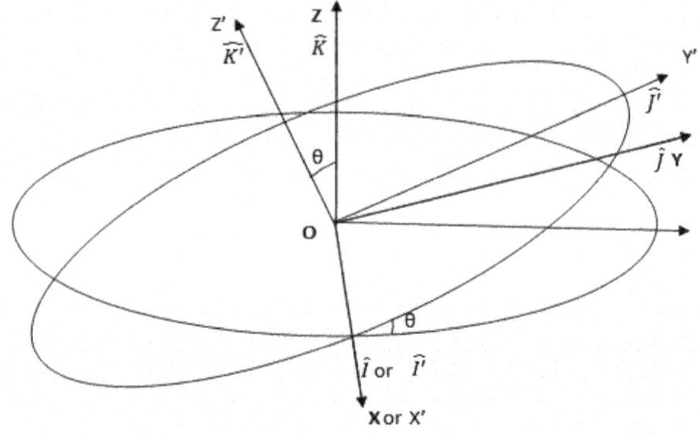

Fig. 8.7 Rotation of coordinate system about the new X-axis through an angle θ

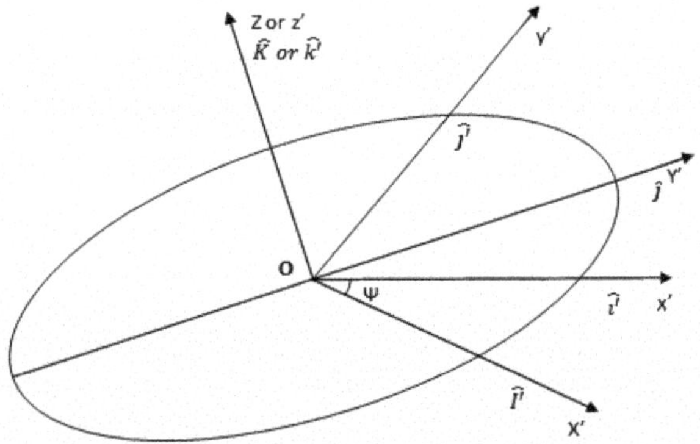

Fig. 8.8 Rotation of coordinate system about the new Z-axis through an angle ψ

From Fig. 8.8:

$$\left.\begin{aligned} \hat{I}' &= \hat{i}' \cos\psi + \hat{j}' \cos\left(90^0 + \psi\right) + \hat{k}' \cos 90^0 = \hat{i}' \cos\psi - \hat{j}' \sin\psi \\ \hat{J}' &= \hat{i}' \cos\left(90^0 - \psi\right) + \hat{j}' \cos\psi + \hat{k}' \cos 90^0 = \hat{i}' \sin\psi + \hat{j}' \cos\psi \\ \hat{K}' &= \hat{k}' \end{aligned}\right\}. \tag{8.12.3}$$

We have

$$\boldsymbol{\omega} = \omega_1 \hat{i}' + \omega_2 \hat{j}' + \omega_3 \hat{k}'. \tag{8.12.4}$$

Again

$$\boldsymbol{\omega} = \omega_\varphi \hat{K} + \omega_\theta \hat{I}' + \omega_\psi \hat{k}' = \dot{\phi}\left(\hat{J}' \sin\theta + \hat{K}' \cos\theta\right) + \dot{\theta}\left(\hat{i}' \cos\psi - \hat{j}' \sin\psi\right) + \dot{\psi}\hat{k}',$$

i.e., $\boldsymbol{\omega} = \dot{\phi}\left\{\sin\theta\left(\hat{i}' \sin\psi + \hat{j}' \cos\psi\right) + \hat{k}' \cos\theta\right\} + \dot{\theta}\left(\hat{i}' \cos\psi - \hat{j}' \sin\psi\right) + \dot{\psi}\hat{k}',$

$$\text{i.e.,}\quad \boldsymbol{\omega} = \hat{i}'\left(\dot{\phi}\sin\theta\sin\psi + \dot{\theta}\cos\psi\right) + \hat{j}'\left(\dot{\phi}\sin\theta\cos\psi - \dot{\theta}\sin\psi\right) + \hat{k}'\left(\dot{\phi}\cos\theta + \dot{\psi}\right). \tag{8.12.5}$$

Equations (8.12.4) and (8.12.5) give

$$\left.\begin{array}{l} \omega_1 = \dot{\phi}\sin\theta\sin\psi + \dot{\theta}\cos\psi \\ \omega_2 = \dot{\phi}\sin\theta\cos\psi - \dot{\theta}\sin\psi \\ \omega_3 = \dot{\phi}\cos\theta + \dot{\psi} \end{array}\right\}. \tag{8.12.6}$$

The set of equations (8.12.6) is called **Euler's Geometrical Equations**.

Deduction: If θ is replaced by $-\theta$, ϕ by $90^0 + \psi$, and ψ by $90^0 + \phi$, Eq. (8.12.6) transforms to

$$\left.\begin{array}{l} \omega_1 = -\dot{\psi}\sin\theta\cos\phi + \dot{\theta}\sin\phi \\ \omega_2 = \dot{\psi}\sin\theta\sin\phi + \dot{\theta}\cos\phi \\ \omega_3 = \dot{\phi} + \dot{\psi}\cos\theta \end{array}\right\}. \tag{8.12.7}$$

8.13 Illustrations

Example 8.1 A rigid body is moving freely about a point O under no external forces. Prove that if two of the principal moments of inertia at O are equal, the instantaneous axis of rotation describes a circular cone, with a vertex at O, fixed in the body.

Solution The integrals of energy and angular momentum of a rigid body rotating about a fixed point O under no external forces with usual notations are

$$A\omega_1^2 + B\omega_2^2 + C\omega_3^2 = 2T = \text{a constant}, \tag{8.13.1}$$

$$A^2\omega_1^2 + B^2\omega_2^2 + C^2\omega_3^2 = \Omega^2 = \text{a constant}. \tag{8.13.2}$$

Equations of the instantaneous axis of rotation are

$$\frac{x}{\omega_1} = \frac{y}{\omega_2} = \frac{z}{\omega_3}. \tag{8.13.3}$$

Now (8.13.1) ÷ (8.13.2) yields

$$\frac{A\omega_1^2 + B\omega_2^2 + C\omega_3^2}{A^2\omega_1^2 + B^2\omega_2^2 + C^2\omega_3^2} = \frac{2T}{\Omega^2}$$

$$\text{i.e., } \Omega^2\left(A\omega_1^2 + B\omega_2^2 + C\omega_3^2\right) = 2T\left(A^2\omega_1^2 + B^2\omega_2^2 + C^2\omega_3^2\right). \tag{8.13.4}$$

The locus of the line (8.13.3) subject to the condition (8.13.4) is

$$\Omega^2\left(Ax^2 + By^2 + Cz^2\right) = 2T\left(A^2x^2 + B^2y^2 + C^2z^2\right). \tag{8.13.5}$$

In the present case, $A = B = K$ (say).

Thus, Eq. (8.13.5) reduces to

$$\left(\Omega^2K - 2TK^2\right)x^2 + \left(\Omega^2K - 2TK^2\right)y^2 = \left(2TC^2 - \Omega^2C\right)z^2,$$

$$\text{i.e., } \lambda x^2 + \lambda y^2 = \mu z^2, \tag{8.13.6}$$

where $\lambda = \Omega^2K - 2TK^2$, $\mu = 2TC^2 - \Omega^2C$.

Equation (8.13.6) can be written as

$$x^2 + y^2 = \frac{\mu}{\lambda}z^2 = \nu z^2$$

which represents a right circular cone with O as the vertex and OZ as the axis of rotation.

Example 8.2 A circular disc of radius a and mass m is supported on a needle point at its center. It is set spinning with angular velocity ω_0 about a line making an angle α with the normal to the disc. Find the angular velocity of the disc at any subsequent time.

Solution In Fig. 8.9, take the center O as the origin, the Z-axis along the normal, Y-axis and X-axis on the plane of the disc in such a way that the axis of rotation OQ lies on the YZ-plane. Let $\hat{i}, \hat{j}, \hat{k}$ stipulate the unit vectors along OX, OY, OZ, respectively. It is to be noted that OX, OY, OZ are the principal axes for the disc at O. Let $\boldsymbol{\omega} = (\omega_1, \omega_2, \omega_3)$ be the angular velocity of the disc about OQ at time t.

The initial angular velocity of the disc is given by

$$\boldsymbol{\omega}_0 = \omega_0\hat{e} = \omega_0\left(\lambda\hat{i} + \mu\hat{j} + \nu\hat{k}\right) = \omega_0\left\{\hat{i}\cos\frac{\pi}{2} + \hat{j}\cos\left(\frac{\pi}{2} - \alpha\right) + \hat{k}\cos\alpha\right\},$$

$$\text{i.e., } \boldsymbol{\omega}_0 = (\omega_0\sin\alpha)\,\hat{j} + (\omega_0\cos\alpha)\,\hat{k}.$$

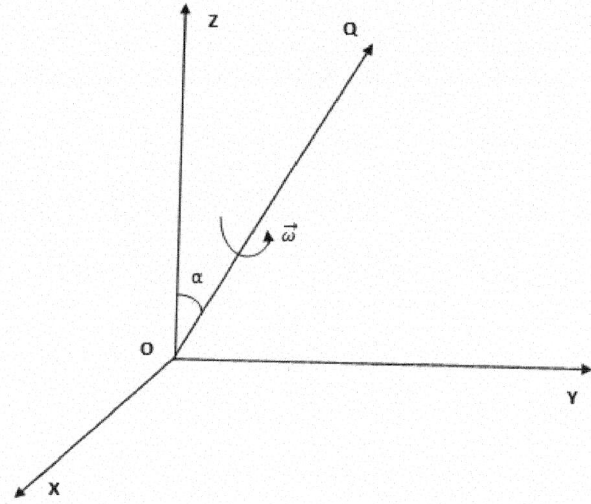

Fig. 8.9 Illustration for Example 8.2

$\therefore$ At time $t = 0$; $\omega_1 = 0,\ \omega_2 = \omega_0 \sin\alpha,\ \omega_3 = \omega_0 \cos\alpha = \xi$ (say) .
Euler's equations of motion are

$$A\dot{\omega}_1 - (B - C)\ \omega_2\omega_3 = L, \tag{8.13.7}$$

$$B\dot{\omega}_2 - (C - A)\ \omega_3\ \omega_1 = M, \tag{8.13.8}$$

$$C\dot{\omega}_3 - (A - B)\ \omega_1\ \omega_2 = N. \tag{8.13.9}$$

The external forces acting on the disc are

(I) The weight W through O in a downward vertical direction.
(II) The reaction at O.

$$\therefore\ \ L = M = N = 0. \tag{8.13.10}$$

Now

$$\text{A} = \text{M.I of the disc about OX} = \frac{1}{4}Ma^2,$$

$$\text{B} = \text{M.I of the disc about OY} = \frac{1}{4}Ma^2,$$

$$\text{C} = \text{M.I of the disc about OZ} = A + B = \frac{1}{2}Ma^2.$$

Equation (8.13.9) gives $\dot{\omega}_3 = 0$.
Integrating the above equation, we get ω_3 = a constant= C_3 (say).
Initially $\omega_3 = \lambda$ at $t = 0$, $\therefore C_3 = \lambda$ and it gives

$$\omega_3 = \lambda. \tag{8.13.11}$$

Equation (8.13.7) yields

$$\frac{1}{4}ma^2\dot{\omega}_1 - \left(\frac{1}{4}ma^2 - \frac{1}{2}ma^2\right)\omega_2\omega_3 = 0,$$

$$\text{i.e.,}\quad \dot{\omega}_1 = -\lambda\omega_2. \tag{8.13.12}$$

Equation (8.13.8) reduces to

$$\frac{1}{4}ma^2\dot{\omega}_2 - \left(\frac{1}{2}ma^2 - \frac{1}{4}ma^2\right)\omega_3\omega_1 = 0,$$

$$\text{i.e.,}\quad \dot{\omega}_2 = -\lambda\omega_1. \tag{8.13.13}$$

Differentiating (8.13.12) w.r.t. t, we get

$$\ddot{\omega}_1 = -\lambda\dot{\omega}_2 = -\lambda^2\omega_1$$

$$\text{i.e.,}\quad \left(D^2 + \lambda^2\right)\omega_1 = 0. \tag{8.13.14}$$

Solving the d.e. (8.13.14), we obtain

$$\omega_1 = A_1 \cos \lambda t + B_1 \sin \lambda t. \tag{8.13.15}$$

Equation (8.13.15) gives

$$\dot{\omega}_1 = -A_1\lambda \sin \lambda t + B_1\lambda \cos \lambda t$$

$$\text{i.e.,}\quad -\lambda\omega_2 = -A_1\lambda \sin \lambda t + B_1\lambda \cos \lambda t,$$

$$\text{i.e.,}\quad \omega_2 = A_1 \sin \lambda t - B_1 \cos \lambda t. \tag{8.13.16}$$

Initially $\omega_1 = 0, \quad \omega_2 = \omega_0 \sin\alpha, \quad \text{at } t = 0.$
Therefore, Eqs. (8.13.15) and (8.13.16) give

$$A_1 = 0, \quad B_1 = -\omega_0 \sin\alpha. \tag{8.13.17}$$

By the use of Eq. (8.13.17), Eqs. (8.13.15) and (8.13.16) take the following forms:

$$\omega_1 = -\omega_0 \sin\alpha \sin\lambda t, \tag{8.13.18}$$

$$\omega_2 = -\omega_0 \sin\alpha \cos\lambda t. \tag{8.13.19}$$

Thus, the angular velocity of the disc at time t is given by

$$\boldsymbol{\omega} = \omega_1\hat{i} + \omega_2\hat{j} + \omega_3\hat{k},$$

where

$$\left.\begin{aligned} \omega_1 &= -\omega_0 \sin\alpha \sin\lambda t \\ \omega_2 &= -\omega_0 \sin\alpha \cos\lambda t \\ \omega_3 &= \lambda = \omega_0 \cos\alpha \end{aligned}\right\}.$$

Example 8.3 A rectangular lamina of sides $2a$ and $2b$ rotates about a diagonal with constant angular velocity $\boldsymbol{\omega}$. Use Euler's equation to determine the components of the moment of the forces about the principal axes of the lamina.

Solution Let ABCD be the rectangular lamina with center O. Consider the X-axis, Y-axis, and Z-axis through O $\parallel$ to **AB**, **BC** and $\perp^r$ to the plane of the lamina, respectively.

Now

$$A = \text{M.I. of the lamina about OX} = \frac{1}{3}mb^2,$$

$$B = \text{M.I. of the lamina about OY} = \frac{1}{3}ma^2,$$

$$C = \text{M.I. of the lamina about OZ} = \frac{1}{3}m(a^2 + b^2),$$

$$D = \text{PI of the lamina about OY-OZ} = 0,$$

$$E = \text{PI of the lamina about OZ-OX} = 0,$$

$$D = \text{PI of the lamina about OX-OY} = 0 \text{ (due to symmetry)}.$$

Thus $D = E = F = O$ and hence OX, OY, and OZ are the principal axes for the body at O. In Fig. 8.10, let $\angle\text{XOC} = \theta$, and so

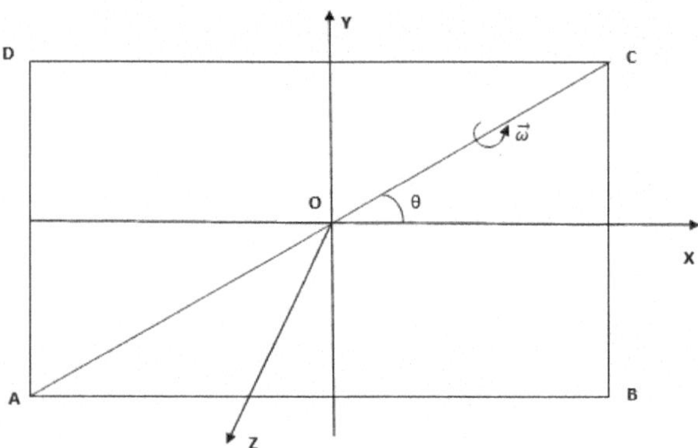

Fig. 8.10 Illustration for Example 8.3

$$\tan\theta = \frac{b}{a}. \tag{8.13.20}$$

Now

$\omega_1 =$ component of the angular velocity about OX $= \omega\cos\theta$,

$\omega_2 =$ that of the body about OY $= \omega\sin\theta$,

$\omega_3 =$ component of $\boldsymbol{\omega}$ about OZ $= 0$.

Euler's equations of motion are

$$\left.\begin{aligned} A\dot{\omega}_1 - (B-C)\,\omega_2\omega_3 = L \\ B\dot{\omega}_2 - (C-A)\,\omega_3\omega_1 = M \\ C\dot{\omega}_3 - (A-B)\,\omega_1\omega_2 = N \end{aligned}\right\}. \tag{8.13.21}$$

Equation (8.13.21) determines

$$L = 0,\ \ M = 0,\ \ N = (B-A)\omega_1\,\omega_2.$$

On simplification

$$N = \left(\frac{1}{3}ma^2 - \frac{1}{3}mb^2\right)\omega\cos\theta\,\omega\,\sin\theta = \frac{m}{6}\left(a^2-b^2\right)\omega^2\sin 2\theta,$$

$$\text{i.e.,}\ \ N = \frac{m}{6}\left(a^2-b^2\right)\omega^2\frac{2\tan\theta}{1+\tan^2\theta} = \frac{m}{3}\left(a^2-b^2\right)\omega^2\frac{\frac{b}{a}}{1+\frac{b^2}{a^2}},$$

$$\text{i.e.,}\ \ N = \frac{m}{3}\left(a^2-b^2\right)\omega^2\frac{ab}{a^2+b^2} = \frac{mab\omega^2\left(a^2-b^2\right)}{3\left(a^2+b^2\right)}.$$

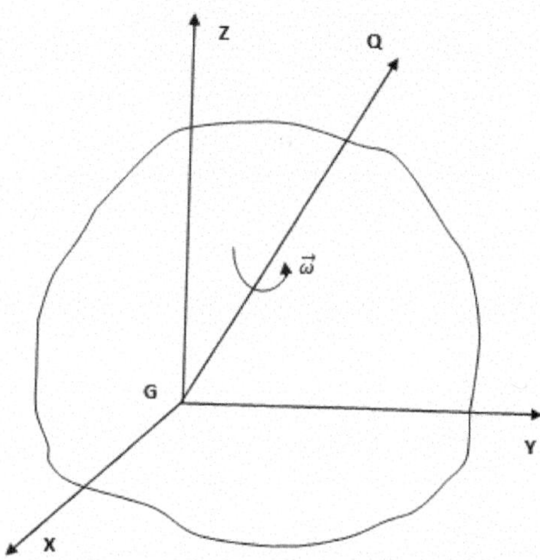

Fig. 8.11 Illustration for Example 8.4

Hence, the moments of the forces about the principal axes are 0, 0, and $\frac{mab\omega^2(a^2-b^2)}{3(a^2+b^2)}$.

Example 8.4 A rigid body is free to rotate about its centroid G, the principal moments of inertia at which are 7, 25, and 32 units, respectively. The body is given an angular velocity $\mathbf{\Omega}$ about a line through G whose direction ratios are 4:0:3. Show that after time t, the components of angular velocity about the principal axes of inertia at G are $\frac{4}{5}\Omega\cos\varphi,\ \frac{4}{5}\Omega\sin\varphi,\ \frac{3}{5}\Omega\cos\varphi$, where $\tan\left(\frac{\varphi}{2}\right) = \tanh\left(\frac{3\Omega t}{10}\right)$.

Solution Let GX, GY, and GZ stipulate the principal axes for the body at Q, and let OQ be the instantaneous axis of rotation (see Fig. 8.11). Let $\boldsymbol{\omega} = (\omega_1, \omega_2, \omega_3)$ denote the angular velocity of the body about OQ.

Given

$$A = \text{M.I. of the body about GX} = 7,$$
$$B = \text{M.I. of the body about GY} = 25,$$
$$C = \text{M.I. of the body about GZ} = 32.$$

The direction ratios of OQ are 4:0:3 and hence the direction cosines of the line OQ are $l = \frac{4}{5},\ m = 0,\ n = \frac{3}{5}$.

External forces acting on the body are (I) the weight of the body through G in a downward vertical direction and (II) the reaction of the body on the axis of rotation at G.

$$\therefore\quad \mathbf{\Lambda} = \mathbf{0},\ \text{ which shows that } L = M = N = 0.$$

Euler's dynamical equations are

$$\left.\begin{aligned} A\dot{\omega}_1 - (B - C)\ \omega_2\omega_3 = 0 \\ B\dot{\omega}_2 - (C - A)\ \omega_3\omega_1 = 0 \\ C\dot{\omega}_3 - (A - B)\ \omega_1\omega_2 = 0 \end{aligned}\right\}. \tag{8.13.22}$$

Equation (8.13.22) gives

$$\dot{\omega}_1 = -\omega_2\omega_3, \tag{8.13.23}$$

$$\dot{\omega}_2 = \omega_3\omega_1, \tag{8.13.24}$$

$$16\dot{\omega}_3 = -9\omega_1\omega_2. \tag{8.13.25}$$

Now (8.13.23) ÷ (8.13.24) yields

$$\frac{\dot{\omega}_1}{\dot{\omega}_2} = -\frac{\omega_2}{\omega_1},$$

i.e., $\omega_1\dot{\omega}_1 + \omega_2\dot{\omega}_2 = 0.$

On the integration of the above equation, we get

$$\omega_1^2 + \omega_2^2 = C_1, \tag{8.13.26}$$

where C_1 is the constant of integration.

Initially $\omega_1 = \Omega l = \frac{4}{5}\Omega,\ \omega_2 = \Omega m = 0$ at t = 0, and so (8.13.26) gives

$$C_1 = \left(\frac{4}{5}\Omega\right)^2 = a^2; \quad a = \frac{4}{5}\Omega.$$

Hence (8.13.26) reduces to

$$\omega_1^2 + \omega_2^2 = a^2$$

i.e., $\omega_1 = \sqrt{a^2 - \omega_2^2}.$ (8.13.27)

Again (8.13.24)÷ (8.13.25) yields

$$\frac{\dot{\omega}_2}{16\dot{\omega}_3} = -\frac{\omega_3}{9\omega_2},$$

i.e., $9\omega_2\dot{\omega}_2 + 16\omega_3\dot{\omega}_3 = 0.$

Integrating the above equation, we obtain

$$9\omega_2^2 + 16\omega_3^2 = C_2, \tag{8.13.28}$$

C_2 being the constant of integration.

Initially $\omega_2 = \Omega m = 0,\ \omega_3 = \Omega n = \frac{3\Omega}{5};\ t = 0$. So

$$C_2 = \frac{144}{25}\Omega^2.$$

Hence Eq. (8.13.28) gets reduced to

$$9\omega_2^2 + 16\omega_3^2 = \frac{144}{25}\Omega^2,$$

$$\text{i.e.,}\quad 16\omega_3^2 = \frac{144}{25}\Omega^2 - 9\omega_2^2,$$

$$\text{i.e.,}\quad 16\omega_3^2 = 9\left(a^2 - \omega_2^2\right),$$

$$\text{i.e.,}\quad \omega_3 = \frac{3}{4}\sqrt{a^2 - \omega_2^2}. \tag{8.13.29}$$

On substitution of Eqs. (8.13.27) and (8.13.29), Eq. (8.13.24) reduces to

$$\dot{\omega}_2 = \frac{3}{4}\left(a^2 - \omega_2^2\right),$$

$$\text{i.e.,}\quad \frac{d\omega_2}{a^2 - \omega_2^2} = \frac{3}{4}dt. \tag{8.13.30}$$

Integrating Eq. (8.13.30) and using the initial condition $\omega_2 = 0$ at $t = 0$, we obtain

$$\frac{1}{a}\tanh^{-1}\left(\frac{\omega_2}{a}\right) = \frac{3}{4}t,$$

$$\text{i.e.,}\quad \tanh^{-1}\left(\frac{\omega_2}{a}\right) = \frac{3a}{4}t,$$

$$\text{i.e.,}\quad \omega_2 = a\tanh\left(\frac{3a}{4}t\right) = a\tanh\left(\frac{3\Omega t}{5}\right),$$

$$\text{i.e.,}\quad \omega_2 = a\tanh(2\theta)\,;\quad \theta = \frac{3\Omega t}{10},$$

$$\text{i.e.,}\quad \omega_2 = a\frac{2\tanh\theta}{1+\tanh^2\theta}. \tag{8.13.31}$$

Given

$$\tan h\left(\frac{3\Omega t}{10}\right) = \tan\frac{\varphi}{2},$$

$$\text{i.e.,}\quad \tan h\theta = \tan\frac{\varphi}{2}. \tag{8.13.32}$$

By the use of (8.13.32), Eq. (8.13.31) takes the form:

$$\omega_2 = a\frac{2\tan\frac{\varphi}{2}}{1+\tan^2\frac{\varphi}{2}} = a\sin\varphi. \tag{8.13.33}$$

Substituting (8.13.33) in (8.13.27) and (8.13.29), we derive

$$\omega_1 = a\cos\varphi,\quad \omega_3 = \frac{3a}{4}\cos\varphi.$$

Thus, the components of the angular velocity are

$$\left.\begin{aligned} \omega_1 &= a\cos\varphi = \tfrac{4}{5}\Omega\cos\varphi \\ \omega_2 &= a\sin\varphi = \tfrac{4}{5}\Omega\sin\varphi \\ \omega_3 &= \tfrac{3a}{4}\cos\varphi = \tfrac{3}{5}\Omega\cos\varphi \end{aligned}\right\}.$$

Example 8.5 A uniform square lamina is rotating under no impressed forces about one corner O, which is fixed. If $\omega_1,\ \omega_2,\ \omega_3$ are the components of the angular velocity about the principal axes at O (in order of increasing moments of inertia), show that $\omega_1^2+\omega_2^2$ and $3\omega_2^2+4\omega_3^2$ are constant.

Solution Consider a square lamina OPQR as shown in Fig. 8.12, with mass M and side $2a$ having its corner O fixed. Take O as the origin, X-axis along **OQ**, Y-axis $\parallel$ to **PR**, and Z-axis $\perp^r$ to the plane of the lamina. The plane of the lamina is the z-plane and hence $D = E = 0$. Further $F = 0$ due to symmetry and so OX, OY, OZ are the principal axes for the lamina at O. Draw lines GA, GB, GC such that **GA $\parallel$ OP, GB $\parallel$ OR, GC $\parallel$ OZ.**

Fig. 8.12 Illustration for Example 8.5

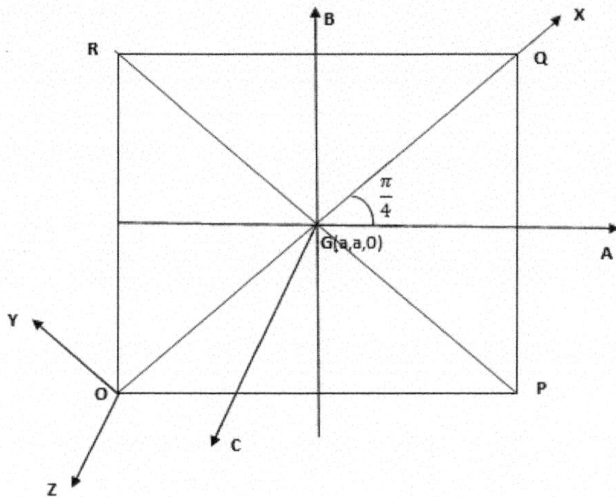

Now

$$A' = \text{M.I. of the lamina about GA} = \frac{1}{3}Ma^2,$$

$$B' = \text{M.I. of the lamina about GB} = \frac{1}{3}Ma^2,$$

$$C' = \text{M.I. of the lamina about GC} = \frac{1}{3}Ma^2 + \frac{1}{3}Ma^2 = \frac{2}{3}Ma^2,$$

$$F' = \text{P.I. of the lamina about GA-GB} = 0.$$

We have

$$A = \text{M.I. of the lamina about OX} = A'\cos^2\theta + B'\sin^2\theta - 2F'\sin\theta\cos\theta\left(\theta = \frac{\pi}{4}\right),$$

$$\text{i.e., } A = \frac{1}{3}Ma^2\left(\cos^2\theta + \sin^2\theta\right) = \frac{1}{3}Ma^2,$$

$$B = \text{M.I of the lamina about OY} = \text{M.I. about PR} + M\text{OG}^2,$$

$$\text{i.e. } B = \frac{1}{3}Ma^2 + M\left(a^2 + a^2\right) = \frac{7}{3}Ma^2,$$

$$C = \text{M.I. of the lamina about OZ} = A + B = \frac{8}{3}Ma^2.$$

As there is no impressed force acting on the lamina, the integrals of energy and momentum give

$$A\omega_1^2 + B\omega_2^2 + C\omega_3^2 = \text{a constant}, \tag{8.13.34}$$

$$A^2\omega_1^2 + B^2\omega_2^2 + C^2\omega_3^2 = \text{a constant}. \tag{8.13.35}$$

(8.13.34) and (8.13.35) yield

$$\omega_1^2 + 7\omega_2^2 + 8\omega_3^2 = \lambda = \text{a constant}, \tag{8.13.36}$$

$$\omega_1^2 + 49\omega_2^2 + 64\omega_3^2 = \mu = \text{a constant}. \tag{8.13.37}$$

(8.13.36) multiplied by 8 gives

$$8\omega_1^2 + 56\omega_2^2 + 64\omega_3^2 = 8\lambda. \tag{8.13.38}$$

(8.13.38)−−(8.13.37) give

$$7\omega_1^2 + 7\omega_2^2 = 8\lambda - \mu,$$

$$\text{i.e.,}\quad \omega_1^2 + \omega_2^2 = \text{a constant}.$$

(8.13.37)−−(8.13.36) give

$$42\omega_2^2 + 56\omega_3^2 = \mu - \lambda,$$

$$\text{i.e.,}\quad 3\omega_2^2 + 4\omega_3^2 = \text{a constant}.$$

Example 8.6 If T is the total KE of rotation of a rigid body with one point fixed, prove that $\frac{dT}{dt} = \boldsymbol{\omega} \cdot \boldsymbol{\Lambda}$, where all quantities refer to the body principal axes.

Solution With the usual meaning of the notations, the KE of the body is given by

$$T = \frac{1}{2}\left(A\omega_1^2 + B\omega_2^2 + C\omega_3^2 - 2D\omega_2\omega_3 - 2E\omega_3\omega_1 - 2F\omega_1\omega_2\right). \tag{8.13.39}$$

As the axes of references fixed in the body are the principal axes,

$$D = E = F = 0. \tag{8.13.40}$$

By the use of Eq. (8.13.40), (8.13.39) reduces to

$$T = \frac{1}{2}\left(A\omega_1^2 + B\omega_2^2 + C\omega_3^2\right).$$

$$\therefore \quad \frac{dT}{dt} = A\omega_1\dot{\omega}_1 + B\omega_2\dot{\omega}_2 + C\omega_3\dot{\omega}_3. \tag{8.13.41}$$

Euler's dynamical equations of motion are

$$\left.\begin{aligned} A\dot{\omega}_1 - (B-C)\,\omega_2\omega_3 &= L \\ B\dot{\omega}_2 - (C-A)\,\omega_3\omega_1 &= M \\ C\dot{\omega}_3 - (A-B)\,\omega_1\omega_2 &= N \end{aligned}\right\}. \tag{8.13.42}$$

Multiplying both sides of Eqs. (8.13.42) by $\omega_1,\ \omega_2,\ \omega_3$, respectively, and adding we obtain

$$A\omega_1\dot{\omega}_1 + B\omega_2\dot{\omega}_2 + C\omega_3\dot{\omega}_3 = L\omega_1 + M\omega_2 + N\omega_3,$$

$$\text{i.e.,}\quad \frac{dT}{dt} = \left(L\hat{i} + M\hat{j} + N\hat{k}\right)\cdot\left(\omega_1\hat{i} + \omega_2\hat{j} + \omega_3\hat{k}\right) = \mathbf{\Lambda}\cdot\boldsymbol{\omega}.$$

Example 8.7 A rigid body is moving freely about a fixed point O under no external force. Prove that if two of the principal moments of inertia at O are equal, the instantaneous axis of rotation describes a circular cone, with a vertex at O, fixed in space.

Solution Given that the rigid body with one point O fixed moves freely under no external forces.

With usual rotations, the integrals of energy and angular momentum give

$$A\omega_1^2 + B\omega_2^2 + C\omega_3^2 = 2T, \tag{8.13.43}$$

$$A^2\omega_1^2 + B^2\omega_2^2 + C^2\omega_3^2 = \Omega^2. \tag{8.13.44}$$

Suppose that $A = B = K$. Then Eqs. (8.13.43) and (8.13.44) reduce to

$$\omega_1^2 + \omega_2^2 + \frac{C}{K}\omega_3^2 = \frac{2T}{K}, \tag{8.13.45}$$

$$\omega_1^2 + \omega_2^2 + \frac{C^2}{K^2}\omega_3^2 = \frac{\Omega^2}{K^2}. \tag{8.13.46}$$

Now (8.13.46) – (8.13.45) gives

$$\left(\frac{C^2}{K^2} - \frac{C}{K}\right)\omega_3^2 = \frac{\Omega^2}{K^2} - \frac{2T}{K},$$

$$\text{i.e.,}\quad \left(C^2 - CK\right)\omega_3^2 = \Omega^2 - 2TK,$$

$$\text{i.e.,}\quad \omega_3^2 = \frac{\Omega^2 - 2TK}{C^2 - CK} = \text{a constant} = \lambda^2\ (\text{say})\,.$$

$$\therefore\quad \omega_3 = \lambda. \tag{8.13.47}$$

On substitution of (8.13.47) in (8.13.45) we get

$$\omega_1^2 + \omega_2^2 + \frac{C}{K}\lambda^2 = \frac{2T}{k}$$

$$\text{i.e.,}\quad \omega_1^2 + \omega_2^2 = \frac{2T}{K} - \frac{C}{K}\lambda^2 = \text{a constant} = \mu^2\ \ (\text{say})\,. \tag{8.13.48}$$

Now

$$\omega^2 = \omega_1^2 + \omega_2^2 + \omega_3^2 = \mu^2 + \lambda^2 = \nu^2$$

$$\therefore\quad \omega = |\boldsymbol{\omega}| = \nu = \text{a constant}. \tag{8.13.49}$$

The instantaneous axis of rotation is given by

$$\frac{x}{\omega_1} = \frac{y}{\omega_2} = \frac{z}{\omega_3}. \tag{8.13.50}$$

The invariable line is given by

$$\frac{x}{A\omega_1} = \frac{y}{B\omega_2} = \frac{z}{C\omega_3}. \tag{8.13.51}$$

Let θ be angle between the lines (8.13.50) and (8.13.51)

$$\therefore\quad \cos\theta = \frac{\omega_1 A\omega_1 + \omega_2 B\omega_2 + \omega_3 C\omega_3}{\sqrt{\omega_1^2 + \omega_2^2 + \omega_3^2}\sqrt{A^2\omega_1^2 + B^2\omega_2^2 + C^2\omega_3^2}},$$

i.e., $\cos\theta = \dfrac{A\omega_1^2 + B\omega_2^2 + C\omega_3^2}{\sqrt{\omega^2}\sqrt{\Omega^2}} = \dfrac{2T}{\omega\Omega} = \text{a constant} = \cos\alpha \text{ (say)}$

$\therefore \quad \theta = \alpha = a \text{ Constant.}$

Hence the variable line (8.13.50) always makes a constant angle α with the invariable line (8.13.51). Thus, the instantaneous axis of rotation describes a circular cone fixed in space, with the invariable line as the axis, O as the vertex, and α as the semi-vertical angle.

Example 8.8 A rigid body having one point O fixed and no external torque about O has two equal principal moments of inertia. Prove that it must rotate with an angular velocity of constant magnitude.

Solution With usual notations, Euler's equations of motion are

$$A\dot{\omega}_1 - (B - C)\,\omega_2\omega_3 = 0, \tag{8.13.52}$$

$$B\dot{\omega}_2 - (C - A)\,\omega_3\omega_1 = 0, \tag{8.13.53}$$

$$C\dot{\omega}_3 - (A - B)\,\omega_1\omega_2 = 0. \tag{8.13.54}$$

Suppose that $A = B = K$ (say).
Therefore, Eq. (8.13.54) gives

$$\dot{\omega}_3 = 0,$$

$$\text{i.e.,} \quad \omega_3 = \lambda = \text{a constant.} \tag{8.13.55}$$

Equations (8.13.52) and (8.13.53) give

$$K\dot{\omega}_1 = (K - C)\,\omega_2\lambda, \tag{8.13.56}$$

$$K\dot{\omega}_2 = (C - K)\,\lambda\omega_1. \tag{8.13.57}$$

(8.13.56) ÷ (8.13.57) gives

$$\frac{\dot{\omega}_1}{\dot{\omega}_2} = -\frac{\omega_2}{\omega_1},$$

$$\text{i.e.,}\quad \omega_1\dot{\omega}_1 + \omega_2\dot{\omega}_2 = 0. \tag{8.13.58}$$

Integrating (8.13.58), we get

$$\omega_1^2 + \omega_2^2 = \text{a constant} = \mu^2.$$

But $\boldsymbol{\omega} = (\omega_1, \omega_2, \omega_3)$,

$$\therefore\quad \omega^2 = \omega_1^2 + \omega_2^2 + \omega_3^2,$$

$$\therefore\quad |\boldsymbol{\omega}|^2 = \mu^2 + \lambda^2 = \nu^2,$$

$$\text{i.e.,}\quad |\boldsymbol{\omega}| = \nu = \text{a constant}.$$

Thus, the angular velocity of the body is of constant magnitude.

Example 8.9 A rigid lamina moves under no forces. A rigid lamina moves under no forces. The principal axes at the center of mass O are Ox, Oy, Oz, the last being normal to the lamina, and the principal moments of inertia about Ox, Oy, Oz are A, B, and C, respectively, with ($B > A$). Prove that the resolved parts of angular velocity vector about Ox, Oy, Oz are, respectively, given by $\omega\cos\lambda,\ \omega\sin\lambda,\ \dot{\lambda}$ where ω is a constant and λ satisfies the differential equation:

$$2\ddot{\lambda} + K^2\sin 2\lambda = 0,\quad K^2 = \frac{\omega^2(B - A)}{B + A}.$$

Solution With usual notations, Euler's equations of motion under no forces are

$$A\dot{\omega}_1 - (B - C)\,\omega_2\omega_3 = 0, \tag{8.13.59}$$

$$B\dot{\omega}_2 - (C - A)\,\omega_3\omega_1 = 0, \tag{8.13.60}$$

$$C\dot{\omega}_3 - (A - B)\,\omega_1\omega_2 = 0. \tag{8.13.61}$$

As Oz is $\perp^r$ to the plane of the lamina,

$$\therefore \quad C = A + B. \tag{8.13.62}$$

Equation (8.13.59) gives

$$A\dot{\omega}_1 - (B - A - B)\,\omega_2\omega_3 = 0,$$

i.e., $\dot{\omega}_1 + \omega_2\omega_3 = 0,$

$$\text{i.e.,} \quad \dot{\omega}_1 = -\omega_2\omega_3. \tag{8.13.63}$$

Equation (8.13.60) transforms to

$$B\dot{\omega}_2 - (A + B - A)\,\omega_3\omega_1 = 0$$

$$\text{i.e.,} \quad \dot{\omega}_2 = \omega_3\omega_1. \tag{8.13.64}$$

Equation (8.13.63) divided by Eq. (8.13.64) gives

$$\frac{\dot{\omega}_1}{\dot{\omega}_2} = -\frac{\omega_2}{\omega_1},$$

i.e., $\omega_1\dot{\omega}_1 + \omega_2\dot{\omega}_2 = 0,$

$$\text{i.e.,} \quad \omega_1^2 + \omega_2^2 = \omega^2, \tag{8.13.65}$$

where ω is a constant.

Equation (8.13.65) is satisfied by the substitution

$$\left.\begin{aligned} \omega_1 &= \omega\cos\lambda \\ \omega_2 &= \omega\sin\lambda \end{aligned}\right\}. \tag{8.13.66}$$

Equation (8.13.63) on substitution of (8.13.66) gets reduced to

$$-\,\omega\sin\lambda\,\dot{\lambda} + \omega\sin\lambda\,\omega_3 = 0,$$

$$\text{i.e.,} \quad \omega_3 = \dot{\lambda}. \tag{8.13.67}$$

Substitution of (8.13.62), (8.13.66), (8.13.67) in (8.13.61) gives

$$(A+B)\ddot{\lambda}-(A-B)\omega^2\sin\lambda\cos\lambda=0,$$

$$\text{i.e., }\ \ddot{\lambda}=\frac{A-B}{A+B}\omega^2\sin\lambda\cos\lambda,$$

$$\text{i.e., }\ \ddot{\lambda}=-\left(\frac{B-A}{B+A}\omega^2\right)\sin\lambda\cos\lambda=-K^2\sin\lambda\cos\lambda, \qquad (8.13.68)$$

where $K^2=\frac{B-A}{B+A}\,\omega^2$.

On simplification, Eq. (8.13.68) takes the form

$$2\ddot{\lambda}=-K^2\sin 2\lambda,$$

$$\text{i.e., }\ 2\ddot{\lambda}+K^2\sin 2\lambda=0.$$

Thus, the components of the angular velocity about Ox, Oy, Oz are $\omega\cos\lambda$, $\omega\sin\lambda$, $\dot{\lambda}$, where ω is a constant and λ satisfies the differential equation

$$2\ddot{\lambda}+K^2\sin 2\lambda=0;\quad K^2=\frac{B-A}{B+A}\omega^2.$$

Example 8.10 A uniform rectangular plate ABCD of mass M, for which AB=$2a$, BC=$2b$, is at rest and is then given a blow of impulse $J\perp^r$ to the plate at C. Show that, if A is fixed but the plate is free to turn about it, the KE generated by the blow is $\frac{12J^2}{7M}$.

Solution Consider a rectangular plate ABCD with mass M having sides $2a$ and $2b$ as shown in Fig. 8.13. Let the origin O be the center of the plate, and the X-axis and Y-axis through O parallel to AB and AD, respectively. Let ALQ be the axis of the situation through A and parallel to BD.

Let the equation of the diagonal BD be

$$y=mx. \qquad (8.13.69)$$

The line (8.13.69) passes through $B\,(a,-b)$ and so $-b=am$ i.e., $m=-\frac{b}{a}$. Hence the equation of the diagonal BD is

$$y=-\frac{b}{a}x,$$

$$\text{i.e., }\ bx+ay=0. \qquad (8.13.70)$$

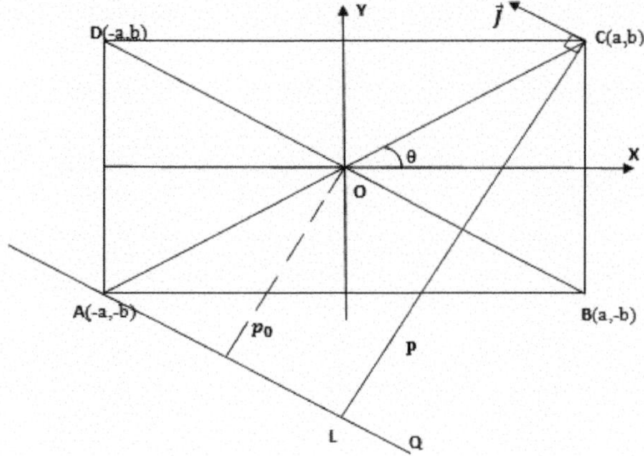

Fig. 8.13 Illustration for Example 8.10

As QA || BD, let its equation be

$$bx + ay = k. \tag{8.13.71}$$

The line (8.13.71) passes through $(-a, -b)$.

$$\therefore \quad k = -ab - ab = -2ab.$$

Thus, the equation of the axis of rotation is

$$bx + ay + 2ab = 0. \tag{8.13.72}$$

Let p be the $\perp^r$ distance of the point $C\ (a, b)$ from QA.

$$\therefore \quad p = \frac{ab + ab + 2ab = 0}{\sqrt{b^2 + a^2}} = \frac{4ab}{\sqrt{a^2 + b^2}}. \tag{8.13.73}$$

Let I denote the MI of the lamina about QA and ω be the angular velocity of the lamina just after impulse J is applied at point C. Equation of the rotational motion of a rigid body under impulsive forces gives

$$I\ (\omega - 0) = pJ,$$

i.e., $I\omega = pJ$,

$$\therefore \quad \omega = \frac{pJ}{I}. \tag{8.13.74}$$

If p_0 is $\perp^r$ distance of O(0,0) from QA, then

$$p_0 = \frac{2ab}{\sqrt{a^2+b^2}}.$$

Now

$$\begin{aligned} I &= \text{MI of the lamina about QA} = \text{MI of the lamina about BD} + Mp_0^2, \\ &= \text{MI of the lamina about diagonal AC} + Mp_0^2 \\ &= A\cos^2\theta + B\sin^2\theta + Mp_0^2, \\ &= \frac{1}{3}Mb^2\frac{a^2}{a^2+b^2} + \frac{1}{3}Ma^2\frac{b^2}{a^2+b^2} + M\frac{4a^2b^2}{a^2+b^2}, \\ &= \frac{2}{3}\frac{Ma^2b^2}{a^2+b^2} + \frac{4Ma^2b^2}{a^2+b^2}, \\ &= \frac{2Ma^2b^2 + 12Ma^2b^2}{3\left(a^2+b^2\right)}, \\ &= \frac{14Ma^2b^2}{3\left(a^2+b^2\right)}. \end{aligned}$$

The KE of the lamina generated by the blow is given by

$$T = \frac{1}{2}I\omega^2 = \frac{1}{2}I\frac{p^2J^2}{I^2} = \frac{p^2J^2}{2I} = \frac{\frac{16a^2b^2}{a^2+b^2}J^2}{2\frac{14Ma^2b^2}{3(a^2+b^2)}} = \frac{12J^2}{7M}.$$

Example 8.11 A uniform solid rectangular parallelepiped is free to turn about its center of gravity, which is fixed. The edges of the solid have lengths $2a$, $2a$, a. Initially, the solid has angular velocity Ω about a diagonal. Find the components, parallel to the edges of the solid, of its angular velocity after a time t. Hence or otherwise show that the resolute of the angular velocity of the solid about the same diagonal after the t is

$$\frac{1}{9}\Omega\left[1 + 8\cos\left(\frac{\Omega t}{5}\right)\right].$$

Solution Consider a rectangular parallelepiped OABCA′B′C′O′, with its center of gravity G taken as the origin, as illustrated in Fig. 8.14. Take the X-axis ∥ **OA**, Y-axis ∥ **OB**, and Z-axis ∥ **OC**. Let the diagonal OO′ be the initial axis of rotation. Now $\mathbf{GO} = a\hat{i} + a\hat{j} + \frac{a}{2}\hat{k}$.

The unit vector in the direction of $\mathbf{GO} = \hat{e} = \frac{a\hat{i}+a\hat{j}+\frac{a}{2}\hat{k}}{\sqrt{a^2+a^2+\frac{a^2}{4}}} = \frac{2\hat{i}+2\hat{j}+\hat{k}}{3}$.

Let $\boldsymbol{\omega} = \omega_1\hat{i} + \omega_2\hat{j} + \omega_3\hat{k}$ be the angular velocity of the body at time t. The initial angular velocity of the body about OO′ is given by

Fig. 8.14 Schematic diagram for Example 8.11

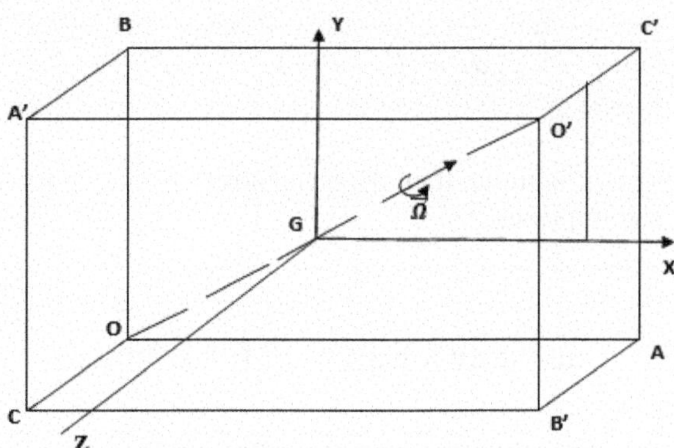

$$\mathbf{\Omega} = \Omega \hat{e} = \Omega \frac{2\hat{i} + 2\hat{j} + \hat{k}}{3} = \left(\frac{2\Omega}{3}, \frac{2\Omega}{3}, \frac{\Omega}{3} \right).$$

$$\text{A} = \text{MI of the body about GX} = \frac{1}{3} M \left(a^2 + \frac{a^2}{4} \right) = \frac{5}{12} M a^2,$$

$$\text{B} = \text{MI of the body about GY} = \frac{1}{3} M \left(\frac{a^2}{4} + a^2 \right) = \frac{5}{12} M a^2,$$

$$\text{C} = \text{MI of the body about GZ} = \frac{2}{3} M a^2,$$

$$D = E = F = 0.$$

Therefore, GX, GY, and GZ are the principal axes for the body at G.

$\mathbf{\Lambda}$ = torque of the forces about G = $\mathbf{0}$,

$$\therefore \quad L = M = N = 0.$$

Euler's dynamical equations are

$$A\dot{\omega}_1 - (B - C)\,\omega_2\omega_3 = 0, \tag{8.13.75}$$

$$B\dot{\omega}_2 - (C - A)\,\omega_3\omega_1 = 0, \tag{8.13.76}$$

$$C\dot{\omega}_3 - (A - B)\,\omega_1\omega_2 = 0. \tag{8.13.77}$$

Substituting the values of A, B, and C in (7.13.75), (7.13.76), and (7.13.77), we obtain

$$\frac{5}{12}Ma^2\dot{\omega}_1 = \left(\frac{5}{12}Ma^2 - \frac{2}{3}Ma^2\right)\omega_2\omega_3,$$

$$\text{i.e.,}\quad 5\dot{\omega}_1 = (5-8)\,\omega_2\omega_3,$$

$$\text{i.e.,}\quad 5\dot{\omega}_1 = -3\omega_2\omega_3. \tag{8.13.78}$$

$$\frac{5}{12}Ma^2\dot{\omega}_2 = \left(\frac{2}{3}Ma^2\frac{5}{12}Ma^2\right)\omega_3\omega_1,$$

$$\text{i.e.,}\quad 5\dot{\omega}_2 = (8-5)\,\omega_3\omega_1,$$

$$\text{i.e.,}\quad 5\dot{\omega}_2 = 3\omega_3\omega_1. \tag{8.13.79}$$

$$\frac{2}{3}Ma^2\dot{\omega}_3 = 0,$$

$$\dot{\omega}_3 = 0,$$

$$\omega_3 = C_1. \tag{8.13.80}$$

Initially

$$\omega_3 = \frac{\Omega}{3} \quad \text{at } t = 0.$$

Therefore, Eq. (8.13.80) gives

$$\frac{\Omega}{3} = C_1.$$

Thus, (8.13.80) reduces to

$$\omega_3 = \frac{\Omega}{3}. \tag{8.13.81}$$

Equation (8.13.78) gives

$$5\dot{\omega}_1 = -3\omega_2\omega_3,$$

$$5\ddot{\omega}_1 = -3\dot{\omega}_2\omega_3 = -3\frac{3\omega_3\omega_1}{5}\omega_3 = -\frac{9}{5}\omega_1\frac{\Omega^2}{9} = -\frac{\Omega^2\omega_1}{5},$$

$$\text{i.e.,}\quad \ddot{\omega}_1 = -\frac{\Omega^2\omega_1}{25},$$

$$\text{i.e.,}\quad \left(D^2 + \frac{\Omega^2}{25}\right)\omega_1 = 0,$$

$$\therefore \quad \omega_1 = \lambda \cos \frac{\Omega}{5} t + \mu \sin \frac{\Omega}{5} t. \tag{8.13.82}$$

(8.13.82) gives

$$\dot{\omega}_1 = -\lambda \sin\left(\frac{\Omega t}{5}\right)\frac{\Omega}{5} + \mu \cos\left(\frac{\Omega t}{5}\right)\frac{\Omega}{5},$$

$$\text{i.e.,} \quad \frac{-3\omega_2\omega_3}{5} = -\lambda\frac{\Omega}{5}\sin\left(\frac{\Omega t}{5}\right) + \mu\frac{\Omega}{5}\cos\left(\frac{\Omega t}{5}\right),$$

$$\text{i.e.,} \quad -3\omega_2\frac{\Omega}{3} = -\lambda\Omega\sin\left(\frac{\Omega t}{5}\right) + \mu\Omega\cos\left(\frac{\Omega t}{5}\right),$$

$$\text{i.e.,} \quad \omega_2\Omega = \lambda\Omega\sin\left(\frac{\Omega t}{5}\right) - \mu\Omega\cos\left(\frac{\Omega t}{5}\right),$$

$$\text{i.e.,} \quad \omega_2 = \lambda\sin\left(\frac{\Omega t}{5}\right) - \mu\cos\left(\frac{\Omega t}{5}\right). \tag{8.13.83}$$

Initially $\omega_1 = \frac{2\Omega}{3}$, $\omega_2 = \frac{2\Omega}{3}$, at $t = 0$.
Therefore, Eqs. (8.13.82) and (8.13.83) give

$$\lambda = \frac{2\Omega}{3}, \quad \mu = -\frac{2\Omega}{3}.$$

Thus, Eqs. (8.13.81), (8.13.82), and (8.13.83) lead to

$$\left.\begin{aligned} \omega_1 &= \tfrac{2\Omega}{3}\cos\tfrac{\Omega t}{5} - \tfrac{2\Omega}{3}\sin\tfrac{\Omega t}{5} \\ \omega_2 &= \tfrac{2\Omega}{3}\sin\left(\tfrac{\Omega t}{5}\right) + \tfrac{2\Omega}{3}\cos\left(\tfrac{\Omega t}{5}\right) \\ \omega_3 &= \tfrac{\Omega}{3} \end{aligned}\right\}. \tag{8.13.84}$$

Equation (8.13.84) gives

$$\left.\begin{aligned} \omega_1 &= \tfrac{2\Omega}{3}(\cos\theta - \sin\theta) \\ \omega_2 &= \tfrac{2\Omega}{3}(\sin\theta + \cos\theta) \\ \omega_3 &= \tfrac{\Omega}{3}; \quad \theta = \tfrac{\Omega t}{5} \end{aligned}\right\}. \tag{8.13.85}$$

From Fig. 8.15, we have

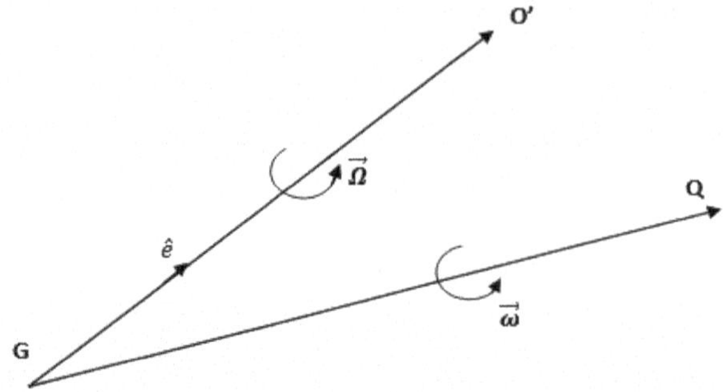

Fig. 8.15 Illustration for the resolved part of ω

The resolved part ofω along $\mathbf{OO}' = \omega \cdot \hat{e}$

$$
\begin{aligned}
&= \left(\omega_1\hat{i} + \omega_2\hat{j} + \omega_3\hat{k}\right) \cdot \left(\frac{2}{3}\hat{i} + \frac{2}{3}\hat{j} + \frac{1}{3}\hat{k}\right), \\
&= \frac{2\omega_1}{3} + \frac{2\omega_2}{3} + \frac{\omega_3}{3}, \\
&= \frac{2}{3}\frac{2\Omega}{3}(\cos\theta - \sin\theta) + \frac{2}{3}\frac{2\Omega}{3}(\sin\theta + \cos\theta) + \frac{1}{3}\frac{\Omega}{3}, \\
&= \frac{4\Omega}{9} 2\cos\theta + \frac{\Omega}{9}, \\
&= \frac{8\Omega}{9}\cos\theta + \frac{\Omega}{9}, \\
&= \frac{\Omega}{9}[1 + 8\cos\theta] = \frac{\Omega}{9}\left[1 + 8\cos\frac{\Omega t}{5}\right], \\
&= \frac{\Omega}{9}\left[1 + 8\ \cos\frac{\Omega t}{5}\right].
\end{aligned}
$$

Example 8.12 Prove that, if a rectangular parallelepiped (edges $2a,\ 2a,\ 2b$) rotates about its center of gravity, its angular velocity about one principal axis is constant and about the other principal axes is periodic, the period being to the period about the first mentioned principal axis as $b^2 + a^2\ :\ b^2 - a^2$.

Solution Consider a rectangular parallelepiped OABCA′B′C′O′, with G as the center of gravity, and OA $= 2a$, OB $= 2a$, OC $= 2b$, as shown in Fig. 8.16. Take G as the origin and mass m of the parallelepiped, and the X-axis is parallel to OA, Y-axis parallel to OB, and the Z-axis is parallel to OC.

The external forces acting on the body are

(I) the weight through G,

(II) the reaction of the axis of rotation at G.

$$\therefore \quad \mathbf{\Lambda} = \text{external torque about G} = \mathbf{0},$$

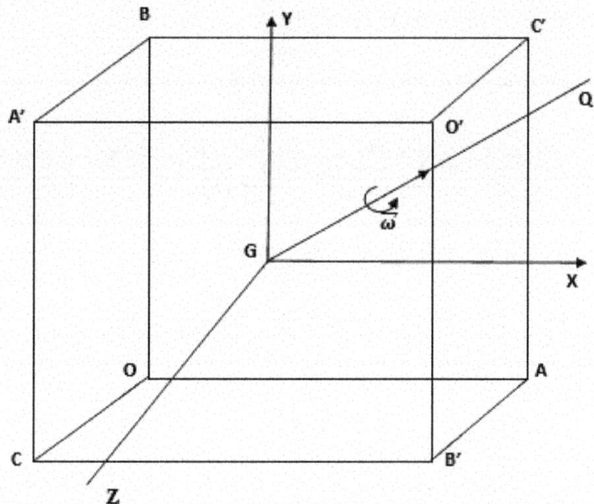

Fig. 8.16 Illustration for Example 8.12

and so, $L = M = N = 0$.

Now

$$A = \text{MI of the body about GX} = \frac{1}{3}m\left(a^2 + b^2\right),$$

$$B = \text{MI of the body about GY} = \frac{1}{3}m\left(a^2 + b^2\right),$$

$$C = \text{MI of the body about GZ} = \frac{1}{3}m\left(a^2 + a^2\right) = \frac{2}{3}ma^2.$$

$D = E = F = 0$ (due to symmetry),

$\therefore$ GX, GY, GZ are the principal axes for the body at G.

Euler's equations are

$$\left.\begin{aligned} A\dot{\omega}_1 - (B - C)\,\omega_2\omega_3 = 0 \\ B\dot{\omega}_2 - (C - A)\,\omega_3\omega_1 = 0 \\ C\dot{\omega}_3 - (A - B)\,\omega_1\omega_2 = 0 \end{aligned}\right\}. \tag{8.13.86}$$

The last equation of (8.13.86) gives

$$C\dot{\omega}_3 = 0,$$

$$\text{i.e.,}\quad \omega_3 = \nu = \text{a constant.} \tag{8.13.87}$$

First equation of (8.13.86) gives

$$\frac{1}{3}m\left(a^2+b^2\right)\dot{\omega}_1 - \left\{\frac{1}{3}m\left(a^2+b^2\right) - \frac{2}{3}ma^2\right\}\omega_2\omega_3 = 0,$$

$$\text{i.e.,}\quad \left(a^2+b^2\right)\dot{\omega}_1 - \left(a^2+b^2-2a^2\right)\omega_2\omega_3 = 0,$$

$$\text{i.e.,}\quad \left(a^2+b^2\right)\dot{\omega}_1 = -\left(a^2-b^2\right)\omega_2\omega_3,$$

$$\text{i.e.,}\quad \left(a^2+b^2\right)\dot{\omega}_1 = \left(b^2-a^2\right)\omega_2\nu. \tag{8.13.88}$$

Second equation of (8.13.86) gives

$$\frac{1}{3}m\left(a^2+b^2\right)\dot{\omega}_2 - \left(\frac{1}{3}ma^2 - \frac{1}{3}ma^2 - \frac{1}{3}mb^2\right)\omega_3\omega_1 = 0,$$

$$\text{i.e.,}\quad \left(a^2+b^2\right)\dot{\omega}_2 - \left(2a^2-a^2-b^2\right)\omega_3\omega_1 = 0,$$

$$\text{i.e.,}\quad .\ \left(a^2+b^2\right)\dot{\omega}_2 - \left(a^2-b^2\right)\omega_3\omega_1 = 0,$$

$$\text{i.e.,}\quad \left(a^2+b^2\right)\dot{\omega}_2 = \left(a^2-b^2\right)\nu\omega_1. \tag{8.13.89}$$

Equation (8.13.88) gives

$$\left(a^2+b^2\right)\ddot{\omega}_1 = \left(b^2-a^2\right)\nu\dot{\omega}_2,$$

$$\text{i.e.,}\quad \left(a^2+b^2\right)\ddot{\omega}_1 = \left(b^2-a^2\right)\nu\frac{a^2-b^2}{a^2+b^2}\nu\omega_1,$$

$$\text{i.e.,}\quad \left(a^2+b^2\right)^2\ddot{\omega}_1 = -\left(b^2-a^2\right)^2\nu^2\omega_1,$$

$$\text{i.e.,}\quad \ddot{\omega}_1 + \frac{\left(b^2-a^2\right)^2}{\left(a^2+b^2\right)^2}\nu^2\omega_1 = 0,$$

$$\text{i.e.,}\quad \ddot{\omega}_1 + K^2\omega_1 = 0, \tag{8.13.90}$$

where $K = \frac{b^2-a^2}{a^2+b^2}\nu$. Similarly

$$\ddot{\omega}_2 + K^2\omega_2 = 0. \tag{8.13.91}$$

Equations (8.13.90) and (8.13.91) represent SHM each of period

$$T = \frac{2\pi}{\sqrt{K^2}} = \frac{2\pi}{K} = \frac{2\pi\left(a^2+b^2\right)}{\left(b^2-a^2\right)\nu}. \tag{8.13.92}$$

Equation (8.13.87) shows that angular velocity about the third principal axis is ν i.e., the angle described per unit time about the third principal axis is ν rad. Therefore, the period of rotation about the third principal axis is given by

$$T' = \frac{2\pi}{\nu}. \tag{8.13.93}$$

(8.13.92)÷ (8.13.93) gives

$$\frac{T}{T'} = \frac{2\pi\left(a^2+b^2\right)}{\left(b^2-a^2\right)\nu} \times \frac{\nu}{2\pi} = \frac{b^2+a^2}{b^2-a^2},$$

$$\therefore \quad T : T' = b^2 + a^2 : b^2 - a^2.$$

Example 8.13 A plane lamina, under the action of no external forces, is set rotating with angular velocity Ω_0 about an axis through its center of inertia whose angular coordinates are $\theta = \alpha,\ \varphi = \beta$ referred to the principal axes at the center of inertia. Prove that at time t the component angular velocities are given by

$$\omega_1 = \Omega_0 \sin\alpha\, cos\varphi,$$

$$\omega_2 = \Omega_0 \sin\alpha\ \sin\varphi,$$

$$\omega_3 = \dot{\varphi} = \Omega_0\left[\cos^2\alpha - \frac{A-B}{A+B}\sin^2\alpha\left(\sin^2\beta - \sin^2\varphi\right)\right]^{1/2},$$

where A and B are the moments of inertia of the lamina about the principal axes in its plane.

Solution In Fig. 8.17, let O be the center of inertia, and OX, OY, and OZ be the three principal axes of the lamina at O, where OX and OY lie on the plane of the lamina. Let OL be the initial axis of rotation of the lamina. Let (x_0, y_0, z_0) be the coordinates of L.

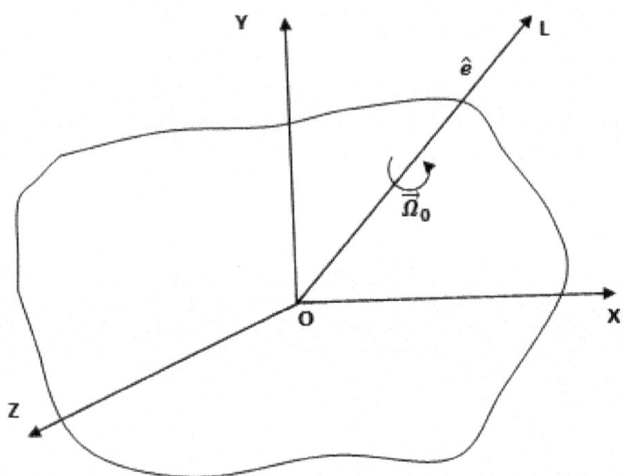

Fig. 8.17 Illustration for Example 8.13

Now

$$\left.\begin{aligned} x_0 &= r\sin\theta\cos\varphi = r_0\sin\alpha\cos\beta \\ y_0 &= r\sin\theta\sin\varphi = r_0\sin\alpha\sin\beta \\ z_0 &= r\cos\theta = r\cos\alpha \end{aligned}\right\} .$$

Let $\mathbf{r}_0$ be the p.v. of L relative to O.

$$\therefore \quad \mathbf{r}_0 = x_0\hat{i} + y_0\hat{j} + z_0\hat{k} = r_0\sin\alpha\cos\beta\hat{i} + r_0\sin\alpha\sin\beta\hat{j} + z_0\cos\alpha\hat{k},$$

$$\therefore \quad \hat{e} = \text{unit vector along } \mathbf{OL} = \frac{\mathbf{r}_0}{|\mathbf{r}_0|} = \sin\alpha\cos\beta\hat{i} + \sin\alpha\sin\beta\hat{j} + \cos\alpha\hat{k}.$$

$$\therefore \quad \boldsymbol{\Omega}_0 = \Omega_0\hat{e} = (\Omega_0\sin\alpha\cos\beta,\ \Omega_0\sin\alpha\sin\beta,\ \Omega_0\cos\alpha)\,.$$

Euler's equations under no external forces are

$$A\dot{\omega}_1 = (B - C)\,\omega_2\omega_3, \tag{8.13.94}$$

$$B\dot{\omega}_2 = (C - A)\,\omega_3\omega_1, \tag{8.13.95}$$

$$C\dot{\omega}_3 = (A - B)\,\omega_1\omega_2. \tag{8.13.96}$$

We have $C = A + B$.

Equation (8.13.94) gives

$$A\dot{\omega}_1 = (B - A - B)\,\omega_2\omega_3 = -A\omega_2\omega_3,$$

$$\text{i.e.,}\quad \dot{\omega}_1 = -\omega_2\omega_3. \tag{8.13.97}$$

Equation (8.13.95) gives

$$B\dot{\omega}_2 = B\omega_3\omega_1,$$

$$\text{i.e.,}\quad \dot{\omega}_2 = \omega_3\omega_1. \tag{8.13.98}$$

Equation (8.13.96) gives

$$(A + B)\,\dot{\omega}_3 = (A - B)\,\omega_1\omega_2. \tag{8.13.99}$$

(8.13.97) ÷ (8.13.98) yields

$$\frac{\dot{\omega}_1}{\dot{\omega}_2} = -\frac{\omega_2}{\omega_1},$$

$$\text{i.e.,}\quad \omega_1\dot{\omega}_1 + \omega_2\dot{\omega}_2 = 0,$$

$$\text{i.e.,}\quad \omega_1^2 + \omega_2^2 = C_1. \tag{8.13.100}$$

Initially

$$\left.\begin{aligned} \omega_1 &= \Omega_0 \sin\alpha \cos\beta \\ \omega_2 &= \Omega_0 \sin\alpha \sin\beta \end{aligned}\right\} \text{ at } t = 0. \tag{8.13.101}$$

∴ Equation (8.13.100) shows that

$$C_1 = \Omega_0^2 \sin^2\alpha.$$

From (8.13.100), we get

$$\left.\begin{aligned} \omega_1 &= \Omega_0 \sin\alpha \cos\varphi \\ \omega_2 &= \Omega_0 \sin\alpha \sin\varphi \end{aligned}\right\}. \tag{8.13.102}$$

(8.13.99) ÷ (8.13.98) gives

$$\frac{(A+B)\,\dot{\omega}_3}{\dot{\omega}_2} = \frac{(A-B)\,\omega_1\omega_2}{\omega_3\omega_1} = \frac{(A-B)\,\omega_2}{\omega_3},$$

i.e., $(A+B)\,\omega_3\dot{\omega}_3 + (B-A)\,\omega_2\dot{\omega}_2 = 0,$

i.e., $(A+B)\,\omega_3^2 + (B-A)\,\omega_2^2 = C_2.$ (8.13.103)

Initially

$$\left.\begin{aligned}\omega_3 &= \Omega_0 \cos\alpha \\ \omega_2 &= \Omega_0 \sin\alpha \sin\beta\end{aligned}\right\} \text{ at } t = 0.$$

$$\therefore \quad C_2 = (A+B)\,\Omega_0^2 \cos^2\alpha + (B-A)\,\Omega_0^2 \sin^2\alpha \, \sin^2\beta.$$

Therefore, Equation (8.13.103) gives

$$(A+B)\,\omega_3^2 + (B-A)\,\omega_2^2 = \Omega_0^2\left[(A+B)\cos^2\alpha + (B-A)\sin^2\alpha\,\sin^2\beta\right],$$

i.e., $(A+B)\,\omega_3^2 + (B-A)\,\Omega_0^2 \sin^2\alpha\,\sin^2\varphi = \Omega_0^2\left[(A+B)\cos^2\alpha + (B-A)\sin^2\alpha\,\sin^2\beta\right],$

i.e., $(A+B)\,\omega_3^2 = \Omega_0^2\left[(A+B)\cos^2\alpha + (A-B)\sin^2\alpha\left(\sin^2\varphi - \sin^2\beta\right)\right],$

i.e., $\omega_3^2 = \Omega_0^2\left[\cos^2\alpha - \dfrac{A-B}{A+B}\sin^2\alpha\left(\sin^2\beta - \sin^2\varphi\right)\right],$

$$\therefore \quad \omega_3 = \Omega_0\left[\cos^2\alpha - \frac{A-B}{A+B}\sin^2\alpha\left(\sin^2\beta - \sin^2\varphi\right)\right]^{\frac{1}{2}}.$$

Equation (8.13.98) gives

$$\Omega_0 \sin\alpha\cos\varphi\,\dot{\varphi} = \omega_3\Omega_0 \sin\alpha\cos\varphi,$$

$$\therefore \quad \dot{\varphi} = \omega_3,$$

i.e., $\omega_3 = \dot{\varphi}.$

Thus, the components of the angular velocity are

$$\left.\begin{aligned}\omega_1 &= \Omega_0 \sin\alpha\cos\varphi \\ \omega_2 &= \Omega_0 \sin\alpha\sin\varphi \\ \omega_3 = \dot{\varphi} &= \Omega_0\left[\cos^2\alpha - \tfrac{A-B}{A+B}\sin^2\alpha\left(\sin^2\beta - \sin^2\varphi\right)\right]^{\frac{1}{2}}\end{aligned}\right\}.$$

Example 8.14 A uniform heavy square lamina, free to turn about its center which is fixed, is set rotating with angular velocity Ω about an axis inclined at 45° to its plane. Show that, in the subsequent motion, the axis of the figure of the lamina describes, with uniform angular velocity, a right circular cone whose axis is inclined at $\tan^{-1}\frac{1}{3}$ with the invariable line. Find the semi-angle of the cone and the uniform angular velocity with which it is described.

Consider a square lamina PQRS shown in Fig. 8.18, having sides $2a$ and mass m. The center of gravity G is taken as the origin, X-axis $\parallel$ **PQ**, Y-axis $\parallel$ **QR**, and Z-axis along the normal to the lamina at G. Here, GZ (normal) is the axis of Fig. 8.18.

Let α, β, γ be the direction angles of GL.

$\therefore\quad \gamma = 45°$, where GL is the initial axis of rotation.

Unit vector along **GL** is given by

$$\hat{e} = l\hat{i} + m\hat{j} + n\hat{k} = l\hat{i} + m\hat{j} + \cos 45°\hat{k} = l\hat{i} + m\hat{j} + \frac{1}{\sqrt{2}}\hat{k}.$$

The initial angular velocity of the lamina is given by

$$\boldsymbol{\Omega} = \Omega\hat{e} = \Omega\left(l,\ m,\ \frac{1}{\sqrt{2}}\right) = \left(\Omega l,\ \Omega m,\ \frac{\Omega}{\sqrt{2}}\right).$$

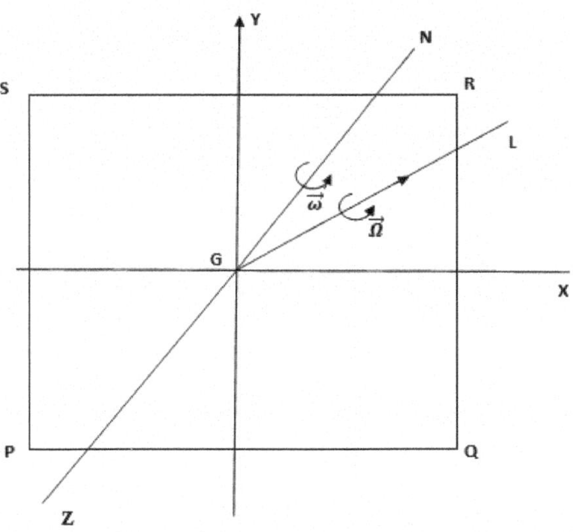

Fig. 8.18 Illustration for Example 8.14

Let $\boldsymbol{\omega} = (\omega_1, \omega_2, \omega_3)$ be the angular velocity of the lamina at time t, about the instantaneous axis of rotation GN. Now

$$A = \text{MI of the lamina about GX} = \frac{1}{3}mb^2 = \frac{1}{3}ma^2,$$

$$B = \text{MI of the lamina about GY} = \frac{1}{3}ma^2,$$

$$C = \text{MI of the lamina about GZ} = \frac{2}{3}ma^2,$$

$D = E = F = 0$ (due to symmetry).

Therefore, GX, GY, GZ are the principal axes for the body at G.

The external forces acting on the lamina are

(I) the weight of the lamina through G,

(II) the reaction of the axis of rotation at G.

$$\therefore \quad \boldsymbol{\Lambda} = \text{external torque} = \mathbf{0},$$

$$\therefore \quad L = M = N = 0.$$

Integrals of energy and angular momentum give

$$A\omega_1^2 + B\omega_2^2 + C\omega_3^2 = \text{Constant}, \tag{8.13.104}$$

$$A^2\omega_1^2 + B^2\omega_2^2 + C^2\omega_3^2 = \text{Constant}. \tag{8.13.105}$$

Equation (8.13.104) gives

$$\frac{1}{3}ma^2\omega_1^2 + \frac{1}{3}ma^2\omega_2^2 + \frac{2}{3}ma^2\omega_3^2 = \text{Constant},$$

i.e., $\omega_1^2 + \omega_2^2 + 2\omega_3^2 = C_1$.

Initially $\omega_1 = \Omega l,\ \omega_2 = \Omega m,\ \omega_3 = \frac{\Omega}{\sqrt{2}}$ at $t = 0$.

$$\therefore \quad C_1 = \Omega^2\left(l^2 + m^2 + 1\right) = \Omega^2\left(2 - n^2\right) = \Omega^2\left(2 - \frac{1}{2}\right) = \frac{3}{2}\Omega^2.$$

$$\therefore \quad \omega_1^2 + \omega_2^2 + 2\omega_3^2 = \frac{3}{2}\Omega^2. \tag{8.13.106}$$

Equation (8.13.105) gives

$$\frac{1}{9}m^2a^4\omega_1^2 + \frac{1}{9}m^2a^4\omega_2^2 + \frac{4}{9}m^2a^4\omega_3^2 = \text{Constant},$$

i.e., $\omega_1^2 + \omega_2^2 + 4\omega_3^2 = C_2$.

Initially $\omega_1 = \Omega l, \ \omega_2 = \Omega m, \ \omega_3 = \frac{\Omega}{\sqrt{2}}$ at $t = 0$.

$$\therefore \quad C_2 = \Omega^2 l^2 + \Omega^2 m^2 + 4\frac{\Omega^2}{2} = \Omega^2\left(l^2 + m^2 + 2\right) = \Omega^2\left(3 - \frac{1}{2}\right) = \frac{5}{2}\Omega^2.$$

$$\text{i.e., } \omega_1^2 + \omega_2^2 + 4\omega_3^2 = \frac{5}{2}\Omega^2. \tag{8.13.107}$$

(8.13.107) − −(8.13.106) give

$$2\omega_3^2 = \frac{\Omega^2}{2}2 = \Omega^2,$$

$$\text{i.e., } \omega_3^2 = \frac{\Omega^2}{2}. \tag{8.13.108}$$

$\therefore$ (8.13.106) gives

$$\omega_1^2 + \omega_2^2 + \Omega^2 = \frac{3}{2}\Omega^2,$$

$$\text{i.e., } \omega_1^2 + \omega_2^2 = \frac{\Omega^2}{2}.$$

Now

$$|\boldsymbol{\omega}|^2 = \omega^2 = \omega_1^2 + \omega_2^2 + \omega_3^2 = \frac{\Omega^2}{2} + \frac{\Omega^2}{2} = \Omega^2. \tag{8.13.109}$$

Thus, the lamina rotates about G with an angular velocity $\boldsymbol{\omega}$ of constant magnitude, i.e., the axis of the figure (normal) also rotates about the instantaneous axis of rotation GN with uniform angular velocity.

Equations of the axis of the figure are

$$\frac{x}{0}=\frac{y}{0}=\frac{z}{1}. \tag{8.13.110}$$

Equations of the instantaneous axis of rotation GN are

$$\frac{x}{\omega_1}=\frac{y}{\omega_2}=\frac{z}{\omega_3}. \tag{8.13.111}$$

Let θ be the angle between (8.13.110) and (8.13.111)

$$\therefore \quad \cos\theta=\frac{\omega_3}{\sqrt{\omega_1^2+\omega_2^2+\omega_3^2}}=\frac{\frac{\Omega}{\sqrt{2}}}{\Omega}=\frac{1}{\sqrt{2}},$$

$$\therefore \quad \theta=\frac{\pi}{4}=\text{Constant}.$$

$\therefore$ The axis of the figure describes a right circular cone about GN having semi-vertical angle $\frac{\pi}{4}$ with uniform angular velocity $\boldsymbol{\omega}=\boldsymbol{\Omega}$.

Thus, the axis of the figure describes a right cone with the instantaneous axis of rotation GN as an axis.

The invariable line (fixed in space) is given by

$$\frac{x}{A\omega_1}=\frac{y}{B\omega_2}=\frac{z}{C\omega_3},$$

$$\text{i.e.,} \quad \frac{x}{\frac{1}{3}ma^2\omega_1}=\frac{y}{\frac{1}{3}ma^2\omega_2}=\frac{z}{\frac{2}{3}ma^2\omega_3},$$

$$\text{i.e.,} \quad \frac{x}{\omega_1}=\frac{y}{\omega_2}=\frac{z}{2\omega_3}. \tag{8.13.112}$$

Let φ be the angle of inclination of the axis of the cone GN to the invariable line (8.13.112)

$$\therefore \quad \cos\varphi=\frac{\omega_1^2+\omega_2^2+2\omega_3^2}{\sqrt{\omega_1^2+\omega_2^2+\omega_3^2}\sqrt{\omega_1^2+\omega_2^2+4\omega_3^2}}=\frac{\frac{3}{2}\Omega^2}{\sqrt{\Omega^2}\sqrt{\frac{5}{2}\Omega^2}}=\frac{3}{2}\times\frac{\sqrt{2}}{\sqrt{5}}=\frac{3}{\sqrt{10}}.$$

[by (8.13.106), (8.13.107), (8.13.109)]

Now

$$\tan\varphi=\sqrt{\sec^2\varphi-1}=\sqrt{\frac{10}{9}-1}=\frac{1}{3}.$$

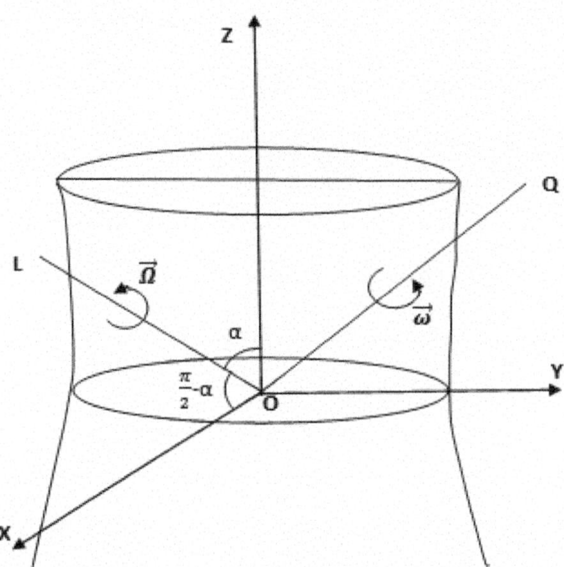

Fig. 8.19 Illustration for Example 8.15

$$\therefore \quad \varphi = \tan^{-1}\frac{1}{3}.$$

Example 8.15 A solid of revolution whose principal moments at the center of inertia are A, A, C ($C > A$) is set spinning with angular velocity $\boldsymbol{\Omega}$ about an axis passing through the center of inertia and making an angle α with the axis of the figure. Prove that the instantaneous axis of rotation describes in the body a right cone of semi-vertical angle α in period $\dfrac{2\pi A \sec\alpha}{(C-A)\Omega}$.

Solution Let O be the center of inertia and OZ be the axis of revolution of the solid. Let OL be the initial axis of rotation of the body, inclined at an angle α to OZ. Take the X-axis on the plane passing through OZ and OL.

From Fig. 8.19, the direction cosines of OL are

$$\cos\left(\frac{\pi}{2}-\alpha\right), \quad \cos 90^\circ, \quad \cos\alpha;$$

$$\text{i.e.,} \quad l = \sin\alpha, \quad m = 0, \quad n = \cos\alpha.$$

Let $\hat{e}$ be the unit vector along **OL**.

$$\therefore \quad \hat{e} = l\hat{i} + m\hat{j} + n\hat{k} = (l,\ m,\ n) = (\sin\alpha,\ 0,\ \cos\alpha)\,.$$

$\therefore$ The initial angular velocity of the body is given by

$$\boldsymbol{\Omega} = \Omega\hat{e} = \Omega\,(\sin\alpha,\ 0,\ \cos\alpha) = (\Omega\sin\alpha,\ 0,\ \Omega\cos\alpha)\,.$$

The external forces acting on the body are

(I) the weight of the body through O.

(II) the reaction of the axis of rotation at O.

$\therefore$ $\mathbf{\Lambda}$ = external torque about O $= \mathbf{0}$, and so, $L = M = N = 0$.
Euler's equation for the motion of a rigid body with one point fixed gives

$$\left.\begin{aligned} A\dot{\omega}_1 - (B - C)\,\omega_2\omega_3 = 0 \\ B\dot{\omega}_2 - (C - A)\,\omega_3\omega_1 = 0 \\ C\dot{\omega}_3 - (A - B)\,\omega_1\omega_2 = 0 \end{aligned}\right\}. \tag{8.13.113}$$

In the present case, $B = A$.
Thus, third equation of (8.13.113) gives

$$C\dot{\omega}_3 = 0,$$

i.e., $\omega_3 = C_3 =$ a constant.

Initially

$$\omega_3 = \Omega\cos\alpha \text{ at } t = 0.$$

$$\therefore \quad \Omega\cos\alpha = C_3,$$

$$\text{i.e., } \omega_3 = \Omega\cos\alpha. \tag{8.13.114}$$

First equation of (8.13.113) gives

$$A\dot{\omega}_1 - (A - C)\,\omega_2\omega_3 = 0,$$

$$\text{i.e., } A\dot{\omega}_1 = (A - C)\,\omega_2\omega_3. \tag{8.13.115}$$

Second equation of (8.13.113) gives

$$A\dot{\omega}_2 = (C - A)\,\omega_3\omega_1. \tag{8.13.116}$$

(8.13.115) $\div$ (8.13.116) yields

$$\frac{A\dot{\omega}_1}{A\dot{\omega}_2} = \frac{(A - C)\,\omega_2\omega_3}{(C - A)\,\omega_3\omega_1}$$

i.e., $\frac{\dot{\omega}_1}{\dot{\omega}_2} = -\frac{\omega_2}{\omega_1}$,

i.e., $\omega_1\dot{\omega}_1 + \omega_2\dot{\omega}_2 = 0$,

i.e., $\omega_1^2 + \omega_2^2 = C_1$.

Initially

$\omega_1 = \Omega \sin\alpha, \ \omega_2 = 0 \ \text{at } t = 0.$

$\therefore \quad \Omega^2 \sin^2\alpha = C_1,$

$$\text{i.e.,} \quad \omega_1^2 + \omega_2^2 = \Omega^2 \sin^2\alpha. \tag{8.13.117}$$

Equation (8.13.114) gives

$$\omega_3^2 = \Omega^2 \cos^2\alpha. \tag{8.13.118}$$

(8.13.117) ÷ (8.13.118) gives

$$\frac{\omega_1^2 + \omega_2^2}{\omega_3^2} = \tan^2\alpha,$$

$$\text{i.e.,} \quad \omega_1^2 + \omega_2^2 = \omega_3^2 \tan^2\alpha. \tag{8.13.119}$$

Let OQ be instantaneous of rotation.

Direction ratios of OQ = Direction ratios of $\boldsymbol{\omega}$ are $\omega_1, \ \omega_2, \ \omega_3$.

Therefore, the equations of the instantaneous axis of rotation OQ are

$$\frac{x}{\omega_1} = \frac{y}{\omega_2} = \frac{z}{\omega_3}. \tag{8.13.120}$$

Therefore, the locus of (8.13.120) subject to the condition (8.13.119) is

$x^2 + y^2 = z^2 \tan^2\alpha,$

which represents a right circular cone with vertex at O, Z-axis as the axis, and α as the semi-vertical angle. Thus, the instantaneous axis of rotation describes in the body (as Z-axis is fixed in the body) a right cone of semi-vertical angle α.

Equation (8.13.115) gives

$$A\ddot{\omega}_1 = (A - C)\,\dot{\omega}_2\dot{\omega}_3 = (A - C)\,\dot{\omega}_2\Omega\cos\alpha = (A - C)\,\Omega\cos\alpha\frac{C - A}{A}\omega_3\omega_1,$$

$$\text{i.e.,}\quad A\ddot{\omega}_1 = (A - C)\,\Omega\cos\alpha\frac{C - A}{A}\Omega\omega_1\cos\alpha = -\frac{(C - A)^2\omega_1\Omega^2\cos^2\alpha}{A},$$

$$\text{i.e.,}\quad \ddot{\omega}_1 = -\frac{(C - A)^2\Omega^2\cos^2\alpha}{A^2}\omega_1 = -\lambda^2\omega_1;$$

$$\lambda = \frac{(C - A)\,\Omega\cos\alpha}{A}.$$

Similarly $\ddot{\omega}_2 = -\lambda^2\omega_2$, and $\ddot{\omega}_3 = 0$.
We have

$$\boldsymbol{\omega} = \omega_1\hat{i} + \omega_2\hat{j} + \omega_3\hat{k},$$

$$\therefore\quad \ddot{\boldsymbol{\omega}} = \ddot{\omega}_1\hat{i} + \ddot{\omega}_2\hat{j} = -\lambda^2\omega_1\hat{i} - \lambda^2\omega_2\hat{j} = -\lambda^2\left(\omega_1\hat{i} + \omega_2\hat{j}\right),$$

$$\text{i.e.,}\quad \ddot{\boldsymbol{\omega}} = -\lambda^2\left(\omega_1\hat{i} + \omega_2\hat{j} + \omega_3\hat{k}\right) + \lambda^2\omega_3\hat{k} = -\lambda^2\boldsymbol{\omega} + \lambda^2\omega_3\hat{k},$$

which represents a periodic motion of period

$$T = \frac{2\pi}{\sqrt{\mu}} = \frac{2\pi}{\sqrt{\lambda^2}} = \frac{2\pi}{\lambda} = \frac{2\pi A}{(C - A)\,\Omega\cos\alpha}.$$

Hence, the instantaneous axis of rotation describes in the body a right cone of semi-vertical angle in period $\frac{2\pi A\sec\alpha}{(C-A)\Omega}$.

8.14 Exercise-VIII

1. If a rigid body with one point fixed rotates with angular velocity $\boldsymbol{\omega}$ and has angular momentum $\boldsymbol{\Omega}$, prove that the KE of the body is given by $T = \frac{1}{2}\boldsymbol{\omega}\cdot\boldsymbol{\Omega}$. Hence show that $T = \frac{1}{2}\left(A\omega_1^2 + B\omega_2^2 + C\omega_3^2 - 2D\omega_2\omega_3 - 2E\omega_3\omega_1 - 2F\omega_1\omega_2\right)$.
2. A rigid body moves freely about a point O under no external forces. Prove that the instantaneous axis of rotation describes a cone, with a vertex at O, fixed in the body.
3. A body moves under no forces about a point O, the principal moments of inertia at O being 6λ, 3λ, λ. Intially, the components of the angular velocity are $\omega_1 = n$,

$\omega_2 = 0$, and $\omega_3 = 3n$ about the principal axes. Show that at any time t, $\omega_2 = -\sqrt{5}\,n \tanh\left(\sqrt{5}\,nt\right)$ [**Hints**: Similar to Example 8.4].

4. A rigid body can rotate about a fixed point O. The coordinates of C.G. referred to the principal axes at O are (a, b, c). If l, m, n are the direction cosines of the upward vertical with reference to the principal axes, prove that the body will rotate about the vertical through the point O steadily if $(B - C)\frac{a}{l} + (C - A)\frac{b}{m} + (A - B)\frac{c}{n} = 0$. Find also the angular velocity for the steady motion.
5. A rigid body is free in space under no external forces and its motion consists of a pure rotation about an axis fixed in the body. The principal moments of inertia A, B, and C at the center of mass are all unequal. Establish that (I) The angular velocity is constant in magnitude. (II) The axis of rotation is fixed in space. (III) The center of mass lies on the axis of rotation.
6. A uni-axial body is supported at its center of mass and is rotating initially with angular velocity about an axis $\perp^r$ to the axis of symmetry. Prove that if a couple of constant moments are applied about the axis of symmetry, the instantaneous axis will describe a cone whose equation referred to the axis fixed in the body of which that of the z-axis coincides with the axis of symmetry is

$$2AN\left(x^2 + y^2\right)\tan^{-1}\frac{y}{x} = C\,(C - A)\,\omega^2 z^2.$$

7. Find the torque needed to rotate a rectangular plate of sides a and b about diagonal with constant angular velocity !. [Hint: See Example No.3.]
   Ans.: $\mathbf{\Lambda} = \frac{M(b^2-a^2)ab\omega^2}{12(a^2+b^2)}\hat{k}$.
8. Derive Euler's equations for the motion of a rigid body with one point fixed, in the case when the axes are not necessarily the principal axes. [**Hint**: Sect. 8.9.]
9. Establish Euler's equations of motion of a rigid body with one point fixed. [**Hint**: Sect. 8.8.]
10. An ellipsoid, free to move about its center, is set in rotation at time $t = 0$ with an angular velocity $\mathbf{\Omega}$. If $n, o, 3n$ are the components of $\mathbf{\Omega}$ along the principal axes at the center, and 6A,3A,A the moments of inertia about three axes, find the components of the angular velocity at time t, and show that for large values of t, the magnitude of the angular velocity is practically $n\sqrt{5}$.
11. Solve the problem of a compound pendulum by using Euler's equations.
    Hints: $\omega_1 = \dot{\theta}$, $\omega_2 = \omega_3 = 0$, $L = mgh\sin\theta$, $M = N = 0$.-component Euler's equation is $A\dot{\omega}_1 - (B - C)\,\omega_2\omega_3 = L$, $mK^2\ddot{\theta} = -mgh\sin\theta$, or $\ddot{\theta} = -\frac{gh}{K^2}\sin\theta$.
12. A uniform thin circular disc is set rotating with an angular velocity $\mathbf{\Omega}$ about an axis through the center making an angle α with the normal. Prove that the semi-vertical angle θ of the cone described by the axis of the disc about the invariable line is given by $\tan\theta = \frac{1}{2}\tan\alpha$.
13. Prove that if a body for which $A = B$ is set rotating with angular velocity η about an axis which makes an angle α with its axis of figure, then its axis of

figure describes in space a cone of semi-angle γ where $\tan\gamma = \frac{A}{C}\tan\alpha$, in time $\frac{2\pi}{\Omega}$, where $\Omega\sin\gamma = \eta\sin\alpha$.

14. A uniform square lamina is rotating under no impressed forces about the principal axis at O (in the order of increasing moments of inertia), show that $\omega_1^2 + \omega_2^2$ and $3\omega_2^2 + 4\omega_3^2$ are constant.
15. A body moves under no forces about a point O, the principal moments of inertia at O being $6\lambda,\ 3\lambda,\ \lambda\,(\lambda > 0)$. Initially the angular velocity of the body has components $\omega_1 = n, \omega_2 = 0, \omega_3 = 3n$ about the principal axes. Show that at any time $t, \omega_2 = -\sqrt{5}n\tanh\left(\sqrt{5}nt\right)$.
16. A rigid body having one point O fixed and no external torque about O has two equal principal moments of inertia. Prove that it must rotate with an angular velocity of constant magnitude.
17. An ellipsoid free to move about its center is set in rotation at time t =0 with an angular velocity $\boldsymbol{\Omega}$. If n, 0, $3n$ are the components of $\boldsymbol{\Omega}$ along the principal axes at the center, and 6A, 3A, A the moments of inertia about these axes, find the components of the angular velocity at time t, and show that, for large values of t, the magnitude of the angular velocity is practically $n\sqrt{5}$.
18. A solid cube is in motion about an angular point which is fixed. If there are no external forces and $\omega_1,\ \omega_2,\ \omega_3$ are the angular velocities about the edges through the fixed point, prove that $\omega_1 + \omega_2 + \omega_3$ and $\omega_1^2 + \omega_2^2 + \omega_3^2$ are each constant. [**Hints**: $L = M = N = 0,\ \frac{d}{dt}\left(\frac{\partial T}{\partial\omega_1}\right) - \omega_3\frac{\partial T}{\partial\omega_2} + \omega_2\frac{\partial T}{\partial\omega_3} = 0$, etc.]
19. An elliptical lamina acted on by no external forces is set rotating about a line through its center which makes equal angles with the major axis and minor axis and with the normal to its plane. Prove that if e be the eccentricity of the ellipse, and $\omega_1,\ \omega_2,\ \omega_3$ be the angular velocities about the major and minor axes and the normal to the plane, then $\omega_1 = F\cos\theta,\ \omega_2 = F\sin\theta,\ \omega_3 = \dot{\theta} = F\sqrt{\frac{1-e^2\sin^2\theta}{2-e^2}}$, where F is a constant.
20. A body moves about a point O under no forces, the principal moments of inertia at O being 3A, 5A, and 6A. Initially the angular velocity has components $\omega_1 = n,\ \omega_2 = 0,\ \omega_3 = n$ about the corresponding principal axis. Show that at any later time $t, \omega_2 = \frac{3n}{\sqrt{5}}\tanh\frac{nt}{\sqrt{5}}$, and that the body ultimately rotates about its mean axis.

Chapter 9
Langrange's Equation of Motion

9.1 Basic Concepts

9.1.1 Constraints

The limitations on motion are often called constraints. For example, when a particle is constrained (restricted) to move on a curve or on a surface.

9.1.2 Generalized Coordinates

Suppose that a system of N particles of masses m_ν $(\nu = 1, 2, 3, \ldots, N)$ moves subject to possible constraints. Then at time t, the position of each particle may be specified by means of n independent variables q_α $(\alpha = 1, 2, \ldots, n)$. These variables $q_1, q_2, q_3, \ldots, q_n$ are termed as the generalized coordinates of the system.

9.1.3 Degree of Freedom

The number of coordinates required to specify the position of a system of one or more particles is called the degree of freedom of the system.

Illustrations:

- Degree of freedom of a particle moving in space $= 3$.
- Degree of freedom of a particle moving on a surface $z = f(x, y)$ is 2.
- Degree of freedom of a rigid body freely moving in space $= 6$.
- Degree of freedom of a rigid body moving with one pt. fixed $= 3$.

Note: Degree of freedom is reduced by constraints.

N. Ahmed et al., *Classical Dynamics*, University Texts in the Mathematical Sciences,
https://doi.org/10.1007/978-981-95-6394-4_9

9.1.4 Notation

The subscript α ranges from 1 to n, the degree of freedom and the subscript ν ranges from 1 to N, the no. of particles in the system.

9.1.5 Holonomic and Non-holonomic Systems

Let a dynamical system be specified by the n generalized coordinates $q_\alpha\,(\alpha = 1, 2, \ldots, n)$. If q_α can be changed to $q_\alpha + \delta q_\alpha$, without making any change in the remaining $(n - 1)$ variables, then the dynamical system is said to be holonomic. Otherwise, it is non-holonomic.

9.1.6 Transformation Equations

Let $\mathbf{r}_\nu = x_\nu \hat{i} + y_\nu \hat{j} + z_\nu \hat{k}$ be the position vector of the ν-th particle w.r.t. X, Y, Z coordinate system. The relationships of the generalized coordinates q_α $(\alpha = 1, 2, 3, \ldots, n)$ to the position coordinates (x_ν, y_ν, z_ν) are given by

$$\left.\begin{array}{l} x_\nu = x_\nu\,(q_1, q_2, q_3, \ldots, q_n, t) \\ y_\nu = y_\nu\,(q_1, q_2, q_3, \ldots, q_n, t) \\ z_\nu = z_\nu\,(q_1, q_2, q_3, \ldots, q_n, t) \end{array}\right\}. \tag{9.1.1}$$

In vector form, (9.1.1) can be written as

$$\mathbf{r}_\nu = \mathbf{r}_\nu\,(q_1, q_2, q_3, \ldots, q_n, t)\,. \tag{9.1.2}$$

9.1.7 Generalized Velocities

Let a dynamical system be composed of N particles of masses m_ν $(v = 1,\ 2,\ 3,\ \ldots,\ N)$ and suppose that the system is specified by n generalized coordinates q_α $(\alpha = 1, 2, 3, \ldots, n)$. Then the quantities $q_\alpha = \frac{dq_\alpha}{dt}$ $(\alpha = 1, 2, \ldots n)$ are called the generalized velocities of the system. Let $\mathbf{r}_\nu$ be the position vector of the ν-th particle at time t relative to some origin O.

$$\therefore\quad \mathbf{r}_\nu = \mathbf{r}_\nu\,(q_1, q_2, q_3, \ldots, q_n, t)\,.$$

Thus,

$$\dot{\mathbf{r}}_\nu = \frac{d\mathbf{r}_\nu}{dt} = \frac{\partial \mathbf{r}_\nu}{\partial q_1}\dot{q}_1 + \frac{\partial \mathbf{r}_\nu}{\partial q_2}\dot{q}_2 + \cdots + \frac{\partial \mathbf{r}_\nu}{\partial q_n}\dot{q}_n + \frac{\partial \mathbf{r}_\nu}{\partial t}. \tag{9.1.3}$$

Now, we regard $\dot{q}_1, \dot{q}_2, \ldots \dot{q}_n$ as new independent variables.
Therefore, Eq. (9.1.3) gives

$$\frac{\partial \dot{\mathbf{r}}_\nu}{\partial \dot{q}_\alpha} = \frac{\partial \mathbf{r}_\nu}{\partial q_\alpha}, \quad \alpha = 1, 2, \ldots n.$$

9.1.8 Generalized Forces

Let the ν-th particle of mass m_ν at the position $\mathbf{r}_\nu$ at time t undergo a virtual displacement to position $\mathbf{r}_\nu + \delta\mathbf{r}_\nu$. Let $\mathbf{F}_\nu$, $\mathbf{F}_\nu{}'$ be the external and internal forces acting on m_ν. Then the virtual work done on m_ν in this displacement is $\left(\mathbf{F}_\nu + \mathbf{F}_\nu{}'\right) \cdot \delta\mathbf{r}_\nu$ and so the total virtual work alone on all particles of the system for similar displacement is

$$\delta W = \sum_{\nu=1}^{N} \left(\mathbf{F}_\nu + \mathbf{F}_\nu{}'\right) \cdot \delta\mathbf{r}_\nu = \sum_{\nu=1}^{N} \mathbf{F}_\nu \cdot \delta\mathbf{r}_\nu + \sum_{\nu=1}^{N} \mathbf{F}_\nu{}' \cdot \delta\mathbf{r}_\nu. \tag{9.1.4}$$

But $\sum_{\nu=1}^{N} \mathbf{F}_\nu{}' \cdot \delta\mathbf{r}_\nu$ =total virtual work done by the internal forces of the system=0.
Hence, equation (9.1.4) reduces to

$$\delta W = \sum_{\nu=1}^{N} \mathbf{F}_\nu \cdot \delta\mathbf{r}_\nu. \tag{9.1.5}$$

Let $\delta q_1, \delta q_2, \ldots \delta q_n$ be the increments of the generalized coordinates. Then the ν-th in particle undergoes a displacement given by

$$\delta\mathbf{r}_\nu = \mathbf{r}_\nu\left(q_1 + \delta q_1, q_2 + \delta q_2, \ldots, q_n + \delta q_n\right) - \mathbf{r}_\nu\left(q_1, q_2, \ldots, q_n\right)$$

$$\text{i.e.,}\quad \delta\mathbf{r}_\nu = \delta q_1 \frac{\partial \mathbf{r}_\nu}{\partial q_1} + \delta q_2 \frac{\partial \mathbf{r}_\nu}{\partial q_2} \ldots\ldots + \delta q_n \frac{\partial \mathbf{r}_\nu}{\partial q_n} = \sum_{\alpha=1}^{n} \frac{\partial \mathbf{r}_\nu}{\partial q_\alpha}\delta q_\alpha. \tag{9.1.6}$$

[neglecting the higher powers of δq_α]

By the use of (9.1.6), (9.1.5) takes the form

$$\delta W = \sum_{\nu=1}^{N} \mathbf{F}_\nu \cdot \sum_{\alpha=1}^{n} \frac{\partial \mathbf{r}_\nu}{\partial q_\alpha} \delta q_\alpha = \sum_{\alpha=1}^{n} \left\{ \sum_{\nu=1}^{N} \mathbf{F}_\nu \cdot \frac{\partial \mathbf{r}_\nu}{\partial q_\alpha} \right\} \delta q_\alpha = \sum_{\alpha=1}^{n} \varphi_\alpha \delta q_\alpha, \tag{9.1.7}$$

where $\varphi_\alpha = \sum_{\nu=1}^{N} \mathbf{F}_\nu \cdot \frac{\partial \mathbf{r}_\nu}{\partial q_\alpha}$; $\varphi_\alpha \, (\alpha = 1, 2, \ldots, n)$ are known as the generalized forces associated with the generalized coordinates $q_\alpha \, (\alpha = 1, 2, \ldots, n)$.

9.1.9 *Conservative Systems*

If the forces acting on each particle of a system are conservative, then the system is termed as conservative. Otherwise, it is non-conservative.

9.1.10 *Kinetic Energy*

The total KE of the system is given by

$$T = \frac{1}{2} \sum_{\nu=1}^{N} m_\nu \dot{\mathbf{r}}_\nu^2.$$

9.2 Lagrange's Equations for a Holonomic System

Let $\mathbf{F}_\nu$ be the net external force acting on the ν-th particle of mass m_ν of a holonomic system.

By Newton's 2nd law of motion, we have

$$m_\nu \ddot{\mathbf{r}}_\upsilon = \mathbf{F}_\nu,$$

$$\text{i.e.,} \quad m_\nu \ddot{\mathbf{r}}_\nu \cdot \frac{\partial \mathbf{r}_\nu}{\partial q_\alpha} = \mathbf{F}_\nu \cdot \frac{\partial \mathbf{r}_\nu}{\partial q_\alpha}. \tag{9.2.1}$$

Now

$$\frac{d}{dt}\left(\dot{\mathbf{r}}_\nu \cdot \frac{\partial \mathbf{r}_\nu}{\partial q_\alpha} \right) = \ddot{\mathbf{r}}_\nu \cdot \frac{\partial \mathbf{r}_\nu}{\partial q_\alpha} + \dot{\mathbf{r}}_\nu \cdot \frac{\partial \dot{\mathbf{r}}_\nu}{\partial q_\alpha},$$

i.e., $\ddot{\mathbf{r}}_\nu \cdot \dfrac{\partial \mathbf{r}_\nu}{\partial q_\alpha} = \dfrac{d}{dt}\left(\dot{\mathbf{r}}_\nu \cdot \dfrac{\partial \mathbf{r}_\nu}{\partial q_\alpha}\right) - \dot{\mathbf{r}}_\nu \cdot \dfrac{\partial \dot{\mathbf{r}}_\nu}{\partial q_\alpha}$,

i.e., $$m_\nu \ddot{\mathbf{r}}_\nu \cdot \frac{\partial \mathbf{r}_\nu}{\partial q_\alpha} = \frac{d}{dt}\left(m_\nu \dot{\mathbf{r}}_\nu \cdot \frac{\partial \mathbf{r}_\nu}{\partial q_\alpha}\right) - m_\nu \dot{\mathbf{r}}_\nu \cdot \frac{\partial \dot{\mathbf{r}}_\nu}{\partial q_\alpha}. \tag{9.2.2}$$

By the use of (9.2.2), (9.2.1) reduces to

$$\frac{d}{dt}\left(m_\nu \dot{\mathbf{r}}_\nu \cdot \frac{\partial \mathbf{r}_\nu}{\partial q_\alpha}\right) - m_\nu \dot{\mathbf{r}}_\nu \cdot \frac{\partial \dot{\mathbf{r}}_\nu}{\partial q_\alpha} = \mathbf{F}_\nu \cdot \frac{\partial \mathbf{r}_\nu}{\partial q_\alpha},$$

i.e., $$\frac{d}{dt}\sum_{\nu=1}^{N} m_\nu \dot{\mathbf{r}}_\nu \cdot \frac{\partial \mathbf{r}_\nu}{\partial q_\alpha} - \sum_{\nu=1}^{N} m_\nu \dot{\mathbf{r}}_\nu \cdot \frac{\partial \dot{\mathbf{r}}_\nu}{\partial q_\alpha} = \sum_{\nu=1}^{N} \mathbf{F}_\nu \cdot \frac{\partial \mathbf{r}_\nu}{\partial q_\alpha}. \tag{9.2.3}$$

The KE T of the system is given by

$$T = \frac{1}{2}\sum_{\nu=1}^{N} m_\nu \dot{\mathbf{r}}_\nu^2 = \frac{1}{2}\sum_{\nu=1}^{N} m_\nu \dot{\mathbf{r}}_\nu \cdot \dot{\mathbf{r}}_\nu. \tag{9.2.4}$$

Equation (9.2.4) gives

$$\frac{\partial T}{\partial q_\alpha} = \sum_{\nu=1}^{N} m_\nu \dot{\mathbf{r}}_\nu \cdot \frac{\partial \dot{\mathbf{r}}_\nu}{\partial q_\alpha} \tag{9.2.5}$$

and

$$\frac{\partial T}{\partial \dot{q}_\alpha} = \sum_{\nu=1}^{N} m_\nu \dot{\mathbf{r}}_\nu \cdot \frac{\partial \mathbf{r}_\nu}{\partial q_\alpha}. \tag{9.2.6}$$

Using (9.2.5) and (9.2.6) in (9.2.3) , we obtain

$$\frac{d}{dt}\left(\frac{\partial T}{\partial \dot{q}_\alpha}\right) - \frac{\partial T}{\partial q_\alpha} = \sum_{\nu=1}^{N} \mathbf{F}_\nu \cdot \frac{\partial \mathbf{r}_\nu}{\partial q_\alpha} = \varphi_\alpha, \tag{9.2.7}$$

where

$$\varphi_\alpha = \sum_{\nu=1}^{N} \mathbf{F}_\nu \cdot \frac{\partial \mathbf{r}_\nu}{\partial q_\alpha} = \text{generalized force}.$$

Thus, we have

$$\frac{d}{dt}\left(\frac{\partial T}{\partial \dot{q}_\alpha}\right) - \frac{\partial T}{\partial q_\alpha} = \varphi_\alpha \ (\alpha = 1, 2, 3, \ldots, n)\,. \tag{9.2.8}$$

Equation (9.2.8) is known as Lagrange's equation for a holonomic system.

9.3 Lagrange's Equation for Conservative Field of Forces

Lagrange's Equation for a holonomic system is

$$\frac{d}{dt}\left(\frac{\partial T}{\partial \dot{q}_\alpha}\right) - \frac{\partial T}{\partial q_\alpha} = \varphi_\alpha \, (\alpha = 1, 2, 3, \ldots, n)\,, \tag{9.3.1}$$

where $\varphi_\alpha = \sum_{\nu=1}^{N} \mathbf{F}_\nu \cdot \frac{\partial \mathbf{r}_\nu}{\partial q_\alpha}$.

We have

$$dW = \sum_{\alpha=1}^{n} \varphi_\alpha dq_\alpha, \tag{9.3.2}$$

and

$$dW = \sum_{\upsilon=1}^{N} \mathbf{F}_\nu \cdot d\mathbf{r}_\nu. \tag{9.3.3}$$

Since $\mathbf{F}_\nu$ is conservative with force potential V_ν (say) ,

$$\therefore \quad \mathbf{F}_\nu = -\nabla V_\upsilon. \tag{9.3.4}$$

Thus (9.3.3) reduces to

$$dW = -\sum_{\nu=1}^{N} \nabla V_\nu \cdot d\mathbf{r}_\nu = -\sum_{\nu=1}^{N} dV_\nu = -d\sum_{\nu=1}^{N} V_\nu = -dV,$$

where

$$V = \sum_{\nu=1}^{N} V_\nu = V_1 + V_2 + \cdots + V_N = \text{PE of the system}$$

$$\therefore \quad W = -V. \tag{9.3.5}$$

Hence, Eq. (9.3.2) gives

$$-\,dV = \sum_{\alpha=1}^{n} \varphi_\alpha dq_\alpha. \tag{9.3.6}$$

But

$$V = V\left(q_1, q_2, q_3, \ldots, q_n\right),$$

$$\therefore \quad dV = \frac{\partial V}{\partial q_1} dq_1 + \frac{\partial V}{\partial q_2} dq_2 + \cdots + \frac{\partial V}{\partial q_n} dq_n = \sum_{\alpha=1}^{n} \frac{\partial V}{\partial q_\alpha} dq_\alpha. \tag{9.3.7}$$

Using (9.3.7) in (9.3.6) we obtain

$$\sum_{\alpha=1}^{n} \varphi_\alpha dq_\alpha = -\sum_{\alpha=1}^{n} \frac{\partial V}{\partial q_\alpha} dq_\alpha. \tag{9.3.8}$$

By equating the coefficient of dq_α in (9.3.8), we get

$$\varphi_\alpha = -\frac{\partial V}{\partial q_\alpha}. \tag{9.3.9}$$

Using (9.3.9) in (9.3.1), we get

$$\frac{d}{dt}\left(\frac{\partial T}{\partial \dot{q}_\alpha}\right) - \frac{\partial T}{\partial q_\alpha} = -\frac{\partial V}{\partial q_\alpha},$$

$$\text{i.e.,} \quad \frac{d}{dt}\left(\frac{\partial T}{\partial \dot{q}_\alpha}\right) = \frac{\partial}{\partial q_\alpha}\left(T - V\right). \tag{9.3.10}$$

Let us write

$$L = T - V, \tag{9.3.11}$$

$$\text{i.e.,} \quad T = L + V,$$

$$\therefore \quad \frac{\partial T}{\partial \dot{q}_\alpha} = \frac{\partial L}{\partial \dot{q}_\alpha} + \frac{\partial V}{\partial \dot{q}_\alpha} = \frac{\partial L}{\partial \dot{q}_\alpha} \quad \left[\because \quad V = V\left(q_\alpha\right),\ \alpha = 1, \ldots, n\right]. \tag{9.3.12}$$

Using (9.3.11) and (9.3.12) in (9.3.10), we obtain

$$\frac{d}{dt}\left(\frac{\partial L}{\partial \dot{q}_\alpha}\right) = \frac{\partial L}{\partial q_\alpha}; \quad \alpha = 1,\ 2,\ 3\ldots, n. \tag{9.3.13}$$

Equation (9.3.13) is Lagrange's equation for a conservative holonomic system. In (9.3.13), $L = T - V$ is called the Lagrangian of the system.

9.4 Linear Displacement and Angular Displacement

In Fig. 9.1, let O and P be two fixed points in a rigid body and $\mathbf{OP} = \mathbf{r}$. Let the rigid body turn through a small angle $\delta\theta$ in the positive sense about an axis through O specified by the unit vector $\hat{a}$.

Then the movement of the point P is $\hat{a}\delta\theta \times \mathbf{r} = \mathbf{e} \times \mathbf{r}$; $\mathbf{e} = \hat{a}\delta\theta =$ rotation.

Let the rigid body be displaced by a small quantity, then this displacement is the sum of a constant translation $\mathbf{u}$ and a small rotation $\mathbf{e}$. If $\mathbf{r}_\alpha$ is displaced to $\mathbf{r}_\alpha + \delta\mathbf{r}_\alpha$ then

$$\delta\mathbf{r}_\alpha = \mathbf{u} + \mathbf{e} \times \mathbf{r}_\alpha; \quad \alpha = 1, 2\ldots, n.$$

9.5 Work Done by a Couple

Consider a couple $\mathbf{G}$ consisting of the force $-\mathbf{F}$, $\mathbf{F}$ acting at the points A ($\mathbf{r}_1$) and B ($\mathbf{r}_2$) of a rigid body, respectively (see Fig. 9.2). Then the moment of the couple is given by

$$\mathbf{G} = \mathbf{AB} \times \mathbf{F} = (\mathbf{r}_2 - \mathbf{r}_1) \times \mathbf{F}.$$

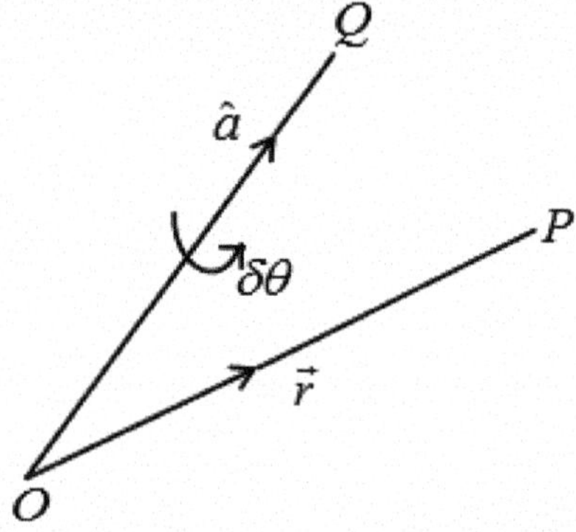

Fig. 9.1 Illustrative diagram for the angular displacement of a rigid body

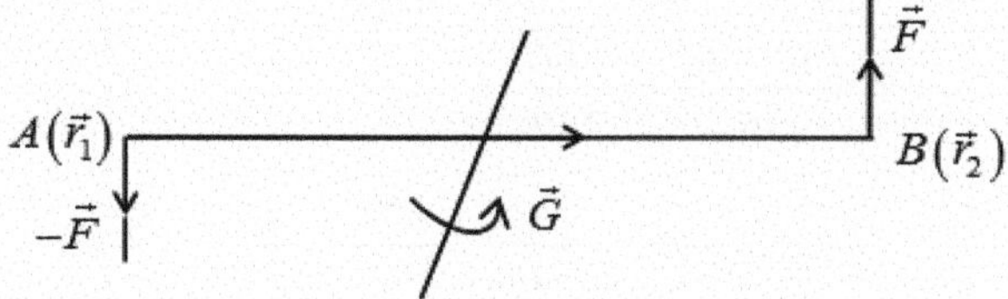

Fig. 9.2 Illustrative diagram for the work done by a couple

Let the body be given a small displacement which is equivalent to translation **u** and rotation **e**.

Then

$$\left.\begin{aligned}\delta\mathbf{r}_1 &= \mathbf{u} + \mathbf{e} \times \mathbf{r}_1 \\ \delta\mathbf{r}_2 &= \mathbf{u} + \mathbf{e} \times \mathbf{r}_2\end{aligned}\right\}. \tag{9.5.1}$$

Let δW be the work done by the couple for this displacement.

$$\therefore \quad \delta W = \mathbf{F} \cdot \delta\mathbf{r}_2 + (-\mathbf{F}) \cdot \delta\mathbf{r}_1 = \mathbf{F} \cdot (\mathbf{u} + \mathbf{e} \times \mathbf{r}_2) - \mathbf{F} \cdot (\mathbf{u} + \mathbf{e} \times \mathbf{r}_1),$$

$$\text{i.e.,} \quad \delta W = \mathbf{F} \cdot (\mathbf{e} \times \mathbf{r}_2) - \mathbf{F} \cdot (\mathbf{e} \times \mathbf{r}_1) = \mathbf{e} \times \mathbf{r}_2 \cdot \mathbf{F} - \mathbf{e} \times \mathbf{r}_1 \cdot \mathbf{F},$$

$$\text{i.e.,} \quad \delta W = \mathbf{e} \cdot (\mathbf{r}_2 \times \mathbf{F}) - \mathbf{e} \cdot (\mathbf{r}_1 \times \mathbf{F}) = \mathbf{e} \cdot (\mathbf{r}_2 \times \mathbf{F} - \mathbf{r}_1 \times \mathbf{F}),$$

$$\text{i.e.,} \quad \delta W = \mathbf{e} \cdot \{(\mathbf{r}_2 - \mathbf{r}_1) \times \mathbf{F}\} = \mathbf{e} \cdot \mathbf{AB} \times \mathbf{F} = \mathbf{e} \cdot \mathbf{G}. \tag{9.5.2}$$

9.6 Generalized Momentum and Impulse

Consider a holonomic system specified by n-generalized coordinates q_α at time t, where $\alpha = 1, 2, 3, \ldots, n$.

Let $T = T(q_1, q_2, \ldots q_n, \dot{q}_1, \dot{q}_2, \ldots, \dot{q}_n, t)$ be the KE of the system.

Then the quantities

$$p_\alpha = \frac{\partial T}{\partial \dot{q}_\alpha} \quad (\alpha = 1, 2, \ldots, n) \tag{9.6.1}$$

are termed as the generalized components of momentum. Let φ_α $(\alpha = 1, 2, \ldots, n)$ be the generalized forces associated with the generalized coordinates q_α $(\alpha = 1, 2, \ldots, n)$.

If

$$\lim_{\substack{\tau \to 0 \\ \varphi_\alpha \to \infty}} \int_0^\tau \varphi_\alpha dt = J_\alpha \ (\alpha = 1,\ 2,\ \ldots,\ n), \tag{9.6.2}$$

where J_α is finite, then the n-quantities J_α $(\alpha = 1,\ 2,\ \ldots,\ n)$ are termed as generalized impulses.

9.7 Lagrange's Equation for Impulsive Forces

Lagrange's equations for holonomic systems are

$$\frac{d}{dt}\left(\frac{\partial T}{\partial \dot{q}_\alpha}\right) - \frac{\partial T}{\partial q_\alpha} = \varphi_\alpha \, (\alpha = 1, 2 \ldots, n)\,,$$

$$\text{i.e.,}\quad \frac{d}{dt} p_\alpha - \frac{\partial T}{\partial q_\alpha} = \varphi_\alpha,$$

$$\text{i.e.,}\quad d p_\alpha - \frac{\partial T}{\partial q_\alpha} dt = \varphi_\alpha dt. \tag{9.7.1}$$

Integrating (9.7.1) from $t = 0$ to $t = \tau$, we obtain

$$\int_0^\tau d p_\alpha - \int_0^\tau \frac{\partial T}{\partial q_\alpha} dt = \int_0^\tau \varphi_\alpha dt,$$

$$\text{i.e.,}\quad (p_\alpha)_\tau - (p_\alpha)_0 - \int_0^\tau \frac{\partial T}{\partial q_\alpha} dt = \int_0^\tau \varphi_\alpha dt,$$

$$\text{i.e.,}\quad \lim_{\substack{\tau \to 0 \\ \varphi_\alpha \to \infty}} \left[(p_\alpha)_\tau - (p_\alpha)_0\right] - \lim_{\substack{\tau \to 0 \\ \varphi_\alpha \to \infty}} \int_0^\tau \frac{\partial T}{\partial q_\alpha} dt = \lim_{\substack{\tau \to 0 \\ \varphi_\alpha \to \infty}} \int_0^\tau \varphi_\alpha dt. \tag{9.7.2}$$

We note that the coordinates q_α do not change abruptly.
i.e., $\left.q_\alpha\right]_{t=0} \simeq \left.q_\alpha\right]_{t=\tau}$ for small τ.

So,

$$\lim_{\substack{\tau\to 0\\ \varphi_\alpha\to\infty}} \int_0^\tau \frac{\partial T}{\partial q_\alpha} dt = \lim_{\substack{\tau\to 0\\ \varphi_\alpha\to\infty}} \frac{\partial T}{\partial q_\alpha}\tau = 0. \tag{9.7.3}$$

Therefore, Eq. (9.7.2) reduces to

$$(p_\alpha)_\tau - (p_\alpha)_0 = J_\alpha,$$

$$\text{i.e.,}\quad (p_\alpha)_2 - (p_\alpha)_1 = J_\alpha\,(\alpha = 1, 2, \ldots n). \tag{9.7.4}$$

Equation (9.7.4) is Lagrange's equation for impulsive forces.

9.8 KE as a Quadratic Function of Velocities

Let $\mathbf{r}_\nu$ be the position vector of the ν-th particle of mass m_ν, relative to some origin O. The KE of the holonomic dynamical system is given by

$$T = \sum_{\nu=1}^{N} \frac{1}{2} m_\nu \dot{\mathbf{r}}_\nu^2, \tag{9.8.1}$$

where N is the no. of particles in the system. Suppose that the system is specified by the n-generalized coordinates q_α $(\alpha = 1,\ 2,\ 3, \ldots, n)$.

Then

$$\mathbf{r}_\nu = \mathbf{r}_\nu\,(q_1, q_2, \ldots, q_n;\ t)\ \ (\nu = 1, 2, \ldots N). \tag{9.8.2}$$

Now

$$\dot{\mathbf{r}}_\nu = \frac{d\mathbf{r}_\nu}{dt} = \frac{\partial \mathbf{r}_\nu}{\partial q_1}\dot{q}_1 + \frac{\partial \mathbf{r}_\nu}{\partial q_2}\dot{q}_2 + \cdots + \frac{\partial \mathbf{r}_\nu}{\partial q_n}\dot{q}_n + \frac{\partial \mathbf{r}_\nu}{\partial t}.$$

$$\therefore\quad T = \frac{1}{2}\sum_{\nu=1}^{N} m_\nu \dot{\mathbf{r}}_\nu^2 = \frac{1}{2}\sum_{\nu=1}^{N} m_\nu \left[\left(\dot{q}_1 \frac{\partial \mathbf{r}_\nu}{\partial q_1} + \dot{q}_2 \frac{\partial \mathbf{r}_\nu}{\partial q_2} + \cdots + \dot{q}_n \frac{\partial \mathbf{r}_\upsilon}{\partial q_n}\right) + \frac{\partial \mathbf{r}_\upsilon}{\partial t}\right]^2$$

$$\text{i.e.,}\quad T = \frac{1}{2}\sum_{\upsilon=1}^{N} m_\upsilon \left[\left(\dot{q}_1 \frac{\partial \mathbf{r}_\nu}{\partial q_1} + \cdots + \dot{q}_n \frac{\partial \mathbf{r}_\nu}{\partial q_n}\right)^2 + 2\frac{\partial \mathbf{r}_\nu}{dt}\cdot\left(\dot{q}_1 \frac{\partial \mathbf{r}_\nu}{\partial q_1} + \cdots + \dot{q}_n \frac{\partial \mathbf{r}_\nu}{\partial q_n}\right) + \left(\frac{\partial \mathbf{r}_\upsilon}{\partial t}\right)^2\right]$$

$$\text{i.e.,}\quad T = \frac{1}{2}\left[\left(a_{11}\dot{q}_1^2 + \cdots + a_{nn}\dot{q}_n^2\right) + (2a_{12}\dot{q}_1\dot{q}_2 + \cdots) + 2\,(a_1\dot{q}_1 + \cdots) + a\right], \tag{9.8.3}$$

where

$$a_{11} = \sum_{\nu=1}^{N} m_\nu \left(\frac{\partial \mathbf{r}_\nu}{\partial q_1}\right)^2,\quad a_{22} = \sum_{\nu=1}^{N} m_\nu \left(\frac{\partial \mathbf{r}_\nu}{\partial q_2}\right)^2,\ \ldots,\ a_{nn} = \sum_{\nu=1}^{N} m_\nu \left(\frac{\partial \mathbf{r}_\nu}{\partial q_n}\right)^2,$$

$$a_{12} = \sum_{\nu=1}^{N} m_\nu \frac{\partial \mathbf{r}_\nu}{\partial q_1}\cdot\frac{\partial \mathbf{r}_\nu}{\partial q_2}, \cdots, \text{etc.},\ a_1 = \sum_{\nu=1}^{N} m_\nu \frac{\partial \mathbf{r}_\nu}{\partial t_1}\cdot\frac{\partial \mathbf{r}_\nu}{\partial q_1}, \cdots, \text{etc.},\ a = \sum_{\upsilon=1}^{N} m_\nu \left(\frac{\partial \mathbf{r}_\nu}{\partial t}\right)^2.$$

Thus, we see that T is a quadratic function of the generalized velocities.
If $\frac{\partial \mathbf{r}_\nu}{\partial t} = \mathbf{0}$, then T is given by

$$T = \frac{1}{2}\left[a_{11}\dot{q}_1^2 + \cdots + 2a_{12}\dot{q}_1\dot{q}_2 + \cdots\right] = \frac{1}{2}\sum_{r=1}^{n}\sum_{s=1}^{n} a_{rs}\dot{q}_r\dot{q}_s\ [a_{rs} = a_{sr}], \tag{9.8.4}$$

which is a homogeneous function of degree 2 in $\dot{q}_1, \dot{q}_2, \ldots, \dot{q}_n$.

9.9 Energy Equation from Lagrange's Equation

In case of a conservative holonomic system, specified by n-generalized coordinates $q_\alpha\,(\alpha = 1, 2, \ldots, n)$,

Lagrange's Equations are

$$\frac{d}{dt}\left(\frac{\partial T}{\partial \dot{q}_\alpha}\right) - \frac{\partial T}{\partial q_\alpha} = -\frac{\partial V}{\partial q_\alpha}\ (\alpha = 1, 2, \ldots, n). \tag{9.9.1}$$

Let $\mathbf{r}_\nu$ be the position vector of the ν-th particle of mass m_ν at time t, relative to some origin O and suppose that $\frac{\partial \mathbf{r}_\nu}{\partial t} = \mathbf{0}\ (\nu = 1, 2, \ldots, N)$.

In this case, the KE of the system is given by

$$T = \frac{1}{2}\sum_{r=1}^{n}\sum_{s=1}^{n} a_{rs}\dot{q}_r\dot{q}_s\ \ [a_{rs} = a_{sr}]. \tag{9.9.2}$$

As T is a homogeneous function of degree 2 in $\dot{q}_1, \dot{q}_2, \ldots, \dot{q}_n$, therefore by Euler's theorem on the homogeneous function it follows that

$$\dot{q}_1 \frac{\partial T}{\partial \dot{q}_1} + \dot{q}_2 \frac{\partial T}{\partial \dot{q}_2} \qquad + \dot{q}_n \frac{\partial T}{\partial \dot{q}_n} = 2T,$$

$$\text{i.e.,} \quad \sum_{\alpha=1}^{n} \dot{q}_\alpha \frac{\partial T}{\partial \dot{q}_\alpha} = 2T. \tag{9.9.3}$$

Equation (9.9.1) gives

$$\dot{q}_\alpha \frac{d}{dt}\left(\frac{\partial T}{\partial \dot{q}_\alpha}\right) - \dot{q}_\alpha \frac{\partial T}{\partial q_\alpha} = -\dot{q}_\alpha \frac{\partial V}{\partial q_\alpha}. \tag{9.9.4}$$

We have

$$\frac{d}{dt}\left(\dot{q}_\alpha \frac{\partial T}{\partial \dot{q}_\alpha}\right) = \dot{q}_\alpha \frac{d}{dt}\left(\frac{\partial T}{\partial \dot{q}_\alpha}\right) + \frac{\partial T}{\partial \dot{q}_\alpha}\ddot{q}_\alpha,$$

$$\therefore \quad \dot{q}_\alpha \frac{d}{dt}\left(\frac{\partial T}{\partial \dot{q}_\alpha}\right) = \frac{d}{dt}\left(\dot{q}_\alpha \frac{\partial T}{\partial \dot{q}_\alpha}\right) - \frac{\partial T}{\partial \dot{q}_\alpha}\ddot{q}_\alpha. \tag{9.9.5}$$

Using (9.9.5) in (9.9.4), we obtain

$$\frac{d}{dt}\left(\dot{q}_\alpha \frac{\partial T}{\partial \dot{q}_\alpha}\right) - \frac{\partial T}{\partial \dot{q}_\alpha}\ddot{q}_\alpha - \dot{q}_\alpha \frac{\partial T}{\partial \dot{q}_\alpha} = -\dot{q}_\alpha \frac{\partial V}{\partial q_\alpha}.$$

This gives

$$\sum_{\alpha=1}^{n} \frac{d}{dt}\left(\dot{q}_\alpha \frac{\partial T}{\partial \dot{q}_\alpha}\right) - \sum_{\alpha=1}^{n} \frac{\partial T}{\partial \dot{q}_\alpha}\ddot{q}_\alpha - \sum_{\alpha=1}^{n} \frac{\partial T}{\partial q_\alpha}\dot{q}_\alpha = -\sum_{\alpha=1}^{n} \dot{q}_\alpha \frac{\partial V}{\partial q_\alpha}$$

$$\text{i.e.,} \quad \frac{d}{dt}\left(\sum_{\alpha=1}^{n} \dot{q}_\alpha \frac{\partial T}{\partial \dot{q}_\alpha}\right) - \sum_{\alpha=1}^{n}\left[\frac{\partial T}{\partial \dot{q}_\alpha}\ddot{q}_\alpha + \frac{\partial T}{\partial q_\alpha}\dot{q}_\alpha\right] = -\sum_{\alpha=1}^{n} \dot{q}_\alpha \frac{\partial V}{\partial q_\alpha},$$

$$\text{i.e.,} \quad \frac{d}{dt}(2T) - \sum_{\alpha=1}^{n}\left[\frac{\partial T}{\partial \dot{q}_\alpha}\ddot{q}_\alpha + \frac{\partial T}{\partial q_\alpha}\dot{q}_\alpha\right] = -\sum_{\alpha=1}^{n} \dot{q}_\alpha \frac{\partial V}{\partial q_\alpha}. \tag{9.9.6}$$

We note that

$$T = T\left(q_\alpha,\ \dot{q}_\alpha\right)\ \ (\alpha = 1,\ 2,\ \ldots,\ n),$$

$$\therefore\quad \frac{dT}{dt} = \sum_{\alpha=1}^{n} \frac{\partial T}{\partial q_\alpha}\dot{q}_\alpha + \sum_{\alpha=1}^{n} \frac{\partial T}{\partial \dot{q}_\alpha}\ddot{q}_\alpha = \sum_{\alpha=1}^{n}\left[\frac{\partial T}{\partial \dot{q}_\alpha}\ddot{q}_\alpha + \frac{\partial T}{\partial q_\alpha}\dot{q}_\alpha\right]. \tag{9.9.7}$$

Again

$$V = V\left(q_\alpha\right) \qquad (\alpha = 1, 2, \ldots n),$$

$$\therefore\quad \frac{dV}{dt} = \sum_{\alpha=1}^{n} \frac{\partial V}{\partial q_\alpha}\dot{q}_\alpha. \tag{9.9.8}$$

By virtue of (9.9.7) and (9.9.8), Eq. (9.9.6) reduces to

$$2\frac{dT}{dt} - \frac{dT}{dt} = -\frac{dV}{dt},$$

i.e., $\dfrac{dT}{dt} = -\dfrac{dV}{dt},$

i.e., $\dfrac{d}{dt}(T + V) = 0,$

i.e., $T + V = \text{Constant} = E\ \ \text{(say)},$

which is the principle of conservation of energy.

9.10 Equilibrium Configuration for Conservative Holonomic Dynamical Systems

Consider a holonomic dynamical system specified by n-generalized coordinates $q_\alpha\,(\alpha = 1, 2, \ldots, n)$ with time explicitly absent. The KE of the system is given by

$$T = \frac{1}{2}\sum_{r=1}^{n}\sum_{s=1}^{n} a_{rs}\,\dot{q}_r\dot{q}_s\,, \tag{9.10.1}$$

where $a_{rs} = a_{sr}$ and $a_{rs} = f(q_1, q_2, \ldots, q_n)$. The PE of the system is given by

$$V = V(q_1, q_2, \ldots q_n). \tag{9.10.2}$$

Lagrange's equations for conservative holonomic system are

$$\frac{d}{dt}\left(\frac{\partial T}{\partial \dot{q}_\alpha}\right) - \frac{\partial T}{\partial q_\alpha} = -\frac{\partial V}{\partial q_\alpha}\ (\alpha = 1, 2, \ldots, n). \tag{9.10.3}$$

Now

$$\frac{\partial T}{\partial q_\alpha} = \frac{1}{2}\sum_{r=1}^{n}\sum_{s=1}^{n}\frac{\partial a_{rs}}{\partial q_\alpha}\dot{q}_r\dot{q}_s. \tag{9.10.4}$$

$$\frac{\partial T}{\partial \dot{q}_\alpha} = \frac{1}{2}\sum_{\alpha=1}^{n}\sum_{s=1}^{n}a_{rs}\left[\dot{q}_r\frac{\partial \dot{q}_s}{\partial \dot{q}_\alpha} + \dot{q}_s\frac{\partial \dot{q}_r}{\partial \dot{q}_\alpha}\right],$$

i.e., $$\frac{\partial T}{\partial \dot{q}_\alpha} = \frac{1}{2}\sum_{r=1}^{n}\left(\sum_{s=1}^{n}a_{rs}\frac{\partial \dot{q}_s}{\partial \dot{q}_\alpha}\right)\dot{q}_r + \frac{1}{2}\sum_{s=1}^{n}\left(\sum_{r=1}^{n}a_{rs}\frac{\partial \dot{q}_r}{\partial \dot{q}_\alpha}\right)\dot{q}_s,$$

i.e., $$\frac{\partial T}{\partial \dot{q}_\alpha} = \frac{1}{2}\sum_{r=1}^{n}a_{r\alpha}\dot{q}_r + \frac{1}{2}\sum_{s=1}^{n}a_{\alpha s}\dot{q}_s,$$

i.e., $$\frac{\partial T}{\partial \dot{q}_\alpha} = \frac{1}{2}\sum_{r=1}^{n}a_{r\alpha}\dot{q}_r + \frac{1}{2}\sum_{r=1}^{n}a_{\alpha r}\dot{q}_r,$$

i.e., $$\frac{\partial T}{\partial \dot{q}_\alpha} = \frac{1}{2}\sum_{r=1}^{n}a_{r\alpha}\dot{q}_r + \frac{1}{2}\sum_{r=1}^{n}a_{r\alpha}\dot{q}_r = \sum_{r=1}^{n}a_{r\alpha}\dot{q}_r. \tag{9.10.5}$$

Now

$$\frac{d}{dt}\left(\frac{\partial T}{\partial \dot{q}_\alpha}\right) = \sum_{r=1}^{n}\left[a_{r\alpha}\ddot{q}_r + \dot{q}_r\frac{d}{dt}a_{r\alpha}\right],$$

i.e., $$\frac{d}{dt}\left(\frac{\partial T}{\partial \dot{q}_\alpha}\right) = \sum_{r=1}^{n}\left[a_{r\alpha}\ddot{q}_r + \dot{q}_r\left\{\frac{\partial a_{r\alpha}}{\partial q_1}\dot{q}_1 + \cdots + \frac{\partial a_{rs}}{\partial q_n}\dot{q}_n\right\}\right]. \tag{9.10.6}$$

Suppose that the system is in equilibrium for a configuration specified by the coordinates

$$q_\alpha = \mu_\alpha \, (\alpha = 1, 2, \ldots, n) . \tag{9.10.7}$$

For these values of q_α,

$$\dot{q}_\alpha = 0, \quad \ddot{q}_\alpha = 0 \quad (\alpha = 1, 2, \ldots, n) . \tag{9.10.8}$$

From (9.10.4) and (9.10.6), it follows that

$$\left.\begin{array}{l} \frac{\partial T}{\partial q_\alpha} = 0 \\ \frac{d}{dt}\left(\frac{\partial T}{\partial \dot{q}_\alpha}\right) = 0 \end{array}\right\} . \tag{9.10.9}$$

Therefore, Eq. (9.10.3) reduces to

$$\frac{\partial V}{\partial q_\alpha} = 0 \text{ for } q_\alpha = \mu_\alpha; \quad \alpha = 1, 2, \cdots, n. \tag{9.10.10}$$

In Eq. (9.10.10), there are n equations, and by solving them, we get the coordinates specifying the equilibrium configuration.

9.11 Theory of Small Oscillations of Conservative Holonomic Dynamical System

Let $q_\alpha = \mu_\alpha; \ \alpha = 1, 2, \ldots, n$ be the generalized coordinates which define an equilibrium configuration. Let $V_0 = V(\mu_1, \mu_2, \mu_3, \cdots, \mu_n)$ be the potential energy in the equilibrium position. Let the system be given a small displacement from the equilibrium position. Let $q_\alpha = \mu_\alpha + \xi_\alpha$ at time t where ξ_α and their time derivatives are small.

We have for equilibrium position $\frac{\partial V}{\partial q_\alpha} = 0$ for $q_\alpha = \mu_\alpha; \ \alpha = 1, 2, \cdots, n.$

Now

$$V = V(q_1, q_2, q_3, \cdots, q_n) = V(\mu_1 + \xi_1, \mu_2 + \xi_2, \mu_3 + \xi_3, \cdots, \mu_n + \xi_n) . \tag{9.11.1}$$

Expanding (9.11.1) in Taylor's series and neglecting the higher powers of ξ_α; $\alpha = 1, 2, \ldots, n$, we obtain

$$V = V(\mu_1, \mu_2, \mu_3, \cdots, \mu_n) + \left[\left(\sum_{\alpha=1}^{n} \xi_\alpha \frac{\partial}{\partial q_\alpha}\right) V\right]_0 + \frac{1}{2}\left[\left(\sum_{\alpha=1}^{n} \xi_\alpha \frac{\partial}{\partial q_\alpha}\right)^2 V\right]_0 ,$$

$$\text{i.e.,}\quad V = V_0 + \frac{1}{2}\left[\left(\sum_{\alpha=1}^{n} \xi_\alpha \frac{\partial}{\partial q_\alpha}\right)^2 V\right]_0,$$

$$\text{i.e.,}\quad V = V_0 + \frac{1}{2}\sum_{r=1}^{n}\sum_{s=1}^{n} C_{rs}\xi_r\xi_s, \tag{9.11.2}$$

where $C_{rs} = C_{sr} = \left(\frac{\partial^2 V}{\partial q_r \partial q_s}\right)_0$.

Again

$$T = \frac{1}{2}\sum_{r=1}^{n}\sum_{s=1}^{n} a_{rs}\dot{q}_r\dot{q}_s = \frac{1}{2}\sum_{r=1}^{n}\sum_{s=1}^{n} a_{rs}\dot{\xi}_r\dot{\xi}_s. \tag{9.11.3}$$

Let a^*_{rs} denote the value of each a_{rs} in the equilibrium configuration $q_\alpha = \mu_\alpha;\ \alpha = 1, 2, \cdots, n$.

Now

$$a_{rs} = a_{rs}\left(q_1, q_2, \cdots, q_n\right) = a_{rs}\left(\mu_1 + \xi_1, \mu_2 + \xi_2, \cdots, \mu_n + \xi_n\right),$$

$$\text{i.e.,}\quad a_{rs} = a_{rs}\left(\mu_1, \mu_2, \cdots, \mu_n\right) + \left[\left(\xi_1\frac{\partial}{\partial q_1} + \xi_2\frac{\partial}{\partial q_2} + \cdots + \xi_n\frac{\partial}{\partial q_n}\right)a_{rs}\right]_0 + \cdots$$

$$\text{i.e.,}\quad a_{rs} = a^*_{rs}\ \text{(approximately)}.$$

Thus Eq. (9.11.3) gives

$$T = \frac{1}{2}\sum_{r=1}^{n}\sum_{s=1}^{n} a^*_{rs}\dot{\xi}_r\dot{\xi}_s. \tag{9.11.4}$$

Hence, we have

$$\left.\begin{aligned} V - V_0 &= \tfrac{1}{2}\sum_{r=1}^{n}\sum_{s=1}^{n} C_{rs}\xi_r\xi_s \\ T &= \tfrac{1}{2}\sum_{r=1}^{n}\sum_{s=1}^{n} a^*_{rs}\dot{\xi}_r\dot{\xi}_s \end{aligned}\right\}. \tag{9.11.5}$$

It is clear that $V - V_0$ and T are homogeneous functions of ξ_r & $\dot{\xi}_r$ respectively.

Now suppose that

$$\xi_r = b_{r1}\eta_1 + b_{r2}\eta_2 + \cdots + b_{rn}\eta_n, \tag{9.11.6}$$

where $r = 1, 2, \ldots, n$.

Let us choose $b_{r1},\ b_{r2},\ \ldots,\ b_{rn}$ in such a way that

$$\left.\begin{aligned} T &= \tfrac{1}{2}\left(\dot{\eta}_1^2 + \dot{\eta}_2^2 + \cdots + \dot{\eta}_n^2\right) \\ V - V_0 &= \tfrac{1}{2}\left(\gamma_1\eta_1^2 + \gamma_2\eta_2^2 + \cdots + \gamma_n\eta_n^2\right) \end{aligned}\right\}. \tag{9.11.7}$$

Now

$$L = T - V = \frac{1}{2}\sum_{r=1}^{n}\dot{\eta}_r^2 - \frac{1}{2}\sum_{r=1}^{n}\gamma_r\eta_r^2 - V_0. \tag{9.11.8}$$

Lagrange's Equations for the new generalized coordinates η_r $(r = 1, 2, \ldots, n)$ are

$$\frac{d}{dt}\left(\frac{\partial L}{\partial \dot{\eta}_r}\right) - \frac{\partial L}{\partial \eta_r} = 0,$$

$$\text{i.e.,}\quad \ddot{\eta}_r + \gamma_r\eta_r = 0,$$

$$\text{i.e.,}\quad \ddot{\eta}_r + \omega_r^2\eta_r = 0,$$

$$\text{i.e.,}\quad \left(D^2 + \omega_r^2\right)\eta_r = 0,$$

$$\text{i.e.,}\quad \eta_r = A_r\cos\left(\omega_r t + \varepsilon_r\right);\ \ (r = 1, 2, \ldots, n)\,. \tag{9.11.9}$$

The coordinates η_r $(r = 1, 2, \ldots, n)$ are called the normal coordinates of the system. Substituting (9.11.9) in (9.11.6), we get the solutions for the original coordinates ξ_r $(r = 1, 2, \ldots, n)$. This solution is linear in normal coordinates. If $(n-1)$ normal coordinates $\eta_1,\ \eta_2,\ \ldots,\ \eta_{r-1},\ \eta_{r+1},\ \ldots,\ \eta_n$ are all permanently zero for a system vibrating about a configuration of stable equilibrium and if $\eta_r \neq 0$, then the system is said to execute a normal mode vibration. The motion of each particle is periodic, passing twice through the equilibrium configuration in every complete vibration and the velocity of each particle vanishes at the same instant twice in every complete oscillation.

9.12 Illustrations

Example 9.1 Use Lagrange's Equation to discuss the motion of a particle in a plane under an attractive central force obeying the law of inverse square.

Solution Let a particle P of mass m be moving around the center of force O on the plane of the motion. Let (r, θ) be the polar coordinates of P referred to O as the pole and **OX** as the initial line. Let $\mathbf{r}$ be the position vector of P relative to O. Let $\hat{r}$, $\hat{\theta}$ denote the unit vectors along the increasing directions of r & θ, as shown in Fig. 9.3.

If $\mathbf{v}$ is the velocity of the particle at P at the t, then

$$\mathbf{v} = v_r\hat{r} + v_\theta\hat{\theta} = \dot{r}\hat{r} + r\dot{\theta}\hat{\theta}. \tag{9.12.1}$$

The KE of the particle is given by

$$T = \frac{1}{2}m\mathbf{v}^2 = \frac{1}{2}m\left(v_r^2 + v_\theta^2\right) = \frac{1}{2}m\left(\dot{r}^2 + r^2\dot{\theta}^2\right). \tag{9.12.2}$$

The force acting on the particle at P is given by

$$\mathbf{F} = \frac{\mu m}{r^2}\left(-\hat{r}\right) = -\frac{\mu m}{r^2}\frac{\mathbf{r}}{r} = -\frac{\mu m\mathbf{r}}{r^3}. \tag{9.12.3}$$

Let V be the PE of the particle at P.

$$\therefore \quad \mathbf{F} = -\nabla V,$$

$$\text{i.e.,} \quad -\frac{\mu m\mathbf{r}}{r^3} = -\nabla V,$$

$$\text{i.e.,} \quad \nabla V = \frac{\mu m}{r^3}\mathbf{r},$$

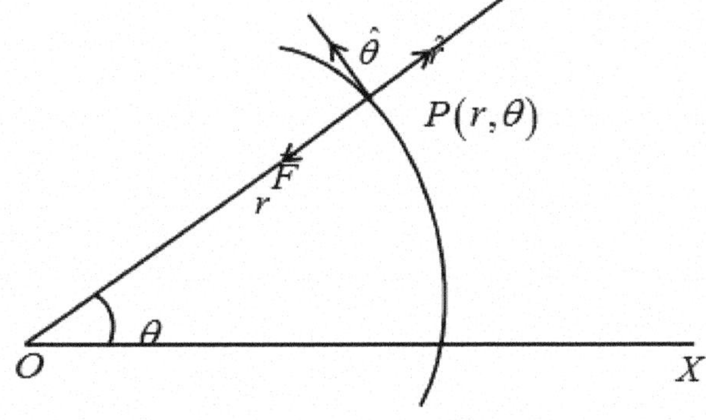

Fig. 9.3 Diagram illustrating the configuration in Example 9.1

which yields

$$\nabla V \cdot d\mathbf{r} = \frac{\mu m}{r^3}(\mathbf{r} \cdot d\mathbf{r}),$$

$$\text{i.e.,}\quad dV = \frac{\mu m}{r^3} r dr = \frac{\mu m}{r^2} dr, \quad \left[\because\ \mathbf{r}^2 = r^2 \Rightarrow \mathbf{r} \cdot d\mathbf{r} = r dr\right]$$

$$\text{i.e.,}\quad V = -\frac{\mu m}{r}. \tag{9.12.4}$$

Therefore, the Lagrangian for the motion is given by

$$L = T - V = \frac{1}{2} m \left(\dot{r}^2 + r^2 \dot{\theta}^2\right) + \frac{\mu m}{r}. \tag{9.12.5}$$

Here r, θ are the only generalized coordinates.
Lagrange's equation of motion for the system is

$$\frac{d}{dt}\left(\frac{\partial L}{\partial \dot{q}_\alpha}\right) = \frac{\partial L}{\partial q_\alpha}; \quad \alpha = 1, 2. \tag{9.12.6}$$

Equation (9.12.6) gives

$$\left.\begin{array}{l} \frac{d}{dt}\left(\frac{\partial L}{\partial \dot{r}}\right) = \frac{\partial L}{\partial r} \\ \frac{d}{dt}\left(\frac{\partial L}{\partial \dot{\theta}}\right) = \frac{\partial L}{\partial \theta} \end{array}\right\}. \tag{9.12.7}$$

1st Equation of (9.12.7) yields

$$\frac{d}{dt}\left[\frac{1}{2} m (2\dot{r})\right] = \frac{1}{2} m \dot{\theta}^2 2r - \frac{\mu m}{r^2},$$

$$\text{i.e.,}\quad m\ddot{r} = mr\dot{\theta}^2 - \frac{\mu m}{r^2},$$

$$\text{i.e.,}\quad \ddot{r} - r\dot{\theta}^2 = -\frac{\mu}{r^2}. \tag{9.12.8}$$

2nd Equation of (9.12.7) gives

$$\frac{d}{dt}\left[\frac{1}{2} m r^2 2\dot{\theta}\right] = 0,$$

$$\text{i.e.,}\quad \frac{d}{dt}\left(r^2\dot{\theta}\right) = 0,$$

$$\text{i.e.,}\quad r^2\dot{\theta} = h = \text{a constant.} \tag{9.12.9}$$

Equations (9.12.8) and (9.12.9) determine the motion of the particle.

Example 9.2 Obtain the equation of motion for the Lagrangian

$$L = a^2(1 - \cos\theta)\,\dot{\theta}^2 - ag(1 + \cos\theta)\,.$$

Solution Given $L = a^2(1 - \cos\theta)\,\dot{\theta}^2 - ag(1 + \cos\theta)\,.$
Here θ is the only generalized coordinate.
Lagrange's equation for conservative force field is

$$\frac{d}{dt}\left(\frac{\partial L}{\partial \dot{q}_\alpha}\right) = \frac{\partial L}{\partial q_\alpha},\quad \alpha = 1. \tag{9.12.10}$$

Equation (9.12.10) gives

$$\frac{d}{dt}\left(\frac{\partial L}{\partial \dot{\theta}}\right) = \frac{\partial L}{\partial \theta}. \tag{9.12.11}$$

Now,

$$\frac{\partial L}{\partial \dot{\theta}} = a^2(1 - \cos\theta)\,2\dot{\theta},\quad \frac{\partial L}{\partial \theta} = a^2\dot{\theta}^2\sin\theta + ag\sin\theta.$$

Therefore, Eq. (9.12.11) gives

$$\frac{d}{dt}\left[2a^2\dot{\theta}(1 - \cos\theta)\right] = a^2\dot{\theta}^2\sin\theta + ag\sin\theta,$$

$$\text{i.e.,}\quad 2a^2\left[(1 - \cos\theta)\,\ddot{\theta} + \dot{\theta}\sin\theta\,\dot{\theta}\right] = a^2\dot{\theta}^2\sin\theta + ag\sin\theta,$$

$$\text{i.e.,}\quad 2a^2\ddot{\theta}(1 - \cos\theta) + a^2\dot{\theta}^2\sin\theta = ag\sin\theta,$$

$$\text{i.e.,}\quad 2\ddot{\theta}(1 - \cos\theta) + \dot{\theta}^2\sin\theta = \frac{g}{a}\sin\theta,$$

$$\text{i.e.,}\quad 2(1 - \cos\theta)\,\dot{\theta}\ddot{\theta} + \sin\theta\dot{\theta}^3 = \frac{g}{a}\sin\theta\,\dot{\theta},$$

$$\text{i.e.,}\quad (1 - \cos\theta)\frac{d}{dt}\dot{\theta}^2 + \dot{\theta}^2\frac{d}{dt}(1 - \cos\theta) = \frac{g}{a}\sin\theta\,\frac{d\theta}{dt},$$

$$\text{i.e.,}\quad \frac{d}{dt}\left\{\dot{\theta}^2(1 - \cos\theta)\right\} = \frac{g}{a}\sin\theta\,\frac{d\theta}{dt},$$

$$\text{i.e.,}\quad d\left\{\dot{\theta}^2(1-\cos\theta)\right\} = \frac{g}{a}\sin\theta\, d\theta,$$

$$\text{i.e.,}\quad \dot{\theta}^2(1-\cos\theta) = -\frac{g}{a}\cos\theta + C,$$

which is the required equation of motion.

Example 9.3 Twice the K.E. of a system is $A\dot{\theta}^2 + 2H\dot{\theta}\omega + B\omega^2$, where $A,\ H,\ B$ are all functions of θ and ω a constant; also the work function of the field of conservative force U, a function of θ alone, show that

$$\frac{1}{2}\left(A\dot{\theta}^2 - B\omega^2\right) = U + C.$$

Solution Given $2T = A\dot{\theta}^2 + 2H\dot{\theta}\omega + B\omega^2$.
Here $A = A(\theta)\,,\ B = B(\theta)\,,\ H = H(\theta)\ \&\omega = \text{a constant}$.
Let V be the force potential,

$$\therefore\quad \mathbf{F} = -\nabla V.$$

We have

$$dW = \mathbf{F}\cdot d\mathbf{r},$$

$$\text{i.e.,}\quad dU = -\nabla V\cdot d\mathbf{r} = -dV,$$

$$\text{i.e.,}\quad dV = -dU,$$

$$\text{i.e.,}\quad V = -U.$$

Therefore, the Lagrangian of the system is given by

$$L = T - V = T + U = \frac{1}{2}\left(A\dot{\theta}^2 + 2H\dot{\theta}\omega + B\omega^2\right) + U.$$

Here θ is the only generalized coordinate. Lagrange's equation is

$$\frac{d}{dt}\left(\frac{\partial L}{\partial\dot{\theta}}\right) = \frac{\partial L}{\partial\theta}. \tag{9.12.12}$$

Now

$$\frac{\partial L}{\partial\theta} = \frac{1}{2}\left[\frac{dA}{d\theta}\dot{\theta}^2 + 2\dot{\theta}\omega\frac{dH}{d\theta} + \frac{dB}{d\theta}\omega^2\right] + \frac{dU}{d\theta},$$

$$\text{i.e.,}\quad \frac{\partial L}{\partial\theta} = \frac{1}{2}\left[\dot{A}\frac{dt}{d\theta}\dot{\theta}^2 + 2\dot{\theta}\omega\frac{dH}{dt}\frac{dt}{d\theta} + \frac{dB}{dt}\frac{dt}{d\theta}\omega^2\right] + \frac{dU}{dt}\frac{dt}{d\theta},$$

i.e., $$\frac{\partial L}{\partial \theta} = \frac{1}{2}\left[\frac{\dot{A}}{\dot{\theta}}\dot{\theta}^2 + 2\dot{\theta}\omega\frac{\dot{H}}{\dot{\theta}} + \frac{\dot{B}}{\dot{\theta}}\omega^2\right] + \frac{\dot{U}}{\dot{\theta}},$$

i.e., $$\frac{\partial L}{\partial \theta} = \frac{1}{2}\left[\dot{A}\dot{\theta} + 2\omega\dot{H} + \frac{\dot{B}\omega^2}{\dot{\theta}}\right] + \frac{\dot{U}}{\dot{\theta}}.$$

Again,

$$\frac{\partial L}{\partial \dot{\theta}} = \frac{1}{2}\left[2A\dot{\theta} + 2H\omega\right] = A\dot{\theta} + H\omega.$$

Thus, (9.12.12) gives

$$\frac{d}{dt}\left(A\dot{\theta} + H\omega\right) = \frac{1}{2}\left[\dot{A}\dot{\theta} + 2\omega\dot{H} + \frac{\dot{B}\omega^2}{\dot{\theta}}\right] + \frac{\dot{U}}{\dot{\theta}},$$

i.e., $$\dot{A}\dot{\theta} + A\ddot{\theta} + \dot{H}\omega = \frac{1}{2}\left[\dot{A}\dot{\theta} + 2\omega\dot{H} + \frac{\dot{B}\omega^2}{\dot{\theta}}\right] + \frac{\dot{U}}{\dot{\theta}},$$

i.e., $$2\dot{A}\dot{\theta}^2 + 2A\dot{\theta}\ddot{\theta} + 2\dot{H}\dot{\theta}\omega = \dot{A}\dot{\theta}^2 + 2\omega\dot{H}\dot{\theta} + \dot{B}\omega^2 + 2\dot{U},$$

i.e., $$\dot{A}\dot{\theta}^2 + 2A\dot{\theta}\ddot{\theta} = \dot{B}\omega^2 + 2\dot{U},$$

i.e., $$\dot{\theta}^2\frac{dA}{dt} + A\frac{d}{dt}\dot{\theta}^2 = \omega^2\frac{dB}{dt} + 2\frac{dU}{dt},$$

i.e., $$\frac{d}{dt}\left(A\dot{\theta}^2\right) = \omega^2\frac{dB}{dt} + 2\frac{dU}{dt},$$

i.e., $$d\left(A\dot{\theta}^2\right) = \omega^2 dB + 2dU,$$

i.e., $$A\dot{\theta}^2 = \omega^2 B + 2U + 2C,$$

i.e., $$\frac{1}{2}\left(A\dot{\theta}^2 - B\omega^2\right) = U + C.$$

Example 9.4 A bead slides without friction on a frictionless wire in the shape of a cycloid with equations

$$x = a\left(\theta - \sin\theta\right), \quad y = a\left(1 + \cos\theta\right); \quad 0 \le \theta \le 2\pi.$$

Using Lagrange's equations, obtain the equation governing the motion.

Solution Consider a bead of mass m that slides without friction along a smooth wire AB, as shown in Fig. 9.4. Let P (x, y) be the position of the bead at time t. The K.E. of the particle is given by

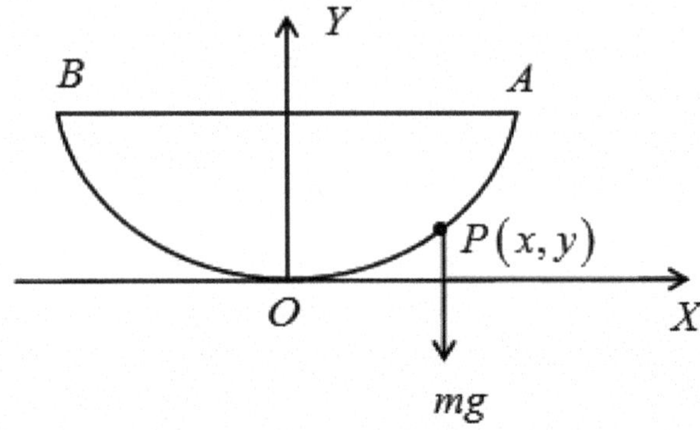

Fig. 9.4 Diagram illustrating Example 9.4

$$T = \frac{1}{2}mv^2 = \frac{1}{2}m\left(\dot{x}^2 + \dot{y}^2\right). \tag{9.12.13}$$

Given

$$x = a\,(\theta - \sin\theta)\,, \quad y = a\,(1 + \cos\theta)\,;$$

$$\therefore \quad \dot{x} = a\dot{\theta}\,(1 - \cos\theta)\,, \quad \dot{y} = -a\dot{\theta}\sin\theta.$$

Now

$$v^2 = \dot{x}^2 + \dot{y}^2 = a^2\dot{\theta}^2\left[(1 - \cos\theta)^2 + \sin^2\theta\right] = 2a^2\dot{\theta}^2\,(1 - \cos\theta)\,. \tag{9.12.14}$$

$$\therefore \quad T = \frac{1}{2}mv^2 = ma^2\dot{\theta}^2\,(1 - \cos\theta)\,. \tag{9.12.15}$$

The P.E. is given by

$$V = mgy = mga\,(1 - \cos\theta)\,. \tag{9.12.16}$$

The Lagrangian for the system is given by

$$L = T - V = ma^2\dot{\theta}^2\,(1 - \cos\theta) - mga\,(1 + \cos\theta)\,.$$

Here θ is the only generalized coordinate.
Lagrange's equation is

$$\frac{d}{dt}\left(\frac{\partial L}{\partial\dot{\theta}}\right) = \frac{\partial L}{\partial\theta}. \tag{9.12.17}$$

Now

$$\frac{\partial L}{\partial \dot{\theta}} = ma^2 (1 - \cos\theta)\, 2\dot{\theta} = 2ma^2 (1 - \cos\theta)\, \dot{\theta},$$

$$\frac{\partial L}{\partial \theta} == ma^2 \dot{\theta}^2 \sin\theta + mga \sin\theta.$$

Therefore, Eq. (9.12.17) gives

$$\frac{d}{dt}\left[2ma^2 (1 - \cos\theta)\, \dot{\theta}\right] = ma^2 \dot{\theta}^2 \sin\theta + mga \sin\theta,$$

$$\text{i.e.,}\quad 2a^2 \left[(1 - \cos\theta)\, \ddot{\theta} + \dot{\theta}^2 \sin\theta\right] = a^2 \dot{\theta}^2 \sin\theta + ga \sin\theta,$$

$$\text{i.e.,}\quad 2\,(1 - \cos\theta)\, \ddot{\theta} + 2\dot{\theta}^2 \sin\theta = \dot{\theta}^2 \sin\theta + \frac{g}{a} \sin\theta,$$

$$\text{i.e.,}\quad 2\,(1 - \cos\theta)\, \ddot{\theta} + \dot{\theta}^2 \sin\theta = \frac{g}{a} \sin\theta,$$

$$\text{i.e.,}\quad 2\,(1 - \cos\theta)\, \dot{\theta}\ddot{\theta} + \dot{\theta}^2 \left(\dot{\theta} \sin\theta\right) = \frac{g}{a} \sin\theta\dot{\theta},$$

$$\text{i.e.,}\quad (1 - \cos\theta)\, \frac{d}{dt}\dot{\theta}^2 + \dot{\theta}^2 \frac{d}{dt}\,(1 - \cos\theta) = \frac{g}{a} \sin\theta \frac{d\theta}{dt},$$

$$\text{i.e.,}\quad \frac{d}{dt}\left[\dot{\theta}^2\, (1 - \cos\theta)\right] = \frac{g}{a} \sin\theta \frac{d\theta}{dt},$$

$$\text{i.e.,}\quad d\left[\dot{\theta}^2\, (1 - \cos\theta)\right] = \frac{g}{a} \sin\theta d\theta,$$

$$\text{i.e.,}\quad \dot{\theta}^2\, (1 - \cos\theta) = -\frac{g}{a} \cos\theta + C,$$

which is the required equation governing the motion.

Example 9.5 A particle on a smooth horizontal table is attached to a string passing through a small hole in the table and carries an equal particle hanging vertically. If at time t, the particle is at a distance r from the hole and the string makes an angle with some fixed line in the table. Obtain Lagrange's equation of motion of the system in terms of r and θ and their time derivatives. The particle is projected at right angles to the string with velocity $\sqrt{2gh}$ at a distance a from the hole. Prove that $\dot{r}^2 = gh\left(1 - \frac{a^2}{r^2}\right) + g\,(a - r)$.

Solution Let O be the hole and OX be a line on the table through O. Let P be the position of the 1st particle at the t with position vector $\mathbf{r}$ relative to O. In Fig. 9.5, let $(r,\ \theta)$ be the polar coordinates of the particle P at time t referred to O as the pole and **OX** as the initial line.

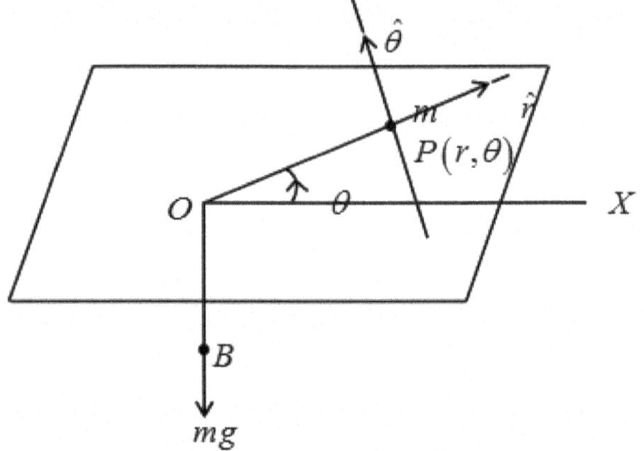

Fig. 9.5 Diagram illustrating Example 9.4

Let $\hat{r},\ \hat{\theta}$ be the unit vectors in the increasing direction of $r,\ \theta$.
Now

$$\mathbf{v}_1 = \text{velocity of P} = \dot{r}\hat{r} + r\dot{\theta}\hat{\theta},$$

$$\mathbf{v}_2 = \text{velocity of } B = \frac{d}{dt}(l-r)\,\hat{k} = -\dot{r}\hat{k};\ l \text{ being the length of the string.}$$

$$T = \text{KE of the system} = \frac{1}{2}mv_1^2 + \frac{1}{2}mv_2^2,$$

$$\text{i.e.,}\quad T = \frac{1}{2}m\left(\dot{r}^2 + r^2\dot{\theta}^2\right) + \frac{1}{2}m\dot{r}^2 = \frac{1}{2}m\left(2\dot{r}^2 + r^2\dot{\theta}^2\right).$$

$$V = \text{PE of the system} = V_1 + V_2 = -mg\,(l-r)\,.$$

$$\therefore\quad L = \text{Lagrangian} = T - V = \frac{1}{2}m\left(2\dot{r}^2 + r^2\dot{\theta}^2\right) + mg\,(l-r)\,. \tag{9.12.18}$$

Here $r,\ \theta$ are the generalized coordinates.

$$\therefore\quad q_1 = r,\ q_2 = \theta.$$

Lagrange's Equations are

$$\frac{d}{dt}\left(\frac{\partial L}{\partial \dot{q}_\alpha}\right) = \frac{\partial L}{\partial q_\alpha};\ \alpha = 1, 2. \tag{9.12.19}$$

Equation (9.12.19) gives

$$\left.\begin{array}{l}\frac{d}{dt}\left(\frac{\partial L}{\partial \dot{r}}\right)=\frac{\partial L}{\partial r}\\ \frac{d}{dt}\left(\frac{\partial L}{\partial \dot{\theta}}\right)=\frac{\partial L}{\partial \theta}\end{array}\right\}. \tag{9.12.20}$$

Equation (9.12.18) gives

$$\left.\begin{array}{l}\frac{\partial L}{\partial r}=mr\dot{\theta}^2-mg,\ \frac{\partial L}{\partial \dot{r}}=2m\dot{r}\\ \frac{\partial L}{\partial \theta}=0,\ \frac{\partial L}{\partial \dot{\theta}}=mr^2\dot{\theta}\end{array}\right\}. \tag{9.12.21}$$

Equation (9.12.20) gives

$$\frac{d}{dt}(2m\dot{r})=mr\dot{\theta}^2-mg,$$

$$\text{i.e.,}\quad 2\ddot{r}-r\dot{\theta}^2+g=0, \tag{9.12.22}$$

and

$$\frac{d}{dt}\left(mr^2\dot{\theta}\right)=0$$

$$\text{i.e.,}\quad r^2\dot{\theta}=C. \tag{9.12.23}$$

Initially $v_\theta = a\dot{\theta} = \sqrt{2gh},\ r = a$ at $t = 0$,
i.e., initially $r = a,\ \dot{\theta} = \frac{\sqrt{2gh}}{a}$ at $t = 0$.
Thus, (9.12.23) gives

$$a^2\frac{\sqrt{2gh}}{a}=C$$

$$\text{i.e.,}\quad C=a\sqrt{2gh},$$

$$\therefore\quad r^2\dot{\theta}=a\sqrt{2gh}. \tag{9.12.24}$$

Using (9.12.24) in (9.12.22), we obtain

$$2\ddot{r}-r\frac{2gha^2}{r^4}+g=0,$$

$$\text{i.e.,}\quad 2\ddot{r}-\frac{2a^2gh}{r^3}+g=0,$$

$$\text{i.e.,}\quad 2\dot{r}\ddot{r} - \frac{2a^2gh}{r^3}\dot{r} + g\dot{r} = 0,$$

$$\text{i.e.,}\quad 2\dot{r}\frac{d\dot{r}}{dt} - 2a^2ghr^{-3}\frac{dr}{dt} + g\frac{dr}{dt} = 0,$$

$$\text{i.e.,}\quad 2\dot{r}d\dot{r} - 2a^2ghr^{-3}dr + gdr = 0,$$

$$\text{i.e.,}\quad \dot{r}^2 + \frac{a^2gh}{r^2} + gr = C_1. \tag{9.12.25}$$

Initially $\dot{r} = 0,\ r = a$ at $t = 0$.
$\therefore$ Eq. (9.12.25) gives

$$gh + ga = C_1$$

Thus, (9.12.25) yields

$$\dot{r}^2 + \frac{a^2gh}{r^2} + gr = gh + ga,$$

$$\text{i.e.,}\quad \dot{r}^2 = gh\left(1 - \frac{a^2}{r^2}\right) + g\,(a - r)\,.$$

Example 9.6 A straight frictionless wire OA fixed at a point O on a downward vertical axis OZ rotates about the axis with constant angular velocity $\boldsymbol{\omega}$. A bead of mass m is constrained to move on the wire. Set up the Lagrangian and find Lagrange's equations.

Solution Let r be the distance of the bead from the point O on the wire. Given $\boldsymbol{\omega}$ is the angular velocity of the wire about OZ, the downward vertical. As illustrated in Fig. 9.6, the point P $(x,\ y,\ z)$ indicates the position of the bead at time t. We have

$$\dot{\phi} = \omega. \tag{9.12.26}$$

Let $(r,\ \theta,\ \phi)$ be the spherical polar coordinates of P.

$$\therefore\quad \left.\begin{aligned} x &= r\sin\theta\cos\phi = r\sin\alpha\cos\phi \\ y &= r\sin\theta\sin\phi = r\sin\alpha\sin\phi \\ z &= r\cos\theta = r\cos\alpha \end{aligned}\right\}, \tag{9.12.27}$$

where $\alpha = \angle AOZ$.

$$T = \text{KE of the bead} = \frac{1}{2}m\left(\dot{x}^2 + \dot{y}^2 + \dot{z}^2\right). \tag{9.12.28}$$

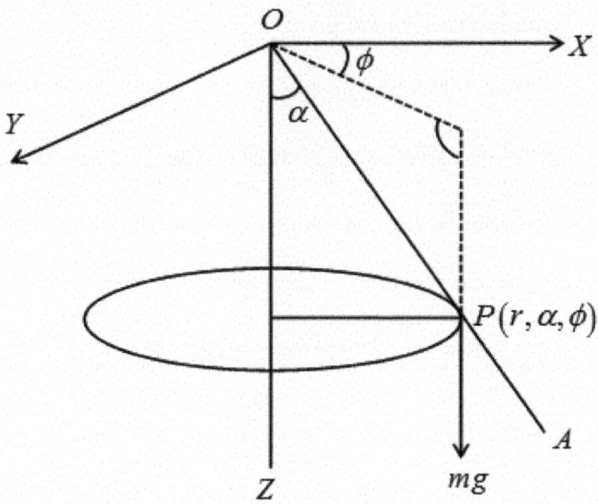

Fig. 9.6 Diagram illustrating Example 9.6

We have

$$\left.\begin{array}{l}\dot{x} = \sin\alpha\left(-r\sin\phi\,\omega + \cos\phi\,\dot{r}\right)\\ \dot{y} = \sin\alpha\left(r\cos\phi\,\omega + \sin\phi\,\dot{r}\right)\\ \dot{z} = \dot{r}\cos\alpha\end{array}\right\}. \tag{9.12.29}$$

$$\therefore\quad \dot{x}^2 + \dot{y}^2 + \dot{z}^2 = \omega^2 r^2 \sin^2\alpha + \dot{r}^2. \tag{9.12.30}$$

Therefore, from Eq. (9.12.28), we obtain

$$T = \frac{1}{2} m \left(\omega^2 r^2 \sin^2\alpha + \dot{r}^2\right).$$

Again,

$$V = \text{PE of the bead} = -mg\,r\cos\alpha.$$

Thus, the Lagrangian for the motion of the bead is given by

$$\begin{aligned} L &= T - V,\\ &= \frac{1}{2} m \left(\omega^2 r^2 \sin^2\alpha + \dot{r}^2\right) + mg\,r\cos\alpha. \end{aligned}$$

Here, r is the only generalized coordinate.
Therefore, Lagrange's equation is

$$\frac{d}{dt}\left(\frac{\partial L}{\partial \dot{r}}\right) = \frac{\partial L}{\partial r},$$

$$\text{i.e.,} \quad \frac{d}{dt}(m\dot{r}) = m\omega^2 r \sin^2\alpha + mg\cos\alpha,$$

$$\text{i.e.,} \quad \ddot{r} = \omega^2 r \sin^2\alpha + g\cos\alpha$$

which is the required equation of motion.

Example 9.7 Set up the Lagrangian for a simple pendulum and obtain an equation describing the motion.

Solution In Fig. 9.7, let OA be the initial position of the string of length l, and OP be its position at time t. Suppose that $\angle\text{AOP} = \theta$. The polar coordinates of P referred to O as the pole and OA as the initial line are $(l,\ \theta)$.
The velocity of the particle is given by

$$\mathbf{v} = v_r\hat{r} + v_\theta\hat{\theta} = \dot{r}\hat{r} + r\dot{\theta}\hat{\theta} = l\dot{\theta}\hat{\theta} \quad \left[\because\ r = l = \text{Constant}\right].$$

$$\therefore \quad T = \text{KE of the particle} = \frac{1}{2}m\mathbf{v}^2 = \frac{1}{2}ml^2\dot{\theta}^2,$$

$$V = \text{PE of the particle} = mgl\,(1 - \cos\theta)\,.$$

Thus the Lagrangian for the motion of the particle is given by

$$\therefore \quad L = T - V = \frac{1}{2}ml^2\dot{\theta}^2 - mgl\,(1 - \cos\theta)\,.$$

Here θ is the only generalized coordinate.
Lagrange's equation is

$$\frac{d}{dt}\left(\frac{\partial L}{\partial \dot{q}_\alpha}\right) = \frac{\partial L}{\partial q_\alpha}; \quad \alpha = 1,$$

.

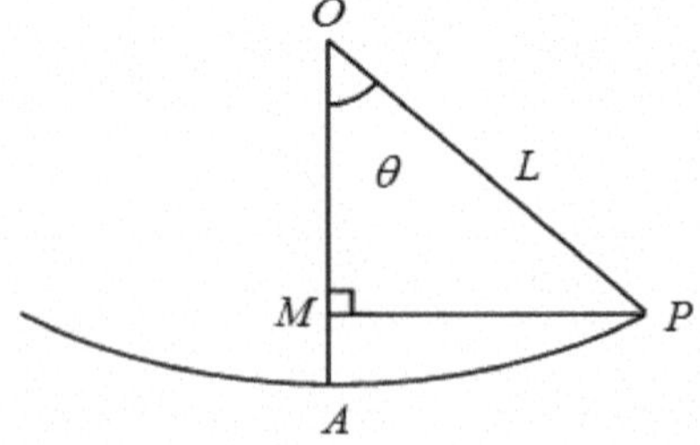

Fig. 9.7 Diagram illustrating Example 9.7

$$\text{i.e.,}\quad \frac{d}{dt}\left(\frac{\partial L}{\partial \dot{\theta}}\right) = \frac{\partial L}{\partial \theta},$$

$$\text{i.e.,}\quad \frac{d}{dt}\left[\frac{1}{2}ml^2 2\dot{\theta}\right] = -mgl\sin\theta,$$

$$\text{i.e.,}\quad l^2\ddot{\theta} = -gl\sin\theta,$$

$$\text{i.e.,}\quad \ddot{\theta} = -\frac{g}{l}\sin\theta,$$

which is the required equation of motion.

Example 9.8 A particle P moves on a smooth horizontal circular wire of radius a, which is free to rotate about a vertical axis through a point O, distant c from the center C. If $\angle$PCO $= \theta$, show that $a\ddot{\theta} + \dot{\omega}\,(a - c\cos\theta) = c\omega^2\sin\theta$, where ω is the angular velocity of the wire.

Solution Take X-axis and Y-axis horizontally on the plane of the wire. Let $\angle$XOC $=$ ϕ. Let P (x, y) be the position of the particle on the wire at time t. Draw $\perp^r$ PL from P on OX. Draw a line CX′ through C $\parallel$ to XO. Produce PL to meet CX′ at M. Draw $\perp^r$ CN from C on OX.

From Fig. 9.8, we have

$$x = \text{OL=ON-LN} = \text{ON-CM} = c\cos\phi - a\cos(\theta+\phi), \tag{9.12.31}$$

$$y = \text{PL} = \text{PM-LM} = \text{CP}\sin(\theta+\phi) - \text{CN} = a\sin(\theta+\phi) - c\sin\phi. \tag{9.12.32}$$

Let T be the KE of the particle at P.

$$\therefore\quad T = \frac{1}{2}mv^2 = \frac{1}{2}m\left(\dot{x}^2 + \dot{y}^2\right). \tag{9.12.33}$$

Now

$$\left.\begin{aligned} \dot{x} &= a\dot{\theta}\sin\psi + \dot{\phi}\,(a\sin\psi - c\sin\phi) \\ \dot{y} &= a\dot{\theta}\cos\psi + \dot{\phi}\,(a\cos\psi - c\cos\phi) \end{aligned}\right\}, \tag{9.12.34}$$

where $\psi = \theta + \phi$.

Therefore, Eq. (9.12.33) gives

$$\therefore\quad T = \frac{1}{2}m\left[a^2\dot{\theta}^2 + \dot{\phi}^2\left\{a^2 + c^2 - 2ac\cos(\psi-\phi)\right\} + 2a\dot{\theta}\dot{\phi}\{a - c\cos(\psi-\phi)\}\right],$$

Fig. 9.8 Diagram illustrating Example 9.8

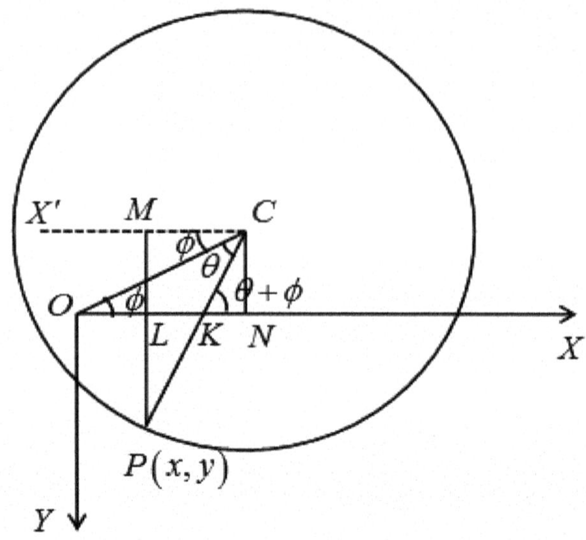

$$\text{i.e.,}\quad T = \frac{1}{2}m\left[a^2\dot{\theta}^2 + \dot{\phi}^2\left(a^2 + c^2 - 2ac\cos\theta\right) + 2a\dot{\theta}\dot{\phi}\left(a - c\cos\theta\right)\right]. \tag{9.12.35}$$

Here θ and ϕ are the only generalized coordinates.
Let **G** be the couple applied in the direction of φ increasing.
Therefore, the work fraction is given by

$$\delta W = \mathbf{G}\cdot\mathbf{e} = G\hat{k}\cdot e\hat{k} = Ge = G\delta\phi = Q_\theta\delta\theta + Q_\phi\delta\phi.$$

$$\therefore\quad Q_\theta = 0,\quad Q_\phi = G.$$

Lagrange's equations are

$$\frac{d}{dt}\left(\frac{\partial T}{\partial \dot{q}_\alpha}\right) - \frac{\partial T}{\partial q_\alpha} = Q_\alpha\ \ (\alpha = 1, 2). \tag{9.12.36}$$

Taking $\alpha = 1$ in (9.12.36), we obtain

$$\frac{d}{dt}\left(\frac{\partial T}{\partial \dot{q}_1}\right) - \frac{\partial T}{\partial q_1} = Q_1$$

$$\text{i.e.,}\quad \frac{d}{dt}\left(\frac{\partial T}{\partial \dot{\theta}}\right) - \frac{\partial T}{\partial \theta} = Q_\theta,$$

$$\text{i.e.,}\quad \frac{d}{dt}\left(\frac{\partial T}{\partial \dot{\theta}}\right) = \frac{\partial T}{\partial \theta}. \tag{9.12.37}$$

Now

$$\frac{\partial T}{\partial \dot{\theta}} = \frac{1}{2}m\left[2a^2\dot{\theta} + 2a\dot{\phi}\,(a - c\cos\theta)\right] = ma\left[a\dot{\theta} + \dot{\phi}\,(a - c\cos\theta)\right],$$

$$\frac{\partial T}{\partial \theta} = \frac{m}{2}\left[\dot{\phi}^2\,(2ac\sin\theta) + 2a\dot{\theta}\dot{\phi}\,(c\sin\theta)\right] = ma\dot{\phi}\left(c\dot{\phi} + c\dot{\theta}\right)\sin\theta.$$

Therefore, Eq. (9.12.37) reduces to

$$\frac{d}{dt}\left[ma\left\{a\dot{\theta} + \dot{\phi}\,(a - c\cos\theta)\right\}\right] = ma\dot{\phi}\left(c\dot{\phi} + c\dot{\theta}\right)\sin\theta,$$

$$\text{i.e.,}\quad \left\{a\ddot{\theta} + \dot{\phi}\left(c\sin\theta\dot{\theta}\right) + (a - c\cos\theta)\,\ddot{\phi}\right\} = \dot{\phi}\left(c\dot{\phi} + c\dot{\theta}\right)\sin\theta,$$

$$\text{i.e.,}\quad a\ddot{\theta} + \omega c\dot{\theta}\sin\theta + (a - c\cos\theta)\,\dot{\omega} = \omega\left(c\omega + c\dot{\theta}\right)\sin\theta,$$

$$\text{i.e.,}\quad a\ddot{\theta} + \omega c\dot{\theta}\sin\theta + (a - c\cos\theta)\,\dot{\omega} = c\omega^2\sin\theta + \omega c\dot{\theta}\sin\theta,$$

$$\text{i.e.,}\quad a\ddot{\theta} + (a - c\cos\theta)\,\dot{\omega} = c\omega^2\sin\theta.$$

Hence the result.

Example 9.9 The ends of a uniform rigid rod of mass m are moving with velocities $\mathbf{u}$ and $\mathbf{v}$. Prove that KE of the rod is $\frac{m}{6}\left(\mathbf{u}^2 + \mathbf{u}\cdot\mathbf{v} + \mathbf{v}^2\right)$.

Solution Consider a rod AB of length $2a$ (say) and center of mass G as shown in Fig. 9.9. Let $\mathbf{GQ}$ be the instantaneous axis of rotation of the rod.

Let $\mathbf{q}$, ω stipulate the velocity of the center of inertia G and angular velocity of the rod AB about $\mathbf{GQ}$ respectively.

We have

$$\mathbf{u} = \mathbf{q} + \omega\times\mathbf{GA} = \mathbf{q} + \omega\times(-\mathbf{a}) = \mathbf{q} - \omega\times\mathbf{a}, \tag{9.12.38}$$

and

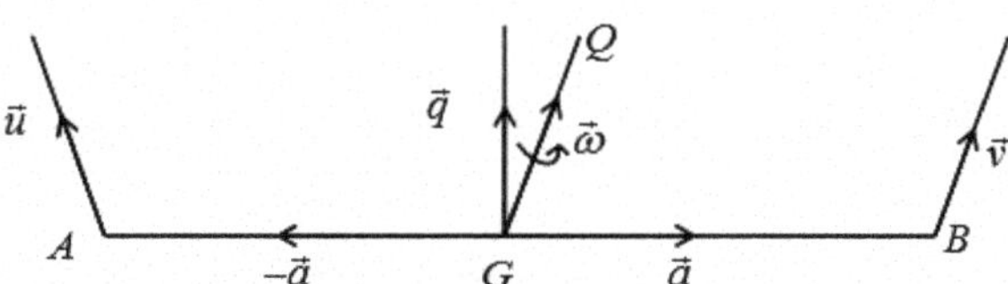

Fig. 9.9 Diagram illustrating Example 9.9

$$\mathbf{v} = \mathbf{q} + \omega \times \mathbf{GB} = \mathbf{q} + \omega \times \mathbf{a}. \tag{9.12.39}$$

(9.12.38)+(9.12.39) gives

$$\mathbf{u} + \mathbf{v} = 2\mathbf{q},$$

$$\therefore \quad \mathbf{q} = \frac{\mathbf{u} + \mathbf{v}}{2}. \tag{9.12.40}$$

(9.12.39)–(9.12.38) gives

$$\mathbf{v} - \mathbf{u} = 2(\omega \times \mathbf{a}),$$

$$\therefore \quad \omega \times \mathbf{a} = \frac{\mathbf{v} - \mathbf{u}}{2}. \tag{9.12.41}$$

In Fig. 9.10, take G as origin, X-axis along **GB**, Y-axis in such a way that GQ lies on the plane XY, and Z-axis through G $\perp^r$ to XY plane.

The direction angles of OQ are $\alpha = \theta,\ \beta = \frac{\pi}{2} - \theta,\ \gamma = \frac{\pi}{2}$.

Therefore, the direction cosines of GQ are $l = \cos\theta,\ m = \sin\theta,\ n = 0$.

Further, GX, GY, and GZ are the principal axis for the rod at G.

$$\therefore \quad \mathrm{D} = \mathrm{E} = \mathrm{F} = 0.$$

Now

$$I = \text{M.I. of the rod about GQ} = \mathrm{A}l^2 + \mathrm{B}m^2 + \mathrm{C}n^2 = \mathrm{A}\cos^2\theta + \mathrm{B}\sin^2\theta. \tag{9.12.42}$$

For the rod, A = M.I. of the rod about GX = 0,

$$\mathrm{B} = \text{M.I. of the rod about GY} = \frac{1}{3}ma^2.$$

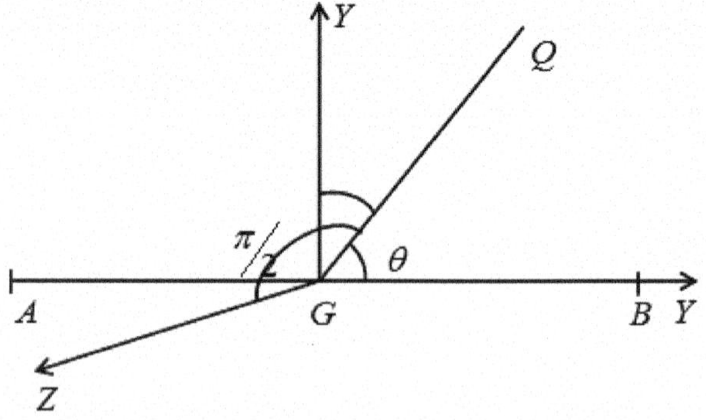

Fig. 9.10 Cartesian system with origin G

Therefore, Eq. (9.12.42) gives

$$I = \frac{1}{3}ma^2 \sin^2\theta. \tag{9.12.43}$$

The KE of the rod is given by

$$T = \frac{1}{2}M\mathbf{V}^2 + \frac{1}{2}I\omega^2 = \frac{1}{2}m\mathbf{q}^2 + \frac{1}{2}I\omega^2 = \frac{1}{2}m\left(\frac{\mathbf{u}+\mathbf{v}}{2}\right)^2 + \frac{1}{2}\frac{1}{3}ma^2\sin^2\theta\,\omega^2,$$

$$\text{i.e.,}\quad T = \frac{1}{8}m(\mathbf{u}+\mathbf{v})^2 + \frac{m}{6}a^2(\omega\sin\theta)^2 = \frac{1}{8}m(\mathbf{u}+\mathbf{v})^2 + \frac{ma^2}{6}\left(\omega\times\hat{i}\right)^2,$$

$$\text{i.e.,}\quad T = \frac{1}{8}m(\mathbf{u}+\mathbf{v})^2 + \frac{m}{6}\left(\omega\times a\hat{i}\right)^2 = \frac{1}{8}m(\mathbf{u}+\mathbf{v})^2 + \frac{m}{6}(\omega\times\mathbf{a})^2,$$

$$\text{i.e.,}\quad T = \frac{1}{8}m(\mathbf{u}+\mathbf{v})^2 + \frac{m}{6}\left(\frac{\mathbf{v}-\mathbf{u}}{2}\right)^2 = \frac{1}{8}m(\mathbf{u}+\mathbf{v})^2 + \frac{m}{24}(\mathbf{v}-\mathbf{u})^2,$$

$$\text{i.e.,}\quad T = \frac{1}{8}m\left(u^2 + 2\mathbf{u}\cdot\mathbf{v} + v^2\right) + \frac{m}{24}\left(v^2 - 2\mathbf{u}\cdot\mathbf{v} + u^2\right)$$

$$\text{i.e.,}\quad T = \frac{m}{24}\left[3\left(u^2 + 2\mathbf{u}\cdot\mathbf{v} + v^2\right) + v^2 - 2\mathbf{u}\cdot\mathbf{v} + u^2\right],$$

$$\text{i.e.,}\quad T = \frac{m}{24}\left(4u^2 + 4\mathbf{u}\cdot\mathbf{v} + 4v^2\right) = \frac{m}{6}\left(\mathbf{u}^2 + \mathbf{u}\cdot\mathbf{v} + \mathbf{v}^2\right).$$

Example 9.10 AB, BC, CD are three equal uniform rods each of mass m, freely jointed at B, C. They lie at rest in a straight line on a smooth horizontal table. If a blow J is administered at B in a horizontal direction $\perp^r$ to the rods, find the initial velocities of A, B, C, and D, and show that the angular velocities of the rods are in the ratio 7 : −6 : 2

Solution Let u_A, u_B, u_C, u_D be the velocities of A, B, C, and D $\perp^r$ to the rods, just after impulse. Let ω_1, ω_2, ω_3 be the anti-clockwise angular velocities of the rods AB, BC, CD, respectively (see Fig. 9.11).

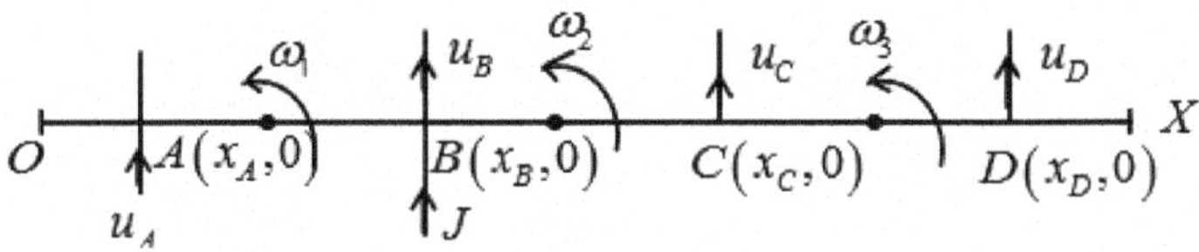

Fig. 9.11 Diagram illustrating Example 9.10

The KE of the rod is given by

$T = \text{KE of AB} + \text{KE of BC} + \text{KE of CD},$

$$\text{i.e., } T = \frac{m}{6}\left(u_A^2 + u_A u_B + u_B^2\right) + \frac{m}{6}\left(u_B^2 + u_B u_C + u_C^2\right) + \frac{m}{6}\left(u_C^2 + u_C u_D + u_D^2\right)$$

$$\text{i.e., } T = \frac{m}{6}\left[u_A^2 + 2u_B^2 + 2u_C^2 + u_D^2 + u_A u_B + u_B u_C + u_C u_D\right]. \tag{9.12.44}$$

Let A, B, C, and D undergo virtual displacements $\delta x_A, \delta x_B, \delta x_C, \delta x_D$ in the direction of J just after impulse.

The impulsive virtual work function is

$$\delta W = J\delta x_B = J_A\delta x_A + J_B\delta x_B + J_C\delta x_C + J_D\delta x_D.$$

$$\therefore \left.\begin{array}{l} J_A = 0 \\ J_B = J \\ J_C = 0 \\ J_D = 0 \end{array}\right\}. \tag{9.12.45}$$

The generalized momenta are given by

$$\left.\begin{array}{l} p_A = \frac{\partial T}{\partial u_A} = \frac{m}{6}\left(2u_A + u_B\right) \\ p_B = \frac{\partial T}{\partial u_B} = \frac{m}{6}\left(4u_B + u_C + u_A\right) \\ p_C = \frac{\partial T}{\partial u_c} = \frac{m}{6}\left(4u_C + u_B + u_D\right) \\ p_D = \frac{\partial T}{\partial u_D} = \frac{m}{6}\left(2u_D + u_C\right) \end{array}\right\}. \tag{9.12.46}$$

Lagrange's equations for impulsive motion are

$$p_A = J_A = 0, \tag{9.12.47}$$

$$p_B = J_B = J, \tag{9.12.48}$$

$$p_C = J_C = 0, \tag{9.12.49}$$

$$p_D = J_D = 0. \tag{9.12.50}$$

Equations (9.12.47), (9.12.48), (9.12.49), (9.12.50) give

$$2u_A + u_B = 0, \tag{9.12.51}$$

$$\frac{m}{6}\left(4u_B + u_A + u_C\right) = J, \tag{9.12.52}$$

$$4u_C + u_B + u_D = 0, \tag{9.12.53}$$

$$2u_D + u_C = 0. \tag{9.12.54}$$

Equation (9.12.51) gives

$$2u_A = -u_B,$$

$$\text{i.e.,} \quad \frac{u_A}{1} = -\frac{u_B}{2} = K \ \text{(say)},$$

$$\therefore \quad u_A = K, u_B = -2K. \tag{9.12.55}$$

Equation (9.12.52) gives

$$\frac{m}{6}(-8K + K + u_C) = J,$$

$$\text{i.e.,} \quad u_C - 7K = \frac{6J}{m},$$

$$\text{i.e.,} \quad u_C = 7K + \frac{6J}{m}. \tag{9.12.56}$$

Equation (9.12.53) gives

$$4\left(7K + \frac{6J}{m}\right) - 2K + u_D = 0,$$

$$\text{i.e.,} \quad u_D = -26K - \frac{24J}{m}. \tag{9.12.57}$$

Equation (9.12.54) yields

$$2\left(-26K - \frac{24J}{m}\right) + \left(7K + \frac{6J}{m}\right) = 0,$$

$$\text{i.e.,} \quad -52mK - 48J + 7mK + 6J = 0,$$

$$\text{i.e.,} \quad K = -\frac{14J}{15m}. \tag{9.12.58}$$

$$\therefore \quad u_A = -\frac{14J}{15m}, \ u_B = \frac{28J}{15m}, \ u_C = -\frac{8J}{15m}, \ u_D = \frac{4J}{15m}. \tag{9.12.59}$$

Now

$$\omega_1 = \frac{u_B - u_A}{2a} = \frac{7J}{5am},$$

$$\omega_2 = \frac{u_C - u_B}{2a} = \frac{-6J}{5am},$$

$$\omega_3 = \frac{u_D - u_C}{2a} = \frac{2J}{5am}.$$

$$\therefore \quad \omega_1 : \omega_2 : \omega_3 = \frac{7J}{5am} : \frac{-6J}{5am} : \frac{2J}{5am} = 7 : -6 : 2.$$

Hence the result is proved.

Example 9.11 Three rods A_1A_2, A_2A_3, A_3A_4, each of mass m, smoothly hinged together at A_2, A_3, hang in equilibrium under gravity from a fixed smooth hinge at A_1. A horizontal impulse is applied to a point P of A_1A_2 . If there is no impulse reaction on the hinge A_1, show that $\frac{A_1P}{PA_2} = \frac{26}{7}$.

Solution Consider three rods A_1A_2, A_2A_3, and A_3A_4 each of length $2a$ and mass m smoothly hinged together at points A_2 and A_3, as shown in Fig. 9.12. Let $A_1P = \lambda$, $PA_2 = \mu$. Let I be the impulse applied at the point P and suppose that I_1 and I_2 are two $\parallel$ components of I applied at A_1 and A_2 respectively.

$$\therefore \quad I_1 + I_2 = I, \tag{9.12.60}$$

and

$$I_1 \cdot A_1P = I_2 \cdot A_2P,$$

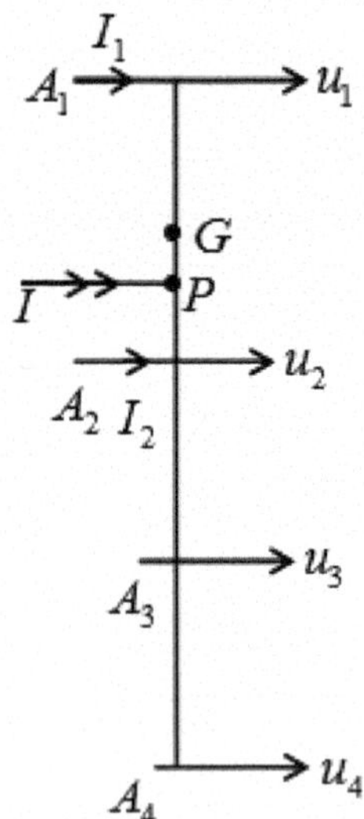

Fig. 9.12 Diagram illustrating Example 9.11

i.e., $I_1\,\lambda = I_2\,\mu$,

i.e., $\dfrac{I_1}{\mu} = \dfrac{I_2}{\lambda} = \xi\ \text{(say)}$,

$$\therefore\quad I_1 = \mu\xi,\ \ I_2 = \lambda\xi. \tag{9.12.61}$$

Equation (9.12.60) gives

$$\mu\xi + \lambda\xi = I,$$

$$\text{i.e.,}\quad \xi = \frac{I}{\lambda+\mu}. \tag{9.12.62}$$

Let u_1, u_2, u_3, u_4 be the velocities of A_1, A_2, A_3, A_4 $\perp^r$ to the rods, just after impulse. Let A_1, A_2, A_3, A_4 undergo virtual displacements δx_1, δx_2, δx_3, δx_4 in the direction of I, i.e., $\perp^r$ to the rods. The impulsive virtual work function is given by

$$\delta W = I_1\delta x_1 + I_2\delta x_2 = J_1\delta x_1 + J_2\delta x_2 + J_3\delta x_3 + J_4\delta x_4.$$

$$\therefore\quad \left.\begin{aligned} J_1 &= I_1 \\ J_2 &= I_2 \\ J_3 &= 0 \\ J_4 &= 0 \end{aligned}\right\}. \tag{9.12.63}$$

The KE of the system of rods is given by

$$\therefore\quad T = \frac{m}{6}\left(u_1^2 + u_1u_2 + u_2^2\right) + \frac{m}{6}\left(u_2^2 + u_2u_3 + u_3^2\right) + \frac{m}{6}\left(u_3^2 + u_3u_4 + u_4^2\right),$$

$$\text{i.e.,}\quad T = \frac{m}{6}\left(u_1^2 + 2u_2^2 + 2u_3^2 + u_4^2 + u_1u_2 + u_2u_3 + u_3u_4\right). \tag{9.12.64}$$

The generalized components of momentum are

$$p_1 = \frac{\partial T}{\partial u_1} = \frac{m}{6}\left(2u_1 + u_2\right),$$

$$p_2 = \frac{\partial T}{\partial u_2} = \frac{m}{6}\left(4u_2 + u_1 + u_3\right),$$

$$p_3 = \frac{\partial T}{\partial u_3} = \frac{m}{6}\left(4u_3 + u_2 + u_4\right),$$

$$p_4 = \frac{\partial T}{\partial u_4} = \frac{m}{6}\left(2u_4 + u_3\right).$$

Lagrange's equations for impulsive forces are

$$p_1 = J_1 = I_1, \tag{9.12.65}$$

$$p_2 = J_2 = I_2, \tag{9.12.66}$$

$$p_3 = J_3 = 0, \tag{9.12.67}$$

$$p_4 = J_4 = 0. \tag{9.12.68}$$

Equation (9.12.68) gives

$$2u_4 + u_3 = 0$$

$$\text{i.e.,}\quad 2u_4 = -u_3,$$

$$\text{i.e.,}\quad u_4 = -\frac{u_3}{2} = k\ (\text{say}),$$

$$\therefore\quad u_4 = k,\quad u_3 = -2k. \tag{9.12.69}$$

Equation (9.12.67) gives

$$4u_3 + u_2 + u_4 = 0$$

$$\text{i.e.,}\quad -8k + u_2 + k = 0,$$

$$\text{i.e.,}\quad u_2 = 7k. \tag{9.12.70}$$

Equation (9.12.66) gives

$$\frac{m}{6}\left(4u_2 + u_1 + u_3\right) = I_2,$$

$$\text{i.e.,}\quad 4u_2 + u_1 + u_3 = \frac{6I_2}{m},$$

$$\text{i.e.,}\quad 28k + u_1 + (-2k) = \frac{6}{m}\lambda\xi,$$

$$\text{i.e.,}\quad 26k + u_1 = \frac{6\lambda}{m}\xi. \tag{9.12.71}$$

Equation (9.12.65) gives

$$\frac{m}{6}\,(2u_1 + u_2) = I_1,$$

$$\text{i.e.,}\quad 2u_1 + u_2 = \frac{6I_1}{m},$$

$$2u_1 + 7k = \frac{6}{m}\mu\xi. \tag{9.12.72}$$

As there is no impulsive reaction at A_1,

$$\therefore\quad u_1 = 0. \tag{9.12.73}$$

Equations (9.12.71) and (9.12.72) give

$$13k = \frac{3\lambda\xi}{m}, \tag{9.12.74}$$

$$7k = \frac{6\mu\xi}{m}. \tag{9.12.75}$$

(9.12.74) $\div$ (9.12.75) gives

$$\frac{13}{7} = \frac{3\lambda\xi}{m} \times \frac{m}{6\mu\xi},$$

$$\text{i.e.,}\quad \frac{13}{7} = \frac{\lambda}{2\mu},$$

$$\text{i.e.,}\quad \frac{\lambda}{\mu} = \frac{26}{7},$$

$$\text{i.e.,}\quad \frac{A_1P}{PA_2} = \frac{26}{7}.$$

Proved.

Example 9.12 Three equal uniform rods AB, BC, CD, each of mass m and length $2a$, are at rest in a straight line smoothly jointed at B and C. A blow I is given to the middle rod at a distance c from its center O in a direction $\perp^r$ to it; show that the initial velocity of O is $\frac{2I}{3m}$, and that the initial angular velocities of the rods are $\frac{(5a+9c)I}{10ma^2}$, $\frac{6cI}{5ma^2}$, $\frac{(5a-9c)I}{10ma^2}$.

Solution Let the parallel components of I be I_1 at B and I_2 at C.

$$\therefore \quad I_1\text{BP} = I_2\text{CP};$$

P being the point of application of the blow.
Thus,

$$I_1(\text{BO} + \text{OP}) = I_2(\text{OC} - \text{OP}),$$

$$\text{i.e.,} \quad I_1\,(a + c) = I_2\,(a - c)\,,$$

$$\text{i.e.,} \quad \frac{I_1}{a - c} = \frac{I_2}{a + c} = k\,(\text{say})\,,$$

$$\text{i.e.,} \quad I_1 = k\,(a - c)\,, I_2 = k\,(a + c)\,.$$

Further

$$I_1 + I_2 = I,$$

$$\text{i.e.,} \quad k\,(a - c) + k\,(a + c) = I,$$

$$\text{i.e.,} \quad k = \frac{I}{2a}.$$

In Fig. 9.13, take the X-axis along **AB** and the Y-axis $\perp^r$ to AB on the plane of motion of the rods. Let $u_A,\ u_B,\ u_C,\ u_D$ be the velocities of A, B, C, and D $\perp^r$ to the rods, just after the impulse. Let A, B, C, and D undergo virtual displacements $\delta x_A,\ \delta x_B,\ \delta x_C,\ \delta x_D$ in the direction of I just after its application.

The impulsive virtual work function is

$$\delta W = T_1\delta x_B + I_2\delta x_C = J_A\,\delta x_A + J_B\delta x_B + J_C\delta x_C + +J_D\delta x_D, \tag{9.12.76}$$

where $J_A,\ J_B,\ J_C,\ J_D$ are the generalized impulses associated with the coordinates $x_A,\ x_B,\ x_C,\ x_D$. Thus

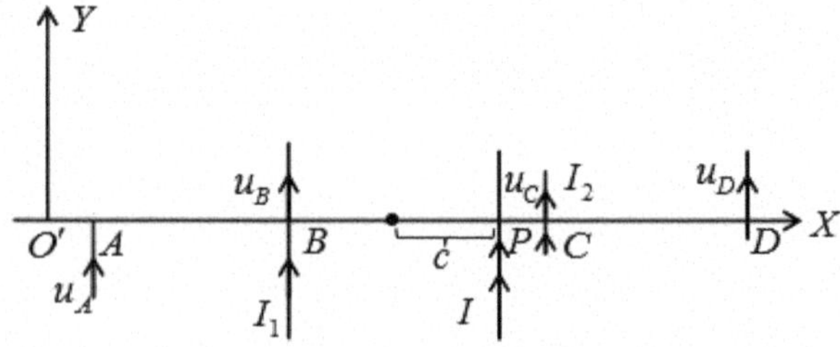

Fig. 9.13 Illustration for Example 9.12

$$J_A = 0,\ J_B = I_1 = k\,(a-c)\,,\ J_C = I_2 = k\,(a+c)\,,\ J_D = 0. \tag{9.12.77}$$

Let T be the KE of the system of rods.

$$\therefore\quad T = \frac{m}{6}\left(u_A^2 + u_A u_B + u_B^2\right) + \frac{m}{6}\left(u_B^2 + u_B u_C + u_C^2\right) + \frac{m}{6}\left(u_C^2 + u_C u_D + u_D^2\right),$$

$$\text{i.e.,}\quad T = \frac{m}{6}\left(u_A^2 + 2u_B^2 + 2u_C^2 + u_D^2 + u_A u_B + u_B u_C + u_C u_D\right). \tag{9.12.78}$$

The generalized momenta are

$$p_A = \frac{\partial T}{\partial u_A} = \frac{m}{6}\,(2u_A + u_B)\,,$$

$$p_B = \frac{\partial T}{\partial u_B} = \frac{m}{6}\,(4u_B + u_A + u_C)\,,$$

$$p_C = \frac{\partial T}{\partial u_C} = \frac{m}{6}\,(4u_C + u_B + u_D)\,,$$

$$p_D = \frac{\partial T}{\partial u_D} = \frac{m}{6}\,(2u_D + u_C)\,.$$

Lagrange's equations of motion for impulsive forces are

$$p_A = J_A = 0, \tag{9.12.79}$$

$$p_B = J_B = k\,(a-c)\,, \tag{9.12.80}$$

$$p_C = J_C = k\,(a+c)\,, \tag{9.12.81}$$

$$p_D = J_D = 0. \tag{9.12.82}$$

Equations (9.12.79), (9.12.80), (9.12.81), (9.12.82) give

$$2u_A + u_B = 0, \tag{9.12.83}$$

$$\frac{m}{6}\,(4u_B + u_A + u_C) = k\,(a-c)\,, \tag{9.12.84}$$

$$\frac{m}{6}\,(4u_C + u_B + u_D) = k\,(a+c)\,, \tag{9.12.85}$$

$$2u_D + u_C = 0. \tag{9.12.86}$$

Equations (9.12.83) and (9.12.86) give

$$u_A = -\frac{u_B}{2}, \quad u_D = -\frac{u_C}{2}.$$

Equations (9.12.84) and (9.12.85) give

$$\frac{m}{6}\left[4u_B - \frac{u_B}{2} + u_C\right] = k\,(a-c)\,,$$

$$\text{i.e.,}\quad m\,(7u_B + 2u_C) = 12k\,(a-c)\,. \tag{9.12.87}$$

and

$$\frac{m}{6}\left(4u_C + u_B - \frac{u_C}{2}\right) = k\,(a+c)\,,$$

$$\text{i.e.,}\quad m\,(7u_C + 2u_B) = 12k\,(a+c)\,. \tag{9.12.88}$$

9.12.87 ÷ 9.12.88 gives

$$\frac{7u_B + 2u_C}{7u_C + 2u_B} = \frac{a-c}{a+c},$$

$$\text{i.e.,}\quad (5a+9c)\,u_B = (5a-9c)\,u_C,$$

$$\text{i.e.,}\quad \frac{u_B}{5a-9c} = \frac{u_c}{5a+9c} = \lambda\,(\text{say})\,,$$

$$\therefore\quad u_B = \lambda\,(5a-9c)\,,\quad u_C = \lambda\,(5a+9c)\,.$$

∴ (9.12.87) gives

$$m\lambda\,[7\,(5a-9c) + 2\,(5a+9c)] = 12k\,(a-c)\,,$$

$$\text{i.e.,}\quad m\lambda\,[45a - 45c] = 12k\,(a-c)\,,$$

$$\text{i.e.,}\quad 15m\lambda = 4k,$$

$$\text{i.e.,}\quad \lambda = \frac{4k}{15m} = \frac{4}{15m}\frac{I}{2a} = \frac{2I}{15am}.$$

$$\text{Initial velocity of O} = \frac{u_B + u_C}{2} = \frac{\lambda}{2}\,[5a - 9c + 5a + 9c] = 5\lambda a,$$

i.e., Initial velocity of O $= 5a\dfrac{2I}{15am} = \dfrac{2I}{3m}$.

The initial angular velocities of the rods AB, BC, and CD are as follows:

$$\omega_1 = \frac{u_B - u_A}{\text{AB}} = \frac{\lambda(5a - ac) + \frac{u_B}{2}}{2a} = \frac{2\lambda(5a - 9c) + \lambda(5a - 9c)}{4a},$$

$$\text{i.e., } \omega_1 = \frac{\lambda[10a - 18c + 5a - 9c]}{4a} = \frac{\lambda(15a - 27c)}{4a} = \frac{(5a - 9c)I}{10ma^2},$$

$$\omega_2 = \text{angular velocity of BC} = \frac{u_C - u_B}{2a} = \frac{\lambda 18C}{2a} = \frac{6IC}{5ma^2},$$

$$\omega_3 = \text{angular velocity of CD} = \frac{u_D - u_C}{2a} = \frac{-\frac{u_C}{2} - u_C}{2a},$$

$$\text{i.e., } \omega_3 = -\frac{3u_C}{4a} = -\frac{3}{4a}\lambda(5a + 9c) = -\frac{(5a + 9c)I}{10ma^2}.$$

Example 9.13 Three uniform rods OA, AB, BC each of mass m are smoothly jointed together at A, B and hang in equilibrium under gravity from a fixed smooth joint O. An impulsive couple G is applied to the rod AB in a vertical plane through O. Prove that the initial KE of the system is $\frac{57G^2}{26ml^2}$, where l is length of AB.

Solution In the Fig. 9.14, take the point O as the origin, X-axis along **OA**, Y-axis $\perp^r$ to **OA** in the plane of the motion of the rod, and Z-axis as shown in the figure.Let $\hat{i}, \hat{j}, \hat{k}$ be the unit vector along +ve X-axis, Y-axis, Z-axis respectively.

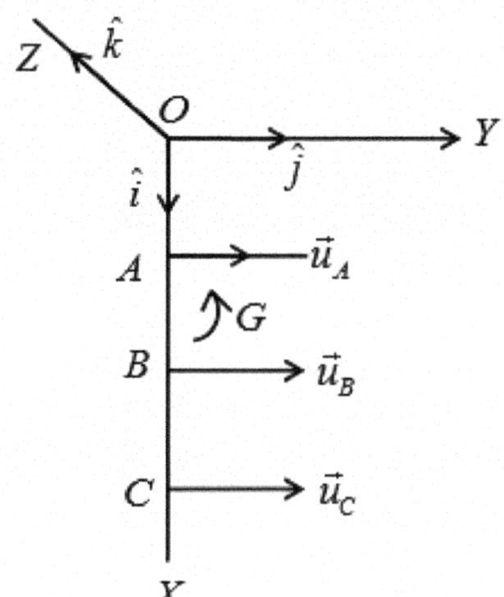

Fig. 9.14 Illustration for Example 9.13

Let $\mathbf{u}_A = u_A\hat{j}$, $\mathbf{u}_B = u_B\hat{j}$, $\mathbf{u}_C = u_C\hat{j}$ be the velocities of A, B, C just after impulse. Let $\delta x_A, \delta x_B, \delta x_C$ be the virtual displacements of A, B, C, respectively, along the Y-axis.
We have

$$\boldsymbol{\delta x}_A = \delta x_A\hat{j},\ \boldsymbol{\delta x}_B = \delta x_B\hat{j},\ \boldsymbol{\delta x}_C = \delta x_C\hat{j},\ \mathbf{G} = G\hat{k}.$$

Let $\mathbf{e} = e\hat{k}$ be the rotation of the rod AB due to the application of the couple $\mathbf{G}$.
We have

$$\left.\begin{aligned}\boldsymbol{\delta x}_A &= \mathbf{u} + \mathbf{e} \times \mathbf{x}_A\\ \boldsymbol{\delta x}_B &= \mathbf{u} + \mathbf{e} \times \mathbf{x}_B\end{aligned}\right\}, \tag{9.12.89}$$

where $\mathbf{u}$ is the translatory displacement of the rod AB.
Equation (9.12.89) gives

$$\left.\begin{aligned}\delta x_A &= u + e x_A\\ \delta x_B &= u + e x_B\end{aligned}\right\}. \tag{9.12.90}$$

Equation (9.12.90) gives

$$\delta x_B - \delta x_A = e\,(x_B - x_A)\,,$$

$$\text{i.e.,}\quad \delta x_B - \delta x_A = e\,(\text{OB} - \text{OA}) = e\text{AB} = e\,l,$$

$$\therefore\quad e = \frac{\delta x_B - \delta x_A}{l}. \tag{9.12.91}$$

The impulsive virtual work fraction is given by

$$\delta W = \mathbf{G}\cdot\mathbf{e} = G\hat{k}\cdot e\hat{k} = Ge = G\frac{\delta x_B - \delta x_A}{l},$$

$$\text{i.e.,}\quad \delta W = J_A\delta x_A + J_B\delta x_B + J_C\delta x_C = \frac{G}{l}\,(\delta x_B - \delta x_A)\,. \tag{9.12.92}$$

Equation (9.12.92) gives

$$\therefore\quad -\frac{G}{l} = J_A,\ \frac{G}{l} = J_B,\ 0 = J_C,$$

$$\text{i.e.,}\quad \left.\begin{aligned} J_A &= -\tfrac{G}{l} \\ J_B &= \tfrac{G}{l} \\ J_C &= 0 \end{aligned}\right\} . \tag{9.12.93}$$

Let T be the KE of the system of rods.

$$\therefore \quad T = \text{K.E. of OA} + \text{K.E. of AB} + \text{K.E. of BC},$$

$$\text{i.e.,}\quad T = \frac{m}{6}\left(u_A^2\right) + \frac{m}{6}\left(u_A^2 + u_A u_B + u_B^2\right) + \frac{m}{6}\left(u_B^2 + u_B u_C + u_C^2\right),$$

$$\text{i.e.,}\quad T = \frac{m}{6}\left(2u_A^2 + 2u_B^2 + u_C^2 + u_A u_B + u_B u_C\right). \tag{9.12.94}$$

The generalized components of momentum are

$$\left.\begin{aligned} p_A &= \tfrac{\partial T}{\partial u_A} = \tfrac{m}{6}\left(4u_A + u_B\right) \\ p_B &= \tfrac{\partial T}{\partial u_B} = \tfrac{m}{6}\left(4u_B + u_A + u_C\right) \\ p_C &= \tfrac{\partial T}{\partial u_C} = \tfrac{m}{6}\left(2u_C + u_B\right) \end{aligned}\right\} . \tag{9.12.95}$$

Lagrange's equations for impulsive motion are

$$p_A = J_A, \tag{9.12.96}$$

$$p_B = J_B, \tag{9.12.97}$$

$$p_C = J_C. \tag{9.12.98}$$

Equations (9.12.96), (9.12.97), (9.12.98) give

$$\frac{m}{6}\left(4u_A + u_B\right) = -\frac{G}{l}, \tag{9.12.99}$$

$$\frac{m}{6}\left(4u_B + u_A + u_C\right) = \frac{G}{l}, \tag{9.12.100}$$

$$2u_C + u_B = 0. \tag{9.12.101}$$

Equation (9.12.101) gives

$$2u_C = -u_B,$$

$$\text{i.e.,}\quad \frac{u_C}{1} = \frac{u_B}{2} = K \text{ (say)},$$

$$\therefore \quad u_C = K, \quad u_B = -2K. \tag{9.12.102}$$

Equation (9.12.99) gives

$$4u_A + u_B = -\frac{6G}{ml},$$

$$\text{i.e.,} \quad 4u_A - 2K = -\frac{6G}{ml},$$

$$\therefore \quad 2u_A = K - \frac{3G}{ml},$$

$$\text{i.e.,} \quad u_A = \frac{K}{2} - \frac{3G}{2ml}. \tag{9.12.103}$$

Equation (9.12.100) gives

$$4u_B + u_A + u_C = \frac{6G}{ml},$$

$$\text{i.e.,} \quad 4(-2K) + \frac{K}{2} - \frac{3G}{2ml} + K = \frac{6G}{ml},$$

$$\text{i.e.,} \quad -16K + K - \frac{3G}{ml} + 2K = \frac{12G}{ml},$$

$$\text{i.e.,} \quad -13k = \frac{15G}{ml},$$

$$\therefore \quad K = -\frac{15G}{13ml}. \tag{9.12.104}$$

Since T is a homogeneous function of degree 2 in u_A, u_B, u_C

$$\therefore \quad u_A\frac{\partial T}{\partial u_A} + u_B\frac{\partial T}{\partial u_B} + u_C\frac{\partial T}{\partial u_C} = 2T,$$

$$\text{i.e.,} \quad u_A p_A + u_B p_B + u_C p_C = 2T,$$

$$\text{i.e.,} \quad 2T = u_A J_A + u_B J_B + u_C J_C,$$

$$\text{i.e.,} \quad 2T = u_A\left(-\frac{G}{L}\right) + u_B\frac{G}{L} = \frac{G}{l}(u_B - u_A) = \frac{G}{2}\left[-2K - \frac{K}{2} + \frac{3G}{2ml}\right],$$

$$\text{i.e.,}\quad 2T = \frac{G}{2l}\left(-4K - K + \frac{3G}{ml}\right) = \frac{G}{2l}\left(\frac{3G}{ml} - 5K\right) = \frac{G}{2l}\left(\frac{3G}{ml} + \frac{75G}{13ml}\right),$$

$$\text{i.e.,}\quad 2T = \frac{G}{26l}\left(\frac{39G}{ml} + \frac{75G}{ml}\right) = \frac{G^2}{26ml^2}114 = \frac{57G^2}{13ml^2},$$

$$\text{i.e.,}\quad T = \frac{57G^2}{26ml^2}.$$

Example 9.14 Uniform rod AB of length $2a$ is suspended from a fixed point O by a string OC of length $\frac{5a}{6}$ attached to a point C of the rod such that AC $= \frac{2a}{3}$. Show that the normal modes of oscillation in a vertical plane are given by $2\pi\sqrt{\frac{a}{2g}}$ and $2\pi\sqrt{\frac{5a}{3g}}$.

Solution Let G be the c.g. of the rod AB.

$\therefore$ AG = GB $= a$. Given OC $= \frac{5a}{6}$, AC $= \frac{2a}{3}$.

Now CG = AG − AC $= a - \frac{2a}{3} = \frac{a}{3}$.

Let OZ be the downward vertical through O, OX the horizontal axis of rotation of the system, θ the angle of inclination of OC to OZ, ϕ that of AB to OZ, and $\hat{\theta}$, $\hat{\phi}$ the unit vectors in the increasing directions of θ and ϕ respectively.

In Fig. 9.15, let **OG** = **r**, **OC** = $\mathbf{r}_1$ &**CG** = $\mathbf{r}_2$.

We have **OG** = **OC** + **CG**,

$$\text{i.e.,}\quad \mathbf{r} = \mathbf{r}_1 + \mathbf{r}_2. \tag{9.12.105}$$

Now

$$\mathbf{v}_0 = \text{velocity of G relative to O} = \frac{d}{dt}\mathbf{OG} = \dot{\mathbf{r}} = \dot{\mathbf{r}}_1 + \dot{\mathbf{r}}_2. \tag{9.12.106}$$

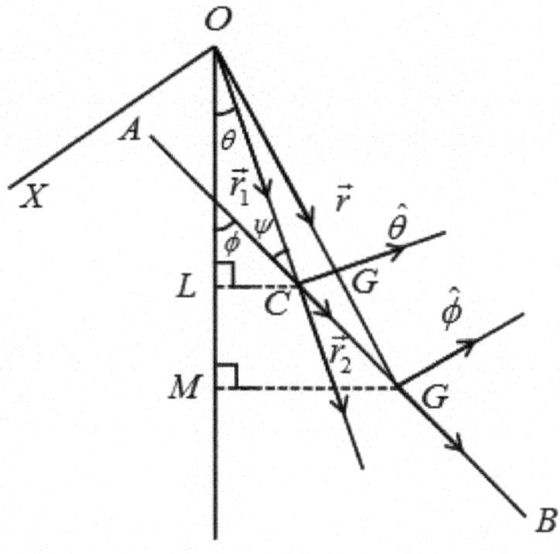

Fig. 9.15 Diagram illustrating Example 9.14

Again

$$\dot{\mathbf{r}}_1 = \frac{d}{dt}\mathbf{OC} = \text{OC}\dot{\theta}\hat{\theta} = \frac{5a}{6}\dot{\theta}\hat{\theta},$$

$$\dot{\mathbf{r}}_2 = \frac{d}{dt}\mathbf{CG} = \text{CG}\dot{\phi}\hat{\phi} = \frac{a}{3}\dot{\phi}\ \hat{\phi}.$$

Equation (9.12.106) gives

$$\mathbf{v}_0 = \frac{5a}{6}\dot{\theta}\hat{\theta} + \frac{a}{3}\dot{\phi}\hat{\phi}. \quad (9.12.107)$$

Let ψ be angle between **OC** & **AB**.
Now

$$\text{T} = \text{KE of the rod AB} = \frac{1}{2}m\mathbf{v}_0^2 + \frac{1}{2}I\omega^2,$$

$$\text{i.e.,}\quad \text{T} = \frac{1}{2}m\left(\frac{5a}{6}\dot{\theta}\hat{\theta} + \frac{a}{3}\dot{\phi}\hat{\phi}\right)^2 + \frac{1}{6}ma^2\dot{\phi}^2,$$

$$\text{i.e.,}\quad \text{T} = \frac{1}{2}m\left(\frac{25a^2}{36}\dot{\theta}^2 + \frac{a^2\dot{\phi}^2}{9} + \frac{5a^2}{9}\dot{\phi}\dot{\theta}\hat{\theta}\cdot\hat{\phi}\right) + \frac{1}{6}ma^2\dot{\phi}^2. \quad (9.12.108)$$

We note that the angle between $\hat{\theta}$ & $\hat{\phi}$ = angle between OC & AB $= \psi$.
From figure,

$$\phi = \theta + \psi,$$

$$\text{i.e.,}\quad \psi = \phi - \theta.$$

Therefore, the angle between $\hat{\theta}$ & $\hat{\phi} = \psi = \phi - \theta \approx 0$.

Now

$$\hat{\theta}\cdot\hat{\phi} = \cos\psi \approx 1.$$

Therefore, Eq. (9.12.108) reduces to

$$T = \frac{1}{2}m\left(\frac{25a^2\dot{\theta}^2}{36} + \frac{a^2\dot{\phi}^2}{9} + \frac{5a^2}{9}\dot{\theta}\dot{\phi}\right) + \frac{1}{6}ma^2\dot{\phi}^2,$$

$$\text{i.e.,}\quad T = \frac{ma^2}{72}\left(25\dot{\theta}^2 + 20\dot{\theta}\dot{\phi} + 16\dot{\phi}^2\right).$$

$$\text{V} = \text{PE of the rod} = -mg\text{OM} + K = -mg\,(\text{OL+LM}) + K,$$

$$\text{i.e.,}\quad \text{V} = -mg\,(\text{OC}\cos\theta + \text{CG}\cos\phi) + K = -mg\left(\frac{5a}{6}\cos\theta + \frac{a}{3}\cos\phi\right) + K,$$

$$\text{i.e.,}\quad \text{V} = -mg\left\{\frac{5a}{6}\left(1 - \frac{\theta^2}{2}\right) + \frac{a}{3}\left(1 - \frac{\phi^2}{2}\right)\right\} + K,$$

$$\text{i.e.,}\quad \text{V} = -mga\left(\frac{5}{6} - \frac{5\theta^2}{12} + \frac{1}{3} - \frac{\phi^2}{6}\right) + K,$$

$$\text{i.e.,}\quad \text{V} = -mga\left(\frac{5\theta^2}{12} + \frac{\phi^2}{6}\right) + \lambda.$$

The Lagrangian for the system is given by

$$L = T - V = \frac{ma^2}{72}\left(25\dot{\theta}^2 + 20\dot{\theta}\dot{\phi} + 16\dot{\phi}^2\right) - mga\left(\frac{5\theta^2}{12} + \frac{\phi^2}{6}\right) - \lambda. \tag{9.12.109}$$

Here θ and ϕ are the only generalized coordinates.
Lagrange's equations are

$$\frac{d}{dt}\left(\frac{\partial L}{\partial \dot{q}_a}\right) = \frac{\partial L}{\partial q_\alpha};\quad \alpha = 1, 2. \tag{9.12.110}$$

Equation (9.12.110) gives

$$\frac{d}{dt}\left(\frac{\partial L}{\partial \dot{q}_1}\right) = \frac{\partial L}{\partial q_1},$$

$$\text{i.e.,}\quad \frac{d}{dt}\left(\frac{\partial L}{\partial \dot{\theta}}\right) = \frac{\partial L}{\partial \theta}, \tag{9.12.111}$$

and

$$\frac{d}{dt}\left(\frac{\partial L}{\partial \dot{\phi}}\right) = \frac{\partial L}{\partial \phi}. \tag{9.12.112}$$

Now

$$\frac{\partial L}{\partial \dot{\theta}} = \frac{ma^2}{72}\left(50\dot{\theta} + 20\dot{\phi}\right), \quad \frac{\partial L}{\partial \dot{\phi}} = \frac{ma^2}{72}\left(20\dot{\theta} + 30\dot{\phi}\right);$$

$$\frac{\partial L}{\partial \theta} = -\frac{5mga}{6}\theta, \quad \frac{\partial L}{\partial \phi} = -mga\frac{\phi}{3}.$$

Equation (9.12.111) gives

$$\frac{d}{dt}\left[\frac{ma^2}{72}\left(50\dot{\theta} + 20\dot{\phi}\right)\right] = -\frac{5mga}{6}\theta,$$

$$\text{i.e.,} \quad \frac{a}{72}\left(50\ddot{\theta} + 20\ddot{\phi}\right) = -\frac{5g\theta}{6},$$

$$\text{i.e.,} \quad 5\ddot{\theta} + 2\ddot{\phi} = -6g\frac{\theta}{a}. \tag{9.12.113}$$

Equation (9.12.112) gives

$$\frac{d}{dt}\left[\frac{ma^2}{72}\left(20\dot{\theta} + 32\dot{\phi}\right)\right] = -\frac{mga\phi}{3},$$

$$\text{i.e.,} \quad 5\ddot{\theta} + 8\ddot{\phi} = -\frac{6g\phi}{a}. \tag{9.12.114}$$

Let

$$\theta = M\cos(pt + \varepsilon), \tag{9.12.115}$$

$$\phi = N\cos(pt + \varepsilon). \tag{9.12.116}$$

$$\dot{\theta} = -M\sin(pt + \varepsilon)p = -Mp\sin(pt + \varepsilon),$$

$$\ddot{\theta} = -p^2M\cos(pt + \varepsilon) = -p^2\theta,$$

Similarly

$$\ddot{\phi} = -p^2\phi.$$

Thus, Eqs. (9.12.113) and (9.12.114) reduce to

$$5\left(-p^2\theta\right)+2\left(-p^2\phi\right)=-\frac{6g\theta}{a},$$

$$\text{i.e.,}\quad 5p^2\theta+2p^2\phi=\frac{6g\theta}{a},$$

$$\text{i.e.}\quad \left(5p^2-\frac{6g}{a}\right)\theta+2p^2\phi=0. \tag{9.12.117}$$

and

$$5\left(-p^2\theta\right)+8\left(-p^2\phi\right)=-6g\frac{\phi}{a},$$

$$\text{i.e.,}\quad 5p^2\theta+8p^2\phi=6g\frac{\phi}{a},$$

$$\text{i.e.,}\quad 5p^2\theta+\left(8p^2-\frac{6g}{a}\right)\phi=0. \tag{9.12.118}$$

By eliminating θ and ϕ from (9.12.117) and (9.12.118), we obtain

$$\begin{vmatrix} 5p^2-\frac{6g}{a} & 2p^2 \\ 5p^2 & 8p^2-\frac{6g}{a} \end{vmatrix}=0$$

$$\text{i.e.,}\quad \left(5p^2-\frac{6g}{a}\right)\left(8p^2-\frac{6g}{a}\right)=10p^4,$$

$$\text{i.e.,}\quad \left(5ap^2-6g\right)\left(8ap^2-6g\right)=10a^2p^4,$$

$$\text{i.e.,}\quad 40a^2p^4-30ap^2g-48ap^2g+36g^2=10a^2p^4,$$

$$\text{i.e.,}\quad 5a^2p^4-13ap^2g+6g^2=0,$$

$$\text{i.e.,}\quad \left(5ap^2-3g\right)\left(ap^2-2g\right)=0,$$

$$\text{i.e.,}\quad p^2=\frac{3g}{5a},\quad p^2=\frac{2g}{a},$$

$$\text{i.e.,}\quad p^2=\frac{2g}{a}=p_1^2,\quad p^2=\frac{3g}{5a}=p_2^2.$$

Therefore, the periods of normal modes are

$$\tau_1 = \frac{2\pi}{p_1} \& \tau_2 = \frac{2\pi}{p_2};$$

$$\text{i.e.,} \quad \tau_1 = \frac{2\pi}{\sqrt{\frac{2g}{a}}}, \tau_2 = \frac{2\pi}{\sqrt{\frac{3g}{5a}}};$$

$$\text{i.e.,} \quad \tau_1 = 2\pi\sqrt{\frac{a}{2g}}, \quad \tau_2 = 2\pi\sqrt{\frac{5a}{3g}}.$$

Example 9.15 A uniform rod, of mass $5m$ and length $2a$, turns freely about one end which is fixed; to its other extremity is attached one end of a light string, of length $2a$, which carries at its other end a particle of mass m; show that the periods of the small oscillations in a vertical plane are the same as those of simple pendulums of lengths $\frac{2a}{3}$ and $\frac{20a}{7}$.

Solution Let OA be the rod of mass $5m$ and length $2a$. Let AB be the string of length $2a$ which is attached to OA at A, and it carries a particle of mass m at B. Let G be the c.g. of the rod OA.

In Fig. 9.16, let $\mathbf{OA} = \mathbf{r}_1$, $\mathbf{AB} = \mathbf{r}_2$ and $\mathbf{OB} = \mathbf{r}$.

$$\therefore \quad \mathbf{r} = \mathbf{r}_1 + \mathbf{r}_2.$$

The velocity of the particle relative to O is given by

$$\mathbf{v} = \dot{\mathbf{r}} = \dot{\mathbf{r}}_1 + \dot{\mathbf{r}}_2. \tag{9.12.119}$$

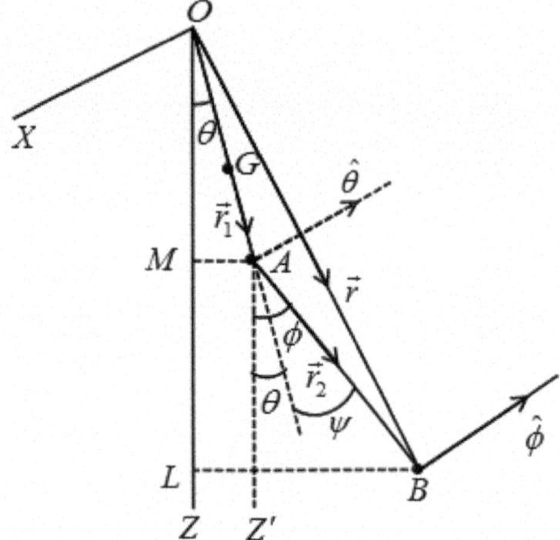

Fig. 9.16 Diagram illustrating Example 9.15

Now

$$\dot{\mathbf{r}}_1 = \frac{d}{dt}\mathbf{OA} = 2a\dot{\theta}\hat{\theta},$$

$$\dot{\mathbf{r}}_2 = \frac{d}{dt}\mathbf{AB} = 2a\dot{\phi}\hat{\phi}.$$

Therefore, (9.12.119) gives

$$\mathbf{v} = 2a\dot{\theta}\hat{\theta} + 2a\dot{\phi}\hat{\phi},$$

where θ and ϕ are angles of inclination of **OA** and **AB** to **OZ** respectively.

$$\therefore \quad v^2 = 4a^2\dot{\theta}^2 + 4a^2\dot{\phi}^2 + 8a^2\dot{\theta}\dot{\phi}\left(\hat{\theta}\cdot\hat{\phi}\right),$$

$$\text{i.e.,} \quad v^2 = 4a^2\dot{\theta}^2 + 4a^2\dot{\phi}^2 + 8a^2\dot{\theta}\dot{\phi}\cos(\phi - \theta),$$

$$\text{i.e.,} \quad v^2 = 4a^2\dot{\theta}^2 + 4a^2\dot{\phi}^2 + 8a^2\dot{\theta}\dot{\phi} \quad \left[\because \quad \cos(\phi - \theta) \approx 1\right],$$

$$\text{i.e.} \quad v^2 = 4a^2\left(\dot{\theta}^2 + \dot{\phi}^2 + 2\dot{\theta}\dot{\phi}\right),$$

$$\text{i.e.,} \quad v^2 = 4a^2\left(\dot{\theta} + \dot{\phi}\right)^2. \tag{9.12.120}$$

Now

$$T_1 = \text{KE of the particle} = \frac{1}{2}m\mathbf{v}^2 = 2ma^2\left(\dot{\theta} + \dot{\phi}\right)^2. \tag{9.12.121}$$

$$\mathbf{v}_0 = \text{Velocity of G relative to O} = \frac{d}{dt}\mathbf{OG} = a\dot{\theta}\hat{\theta}.$$

$$T_2 = \text{KE of the rod OA} = \frac{1}{2}M\mathbf{v}_0^2 + \frac{1}{2}I\omega^2,$$

$$\text{i.e.,} \quad T_2 = \frac{5m}{2}a^2\dot{\theta}^2 + \frac{5m}{6}a^2\dot{\theta}^2 = \frac{10ma^2\dot{\theta}^2}{3}. \tag{9.12.122}$$

Therefore, the KE of the system is given by

$$T = T_1 + T_2 = 2ma^2\left(\dot{\theta} + \dot{\phi}\right)^2 + \frac{10ma^2}{3}\dot{\theta}^2 = \frac{2ma^2}{3}\left[8\dot{\theta}^2 + 6\dot{\theta}\dot{\phi} + 3\dot{\phi}^2\right]. \tag{9.12.123}$$

$$V_1 = \text{PE of the rod OA} = -5mga\cos\theta + C_1,$$

$$V_2 = \text{PE of the particle} = mg\,(-\text{OL}) + C_2 = -mg\,(2a\cos\theta + 2a\cos\phi) + C_2.$$

$$\therefore\quad V = \text{PE of the system} = V_1 + V_2 = -5mga\cos\theta - mg\,(2a\cos\theta + 2a\cos\phi) + \lambda,$$

i.e., $V = -mga\,[5\cos\theta + (2\cos\theta + 2\cos\phi)] + \lambda,$

i.e., $V = -mga\,[7\cos\theta + 2\cos\phi] + \lambda,$

i.e., $V = -mga\left[7\left(1 - \frac{\theta^2}{2!} + \ldots\right) + 2\left(1 - \frac{\phi^2}{2!} + \cdots\right)\right] + \lambda,$

i.e., $$V = \frac{7mga\theta^2}{2} + mga\phi^2 + \mu. \tag{9.12.124}$$

The Lagrangian for the system is given by

$$L = T - V = \frac{2ma^2}{3}\left(8\dot{\theta}^2 + 6\dot{\theta}\dot{\phi} + 3\dot{\phi}^2\right) - \frac{7mga}{2}\theta^2 - mga\phi^2 + \nu. \tag{9.12.125}$$

Here θ and ϕ are the only generalized coordinates. Lagrange's equations are

$$\frac{d}{dt}\left(\frac{\partial L}{\partial \dot{q}_\alpha}\right) = \frac{\partial L}{\partial q_\alpha};\quad \alpha = 1, 2. \tag{9.12.126}$$

Equation (9.12.126) gives

$$\frac{d}{dt}\left(\frac{\partial L}{\partial \dot{q}_1}\right) = \frac{\partial L}{\partial q_1}, \tag{9.12.127}$$

and

$$\frac{d}{dt}\left(\frac{\partial L}{\partial \dot{q}_2}\right) = \frac{\partial L}{\partial q_2}. \tag{9.12.128}$$

Equations (9.12.127) and (9.12.128) give

$$\frac{d}{dt}\left(\frac{\partial L}{\partial \dot{\theta}}\right) = \frac{\partial L}{\partial \theta}, \tag{9.12.129}$$

$$\frac{d}{dt}\left(\frac{\partial L}{\partial \dot{\phi}}\right) = \frac{\partial L}{\partial \phi}. \tag{9.12.130}$$

Now

$$\frac{\partial L}{\partial \dot{\theta}} = \frac{4ma^2}{3}\left(8\dot{\theta} + 3\dot{\phi}\right), \quad \frac{\partial L}{\partial \dot{\phi}} = 4ma^2\left(\dot{\theta} + \dot{\phi}\right).$$

Again

$$\frac{\partial L}{\partial \theta} = -7mga\theta, \quad \frac{\partial L}{\partial \phi} = -2mga\phi.$$

Equation (9.12.129) gives

$$\frac{d}{dt}\left[\frac{4ma^2}{3}\left(8\dot{\theta} + 3\dot{\phi}\right)\right] = -7mga\theta,$$

i.e., $\frac{4}{3}a\left(8\ddot{\theta} + 3\ddot{\phi}\right) = -7g\theta,$

i.e., $$8\ddot{\theta} + 3\ddot{\phi} = -\frac{21g\theta}{4a}. \tag{9.12.131}$$

Equation (9.12.130) gives

$$\frac{d}{dt}\left[4ma^2\left(\dot{\theta} + \dot{\phi}\right)\right] = -2mga\phi,$$

i.e., $2a\left(\ddot{\theta} + \ddot{\phi}\right) = -g\phi,$

i.e., $$\ddot{\theta} + \ddot{\phi} = -\frac{g\phi}{2a}. \tag{9.12.132}$$

Let

$$\theta = M\cos(pt + \varepsilon), \tag{9.12.133}$$

$$\phi = N \cos(pt + \varepsilon). \tag{9.12.134}$$

Equation (8.12.135) gives

$$\dot{\theta} = -Mp \sin(pt + \varepsilon).$$

$$\therefore \quad \ddot{\theta} = -Mp^2 \cos(pt + \varepsilon) = -p^2\theta. \tag{9.12.135}$$

Similarly

$$\ddot{\phi} = -p^2\phi. \tag{9.12.136}$$

Substituting (9.12.135) and (9.12.136) in (9.12.131) and (9.12.132), we get

$$8(-p^2\theta) + 3(-p^2\phi) = -\frac{21g\theta}{4a}$$

$$\text{i.e.,} \quad 8p^2\theta + 3p^2\phi = \frac{21g\theta}{4a},$$

$$\text{i.e.,} \quad \left(8p^2 - \frac{21g}{4a}\right)\theta + 3p^2\phi = 0. \tag{9.12.137}$$

and

$$-p^2\theta - p^2\phi = -\frac{g\phi}{2a},$$

$$\text{i.e.,} \quad p^2\theta + p^2\phi = \frac{g\phi}{2a},$$

$$\text{i.e.,} \quad p^2\theta + \left(p^2 - \frac{g}{2a}\right)\phi = 0. \tag{9.12.138}$$

Eliminating θ and ϕ from (9.12.137) and (9.12.138), we get

$$\begin{vmatrix} 8p^2 - \frac{21g}{4a} & 3p^2 \\ p^2 & p^2 - \frac{g}{2a} \end{vmatrix} = 0,$$

$$\text{i.e.,} \quad \left(p^2 - \frac{g}{2a}\right)\left(8p^2 - \frac{21g}{4a}\right) - 3p^4 = 0,$$

$$\text{i.e.,} \quad (2ap^2 - g)\,(32ap^2 - 21g) = 24a^2p^4,$$

$$\text{i.e.,}\quad 64a^2p^4 - 42a\,p^2g - 32ap^2g + 21g = 24a^2p^4,$$

$$\text{i.e.,}\quad 40a^2p^4 - 74a\,p^2g + 21g^2 = 0,$$

$$\text{i.e.,}\quad (20ap^2 - 7g)\,(2ap^2 - 3g) = 0,$$

$$\text{i.e.,}\quad p^2 = \frac{7g}{20a},\ p^2 = \frac{3g}{2a},$$

$$\text{i.e.,}\quad p^2 = \frac{3g}{2a} = p_1^2,\ p^2 = \frac{7g}{20a} = p_2^2.$$

Thus, the periods of oscillations are

$$\tau_1 = \frac{2\pi}{p_1} \& \tau_2 = \frac{2\pi}{p_2},$$

$$\text{i.e.,}\quad \tau_1 = \frac{2\pi}{\sqrt{\frac{3g}{2a}}} \& \tau_2 = \frac{2\pi}{\sqrt{\frac{7g}{20a}}},$$

$$\text{i.e.,}\quad \tau_1 = 2\pi\sqrt{\frac{2a}{3g}} \& \tau_2 = 2\pi\sqrt{\frac{20a}{7g}},$$

$$\text{i.e.,}\quad \tau_1 = 2\pi\sqrt{\frac{l_1}{g}} \& \tau_2 = 2\pi\sqrt{\frac{l_2}{g}},$$

where

$$l_1 = \frac{2a}{3} \& l_2 = \frac{20a}{7}.$$

Now,

$$\tau_1 = \text{period of oscillation of a simple pendulum of length } l_1 = \frac{2a}{3},$$

and

$$\tau_2 = \text{period of oscillation of a simple pendulum of length } l_2 = \frac{20a}{7}.$$

9.13 Exercise-IX

1. Set up the Lagrangian for a compound pendulum and obtain an equation that describes the motion of the body.
2. A uniform rod of mass m is moving in a plane, its longitudinal velocity being w, and the transverse velocities of its end points being u and v, show that the KE of the rod is $T = \frac{1}{6}m\left(u^2 + v^2 + uv + 3w^2\right)$.
3. The end points of a uniform rod of mass m are moving $\perp^r$ to the line of the rod in the same direction with velocities u, v; prove that the KE of the rod is $\frac{1}{6}m\left(u^2 + uv + v^2\right)$.
4. Three equal uniform rods PQ, QR, RS, each of mass m and length $2a$, are at rest in a straight line smoothly jointed at Q and R. A blow J is given to the middle rod at a distance $\frac{a}{2}$ from its center O in a direction $\perp^r$ to it; show that the initial angular velocities of the rods are as 19 : 12 : 1.
5. Two rods, AB and BC, of length $2a$ and $2b$ and of masses proportional to their lengths, are freely jointed at B and lie in a straight line. A blow is communicated to the end A; show that the resulting KE when the system is free is to the energy when C is fixed as $(4a + 3b)\ (4b + 3c) : 12(a + b)^2$.
6. AB, BC, CD, DE are four equal uniform rods freely jointed at B, C, D, and at rest in a straight line on a smooth horizontal table. If an impulsive couple G is applied to the rod BC in the plane of the table, find the initial velocities of the ends of the rods, taking m to be the mass and $2a$ the length of a rod.
Answer: $u_A = \frac{33G}{56ma},\ u_B = -\frac{33G}{28ma},\ u_C = \frac{9G}{8ma},\ u_D = -\frac{9G}{28ma},\ u_E = \frac{9G}{56ma}$.
7. Three uniform rods AB, BC, CD each of mass m and equal length, are smoothly jointed at B and C and are placed in a st. line. The rods are set in motion from rest by an impulse I applied at A at right angles to AB. Show that the KE generated is $\frac{26I^2}{15m}$ if the rods are free, and is less than this by an amount $\frac{I^2}{390m}$ if the end D is fixed.
8. A uniform rod, of length $2a$ which has one and attached to a fixed point by a light inextensible string, of length $\frac{5a}{12}$, is performing small oscillations in a vertical plane about its position of equilibrium. Shown that the periods of its principal oscillation are $2\pi\sqrt{\frac{5a}{3g}}$ and $\pi\sqrt{\frac{a}{3g}}$.
9. A uniform heavy rod AB of mass m and length $12a$ is suspended from a fixed point by a light string OC of length $5a$ attached to a point C of the rod so that AC=$4a$. The system makes small oscillations about its position of equilibrium in a vertical plane through O, and OC, AB make angles q_1, q_2 respectively, with the vertical at time t. Prove that the KE of the system is $\frac{1}{2}ma^2\left(25\dot{q}_1^2 + 20\dot{q}_1\dot{q}_2 + 16\dot{q}_2^2\right)$ and that the normal periods of oscillation are $\frac{2\pi}{p_1}$ and $\frac{2\pi}{p_2}$, where $3ap_1^2 = g$ and $10ap_2^2 = g$.
10. A homogeneous rod OA, of mass m_1 and length $2a$, is freely hinged at O to a fixed point; at its other end is freely attached another homogeneous rod AB, of mass m_2 and length $2b$; the system moves under gravity; find the Lagrangian and hence obtain Lagrange's equations to determine the motion.

Chapter 10
Theory of Hamilton

10.1 Definition

Let a dynamical system be specified by n-generalized coordinates q_α $(\alpha = 1, 2, 3, \ldots, n)$ with p_α $(\alpha = 1, 2, 3, \ldots, n)$ as the generalized momenta. Then the Hamilton of the system, denoted H, is defined by $H = \sum\limits_{\alpha=1}^{n} p_\alpha \dot{q}_\alpha - L$, where $L = T - V$ is the Lagrangian of the system. On use of Lagrange's equations $\dot{q}_\alpha$ may be expressed in terms of q_α, p_α, t.

Thus,

$$H = H\left(p_\alpha, q_\alpha, t\right), \ \alpha = 1, 2, 3, \ldots, n.$$

10.2 Hamilton's Equations

Statement:

$$\dot{p}_\alpha = -\frac{\partial H}{\partial q_\alpha}, \dot{q}_\alpha = \frac{\partial H}{\partial p_\alpha}, \frac{\partial L}{\partial t} = -\frac{\partial H}{\partial t}; \ \alpha = 1, 2, 3, \ldots, n.$$

Derivation: By definition,

$$H = \sum_{\alpha=1}^{n} p_\alpha \dot{q}_\alpha - L. \tag{10.2.1}$$

N. Ahmed et al., *Classical Dynamics*, University Texts in the Mathematical Sciences,
https://doi.org/10.1007/978-981-95-6394-4_10

Equation (10.2.1) gives

$$dH = \sum_{\alpha=1}^{n} [p_\alpha d\dot{q}_\alpha + \dot{q}_\alpha dp_\alpha] - dL. \tag{10.2.2}$$

We have

$$L = L(q_\alpha, \dot{q}_\alpha, t)\,;\ \alpha = 1, 2, 3, \ldots, n.$$

$$\therefore\quad dL = \sum_{\alpha=1}^{n} \frac{\partial L}{\partial q_\alpha} dq_\alpha + \sum_{\alpha=1}^{n} \frac{\partial L}{\partial \dot{q}_\alpha} d\dot{q}_\alpha + \frac{\partial L}{\partial t} dt. \tag{10.2.3}$$

Equation (10.2.3) leads (10.2.2) to take the form

$$dH = \sum_{\alpha=1}^{n} [p_\alpha d\dot{q}_\alpha + \dot{q}_\alpha dp_\alpha] - \sum_{\alpha=1}^{n} \frac{\partial L}{\partial q_\alpha} dq_\alpha - \sum_{\alpha=1}^{n} \frac{\partial L}{\partial \dot{q}_\alpha} d\dot{q}_\alpha - \frac{\partial L}{\partial t} dt. \tag{10.2.4}$$

Now,

$$\frac{\partial L}{\partial \dot{q}_\alpha} = \frac{\partial}{\partial \dot{q}_\alpha}(T - V) = \frac{\partial T}{\partial \dot{q}_\alpha} - \frac{\partial V}{\partial \dot{q}_\alpha} = \frac{\partial T}{\partial \dot{q}_\alpha} = p_\alpha. \tag{10.2.5}$$

Lagrange's Equations give

$$\frac{d}{dt}\left(\frac{\partial L}{\partial \dot{q}_\alpha}\right) = \frac{\partial L}{\partial q_\alpha};\ \alpha = 1, 2, 3, \ldots, n. \tag{10.2.6}$$

Equation (10.2.6) depicts

$$\frac{\partial L}{\partial q_\alpha} = \dot{p}_\alpha.$$

Replacing $\frac{\partial L}{\partial q_\alpha}$ by $\dot{p}_\alpha$ and $\frac{\partial L}{\partial \dot{q}_\alpha}$ by p_α in (10.2.4), we obtain

$$dH = \sum_{\alpha=1}^{n} p_\alpha d\dot{q}_\alpha + \sum_{\alpha=1}^{n} \dot{q}_\alpha dp_\alpha - \sum_{\alpha=1}^{n} \dot{p}_\alpha dq_\alpha - \sum_{\alpha=1}^{n} p_\alpha d\dot{q}_\alpha - \frac{\partial L}{\partial t} dt,$$

$$\text{i.e.,}\quad dH = \sum_{\alpha=1}^{n} \dot{q}_\alpha dp_\alpha - \sum_{\alpha=1}^{n} \dot{p}_\alpha dq_\alpha - \frac{\partial L}{\partial t} dt. \tag{10.2.7}$$

Again $\because \quad H = H(p_\alpha, q_\alpha, t)\,;\ \alpha = 1, 2, 3, \ldots, n,$

$$\therefore \quad dH = \sum_{\alpha=1}^{n} \frac{\partial H}{\partial p_\alpha} dp_\alpha + \sum_{\alpha=1}^{n} \frac{\partial H}{\partial q_\alpha} dq_\alpha + \frac{\partial H}{\partial t} dt. \tag{10.2.8}$$

Equations (10.2.7) and (10.2.8) give

$$\sum_{\alpha=1}^{n} \frac{\partial H}{\partial p_\alpha} dp_\alpha + \sum_{\alpha=1}^{n} \frac{\partial H}{\partial q_\alpha} dq_\alpha + \frac{\partial H}{\partial t} dt = \sum_{\alpha=1}^{n} \dot{q}_\alpha dp_\alpha - \sum_{\alpha=1}^{n} \dot{p}_\alpha dq_\alpha - \frac{\partial L}{\partial t} dt. \tag{10.2.9}$$

Equating the coefficients of dp_α and dq_α in (10.2.9), we derive,

$$\dot{p}_\alpha = -\frac{\partial H}{\partial q_\alpha}, \dot{q}_\alpha = \frac{\partial H}{\partial p_\alpha}, \frac{\partial L}{\partial t} = -\frac{\partial H}{\partial t}; \ \alpha = 1, 2, 3, \ldots, n. \tag{10.2.10}$$

These equations are termed the Hamiltonian Equations. If H does not contain t explicitly, the Hamiltonian Equations become

$$\dot{p}_\alpha = -\frac{\partial H}{\partial q_\alpha}, \ \dot{q}_\alpha = \frac{\partial H}{\partial p_\alpha}; \quad \alpha = 1, 2, 3, \ldots, n.$$

10.3 Hamiltonian for Conservative Systems

By definition,

$$H = \sum_{\alpha=1}^{n} p_\alpha \dot{q}_\alpha - L = \sum_{\alpha=1}^{n} \frac{\partial T}{\partial \dot{q}_\alpha} \dot{q}_\alpha - L. \tag{10.3.1}$$

Again

$$T = \frac{1}{2} \sum_{r=1}^{n} \sum_{s=1}^{n} a_{rs} \dot{q}_r \dot{q}_s \tag{10.3.2}$$

is a homogeneous function of degree 2 in $\dot{q}_\alpha$ $(\alpha = 1, 2, 3, \ldots, n)$,

$$\text{i.e.,} \quad T = T(\dot{q}_1, \dot{q}_2, \dot{q}_3, \ldots, \dot{q}_n) \tag{10.3.3}$$

is a homogeneous function of degree 2 in $\dot{q}_1, \dot{q}_2, \dot{q}_3, \ldots, \dot{q}_n$. Thus, by Euler's theorem on homogeneous functions, we have

$$\dot{q}_1 \frac{\partial T}{\partial \dot{q}_1} + \dot{q}_2 \frac{\partial T}{\partial \dot{q}_2} + \dot{q}_3 \frac{\partial T}{\partial \dot{q}_3} + \cdots + \dot{q}_n \frac{\partial T}{\partial \dot{q}_n} = 2T,$$

$$\sum_{\alpha=1}^{n} \dot{q}_\alpha \frac{\partial T}{\partial \dot{q}_\alpha} = 2T. \tag{10.3.4}$$

Equation (10.3.1) gives

$$H = 2T - L = 2T - (T - V) = T + V = E = \text{Total energy of the system} = a \text{ constant.}$$

10.4 The Calculus of Variations

10.4.1 Definition

A problem that often arises in Mathematics is that of finding a curve $y = \bar{y}(x)$ joining the points where $x = a$ and $x = b$ such that the integral $\int_a^b F(x, y, y')\,dx$, where $y' = \frac{dy}{dx}$, is a maximum or minimum, also called an extremum or extreme value. The curve itself is often called an extremal.

Theorem 10.1 *A necessary condition for* $I = \int_a^b F(x, y, y')\,dx$ *to be extremum is*

$$\frac{d}{dx}\left(\frac{\partial F}{\partial y'}\right) = \frac{\partial F}{\partial y}.$$

Proof Suppose that the curve $\bar{C}$ which makes I an extremum is given by

$$y = \bar{y}(x) : a \le x \le b. \tag{10.4.1}$$

Consider a neighboring curve C joining A and B given by (see Fig. 10.1)

$$y = \bar{y}(x) + \varepsilon\eta(x), \tag{10.4.2}$$

where ε is a small parameter and $\eta(x)$ is an arbitrary continuous differential function of x satisfying $\eta(a) = 0$ and $\eta(b) = 0$. The value of I for the curve C is

Fig. 10.1 Diagram illustrating Theorem 10.1

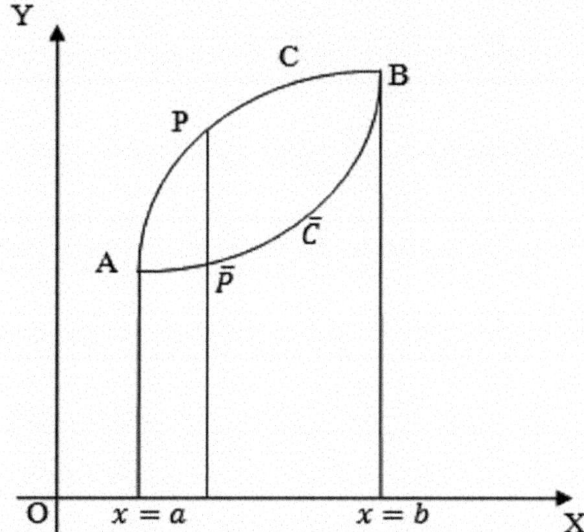

$$I(\varepsilon) = \int_a^b F\left(x, \bar{y} + \varepsilon\eta, \bar{y}' + \varepsilon\eta'\right) dx. \tag{10.4.3}$$

This is an extremum for $\varepsilon = 0$,

$$\therefore \quad \left.\frac{dI}{d\varepsilon}\right]_{\varepsilon=0} = 0,$$

$$\text{i.e.,} \quad I'(0) = 0. \tag{10.4.4}$$

Now

$$I'(\varepsilon) = \int_a^b \left[F_y\left(x, \bar{y} + \varepsilon\eta, \bar{y}' + \varepsilon\eta'\right)\eta + F_{y'}\left(x, \bar{y} + \varepsilon\eta, \bar{y}' + \varepsilon\eta'\right)\eta'\right] dx,$$

$$I'(0) = \int_a^b \left[\eta F_y\left(x, \bar{y}, \bar{y}'\right) + \eta' F_{y'}\left(x, \bar{y}, \bar{y}'\right)\right] dx.$$

$$\int_a^b \left[\eta F_y \left(x, \bar{y}, \bar{y}'\right) + \eta' F_{y'} \left(x, \bar{y}, \bar{y}'\right) \right] dx = 0. \tag{10.4.5}$$

Now

$$\int_a^b \eta' F_{y'} \left(x, \bar{y}, \bar{y}'\right) dx = F_{y'} \left(x, \bar{y}, \bar{y}'\right) \eta\big]_a^b - \int_a^b \eta \frac{d}{dx} F_{y'} \left(x, \bar{y}, \bar{y}'\right) dx = -\int_a^b \eta \frac{d}{dx} F_{y'} \left(x, \bar{y}, \bar{y}'\right) dx.$$

Thus (10.4.5) reduces to

$$\int_a^b \left[\eta F_y - \eta \frac{d}{dx} F_{y'} \right] dx = 0,$$

$$i.e. \quad \int_a^b \eta \left[\frac{d}{dx} F_{y'} - F_y \right] dx = 0. \tag{10.4.6}$$

Since η is arbitrary, Eq. (10.4.6) gives

$$\frac{d}{dx} \left(\frac{\partial F}{\partial y'} \right) = \frac{\partial F}{\partial y}, \tag{10.4.7}$$

which is known as the Euler–Lagrange equation. By using the Taylor series expansion, we find that

$$I(\varepsilon) = I(0) + \varepsilon I'(0) + \text{higher order terms in } \varepsilon^2, \varepsilon^3, \text{etc.,}$$

$$\text{i.e.,} \quad I(\varepsilon) - I(0) = \varepsilon \int_a^b \left[\eta F_y + \eta' F_{y'} \right] dx + O\left(\varepsilon^2\right). \tag{10.4.8}$$

The coefficient of ε in (10.4.8) is often called the variation of the integral and it is denoted by $\delta \int_a^b F\left(x, y, y'\right) dx$,

$$\therefore \quad \delta \int_a^b F\left(x, y, y'\right) dx = \int_a^b \left[\eta F_y + \eta' F_{y'} \right] dx = I'(0).$$

But $\int_a^b F\left(x, y, y'\right)\, dx$ is an extremum if $I'(0) = 0$,

$$\text{i.e.,}\quad \delta \int_a^b F\left(x, y, y'\right)\, dx = 0.$$

□

Remark

The above result may easily be extended to the integral

$$\int_a^b F\left(x, y_1, y'_1, y_2, y'_2, \ldots, y_n, y'_n\right)\, dx$$

and it leads Euler–Lagrange Equations:

$$\frac{d}{dx}\left(\frac{\partial F}{\partial y'_\alpha}\right) = \frac{\partial F}{\partial y_\alpha};\ \alpha = 1, 2, \ldots, n.$$

10.5 Hamilton's Principle

By identifying the function $F\left(x, y, y'\right)$ with the Lagrangian $L\,(t, q, \dot{q})$ where x, y, y' are replaced by $t, q, \dot{q}$ respectively, we see that a necessary and sufficient condition for the integral $\int_{t_1}^{t_2} L dt$ to be extremum is given by

$$\frac{d}{dt}\left(\frac{\partial L}{\partial \dot{q}}\right) = \frac{\partial L}{\partial q}.$$

For systems involving n degree of freedom, we consider the integral

$$\int_{t_1}^{t_2} L\,(t, q_1, \dot{q}_1, q_2, \dot{q}_2, \ldots, q_n, \dot{q}_n)\, dt.$$

The necessary conditions for this integral to be extremum are

$$\frac{d}{dt}\left(\frac{\partial L}{\partial \dot{q}_\alpha}\right) = \frac{\partial L}{\partial q_\alpha};\quad \alpha = 1, 2, \ldots, n,$$

which are Lagrange's equations for a conservative system. Thus,

$$\delta \int_{t_1}^{t_2} L\left(t, q_\alpha, \dot{q}_\alpha\right) dt = 0$$

if

$$\frac{d}{dt}\left(\frac{\partial L}{\partial \dot{q}_\alpha}\right) = \frac{\partial L}{\partial q_\alpha};\ \alpha = 1, 2, \ldots, n.$$

Hence, Lagrange's Equations for a conservative holonomic system can be obtained by considering $\int_{t_1}^{t_2} L dt$ to be extremum. i.e., $\delta \int_{t_1}^{t_2} L dt = 0$.

Thus, we conclude that a conservative system moves from t_1 to t_2 in such a way that $I = \int_{t_1}^{t_2} L dt$ has an extreme value, i.e., $\delta \int_{t_1}^{t_2} L dt = 0$.

This principle is known as Hamilton's principle. Moreover, since the extreme value of I is often minimum, therefore the above principle is also termed Hamilton's principle of least action.

10.6 The Brachistochrone Problem

Statement Suppose A and B are any two points in space but not in the same vertical line. Let A and B be connected by a plane curve C, and suppose A is at a higher level than B. The problem of finding the form of C, which ensures that the time of fall from A to B under the gravity of a smooth particle along it is a minimum, starting from rest at A, is called the Brachistochrone problem.

Solution Consider a two-dimensional Cartesian coordinate system with the X-axis oriented horizontally on the plane and the Y-axis directed vertically downward, with point A as the origin, as shown in Fig. 10.2. Let (x_0, y_0) be the coordinates of B. Let P(x, y) be any position of the particle.

Because gravity is the only external force acting on the particle, the total energy of the particle is conserved.

$$\text{i.e.,}\quad T + V = \text{Constant},$$

$$\text{i.e.,}\quad 0 + mgy_0 = \frac{1}{2}mv^2 + mg\left(y_0 - y\right)\quad \text{(Taking B as the standard position)},$$

$$\text{i.e.,}\quad \frac{1}{2}mv^2 = mgy,$$

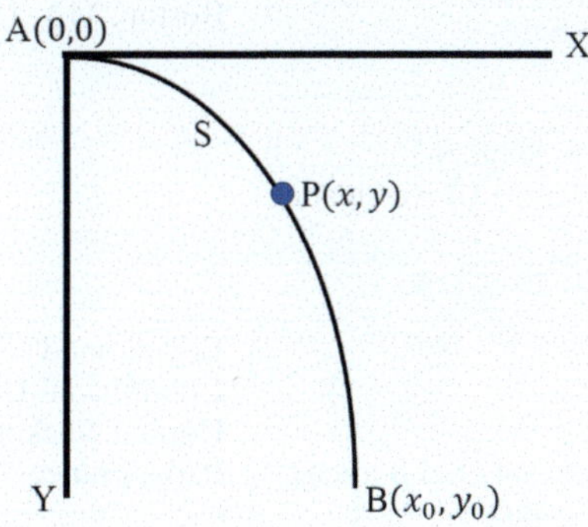

Fig. 10.2 Diagram illustrating the Brachistochrone Problem

$$\text{i.e.,}\quad \frac{1}{2}m\left(\frac{ds}{dt}\right)^2 = mgy,$$

$$\text{i.e.,}\quad \left(\frac{ds}{dt}\right)^2 = 2gy,$$

$$\text{i.e.,}\quad \frac{ds}{dt} = \sqrt{2gy},$$

$$\text{i.e.,}\quad dt = \frac{ds}{\sqrt{2gy}}.$$

Thus, the total time taken to go from A (0, 0) to B (x_0, y_0) along the curve C is given by

$$\tau = \int \frac{ds}{\sqrt{2gy}}. \tag{10.6.1}$$

But $ds^2 = dx^2 + dy^2$,

$$\therefore \quad ds = \sqrt{1 + y'^2}dx. \tag{10.6.2}$$

Equation (10.6.1) gives

$$\tau = \frac{1}{\sqrt{2g}} \int\limits_{x=0}^{x_0} \frac{\sqrt{1 + y'^2}}{\sqrt{y}}dx. \tag{10.6.3}$$

Now to find C so that τ is the least. For the minimum value of τ, we must have

$$\frac{d}{dx}\left(\frac{\partial F}{\partial y'}\right) = \frac{\partial F}{\partial y}, \tag{10.6.4}$$

where

$$F = \sqrt{\frac{(1 + y'^2)}{y}},$$

$$\therefore \quad \frac{\partial F}{\partial y'} = \frac{y'}{\sqrt{y\left(1 + y'^2\right)}},$$

$$\text{and so } \frac{d}{dx}\left(\frac{\partial F}{\partial y'}\right) = \frac{2yy'' - y'^2 - y'^4}{2\{y\left(1 + y'^2\right)\}^{\frac{3}{2}}},$$

$$\frac{\partial F}{\partial y} = -\frac{\sqrt{1 + y'^2}}{2y\sqrt{y}}.$$

Equation (10.6.4) gives

$$\frac{2yy'' - y'^2 - y'^4}{2\{y\left(1 + y'^2\right)\}^{\frac{3}{2}}} = -\frac{\sqrt{1 + y'^2}}{2y\sqrt{y}},$$

$$\text{i.e., } 2yy'' - y'^2 - y'^4 = -\left(1 + y'^2\right)^2,$$

$$\text{i.e., } 2yy'' - y'^2 - y'^4 = -\left(1 + 2y'^2 + y'^4\right),$$

$$\text{i.e., } 1 + y'^2 + 2yy'' = 0. \tag{10.6.5}$$

To solve (10.6.5), put $y' = u$,

$$\therefore \quad y'' = \frac{du}{dx} = \frac{du}{dy}\frac{dy}{dx} = \frac{du}{dy}y' = u\frac{du}{dy}.$$

Hence (10.6.5) reduces to

$$1 + u^2 + 2yu\frac{du}{dy} = 0,$$

$$\text{i.e., } \frac{2udu}{1 + u^2} + \frac{dy}{y} = 0,$$

$$\text{i.e., } \log\left(1 + u^2\right) + \log y = \log b,$$

i.e., $y\left(1+u^2\right)=b,$

i.e., $y'^2=\dfrac{b-y}{y},$

i.e., $y'=\sqrt{\dfrac{b-y}{y}},$

i.e., $\dfrac{dy}{dx}=\sqrt{\dfrac{b-y}{y}},$

i.e., $\sqrt{\dfrac{y}{b-y}}dy=dx,$

i.e., $x=I+c,$ (10.6.6)

where $I=\displaystyle\int\sqrt{\frac{y}{b-y}}dy.$

Put

$$y=b\sin^2\theta, \tag{10.6.7}$$

$\therefore\ dy=2b\sin\theta\cos\theta\,d\theta.$

$$\therefore\ I=\int\sqrt{\frac{b\sin^2\theta}{b\cos^2\theta}}\,2b\sin\theta\cos\theta\,d\theta=2b\int\sin^2\theta\,d\theta=\frac{b}{2}\left(2\theta-\sin 2\theta\right).$$

Let $\theta=\alpha$ at $y=0$.

i.e., $0=b\sin^2\alpha,$

i.e., $\alpha=0.$

i.e., $y=0$ at $\theta=\alpha=0.$

Equation (10.6.6) gives

$$x=\frac{b}{2}\left(2\theta-\sin 2\theta\right)+c. \tag{10.6.8}$$

But $x=0$ when y=0,

$$\text{i.e.,}\quad x = 0 \quad \text{when} \quad \theta = \alpha = 0. \tag{10.6.9}$$

Subjecting (10.6.9) to (10.6.8), we obtain $c = 0$. Hence (10.6.8) gives

$$x = \frac{b}{2}\left(2\theta - \sin 2\theta\right). \tag{10.6.10}$$

Thus, we have

$$\left.\begin{array}{l} x = a\left(\varphi - \sin\varphi\right) \\ y = a\left(1 - \cos\varphi\right) \end{array}\right\}; \quad \varphi = 2\theta,\, a = \frac{b}{2}. \tag{10.6.11}$$

Equations (10.6.11) represent a cycloid. Hence, the curve along which the particle takes the minimum time is a cycloid.

10.7 Illustrations

Example 10.1 A particle moves in the xy-plane under the action of the force $\frac{\mu}{r^2}$ directed towards the origin. Set up the Hamiltonian for the system. Also, write down Hamilton's Equations.

Solution Let O be the origin, which is the center of force. Let P be the position of the particle at time t with polar coordinates (r, θ) referred to O as the pole and **OX** as an initial line, as shown in Fig. 10.3.

The force acting on the particle at P is given by

$$\mathbf{F} = -\frac{\mu}{r^2}\hat{r} = -\frac{\mu}{r^3}\mathbf{r}. \tag{10.7.1}$$

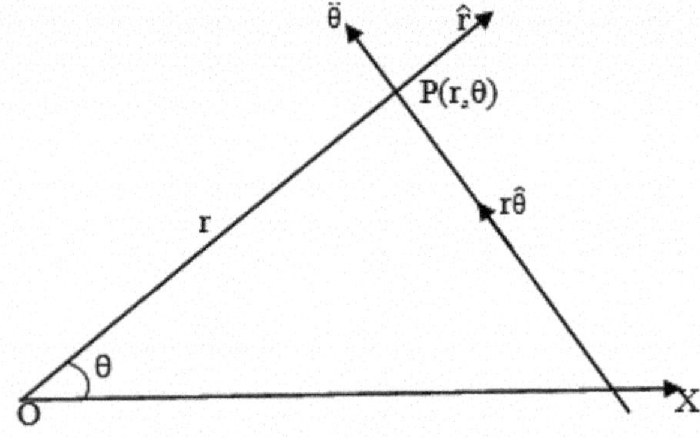

Fig. 10.3 Diagram illustrating Example 10.1

The potential energy of the particle at P is specified by

$$-\nabla V = -\frac{\mu}{r^3}\mathbf{r},$$

$$\text{i.e.,}\quad \nabla V \cdot d\mathbf{r} = \frac{\mu}{r^3}\mathbf{r} \cdot d\mathbf{r},$$

$$\text{i.e.,}\quad dV = \frac{\mu}{r^2}dr,$$

$$\text{i.e.,}\quad V = -\frac{\mu}{r}. \tag{10.7.2}$$

The KE of the particle at time t is quantified by

$$T = \frac{1}{2}m\mathbf{v}^2 = \frac{1}{2}m\left(v_r^2 + v_\theta^2\right) = \frac{1}{2}m\left(\dot{r}^2 + r^2\dot{\theta}^2\right). \tag{10.7.3}$$

The Lagrangian is given by

$$L = T - V = \frac{1}{2}m\left(\dot{r}^2 + r^2\dot{\theta}^2\right) + \frac{\mu}{r}. \tag{10.7.4}$$

Here r and θ are the only generalized coordinates.
We have

$$p_\alpha = \frac{\partial T}{\partial \dot{q}_\alpha}; \quad \alpha = 1, 2. \tag{10.7.5}$$

Equation (10.7.5) give

$$p_r = \frac{\partial T}{\partial \dot{r}} = m\dot{r}, \quad p_\theta = \frac{\partial T}{\partial \dot{\theta}} = mr^2\dot{\theta};$$

$$\text{i.e.,}\quad \dot{r} = \frac{p_r}{m}, \quad \dot{\theta} = \frac{p_\theta}{mr^2}. \tag{10.7.6}$$

As the central force is conservative,

$$\therefore \quad H = T + V = \frac{1}{2}m\left(\dot{r}^2 + r^2\dot{\theta}^2\right) - \frac{\mu}{r}. \tag{10.7.7}$$

By use of (10.7.6), (10.7.7) reduces to

$$H = \frac{1}{2}m\left(\frac{p_r^2}{m^2} + r^2\frac{p_\theta^2}{m^2r^4}\right) - \frac{\mu}{r} = \frac{1}{2m}\left(p_r^2 + \frac{p_\theta^2}{r^2}\right) - \frac{\mu}{r}. \tag{10.7.8}$$

Hamilton's Equations are

$$\left.\begin{aligned} \dot{p}_\alpha &= -\frac{\partial H}{\partial q_\alpha} \\ \dot{q}_\alpha &= \frac{\partial H}{\partial p_\alpha} \end{aligned}\right\}; \quad \alpha = 1, 2. \tag{10.7.9}$$

1st Equation (10.7.9) gives

$$\left.\begin{aligned} \dot{p}_r &= -\frac{\partial H}{\partial r} = \frac{p_\theta^2}{mr^3} - \frac{\mu}{r^2} \\ \dot{p}_\theta &= -\frac{\partial H}{\partial \theta} = 0 \end{aligned}\right\}. \tag{10.7.10}$$

2nd Equation of (10.7.9) gives

$$\left.\begin{aligned} \dot{r} &= \frac{\partial H}{\partial p_r} = \frac{p_r}{m} \\ \dot{\theta} &= \frac{\partial H}{\partial p_\theta} = \frac{p_\theta}{mr^2} \end{aligned}\right\}. \tag{10.7.11}$$

Equations (10.7.10) and (10.7.11) are Hamilton's Equations.

Example 10.2 Write the Hamiltonian function for the motion of a compound pendulum and hence obtain the equation of motion for a compound pendulum.

Solution Let O be the center of suspension, G the position of the c.g., OZ the horizontal axis of rotation, and OA be the downward vertical through O. Let θ be the angle between the line OA fixed in space and line OG fixed in the body.

The KE of the body is given by

$$T = \frac{1}{2}I\omega^2 = \frac{1}{2}MK^2\dot{\theta}^2, \tag{10.7.12}$$

where M is the mass of the body, K is the radius of gyration of the body about OZ, the axis of rotation. The PE of the body is given by

$$V = -Mgh\cos\theta; \quad h = OG. \tag{10.7.13}$$

The Hamiltonian is given by

$$H = T + V = \frac{1}{2}MK^2\dot{\theta}^2 - Mgh\cos\theta. \tag{10.7.14}$$

Here, θ is the only generalized coordinate. The generalized momentum is given by

$$p_\alpha = \frac{\partial T}{\partial \dot{q}_\alpha}; \quad \alpha = 1.$$

$$\therefore \quad p_\theta = \frac{\partial T}{\partial \dot{\theta}} = MK^2\dot{\theta},$$

$$\text{i.e.,} \quad \dot{\theta} = \frac{p_\theta}{MK^2}. \tag{10.7.15}$$

Equations (10.7.14) and (10.7.15) give

$$H = \frac{p_\theta^2}{2MK^2} - Mgh\cos\theta. \tag{10.7.16}$$

Hamilton's Equations are

$$\dot{p}_\alpha = -\frac{\partial H}{\partial q_\alpha}, \quad \dot{q}_\alpha = \frac{\partial H}{\partial p_\alpha}; \quad \alpha = 1. \tag{10.7.17}$$

1st Equation of (10.7.17) gives

$$\dot{p}_\theta = -\frac{\partial H}{\partial \theta} = -Mgh\sin\theta. \tag{10.7.18}$$

2nd Equation of (10.7.17) gives

$$\dot{\theta} = \frac{\partial H}{\partial p_\theta} = \frac{p_\theta}{MK^2}. \tag{10.7.19}$$

Equation (10.7.19) yields

$$\ddot{\theta} = \frac{\dot{p}_\theta}{MK^2} = \frac{-Mgh\sin\theta}{MK^2} = -\frac{gh\sin\theta}{K^2},$$

$$\text{i.e.,} \quad \ddot{\theta} + \frac{gh\sin\theta}{K^2} = 0,$$

which is the equation of motion for a compound pendulum.

Example 10.3 If $H(p, q) = \frac{1}{2}\left(p^2 + \lambda^2 q^2\right)$ and q and p are transformed into Q, P by means of relations $q = (2Q)^{\frac{1}{2}}\lambda^{-\frac{1}{2}}\cos P$, $p = (2Q)^{\frac{1}{2}}\lambda^{\frac{1}{2}}\sin P$, show that the system of Hamilton's in (q, p) is changed again into another such system with the Hamiltonian equal to λQ.

Solution The given Hamiltonian in p, q system is

$$H(p, q) = \frac{1}{2}\left(p^2 + \lambda^2 q^2\right). \tag{10.7.20}$$

Given that

$$p = (2Q)^{\frac{1}{2}} \lambda^{\frac{1}{2}} \sin P, \tag{10.7.21}$$

$$\lambda q = (2Q)^{\frac{1}{2}} \lambda^{\frac{1}{2}} \cos P. \tag{10.7.22}$$

Squaring and adding (10.7.21) and (10.7.22), we get

$$p^2 + \lambda^2 q^2 = 2\lambda Q. \tag{10.7.23}$$

Dividing Eq. (10.7.21) by Eq. (10.7.22), which gives

$$\frac{p}{\lambda q} = \tan P,$$

$$\text{i.e.,} \quad P = \tan^{-1}\left(\frac{p}{\lambda q}\right). \tag{10.7.24}$$

Now

$$H(p, q) = \frac{1}{2}\left(p^2 + \lambda^2 q^2\right) = \lambda Q = H'(P, Q) \text{ (say)}.$$

Therefore, the new Hamiltonian is $H' = \lambda Q$. Hamilton's equations are

$$\left.\begin{array}{l} \dot{p} = -\frac{\partial H}{\partial q} \\ \dot{q} = \frac{\partial H}{\partial p} \end{array}\right\}. \tag{10.7.25}$$

Now

$$\frac{\partial H}{\partial q} = \frac{\partial H'}{\partial q} = \frac{\partial H'}{\partial P}\frac{\partial P}{\partial q} + \frac{\partial H'}{\partial Q}\frac{\partial Q}{\partial q} = -\frac{\lambda p}{p^2 + \lambda^2 q^2}\frac{\partial H'}{\partial P} + \lambda q \frac{\partial H'}{\partial Q}, \tag{10.7.26}$$

$$\frac{\partial H}{\partial p}=\frac{\partial H'}{\partial p}=\frac{\partial H'}{\partial P}\frac{\partial P}{\partial p}+\frac{\partial H'}{\partial Q}\frac{\partial Q}{\partial p}=\frac{\lambda q}{p^2+\lambda^2q^2}\frac{\partial H'}{\partial P}+\frac{p}{\lambda}\frac{\partial H'}{\partial Q}. \tag{10.7.27}$$

Again,

$$\dot{p}=\frac{\partial p}{\partial P}\dot{P}+\frac{\partial p}{\partial Q}\dot{Q}=(2Q)^{\frac{1}{2}}\lambda^{\frac{1}{2}}Cos P\,\dot{P}+\lambda^{\frac{1}{2}}Sin P.\frac{1}{2}(2Q)^{-\frac{1}{2}}.2\dot{Q},$$

$$\text{i.e.,}\quad \dot{p}=(2Q)^{\frac{1}{2}}\lambda^{\frac{1}{2}}\frac{\lambda q}{(2Q)^{\frac{1}{2}}\lambda^{\frac{1}{2}}}\dot{P}+\lambda^{\frac{1}{2}}\frac{p}{(2Q)^{\frac{1}{2}}\lambda^{\frac{1}{2}}}\frac{\dot{Q}}{(2Q)^{\frac{1}{2}}},$$

$$\text{i.e.,}\quad \dot{p}=\lambda q\dot{P}+\frac{p}{2Q}\dot{Q}=\lambda q\dot{P}+\frac{p\lambda}{p^2+\lambda^2q^2}\dot{Q}. \tag{10.7.28}$$

Further,

$$\dot{q}=\frac{\partial q}{\partial P}\dot{P}+\frac{\partial q}{\partial Q}\dot{Q}=-(2Q)^{\frac{1}{2}}\lambda^{-\frac{1}{2}}Sin P\,\dot{P}+\lambda^{-\frac{1}{2}}Cos P\frac{1}{2}(2Q)^{-\frac{1}{2}}2\dot{Q},$$

$$\text{i.e.,}\quad \dot{q}=-(2Q)^{\frac{1}{2}}\lambda^{-\frac{1}{2}}\frac{p}{(2Q)^{\frac{1}{2}}\lambda^{\frac{1}{2}}}\dot{P}+\lambda^{-\frac{1}{2}}\frac{\lambda q}{(2Q)^{\frac{1}{2}}\lambda^{\frac{1}{2}}}\frac{\dot{Q}}{(2Q)^{\frac{1}{2}}},$$

$$\text{i.e.,}\quad \dot{q}=-\frac{p}{\lambda}\dot{P}+\frac{\lambda q}{2\lambda Q}\dot{Q}=-\frac{p}{\lambda}\dot{P}+\frac{\lambda q}{p^2+\lambda^2q^2}\dot{Q}. \tag{10.7.29}$$

1st Equation of (10.7.25) gives

$$\lambda q\dot{P}+\frac{p\lambda}{p^2+\lambda^2q^2}\dot{Q}=\frac{\lambda p}{p^2+\lambda^2q^2}\frac{\partial H'}{\partial P}-\lambda q\frac{\partial H'}{\partial Q}. \tag{10.7.30}$$

2nd Equation of (10.7.25) gives

$$-\frac{p}{\lambda}\dot{P}+\frac{\lambda q}{p^2+\lambda^2q^2}\dot{Q}=\frac{\lambda q}{p^2+\lambda^2q^2}\frac{\partial H'}{\partial P}+\frac{p}{\lambda}\frac{\partial H'}{\partial Q}. \tag{10.7.31}$$

Solving (10.7.30) and (10.7.31), we get

$$\left.\begin{aligned}\dot{P}&=-\tfrac{\partial H'}{\partial q}\\ \dot{Q}&=\tfrac{\partial H'}{\partial p}\end{aligned}\right\},$$

which are again Hamilton's Equations in Q, P system with the Hamiltonian $H'=\lambda Q$.

Example 10.4 Deduce Lagrange's equations from Hamilton's Principle.

Solution By Hamilton's principle, we have

$$\delta \int_{t_1}^{t_2} L\,dt = 0,$$

$$\text{i.e.,} \quad \int_{t_1}^{t_2} \delta L\,dt = 0, \tag{10.7.32}$$

where

$$L = L\left(q_\alpha, \dot{q}_\alpha\right); \quad \alpha = 1, 2, \ldots, n.$$

We have

$$\delta L = \sum_{\alpha=1}^{n} \frac{\partial L}{\partial q_\alpha} \delta q_\alpha + \sum_{\alpha=1}^{n} \frac{\partial L}{\partial \dot{q}_\alpha} \delta \dot{q}_\alpha. \tag{10.7.33}$$

Equation (10.7.32) gives

$$\int_{t_1}^{t_2} \sum_{\alpha=1}^{n} \left(\frac{\partial L}{\partial q_\alpha} \delta q_\alpha + \frac{\partial L}{\partial \dot{q}_\alpha} \delta \dot{q}_\alpha \right) dt = 0,$$

$$\text{i.e.,} \quad \int_{t_1}^{t_2} \left(\sum_{\alpha=1}^{n} \frac{\partial L}{\partial q_\alpha} \delta q_\alpha \right) dt + \int_{t_1}^{t_2} \left(\sum_{\alpha=1}^{n} \frac{\partial L}{\partial \dot{q}_\alpha} \delta \dot{q}_\alpha \right) dt = 0,$$

$$\text{i.e.,} \quad \int_{t_1}^{t_2} \left(\sum_{\alpha=1}^{n} \frac{\partial L}{\partial q_\alpha} \delta q_\alpha \right) dt + \int_{t_1}^{t_2} \left\{ \sum_{\alpha=1}^{n} \frac{\partial L}{\partial \dot{q}_\alpha} \frac{d}{dt} \left(\delta q_\alpha\right) \right\} dt = 0,$$

$$\text{i.e.,} \quad \int_{t_1}^{t_2} \left(\sum_{\alpha=1}^{n} \frac{\partial L}{\partial q_\alpha} \delta q_\alpha \right) dt + \left(\sum_{\alpha=1}^{n} \frac{\partial L}{\partial \dot{q}_\alpha} \delta q_\alpha \right) \Bigg]_{t_1}^{t_2} - \int_{t_1}^{t_2} \left\{ \sum_{\alpha=1}^{n} \frac{d}{dt} \left(\frac{\partial L}{\partial \dot{q}_\alpha} \right) \delta q_\alpha \right\} dt = 0,$$

$$\text{i.e.,} \quad \int_{t_1}^{t_2} \left[\sum_{\alpha=1}^{n} \left\{ \frac{\partial L}{\partial q_\alpha} - \frac{d}{dt} \left(\frac{\partial L}{\partial \dot{q}_\alpha} \right) \right\} \delta q_\alpha \right] dt = 0,$$

$$\text{i.e.,} \quad \sum_{\alpha=1}^{n} \left\{ \frac{\partial L}{\partial q_\alpha} - \frac{d}{dt}\left(\frac{\partial L}{\partial \dot{q}_\alpha}\right) \right\} \delta q_\alpha = 0. \tag{10.7.34}$$

$\because \quad \delta q_\alpha; \quad \alpha = 1, 2, \ldots, n$ are arbitrary and independent of each other, Equation (10.7.34) gives

$$\frac{d}{dt}\left(\frac{\partial L}{\partial \dot{q}_\alpha}\right) - \frac{\partial L}{\partial q_\alpha} = 0; \quad \alpha = 1, 2, \ldots, n$$

which are Lagrange's equations for a holonomic dynamical system specified by n -generalized coordinates q_α; $\alpha = 1, 2, \ldots, n$.

Example 10.5 If $2T = \dot{\theta}^2 + \theta^2 \dot{\varphi}^2$ and $V = \frac{1}{2} n^2 \theta^2$, prove that Hamilton's equations give

$$\theta^2 = a^2 \cos^2(nt + \alpha) + b^2 \sin^2(nt + \alpha) \ \& \ \tan(\varphi + \beta) = \frac{b}{a} \tan(nt + \alpha),$$

where a, b, α, β are constants.

Solution Here θ, φ are the only generalized coordinates.
We have

$$L = T - V = \frac{1}{2}\left(\dot{\theta}^2 + \theta^2 \dot{\varphi}^2\right) - \frac{1}{2} n^2 \theta^2. \tag{10.7.35}$$

The generalized momenta are as follows:

$$\left.\begin{aligned} p_\theta &= \frac{\partial T}{\partial \dot{\theta}} = \dot{\theta} \\ p_\varphi &= \frac{\partial T}{\partial \dot{\varphi}} = \theta^2 \dot{\varphi} \end{aligned}\right\}. \tag{10.7.36}$$

Equation (10.7.36) give

$$\left.\begin{aligned} \dot{\theta} &= p_\theta \\ \dot{\varphi} &= \frac{p_\varphi}{\theta^2} \end{aligned}\right\}. \tag{10.7.37}$$

Further

$$H = T + V = \frac{1}{2}\left(\dot{\theta}^2 + \theta^2 \dot{\varphi}^2\right) + \frac{1}{2} n^2 \theta^2 = \frac{1}{2}\left(p_\theta^2 + \frac{p_\varphi^2}{\theta^2} + n^2 \theta^2\right). \tag{10.7.38}$$

Hamilton's Equations are

$$\left.\begin{aligned} \dot{q}_\alpha &= \frac{\partial H}{\partial p_\alpha} \\ \dot{p}_\alpha &= -\frac{\partial H}{\partial q_\alpha} \end{aligned}\right\}. \tag{10.7.39}$$

1st Equation of (10.7.39) gives

$$\left.\begin{array}{l}\dot{\theta}=\frac{\partial H}{\partial p_\theta}=p_\theta\\ \dot{\varphi}=\frac{\partial H}{\partial p_\varphi}=\frac{p_\varphi}{\theta^2}\end{array}\right\}. \tag{10.7.40}$$

2nd Equation of (10.7.39) gives

$$\left.\begin{array}{l}\dot{p}_\theta=-\frac{\partial H}{\partial \theta}=\frac{p_\varphi^2}{\theta^3}-n^2\theta\\ \dot{p}_\varphi=-\frac{\partial H}{\partial \varphi}=0\end{array}\right\}. \tag{10.7.41}$$

2nd Equation of (10.7.41) gives

$$p_\varphi = cn = \text{a constant.} \tag{10.7.42}$$

1st equations of (10.7.40) and (10.7.41) yield

$$\ddot{\theta}=\dot{p}_\theta=\frac{p_\varphi^2}{\theta^3}-n^2\theta=\frac{c^2n^2}{\theta^3}-n^2\theta,$$

i.e., $\ddot{\theta}+n^2\theta=\dfrac{c^2n^2}{\theta^3}$,

i.e., $2\ddot{\theta}\dot{\theta}+2n^2\theta\dot{\theta}=2\dfrac{c^2n^2}{\theta^3}\dot{\theta}$,

i.e., $\dot{\theta}^2+n^2\theta^2=-\dfrac{c^2n^2}{\theta^2}+2en^2$,

i.e., $\dot{\theta}^2=-\dfrac{c^2n^2}{\theta^2}-n^2\theta^2+2en^2=\dfrac{n^2}{\theta^2}\left[-c^2-\theta^4+2e\theta^2\right]$,

i.e., $\dot{\theta}^2=\dfrac{n^2}{\theta^2}\left[\left(e^2-c^2\right)-\left(\theta^2-e\right)^2\right]$,

i.e., $\dot{\theta}=-\dfrac{n}{\theta}\sqrt{\left(e^2-c^2\right)-\left(\theta^2-e\right)^2}$,

i.e., $-\dfrac{2\theta d\theta}{\sqrt{\left(e^2-c^2\right)-\left(\theta^2-e\right)^2}}=2ndt$,

i.e., $I=2nt+2\alpha. \quad (10.7.43)$

Now

$$I = -\int \frac{2\theta d\theta}{\sqrt{\left(e^2 - c^2\right) - \left(\theta^2 - e\right)^2}}. \tag{10.7.44}$$

Put

$$u = \theta^2 - e,$$

i.e., $du = 2\theta d\theta.$

$$\therefore \quad I = -\int \frac{du}{\sqrt{\lambda^2 - u^2}} = \cos^{-1} \frac{u}{\lambda}. \tag{10.7.45}$$

$\therefore$ (10.7.43) gives

$$2nt + 2\alpha = \cos^{-1} \frac{u}{\lambda},$$

i.e., $u = \lambda \cos(2nt + 2\alpha),$

i.e., $\theta^2 - e = \sqrt{e^2 - c^2} \cos 2(nt + \alpha),$

i.e., $\theta^2 = \dfrac{(a^2 + b^2)}{2} + \dfrac{(a^2 - b^2)}{2} \cos 2(nt + \alpha); \quad e = \dfrac{1}{2}\left(a^2 + b^2\right), \quad c = ab,$

i.e., $\theta^2 = \dfrac{a^2}{2} 2\cos^2(nt + \alpha) + \dfrac{b^2}{2} 2\sin^2(nt + \alpha),$

i.e., $\theta^2 = a^2 \cos^2(nt + \alpha) + b^2 \sin^2(nt + \alpha). \quad (10.7.46)$

Equations (10.7.42) and (10.7.37) gives

$$\theta^2 \dot{\varphi} = c\,n,$$

i.e., $\dot{\varphi} = \dfrac{c\,n}{\theta^2} = \dfrac{c\,n}{a^2 \cos^2(nt + \alpha) + b^2 \sin^2(nt + \alpha)}$ (by (10)). $\quad (10.7.47)$

Equation (10.7.47) gives

$$d\varphi = \frac{c\,n\,dt}{a^2\cos^2(nt+\alpha) + b^2\sin^2(nt+\alpha)},$$

$$\text{i.e.,}\quad d\varphi = \frac{c\,n\sec^2(nt+\alpha)\,dt}{a^2 + b^2\tan^2(nt+\alpha)},$$

$$\text{i.e.,}\quad \varphi = c\,n\int\frac{\sec^2(nt+\alpha)\,dt}{a^2+b^2\tan^2(nt+\alpha)} + \varphi_0,$$

$$\text{i.e.,}\quad \varphi = c\,n\,J + \varphi_0, \tag{10.7.48}$$

where

$$J = \int\frac{\sec^2(nt+\alpha)\,dt}{a^2+b^2\tan^2(nt+\alpha)}. \tag{10.7.49}$$

Consider the substitution

$$\xi = b\tan(nt+\alpha),$$

$$\therefore\quad d\xi = b\sec^2(nt+\alpha)\,n\,dt.$$

Hence (10.7.49) reduces to

$$J = \int\frac{d\xi}{b\,n\,(a^2+\xi^2)} = \frac{1}{abn}\tan^{-1}\frac{\xi}{a} = \frac{1}{c\,n}\tan^{-1}\frac{\xi}{a}. \tag{10.7.50}$$

Equation (10.7.48) gives

$$\varphi = cn\frac{1}{cn}\tan^{-1}\frac{\xi}{a} + \varphi_0 = \tan^{-1}\frac{\xi}{a} + \varphi_0,$$

$$\text{i.e.,}\quad \varphi - \varphi_0 = \tan^{-1}\frac{\xi}{a},$$

$$\text{i.e.,}\quad \varphi + \beta = \tan^{-1}\frac{\xi}{a};\quad \beta = -\varphi_0,$$

$$\text{i.e.,}\quad \tan^{-1}\frac{\xi}{a} = \varphi + \beta,$$

$$\text{i.e.,}\quad \xi = a\tan(\varphi+\beta),$$

i.e., $b \tan(nt + \alpha) = a \tan(\varphi + \beta)$,

i.e., $\tan(\varphi + \beta) = \frac{b}{a} \tan(nt + \alpha)$.

Thus, we have

$$\theta^2 = a^2 \cos^2(nt + \alpha) + b^2 \sin^2(nt + \alpha)$$

and $\tan(\varphi + \beta) = \frac{b}{a} \tan(nt + \alpha)$

10.8 Exercise-X

1. Write the Hamiltonian and equation of motion for a simple pendulum.
2. A particle of mass m moves in a force field of potential V. Write the Hamiltonian and Hamilton's equations in the spherical polar coordinate system (r, θ, φ).
   Hints: $T = \frac{m}{2}\left(\dot{r}^2 + r^2\dot{\theta}^2 + r^2 \sin^2\theta\,\dot{\varphi}^2\right)$, $V = V(r, \theta, \varphi)$.
3. A particle of mass m moves in a force field of potential V. Write the Hamiltonian and Hamilton's equations in the cylindrical polar coordinate system (r, θ, z).
   Hints: $T = \frac{m}{2}\left(\dot{r}^2 + r^2\dot{\theta}^2 + \dot{z}^2\right)$, $V = V(r, \theta, z)$.
4. A particle of mass m moves in a force field of potential V. Write the Hamiltonian and Hamilton's equations in the Cartesian coordinate system (x, y, z).
   Hints: $T = \frac{m}{2}\left(\dot{x}^2 + \dot{y}^2 + \dot{z}^2\right)$, $V = V(x, y, z)$.
5. Use Hamilton's equations to find the equations of motion of a projectile.
   Hints: $T = \frac{m}{2}\left(\dot{x}^2 + \dot{y}^2 + \dot{z}^2\right)$, $V = mgz$.
6. Use Hamilton's equations to find the equations of motion of a particle of mass m moving in a straight line in simple harmonic motion.
   Hints: $\mathbf{F} = -m\mu x\hat{i}$, $\frac{dV}{dx} = m\mu x$, $V = \frac{m\mu x^2}{2}$, $T = \frac{1}{2}m\dot{x}^2$.

Bibliography

1. Chorlton, F.: Text Book of Dynamics. CBS Publishers & Distributors, New Delhi, India (1985)
2. Loney, S.L.: An Elementary Treatise on Dynamics of a Particle and of Rigid Bodies. Surjeet Publications, New Delhi, India (2002)
3. Murray, R.: Spiegel: Theory and Problems of Theoretical Mechanics. Schaum's Outline Series, McGraw-Hill Book Company, Singapore (1986)
4. Ramsey, A.S.: Dynamics. Part I & Part II. CBS Publishers & Distributors, New Delhi, India (1985)

N. Ahmed et al., *Classical Dynamics*, University Texts in the Mathematical Sciences,
https://doi.org/10.1007/978-981-95-6394-4

The manufacturer's authorised representative in the EU is Springer Nature Customer Service Centre GmbH, Europaplatz 3, 69115 Heidelberg, Germany. If you have any concerns regarding our products, please contact ProductSafety@springernature.com

Printed and bound by CPI Group (UK) Ltd, Croydon, CR0 4YY
07/07/2026
02160909-0005